Perspectives on Process Safety from Around the World 2019

AA009072

Held at the 2019 AIChE Spring Meeting and
15th Global Congress on Process Safety

New Orleans, Louisiana, USA
31 March - 4 April 2019

ISBN: 978-1-5108-8690-2

Printed from e-media with permission by:

Curran Associates, Inc.
57 Morehouse Lane
Red Hook, NY 12571

Some format issues inherent in the e-media version may also appear in this print version.

Printed by Curran Associates, Inc. (2019)

For permission requests, please contact AIChE
at the address below.

AIChE
120 Wall Street, FL 23
New York, NY 10005-4020

Phone: (800) 242-4363
Fax: (203) 775-5177

www.aiche.org

Additional copies of this publication are available from:

Curran Associates, Inc.
57 Morehouse Lane
Red Hook, NY 12571 USA
Phone: 845-758-0400
Fax: 845-758-2633
Email: curran@proceedings.com
Web: www.proceedings.com

TABLE OF CONTENTS

Approach of Human Factors in Risk Analysis and Operational Procedures: Stress Level Analysis of Operators in Performing Tasks

Luana Fernanda Marques Santos
Environmental Engineering, PEA/UFRJ
University City, Rio Janeiro
luanafernanda@poli.ufrj.br

Isaac Jose Antonio Luquetti dos Santos
Nuclear Engineering Institute/CNEN and PEA/UFRJ
University City, Rio Janeiro
luquetti@ien.gov.br

Prepared for Presentation at
American Institute of Chemical Engineers
2019Spring Meeting and 15thGlobal Congress on Process Safety
New Orleans, LA
March 31 – April 3, 2019

Approach of Human Factors in Risk Analysis and Operational Procedures: Stress Level Analysis of Operators in Performing Tasks

Luana Fernanda Marques Santos
Environmental Engineering, PEA/UFRJ
University City, Rio Janeiro
luanafernanda@poli.ufrj.br

Isaac Jose Antonio Luquetti dos Santos
Nuclear Engineering Institute/CNEN and PEA/UFRJ
University City, Rio Janeiro
luquetti@ien.gov.br

Keywords: stress, human factors, risk, operational procedure

Abstract

Process risk analyzes generally address operational aspects in their entirety. Human factors are commonly found as root causes of major accidents, which are not mentioned in the process risk analyzes. Include analysis of human factors in the process risks analysis is necessary in an attempt to improve the understanding of major accidents and process safety. This work shows a methodological framework used to include human factor analysis in the process risk analysis, and therefore the inclusion of human factors in the operational procedures, as a point of improvement identified in the risk analysis. For this, the Operator Action Event Tree technique was used to determine the level of stress that operators are subjected to in performing procedures, which can lead to human errors and contribute to the occurrence of major accidents or not to block the effects of an initiating event.

1 Introduction

Human performance, in individual sense, is a series of behaviors executed in order to accomplish specific task objectives. Human performance, in organizational sense, is what the leaders and managers are doing or have done those influence behaviors and eventually affects the physical plant [1]. Human performance improvement is an approach to reducing human errors.

Human error is any mental or physical activity, or failure to perform activity, that leads to either an undesired or unacceptable outcome [2]. Human error is caused by a variety of conditions related to individual behavior, management and leadership practices, and organizational processes. Where human failures are identified, it is necessary to identify the factors that make the failure more or less likely. Performance Shaping Factors (PSFs) are the characteristics of people, tasks and organizations that influence human performance and therefore the likelihood of human failure [3]. PSFs include time pressure, fatigue, design of controls/displays, quality of procedures, stress and so on.

Human factors are a body of scientific factors about human characteristics, covering biomedical, psychological and psychosocial considerations, including principles and applications in the personnel selection, training, job performance aid tools and human performance evaluation [4].

Stress is a term used to cover conditions of a physical, biological, or psychological nature, that typically cannot be controlled by organisms, and that strain organisms often beyond their powers to adapt. Human beings respond to stressors either through extraordinary mental or physical effort or by exhibiting degraded performance [5]. According to [6], depending on the intensity of the stressor and the nature of the task to be performed, stress affects human performance. The stress-related phenomena are often classified as emotional, cognitive and physical. In the situation when human beings are faced with the extreme conditions and complex tasks, performance is crucial and personnel must be prepared to operate under extreme stress, and must have the knowledge to solve problems and take correct decision. Various stressors have been shown to affect human performance, such as crowding, noise, time pressure, workload [7], [8].

Workload is defined as the cost incurred by a human operator to achieve a particular level of performance [9] or the proportion of the capacity an operator spends on task performance [10].

A task is an activity or a set of actions needed to solve a problem or achieve a goal. Task analysis is defined as the study of human beings actions necessary to perform a task, in terms of activities and/or cognitive processes. It provides organization and structure for the description of a set activities performed by a person [11]. A task analysis can be conducted by two modes. The first one is the understanding of sequence or dependency between different activities. The second is the process of representing how tasks can be decomposed into subtasks and the logical relationship between these. A common technique used is called hierarchical decomposition and tasks tabular diagram. Task demands are the physical and mental demands required to perform an activity. When the demands exceed the capabilities or challenge the limitations of the individual, a human error is likely to occur. Task demands include time pressure, simultaneous multiple actions and unclear work standards [11].

The objective of this paper is to present and apply a methodological framework that describes how human factors can be considered in the accident analysis, and therefore the inclusion of human factors in the operational procedures, as a point of improvement identified in the risk analysis. A case study is developed through the use the event tree technique, when an accident is analyzed in order to determine the level of stress that operators were subjected during the performance of operational procedures.

2 Performance modes of the operators

During the performance of their tasks or activities, human beings move among the various performance modes. According to Rasmussen [12], human beings switch among different levels of cognitive control when faced with different situations. Depending on the situation, an individual chooses a performance mode that appears sufficient to control the situation. These cognitive characteristics can result in an unsafe action by interfering with the process of detection (monitoring), situation assessment, response planning and response implementation. The performance modes are skill-based, rule-based and knowledge-based.

The skill-based performance mode is associated with practiced actions under familiar situations with little conscious. The actions required are preprogrammed instructions that have been developed through training and experience. The error mode associated with skill-based performance is inattention. Errors made at this level are referred to as *slips* or *lapses*. Slips and lapses occur as a result of a failure in the execution of an action. For example, the operator started the "A" pump instead of the "B" pump. The lapse is an internalized error involving a memory failure. Slips involve the inadvertent commission, that is, the action was taken, but not correctly or omission, that is, action not taken [12].

The rule-based performance mode is associated with familiar patterns or solutions stored in person memory, in order to solve a problem. The cognitive follows an "if-then" logic. The error mode associated with rule-based performance is misinterpretation, that is, not understand or recognize conditions that require a particular response. Errors in the rule-based performance mode are referred to as mistake. A mistake is a failure in a mental process when selecting a solution to a problem, but it is the wrong solution or a mismatch between what was intended and the intended consequence. Examples are deviating from an approved procedure or applying the correct procedure to the wrong situation [12].

The knowledge-based performance mode is associated with a situation that is unfamiliar, a solution to a problem not seen before. There is a cognitive limitation. The person need additional information for try to solve problem also increases. Errors in the knowledge-based performance mode are referred to as mistake. However, the mechanism for error is different, because result from a cognitive limitation. The error mode is an inaccurate mental model. Decisions are taken with limited information or there is insufficient time to acquire it [12]. IAEA-TECDOC-592 [13] presents a logic diagram for assignment of the type of cognitive processing (Figure 1).

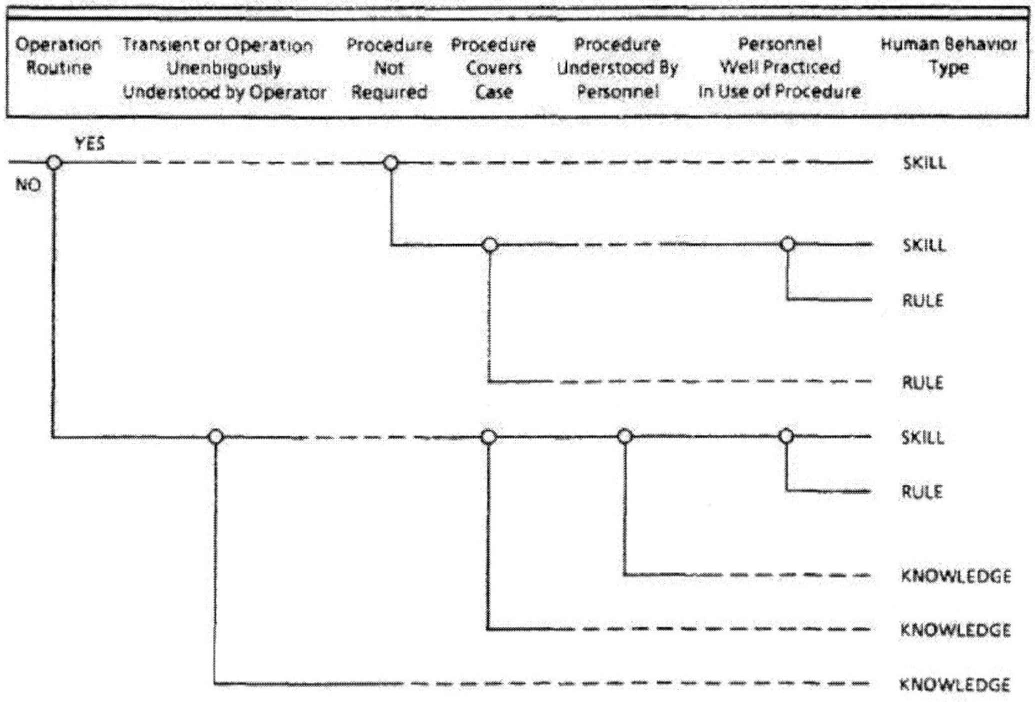

Figure 1. Decision tree for assignment of the type of cognitive processing

3 Methodological framework

This paper deals with the use of a methodological framework that describes how human factors can be considered in the accident analysis, and therefore the inclusion of human factors in the operational procedures, as a point of improvement identified in the risk analysis. The methodology includes the following item: definition and selection of an appropriate event tree technique based on the performance shaping factors (PSFs); description of the accident; analysis of the performance mode of the operators during each action developed; identification and analysis of performance shaping factors and definition of the stress level of the operator.

3.1 Definition and selection of an appropriate event tree technique

An event tree is a graphic that shows scenarios resulting from a specific initiating event (IE) Event trees present the event information in a tree structure that is used to classify scenarios according to their consequences. The headings of the event tree are the IE, which is the starting point, the subsequent events show success or failure of conditions related to mitigating/ aggravating events, and the end state presents the consequence of each scenario.

The EPRI *HRA* Calculator [14] is a software tool designed to facilitate a standardized approach to human reliability analysis (HRA). The software uses different methodologies for HRA and can be possible to integrate the results in probabilistic risk assessment (PRA). It includes modules for pre-initiator events, post-initiator events and dependency analysis. It is possible to addresses issues such as stress level, and identification of rule, skill or knowledge based actions, training, timing information, and clarity of procedures. The HRA Calculator uses decision trees to calculate stress level. The results are a stress level of optimal (low), medium, or high. In this paper, the following decisions tree, shown in figure 2, was used.

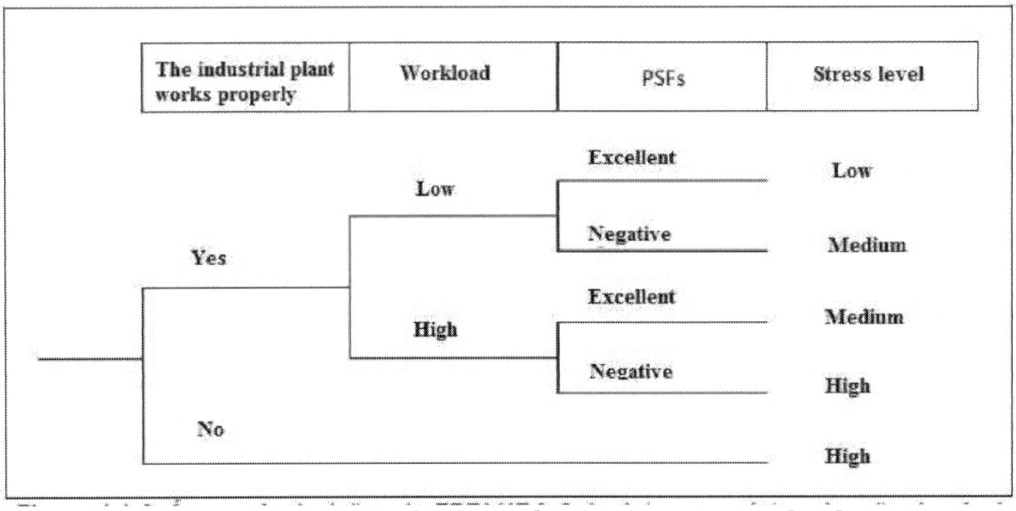

Figure 2. Decision tree for assessment stress level

3.2 Description of the accident

The industrial installation chosen for case study is a petroleum storage terminal. The terminal serves as connection between the oil tankers and pipelines, which are appropriate facilities for transferring products of ships to earth and vice-versa, or between ships. During an operational procedure, the field operator opened the suction valve of the Liquefied petroleum gas (LPG) booster pump. After this action, the field operator opened the vent valve using the right hand. As a result, there was the displacement of the nipple of the vent valve that collided with the fingers of the left hand of the operator. The table 1 shows the actions of the operator through a task tabular diagram.

Table 1. Tasks Tabular Diagram.

Action	How the action was done? What happened?	Who developed the action?	What equipment, component or tools were used?
1. Open the suction valve of LPG booster pump	2. The operator used the two hands	Field operator	The suction pump wheel
2. Due to the floor unlevelling, the operator	2. The operator used the right hand on the drain valve.	Field operator	The drain valve

supported the left hand on suction valve of LPG booster pump, while with the right hand handled the drain valve.	2.1 In the drain valve system there is a prolonged tube (nipple), stationed distant from the operator face where the gas can be drained 2.2 The prolonged tube was not properly welded and was loose		
3. The operator handled the drain valve	3. While the operator was handling the drain valve, a sudden gas leak occurred by the nipple	Field operator	The drain valve
4. The operator handled the drain valve	4. The nipple spun quickly and shocked against the operator's left hand, causing an injury	Field operator	The drain valve

3.3 Analysis of the performance mode of the operator during each action developed

Observations of operator activities, in combination with interviews, are the principal approach of the ergonomist. The ways that data are collected are: watching the performance of the task (direct observation) or hearing a recording or description of the task (indirect observation). These methods can be used in combination with each other in order obtain a broader understanding of the task. In this study, recording and description of the task were used in combination with interviews with the principal actor of the event, the field operator. In this item, the analysis of the performance mode of the operator was based on the decision tree showed in Figure 1.

3.4 Identification and analysis of performance shaping factors

In this item, a set of PSFS (performance shaping factors) was developed based on the following human reliability analysis methods: THERP [15]) and CREAM [3] The PSFs are: Adequacy of equipment, tools and operational support (AETO); availability of procedures (AP); available time (AT); experience (EX); adequacy of training (AT); workplace conditions (WC); adequacy of organization (AO). The identification of the PSFs that affected the performance of the field operator and contributed for the accident was based on this set of PSFs. For each action developed by the field operator one or more PSFs were chosen.

3.5 Definition of stress level of field operator

Figure 2 shows the decision tree used to assess the stress level of field operator. In this phase, it is important to report that all analysis were done by two human factors experts with more than ten years of experience in process safety, ergonomics, risk and human reliability. The results presents in the next item of this study represents the consensus of the judgment and analysis of two experts.

4 Results and Conclusions

Through the use of decision tree (Figure 1) and table 1, the analysis of the operator performance mode was concluded. The results are: The operation was routine; the operation was correctly understood by the field operator; procedure was required; procedure covers the case; procedure was understood; field operator with experience. The performance mode of the field operator is SKILL. The error mode in the skill-based performance selected is slip. There was an attention failure during the operation of the suction valve of LPG booster, but it isn`t the principal factor that contributed for the accident. The field operator knew that the prolonged tube (nipple) was not properly welded and was loose.

Through the use of table 1 and judgment of the human factors experts was possible to identify the PSFs that affected the performance of the field operator and contributed decisively to the event. The results are shown in table 2.

Table 2. Identification of the PSFs

Action	How the action was done? What happened?	Who developed the action?	What equipment, component or tools were used?	PSFs
1. Open the suction valve of LPG booster pump	2. The operator used the two hands	Field operator	The suction pump wheel	AETO; WC; AO
2. Due to the floor unlèvelling, the operator supported the left hand on suction valve of LPG booster pump, while with the right hand handled the drain valve.	2. The operator used the right hand on the drain valve. 2.1 In the drain valve system there is a prolonged tube (nipple), stationed distant from the operator face where the gas can be drained 2.2 The prolonged tube (nipple) was not properly welded and was loose	Field operator	The drain valve	AETO; WC; AO
3. The operator handled the drain valve	3. While the operator was handling the drain valve, a sudden gas leak occurred by the nipple	Field operator	The drain valve	AETO; WC; AO
4. The operator handled the drain valve	4. The nipple spun quickly and shocked against the operator's left hand, causing an injury	Field operator	The drain valve	AETO; WC; AO

Finally, the stress level of the field operator was identified. The decision tree (Figure 2) was used. In this analysis, the plant industrial was working correctly. The workload assessment was performed based on the number of procedures in which the operator is involved during the executive phase, the number of critical steps involved in the implementation of the task and the available time to perform it. The number of procedures was considered according to human limitations. The number of critical steps involved in the implementation of the task was evaluated as low. The available time to perform the task was sufficient. The workload assessment was evaluated as LOW. In the next step, PSFs were assessment. Finally, according to EPRI HRA [14], PSFs can only be considered excellent if none of them present any adverse condition. In this study, the PSFs AETO and WC were assessment as negative. The results show that the stress level of the field operator was MEDIUM.

In this paper, the overall results showed a positive and significant contribution of methodological framework for the inclusion of human factors in the accident analysis and consequently improvement of the risk analysis. The study reflects the importance of having a group of experts for providing decisions and analysis of the PSFs. The study demonstrates the importance of the correct choice of the technique to be used in the analysis carried out, that is, the decision trees. The approach can be used in several industrial plants and constitutes a validated systemic way to assess the stress level of workers.

5 References

[1] EPRI. *Human Performance: Fossil Operations.* Electric Power Research Institute, USA, 2007.

[2] J. Reason. *Human Error*, Cambridge University, 1994.

[3] E. Hollnagel. Cognitive Reliability and Error Analysis Method, Elsevier, 1998.

[4] NUREG. *Human System Interface Design Review Guideline, NUREG 700.* U. S. *Nuclear Regulatory* Commission, 2002.

[5] A. W. K. Gaillard and C. J. E. Wientject. *Mental Load and Work Stress as Two Types of Energy Mobilization.* Work and Stress, 8, pp. 141-152, 1994.

[6] A. W. A. Van Gemmert and G. P. Van Galen. *Stress, Neuro-Motor Noise and Human Performance: A Theoretical Perspective.* Journal of Experimental Psychology: Human Perception and Performance, 23, 1, pp. 299-1313, 1997.

[7] R. N. Williams. *Field Observations and Surveys in Combat Zones.* Social Psychology Quarterly 47, 186-192, 1984.

[8] I. L. Goldstein and P. W. Dorfpan. (1978). *Speed and Load Stress as Determinants of Performance in a Time Sharing Task.* Human Factors, pp. 603-609, 1978.

[9] A. D. Andre. The Value of Workload in the Design and Analysis of Consumer Products. In P.A. Hancock, & P.A. Desmond, *Stress, Workload, and Fatigue,* pp. 373-382. Mahwah, NJ, 2001.

[10] D. Kahneman, D. *Attention and Efsort.* Englewood Cliff, NJ: Prentice-Hall, 1973.

[11] B. Kirwan and L. K. Ainsworth. *A Guide to Task Analysis.* Taylor and Francis, London, 1992.

[12] J. Rasmussen. *Skills, Rules and Knowledge: Signals, Signs and Symbols and others Distinctions in Human Performance Models*. IEEE Transactions on Systems, Man and Cybernetics, 1997.

[13] IAEA-TECDOC-592. Case *Study on the Use of PSA Methods: Human Reliability Analysis*. International Atomic Energy Agency, Vienna, Austria, 1991.

[14] K. D. Kohlhepp. *Evaluation of the Use of Engineering Judgments Applied to Analytical Human Reliablity Analysis (HRA) Methods*. Thesis Master of Science Submitted to the Office of Graduate Studies of Texas A&M University, 2005.

[15] A. D. Swain, and H. E. Guttmann. *Handbook of Human Reliability Analysis with Emphasis on Nuclear Power Plant Applications.* NUREG/CR-1278, U. S. Nuclear Regulatory COmmission, 1983.

546567 Functional Safety Management: The Yellow Brick Road to a Safe Plant. Gerenciamento Da Seguranca Funcional: O Caminho Dourado Para Uma Planta Segura

Gerenciamento da Segurança Funcional:

o caminho dourado para uma planta segura

Monica Hochleitner

SIS SILverstone LLC

11999 Katy Freeway, Suite 600, Houston, TX 77079

Abstract

A experiência observada nos projetos de sistemas instrumentados de segurança (SIS) ao longo dos anos tem mostrado que a indústria cada vez mais busca atender normas e seguir melhores práticas de segurança no processo industrial – seja devido à regulamentação local ou por exigência das seguradoras. O que também se observa, no entanto, é que muitas vezes essas iniciativas abrangem o mínimo necessário dos requerimentos propostos e acabam por se perder ao longo da vida útil das plantas.

Não bastasse a necessidade de identificar, tratar e manter seguros os cenários que ameaçam o risco do processo tolerável pelas organizações, também é preciso proteger essas instalações de invasões cibernéticas, trazendo para o ambiente que tradicionalmente se restringia ao tangível, ameaças que precisam ser endereçadas em um formato completamente novo.

Este trabalho discute a importância em se ter um sistema de gerenciamento da segurança muito bem implementado, que permita gerir potenciais ameaças à segurança do processo e à segurança de informações do processo, garantindo a perpetuação das medidas inicialmente adotadas por meio da integração das fases e atividades de um projeto de SIS, incluindo auditorias e avaliações periódicas de tais sistemas. Por último, é apresentada uma ferramenta que além de integrar e documentar as fases e atividades de um projeto de SIS, funciona como um fluxo de trabalho que conduz o projeto através de uma trilha essencial e, por contemplar todos os requisitos de segurança funcional exigidos pelas normas em vigência, naturalmente resulta em plantas de processo mais seguras muito além de sua etapa de implantação.

Keywords

Sistema de Gerenciamento da Segurança, Inspeção Periódica, Intertravamento & Segurança, Sistemas Instrumentados de Segurança

Process Safety Leading Indicators By Environmental Experts Point of View

Luana Fernanda Marques Santos
Environmental Engineering, PEA/UFRJ
University City, Rio Janeiro
luanafernanda@poli.ufrj.br

Isaac Jose Antonio Luquetti dos Santos
Nuclear Engineering Institute/CNEN and PEA/UFRJ
University City, Rio Janeiro
luquetti@ien.gov.br

Assed Naked Haddad
Environmental Engineering, PEA/UFRJ
University City, Rio Janeiro
assed@poli.ufrj.br

Prepared for Presentation at
American Institute of Chemical Engineers
2019Spring Meeting and 15thGlobal Congress on Process Safety
New Orleans, LA
March 31 – April 3, 2019

Process Safety Leading Indicators By Environmental Experts Point of View

Luana Fernanda Marques Santos
Environmental Engineering, PEA/UFRJ
University City, Rio Janeiro
luanafernanda@poli.ufrj.br

Isaac Jose Antonio Luquetti dos Santos
Nuclear Engineering Institute/CNEN and PEA/UFRJ
University City, Rio Janeiro
luquetti@ien.gov.br

Assed Naked Haddad
Environmental Engineering, PEA/UFRJ
University City, Rio Janeiro
assed@poli.ufrj.br

Keywords: Process safety, leading indicators, environmental experts

Abstract

Process safety accidents can cause severe environmental damage. Process safety indicators are the key elements to improve the safety performance of the process. They can identify factors and causes that may lead to major accidents. Leading indicators are a set of future-oriented metrics that indicate the performance of protection layers that prevent accidents. It is much important to identify leading indicators of process safety in order to verify the weakness and deficiencies in the management system of process safety. This paper shows a methodological framework based on the initial selection of 77 leading indicators; the selection of two groups of experts, one with experience on process safety and another with experience on environmental analysis; the ranking of the leading indicators through the use of a fuzzy model and finally the comparison of the results obtained from the two groups of experts. The results showed the comparison among the opinions of two experts group, the ranking of the most important leading indicators according to opinion of process safety experts and environmental experts and which indicators need to be monitored from the point of view of each group of expert.

1 Introduction

Expert judgment is necessary in many areas of risk analysis, where there isn`t a specific database or the data are insufficient to be used in statistical estimate or for constructing an empirical model. A good way of using expert judgment is based on collecting information from experts to build an expert model with which the desired value may be obtained [1].

The experts who make the judgments must be familiar with the criteria to be judged. If it involves process safety or environmental questions, the experts must have an in-depth knowledge of plant systems, safety standards, investigation and accident analysis. The number of experts needed to make the judgments can be defined. According to [2], six experts would be sufficient for direct estimation and eight experts could be used for paired comparisons.

Each expert perceives different cues and can interpret them differently but the cues that can be perceived are normally interrelated, and therefore a high redundancy will exist. This redundancy reduces the effect of differential selection and makes many judgments rather robust. When there is lack of physical data, one of the key elements of any risk analysis is the use of expert judgment. The application of the expert judgment can vary between methods and degree of subjectivity. In some instances, experts will have to estimate a value based entirely on their experience [3].

Process safety indicators are used to identify events or conditions that could ultimately lead to higher-level consequences. Indicators provide a means to measure performance against requirements. Leading indicators indicate the performance of processes or protective barriers that prevent incidents, giving an indication of potential problems in safety systems early enough for corrective actions may be taken [4].

According to [5], there are six main stages to implement a process safety measurement system. The following stages are defined as:

- Establish the organizational arrangements to implement the indicators: Set up an implementation team.
- Decide on the scope of the measurement system: Identify incident scenarios, the immediate causes of hazard scenarios, review performance and non-conformance.
- Identify the risk control systems in place to prevent major accidents. Decide on the outcomes for each and set a leading indicator: Described the outcome, set a leading indicator, follow up deviations from the outcomes.
- Identify the critical elements of each risk control system and set leading indicators.
- Establish the data collection and reporting system: Collect information. Ensure information/unit of measurement is available or can be established.
- Review performance of process management system and the scope of the indicators.

The management teams of companies work to identify hazards and control the process before something goes wrong to avoid incidents that result in injury. Management control involves establishing and maintaining standards for work performance. This involves evaluating performance, communicating it, correcting deficiencies in performance standards and establishing and tracking performance [6].

The objective of this paper is to present and apply a methodological framework that describes the selection process of leading indicators related to process safety, the analysis of the indicators by two group of experts, one with experience in safety process and another with experience in environmental analysis, the ranking of the leading indicators through a fuzzy model and the comparison of the results obtained.

2 Fuzzy Method Applied to Decision-Making Process

Each expert has their own evaluation, based on their experience, knowledge and convictions on the issue in question. Therefore, finding a consensus function for the group to aggregate the different estimates into a common opinion is an important question [7]. Various decision-making methods with fuzzy logic can be identified in the scientific literature. Fuzzy logic is adapted to the problem by loosening the construction of groups of opinions without defined limits, being adherent to the imprecision and uncertainty that may characterize the lack of consensus [8]. In general, the experts present different levels of importance. Experience, reputation and decision-making power may indicate the necessity to judge the weight of their estimates.

Under certain conditions of extreme uncertainty, even experienced experts may encounter difficulties in their estimates. Human evaluations are subjective and often inconsistent, or even conflicting, when measured at different times. In these cases, the experts may carry out a more subjective evaluation if their choices are restricted to psychometric evaluations (i.e. high relevance, relevant or low relevance variable), without the determination of exact values of the fuzzy number domain. This study proposed adaptations to the Fuzzy-SAM model of [7] for problems with an elevated degree of uncertainty. Initially, the fuzzy sets to be selected through psychometric evaluation of the experts were developed. For simplicity, three triangular fuzzy numbers were elaborated, which represented the options to be selected for each human factor considered in the problem (Figure 1).

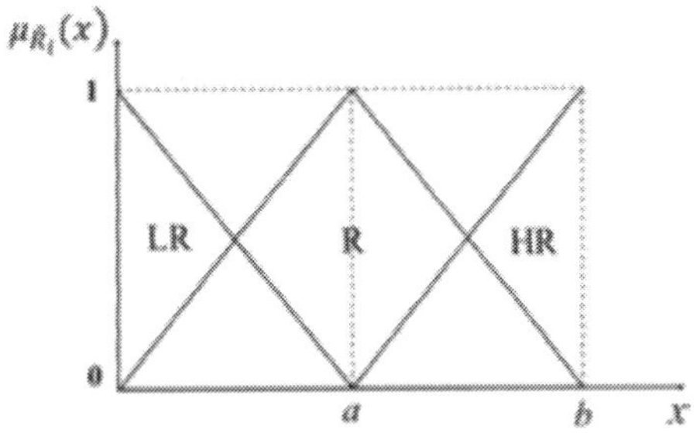

Figure 1 – Fuzzy Triangular Sets

3 Methodological framework

This paper deals with the use of a methodological framework that describes the following item:

- Description of the industrial plant chosen for implementation of leading indicators
- Define criteria to be used for selecting a set of process safety leading indicators
- Identify and select two groups of experts
- Apply a fuzzy method for ranking of each process safety leading indicator

3.1 Description of the industrial plant chosen for implementation of leading indicators

The industrial plant chosen for case study is a petroleum storage terminal. The terminal serves as connection between the oil tankers and pipelines, which are appropriate facilities for transferring products of ships to earth and vice-versa, or between ships.

3.2 Define criteria to be used for selecting a set of process safety leading indicators

A set of 77 leading indicators was selected and grouped by themes. The selection of the indicators was based on safety references guides, reports of major accidents analysis occurred in the terminals and pipelines and recommendations cited by [9]. The criteria used for development of this database can be obtained in [10].

3.3 Identify and select two experts group

Initially, a first group of 15 experts was selected and a questionnaire was used to determine the degree of importance of each expert. The criteria used were based on experience and work years in the area of process safety. Awards were attributed for each criteria. The first group analyzed and validated the 77 leading indicators. After this validation, the pareto principle was used and a set of 44 indicators was selected and grouped by themes

In the second phase, a second group of 9 experts were selected and a questionnaire was used to determine the degree of importance of each expert. The criteria used were based on experience and work years in the area of environmental analysis. Awards were attributed for each criteria. The second group analyzed and validated the 77 leading indicators. After this validation, the pareto principle was used and a set of 43 indicators was selected and grouped by themes This questionnaire is an important part of the fuzzy model applied to indicators ranking.

3.4 Apply a fuzzy method for ranking of each process safety leading indicator

A second questionnaire was developed with the objective to rank the leading indicators. Initially, the fuzzy sets to be selected through psychometric evaluation of the experts were developed. For simplicity, three triangular fuzzy numbers were elaborated, which represented the options to be selected for the relevance of each leading indicator. In Figure 1, the "Low relevance" (LR), "relevance" (R) and "High relevance" (HR) fuzzy sets were represented by fuzzy numbers (0,0,1), (0,1,2) and (0,2,2), respectively.

The phases of the fuzzy method are presented below:

- Verify union areas and intersection areas of fuzzy opinion;
- Calculate degree of opinion;
- Develop agreement array;
- Calculate degree of relative agreement;
- Calculate experts consensus coefficient;
- Identify the fuzzy value of each one of the process safety leading indicators;
- Rank the process safety leading indicators using the indicators fuzzy value.

4 Results and Conclusions

The first group of 15 experts ranked the 44 leading indicators, as showed on Table 1, identifying the indicators as tier 3 or tier 4. The Table 1 results show that the greatest value of triangular fuzzy number occurred in tier 3 indicators called "Demand on safety systems". All tier 3 indicators were among the top eight indicators of hierarchy. This can be justified by the fact that the events monitored by these indicators are very close to the accident occurrence.

Table 1: Ranking of process safety leading indicators by the first experts group

Hierarchy	Tier	Criteria degree
1 - Demand on safety systems: failure on demand or test in instrumented system	3	1.0000
2 - Integrity mechanical, inspection and safety critical equipment maintenance: percentage of inspections of safety critical items of plant and equipment due during the measurement period	4	0.9729
3 - Management of change: number of open management of change	4	0.9676
4 - Demand on safety systems: number of safety instrumented system activations	3	0.8832
5 – Safety culture: percentage of meeting that discuss about process safety	4	0.8600
6 - Demand on safety systems: pressure relief device activations	3	0.8448
7 - Hazard identification and risk analysis: process safety risk analysis progress (number open, number closed, time to closure)	4	0.8380
8 - Demand on safety systems: number of high level and high high level alarms activation in tanks	3	0.8090
9 - Integrity mechanical, inspection and safety critical equipment maintenance: number of inspections or conformance tests indicate that vessels, tanks, pipelines or other equipments was operated with safe limit operating exceeded (results outside acceptable limits)	4	0.8087
10 - Integrity mechanical, inspection and safety critical equipment maintenance: number	4	0.7784

of emergency work orders or corrective maintenance not planned		
11 - Integrity mechanical, inspection and safety critical equipment maintenance: percentage of inspection recommendations completed	4	0.7769
12 - Fatigue risk management: critical alarms per operator hour	4	0.7658
13 - Emergency response plan: emergency response equipment in conformance to preventive or predictive maintenance schedule	4	0.7458
14 - Integrity mechanical, inspection and safety critical equipments maintenance: maintenance backlogs	4	0.7307
15 - Training and competency: training to critical positions in conformance with planned process safety management schedule	4	0.7060
16 - Incident investigation: quality of incident investigation and analysis	4	0.7026
17 - Actions items and audits follow up: number of process safety audits	4	0.7016
18 - Fatigue risk management: number of false alarms and disabled alarms	4	0.6873
19 - Training and competency: percentage of training, efficiency evaluate and grade average	4	0.6872
20 - Fatigue risk management: number of extended shift per operator	4	0.6869
21 - Incident investigation: number of lessons learnt communicated	4	0.6743
22 - Safety culture: number of field inspections by managers	4	0.6703
23 - Management of change: percentage of sampled management of change that satisfied all aspects of procedure requirements	4	0.6604
24 - Workforce involvement: percentage of workforce involved in process safety routine activities	4	0.6600
25 - Procedures: percentage of compliance with critical procedures by observation (safety critical tasks observed where all steps of the relevant safe working procedure were followed)	4	0.6550
26 - Integrity mechanical, inspection and safety critical equipments maintenance: percentage of safety critical equipment operating within specification when inspected	4	0.6235
27 - Emergency response plan: percentage of training to emergency response staff, per shift	4	0.6041
28 - Safety culture: A/(A+B), that: A = frequency that employment was publicly recognized for recording safety events, worried about safety have became lessons learnt or that brought a good contribution for safety; B = frequency that employment was publicly recognized for accomplish production demands or deadlines.	4	0.5975
29 - Permit to work: unsafe conditions or violations by observations	4	0.5945
30 - Incident investigation: percentage of corrective actions completed	4	0.5774
31 - Hazard identification and risk analysis: percentage of process risk analysis current, accuracy and reviewed to plan	4	0.5724
32 - Permit to work: percentage of permit work in conformance with procedure requirements	4	0.5236
33 - Stakeholders engagement: number of complaints received in facility (community, workforce)	4	0.5111
34 - Emergency response plan: emergency simulators completed versus emergency simulators have planned	4	0.4926
35 - Hazard identification and risk analysis: number of trained leaders and participants in the hazard identification and risk analysis	4	0.4630
36 - Actions items and audits follow up: percentage of corrective actions across the spectrum of Process Safety Management Systems completed on time	4	0.4200
37 - Contractors management: frequency of negative performance in safety assessments, field inspections, safe work practice audits and safety related audits	4	0.4165
38 - Emergency response plan: percentage of emergency response plans that are current, reviewed and revised within the designated period	4	0.4129
39 - Incident investigation: average time to closure recommendations	4	0.4064
40 - Contractors management: percentage of contractors with safety training schedule completed before work	4	0.3576
41 - Workforce involvement: rate of suggestions by workforce or report about unsafe conditions or near misses and rate progress over time	4	0.3509
42 - Actions items and audits follow up: average and maximum time to closure recommendations by Process Safety Management Systems	4	0.3367

43 - Incident investigation: rate between accidents occurred and near misses reported	4	0.3233
44 - Contractors management: percentage of attendance at safety meetings	4	0.2475

The second group of 9 experts ranked the 10 first leading indicators, as showed on Table 2, identifying the indicators as tier 3 or tier 4. The Table 2 results show that the greatest value of triangular fuzzy number occurred in tier 3 indicator called "Demand on safety systems" and the second higher value of triangular fuzzy number occurred in tier 4 indicator, called "Integrity mechanical, inspection and safety critical equipment maintenance". This result is similar as the first group. It shows similar opinion between the two groups of experts. All tier 3 indicators were among the top ten indicators of hierarchy. It shows too similar opinion between the two groups. The table 2 results related to criteria degree, showed there is great convergence between experts opinions, resulting in similar consensus coefficient. The table 2 results showed that the tier 4 indicators called "Hazard identification and risk analysis" and "Stakeholders engagement" were ranked in the first positions, demonstrating that the experts evaluated the indicators as high relevance.

Table 2: Ranking of process safety leading indicators by the second experts group

Hierarchy	Tier	Criteria degree
1 - Demand on safety systems: failure on demand or test in instrumented system	3	1.0000
2 - Integrity mechanical, inspection and safety critical equipment maintenance: number of inspections or conformance tests indicate that vessels, tanks, pipelines or other equipment was operated with safe limit operating exceeded (results outside acceptable limits)	4	0,9665
3 - Hazard identification and risk analysis: process safety risk analysis progress (number open, number closed, time to closure)	4	0,9650
4 – Safety culture: percentage of meeting that discuss about process safety	4	0,9544
5 – Stakeholders engagement: number of complaints received in facility (regulators, community, workforce)	4	0,9494
6 - Integrity mechanical, inspection and safety critical equipments maintenance: percentage of inspections of safety critical items of plant and equipment due during the measurement period	4	0.9471
7 - Demand on safety systems: number of safety instrumented system activations	3	0.9370
8 - Demand on safety systems: pressure relief device activations	3	0.9262
9 - Demand on safety systems: number of high level and high high level alarms activation in tanks	3	0.9213
10 - Fatigue risk management: critical alarms per operator hour	4	0.9135

5 References

[1] Podofillini L., Dang. V. N., Zio E., P. Baraldi, M. Librizzi. ***Using Expert Models in Human Reliability Analysis – a Dependence Assessment Method Based on Fuzzy Logic.*** **Risk Analysis 30, 8, pp. 1277-1297, 2010.**

[2] Seaver, D. A., and Stillwell, W. G. ***Procedures for using expert judgment to estimate human error probabilities in nuclear power plant operations (NUREG/CR-2743, SAND82-7054).*** **Albuquerque, NM: Sandia National Laboratories, March 1983.**

[3] Katherine D. Kohlhepp. ***Evaluation of the use of engineering judgments applied to analytical human reliablity analysis (HRA) methods.*** **A Thesis Submitted to the Office of Graduate Studies of Texas A&M University in partial fulfillment of the requirements for the degree of MASTER OF SCIENCE, December 2005.**

[4] ***Process Safety Performance Indicators for the Refining and Petrochemical Industries.*** **ANSI/API recommended practice 754, Second Edition, April 2016.**

[5] ***HSE 254. Developing process safety indicators: A step-by-step guide for chemical and major hazard industries,*** **2006.**

[6] OE/SHE Key Performance Indicators (KPIs). Applied Operational Excellence for the Oil, Gas, and Process Industries, 2015.

[7] Hsu H-M, Chen C-T. ***Aggregation of fuzzy opinions under group decision making. Expert Syst. Appl.*****, 1996.**

[8] Zadeh L. A. ***Information and control. Fuzzy Sets,*** **1965;8:338–53.**

[9] Nolan D. P., Anderson, E. T. ***Applied Operational Excellence for the Oil, Gas, and Process Industries.*** **Elsevier. MA, USA, 2015.**

[10] Luana F. M. Santos. ***Safety Leading Indicators Applied to a Petroleum Storage Terminal: Building an Indicator Panel.*** **A Thesis Submitted to the Federal University of Rio Janeiro, Environmental Engineering Program in partial fulfillment of the requirements for the degree of MASTER OF SCIENCE, December 2018**

Disciplina Total Incondicional. Reduciendo incidentes trabajando con indicadores proactivos Tier-4

Sergio D. Gavin
Pan American Energy
Leandro N. Alem 1180 – Buenos Aires.- Argentina
sergio.gavin@axionenergy.com

Marcelo M. Ledesma
Pan American Energy
Leandro N. Alem 1180 – Buenos Aires.- Argentina
marcelo.m.ledezma@axionenergy.com

Prepared for Presentation at
American Institute of Chemical Engineers
2019 Spring Meeting and 15th Global Congress on Process Safety
New Orleans, LA
March 31 – April 3, 2019

Disciplina Total Incondicional. Reduciendo incidentes trabajando con indicadores proactivos Tier-4

Sergio D. Gavin
Pan American Energy
Leandro N. Alem 1180 – Buenos Aires.- Argentina
sergio.gavin@axionenergy.com

Marcelo M. Ledesma
Pan American Energy

Keywords: disciplina operativa, disciplina impuesta, autodisciplina, cambio cultural.

Abstract

El objetivo de este trabajo es exponer el proceso de desarrollo e implementación exitoso del sistema de gestión de Disciplina Operativa que tuvo lugar dentro de la Gerencia de Producción de Refinería.

Tanto operadores como supervisores de producción eran muy jóvenes y con poca experiencia en la operación de plantas, con lo cual se hizo necesario implementar una manera sistemática de trabajar que redujera la cantidad de incidentes Tier 3, 2 y 1 que estaban ocurriendo. La clave estaba en la disciplina operativa y en crear indicadores proactivos Tier 4 para evitar recurrencia y reducir la tasa de incidentes.

El proceso inició con la definición de una Visión muy simple que pudiera ser fácilmente recordada por todos: "0000" (Cuádruple cero: cero incidentes de seguridad de procesos, de seguridad personal, ambiental y de calidad) y con una premisa de conducta que alcanzara desde la alta gerencia hasta el último operario: "Coherencia entre los que pienso, digo, hago y demuestro".

Fueron dos los factores claves del éxito. El primero considerar este sistema como un sistema vivo que iba creciendo, cambiando y ajustándose. La clave estuvo en mantener un estado permanente de alerta justo. Sin excederse, lo que podría causar desmotivación y parálisis, y sin relajarse, que podría transmitir un mensaje de moda pasajera en cuanto a la importancia de la seguridad. El segundo factor clave estaba basado en el concepto de "intervención". Se generaron herramientas que dieron a la supervisión una excusa para acercarse a hablar de seguridad con los operadores al mismo tiempo que los auditaban. A través de potenciar los encuentros de feedback se logró crear un círculo virtuoso. Dar más feedback mejora el clima laboral, esto genera más compromiso, que resulta en menos incidentes y por lo tanto en mejores resultados para la compañía que contribuyen a seguir mejorando el clima laboral.

La implementación se inició con la aplicación de disciplina externa o impuesta. Posteriormente se incorporaron poco a poco acciones cada vez más blandas como inculcar un alto compromiso por la capacitación, abandonar la postura de víctimas para convertirse en protagonistas, y finalmente desarrollar la disciplina interna o autodisciplina en todos los operadores.

La gestión sobre disciplina operativa está medida con indicadores proactivos focalizados en cómo los operadores están haciendo su trabajo. El objetivo es no tener incidentes donde la causa raíz sea la confiabilidad humana. Son calculados todos los meses y en función de los resultados se orienta el foco de las acciones de mejora del mes siguiente. Estos indicadores son analizados por cada Planta y Guardia generando un Tablero de Control que se revisa todos los meses con la presencia de la Gerencia, Jefatura, Supervisores y Operadores de Campo. Gracias a esta gestión el indicador (Tier1+Tier2)/200.000 hh se redujo un 75% en dos años.

1 Introducción

1.1 Indicadores de gestión de Seguridad de Procesos (API RP 754) [1]

La Gestión de la Seguridad de Procesos es una parte de los sistemas de gestión en las empresas la cual está relacionada con los peligros mayores que impactan en la salud de las personas, el daño al medio ambiente, la integridad de las instalaciones y las pérdidas económicas para el negocio.

La recomendación API RP 754 tiene el propósito de homologar las métricas retrospectivas de seguridad de procesos, es decir una vez ocurrido el incidente clasificarlos según unos criterios establecidos.

La clasificación es de 4 niveles distintos (Tier). Estos son:

Tier 1: Evento catastrófico o consecuencias mayores. LOPC (Loss of primary containment)

Tier 2: Evento de consecuencias menores. PSE (Process Safety Event)

Tier 3: Evento de demanda a los sistemas de seguridad

Tier 4: Eventos asociados a violaciones de la Disciplina Operativa y gerenciamiento de sistemas.

Según la clasificación de incidentes de procesos de CCPS (Center Chemical for Process Safety)[2] los dos primeros Tier son considerados métricas retrospectivas (lagging indicators) esto es, reportan estadísticamente incidentes ya ocurridos, mientras que los Tier 3 y Tier 4 se consideran métricas predictivas (leading indicators) esto es, son indicadores que nos advierten si estamos lejos o cerca de tener un incidente de nivel 1 o 2

Este trabajo describe el sistema de gestión creado para reducir los incidentes de nivel 1, 2 y 3 enfocándose en la creación, seguimiento y acción sobre indicadores que miden el estándar de Disciplina Operativa (indicadores Tier 4) exclusivamente relacionados con las actividades que se realizan dentro del área de operaciones de Refinería.

Como se explicará más adelante en este trabajo, hemos optado por re-definir los indicadores Tier 4 como *"proactivos"* y no *"predictivos"*, tal y como se enuncia en el título de este paper.

1.2 Contexto

Este trabajo expone el sistema de gestión creado para dar solución a un problema de repetición de incidentes cuya causa principal fue la falta de disciplina operativa en la División de Producción de refinería.

Los principales factores que contribuyeron a la pérdida de la disciplina operativa en el personal se considera que fueron los siguientes:

a) Recambio generacional de operadores. Hubo un grupo de operadores expertos y veteranos que se jubilaron y fueron reemplazados por operadores de entre 20 y 25 años sin experiencia en refinería.

b) Dificultad para promocionar operadores de consola a supervisores de turno dado que preferían no ascender ya que el puesto de supervisor de turno es fuera de convenio. Por esta razón esta posición es ocupada en su mayoría por personas ajenas a la refinería y la industria.

c) Alta rotación de aquellos supervisores de turno que no salieron del equipo de producción de Refinería

Esta combinación de operadores jóvenes, sin experiencia, más una supervisión con sólidas competencias genéricas pero con falta de competencias técnicas del mundo del refino propiciaron la pérdida de la disciplina operativa que llevaron a tener repetidos incidentes Tier 1 y 2. Por ejemplo: Error en aislamiento de energía, error en gestión de by pass de dispositivos críticos, incumplimientos de check list de campo, etc. Estos desvíos terminaron ocasionando lesiones al personal y daño a las instalaciones, incluso paradas de planta.

Este trabajo describe acciones concretas llevadas a cabo para mejorar la gestión. Describe "qué se hizo" y "cómo se hizo".

El método se basa en la "intervención" del supervisor de turno sobre los operadores. La misma representa una excusa para conversar y dar feedback no solo de temas de seguridad sino para generar un mayor acercamiento humano entre las partes. La Figura 1 resume este concepto.

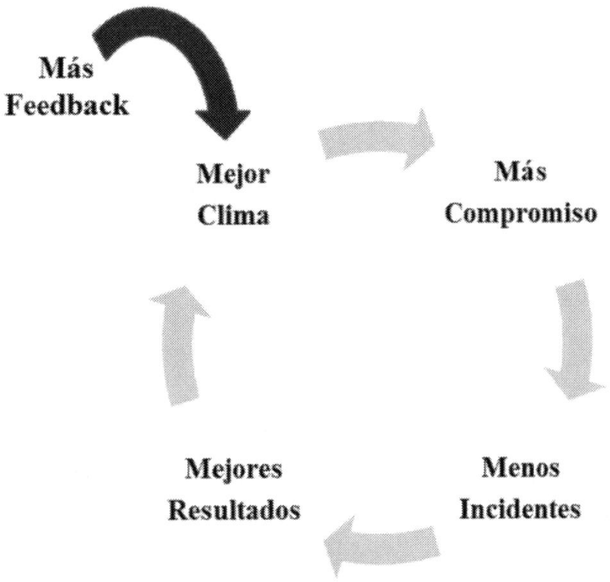

Figura 1

Otra clave del éxito consistió en mantener un estado de "alerta justo" (Figura 2). Esto se consiguió sumando todos los meses una nueva iniciativa de control vinculada a la seguridad. Se procuraba que no agregara mucha más demanda de tiempo a la supervisión de modo tal que pudiera ser cumplible y que a su vez se notara que la seguridad era el principal objetivo. Sin excederse, lo que podría causar desmotivación y parálisis, y sin relajarse, que podría transmitir un mensaje de moda pasajera en cuanto a la importancia de la seguridad.

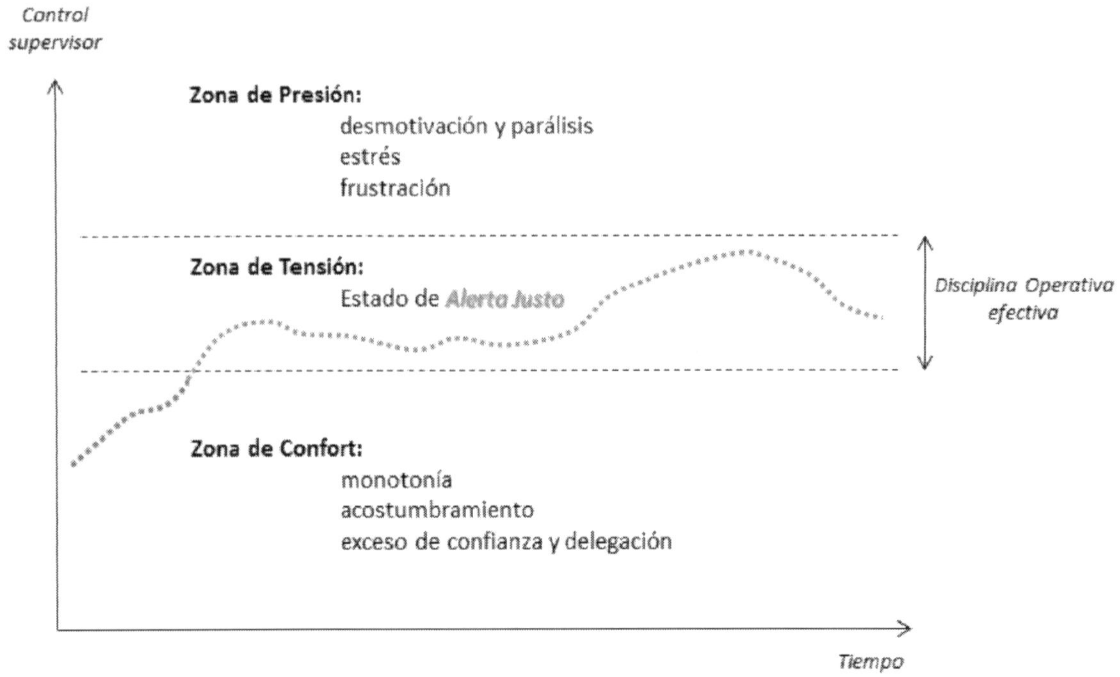

Figura 2

Este sistema de gestión se construyó desde cero, paso a paso, con pequeños nuevos controles que permitían mantener el estado de alerta justo.

A continuación se describen los pasos seguidos en la misma cronología en que se fueron implementando.

Definir la Visión y Expectativas de la División

Reforzar la Disciplina Operativa

Desarrollar Compromisos Totales

Adquirir Responsabilidad Incondicional

Lograr la Autodisciplina

La implementación de todas estas acciones estuvieron orientadas a trabajar sobre la base de la pirámide de los indicadores de seguridad de procesos del CCPS (indicadores Tier 4) relacionados con comportamientos inseguros y la falta de disciplina operativa. Estas métricas las consideramos *'proactivas'* y no *'predictivas'* porque mediante el análisis de qué es necesario controlar y medir en cada momento creamos nuevos controles para reducir su probabilidad de ocurrencia de esta forma abandonamos el concepto de "predictivas" ya que entendemos que es un enfoque pasivo del Sistema de Gestion de la

Seguridad, y pasamos al concepto de "Tier 4 Proactivo", denotando de esta manera una filosofía de acción sobre los desvíos que nos proporcionan los datos que luego de su proceso aportan información para trazar los planes de acción que tenderán a evitar proactivamente la ocurrencia de incidentes del tipo Tier 1 y 2.

La implementación y liderazgo de todas estas acciones se decidió que fuera un proceso cross. Esto es, si bien los supervisores de turno dependían de forma jerárquica del jefe de planta, en el caso de la implementación de todas estas acciones de seguridad reportaban directamente al superintendente de turno.

2 Definir la Visión y Expectativas

En primer lugar definimos una Visión simple, fácil de recordar, fácil de escribir e incluso fácil de representar con un simple gesto. "Cuádruple cero" o "0000" que consistía en cero incidentes personales, ambientales, industriales y de calidad.

También definimos una premisa que debe ser respetada por todos dentro de la División, comenzando por el más alto nivel. "Coherencia entre lo que pienso, digo, hago y demuestro". Si decimos que la seguridad es lo más importante y que así lo representa nuestra Visión, cualquier tema de seguridad es prioridad sobre el resto y así se debe demostrar. Un gerente o jefe no puede faltar nunca a una reunión de seguridad. Debe re agendar cualquier compromiso para cumplir con su presencia. Debe predicar con el ejemplo en todo ya que cualquier desvío o incongruencia entre el discurso y los hechos deteriora la confianza del equipo en su líder.

Sumado a lo anterior definimos las Expectativas que hay dentro de la División de Producción para con los colaboradores. Las mismas se resumen en una sola hoja y abarca los puntos de: presencia en campo, dispositivos críticos, cambio de turno, etc. Estas son repasadas entre supervisores y operadores en dos oportunidades por año siendo su cumplimiento condición de permanencia en la División.

3 Reforzar la Disciplina Operativa

3.1 Principios

Definimos la Disciplina Operativa como la dedicación y compromiso responsable de realizar la tarea de una manera correcta, todas las veces, desde la primera vez. Se basa en tres principios.

Conocimiento: Conocer la manera correcta de realizar las tareas, de acuerdo a los procedimientos y normativas vigentes. Comprender por qué la tarea está siendo realizada de esa manera.

Compromiso: Seguir los procedimientos, aplicarlos y desafiarlos a la mejora continua. Asumir la responsabilidad personal de comprometerse con los resultados. Cuidar la seguridad propia y la de su equipo de trabajo. Mantenerse concentrado en la tarea focalizándose en el cliente.

Estar Alerta: Mantener una visión amplia de lo que está ocurriendo a su alrededor. Reaccionar con rapidez y seguridad ante tareas que pueden desviarse de lo planeado. Pedir ayuda cuando no esté seguro. Si su tarea se ve alterada, revisar lo actuado, los procedimientos y rehacer el plan.

3.2 Aplicación

La forma de llevar a la práctica la disciplina operativa es a través de "Auditorías de campo". De esta manera forzamos la intervención y charla cara a cara entre supervisores y operadores.

Se trata de un control externo a la persona e impuesto por la gerencia de la División.

Con esta práctica no solo se verifican el estado operativo de la planta y los trabajos, y se produce la "intervención" sino que además se produce un "contagio bidireccional". Esto es, del supervisor al grupo y del grupo al supervisor, aprendiendo todos en ese intercambio.

Se diseñaron 18 tipos de auditorías (formularios de entre 5 y 20 preguntas). *'Presenciales'* y *'No Presenciales'*. Las presenciales se realizan con el operador de su propia guardia e implica un recorrido por la unidad. Las No Presenciales consisten en revisar información archivada en la sala de operadores sobre la cual se consultan aspectos de seguridad, cómo hacer la tarea, etc. Esta última generalmente se hace en turno noche o fines de semana y busca también forzar el acercamiento de supervisores y operadores.

Las auditorías diseñadas fueron:

De reuniones de trabajo	**De trabajos en campo**	
1) Cambio de Turno presencial	4) Permisos de Trabajo en ejecución	7) Ronda Operativa presencial
2) Charlar de 15 minutos de seguridad presencial	5) Aislamiento de energía mecánica	8) Operaciones en campo presencial
3) Condiciones no habituales de operación	6) Aislamiento de energía eléctrica	9) Verificación de cumplimiento de procedimientos

10) Operaciones de consola

11) Housekeeping

De verificación y validación

12) Permisos de Trabajo finalizados

13) Lazos críticos

14) Cambio de turno no presencial

15) Charla de 15 minutos de seguridad no presencial

16) Ronda operativa no presencial

17) Operaciones de campo no presencial

18) Evaluación de observaciones para prevenir incidentes

Cada una de estas auditorías se completan con la información de: fecha, supervisor auditor, área, guardia y una serie de preguntas que deben contestarse con una valoración entre 0, 1 o 2 (siendo el 0 muy mal y 2 muy bien). A su vez cada pregunta tiene una ponderación diferente dentro del puntaje final de la auditoría. Por ejemplo. En aislamiento de energía no pondera lo mismo que falte una tarjeta en un candado, a que falte colocar cadena con candado a una válvula, o que la válvula haya quedado abierta cuando debe quedar cerrada. Con todo esto se obtiene un resultado final expresado en porcentaje.

En un principio para calcular "la salud" de cada tipo de control implementado se promediaban los resultados de todas las auditorias del mes de cada tipo y la expectativa era que fuera mayor a 95%. Sin embargo este no era un buen indicador ya que podía tener por ejemplo 90% de auditorías con puntaje 100% y 10% con puntaje 50% que el promedio final me daría 95% y estaríamos considerando que el sistema está bien.

Se decidió medir el porcentaje de auditorías con puntaje menor al 90%, y el objetivo fijado es que sea inferior a 5%.

Una consideración que se tuvo en cuenta en la ponderación del peso relativo de cada pregunta en la auditoría es que aquella pregunta que es muy importante que esté bien respondida tiene mucho peso relativo, de modo tal que si se califica con 0 o 1 el resultado final de la auditoría será menor a 90%

Cada auditoría es cargada en una base de datos por los propios supervisores. Con la información almacenada se obtienen estadísticas de la cantidad de auditorías realizadas por área, por guardia, por supervisor: Ídem con cantidad de desvíos. Ídem con tipo de desvíos

El análisis incluye verificar que se realicen auditorías en todas las salas de operadores y en todas las guardias. No puede quedar ninguna planta sin cubrir por todas las guardias.

Se realizan análisis comparativos. Por ejemplo. ¿Por qué un área detectó más desvíos que otra? ¿Por qué una guardia encontró más desvíos que otra? A priori no hay respuestas correctas o incorrectas a estas preguntas ya que debe analizarse si hay más desvíos porque es una planta (o guardia) con menos disciplina operativa o bien es una planta (o guardia) con un ojo supervisor mejor entrenado.

Por cada uno de los 18 tipos de auditorías está definida una cantidad a realizar por planta y por guardia en el mes. En total se realizan más de 1000 por mes respondiendo alrededor de 15.000 preguntas cuyas respuestas quedan todas almacenadas en la base de datos que permite elaborar los informes mensuales.

A continuación en la Tabla 1 se observa la cantidad de auditorías realizadas en 2016, 2017 y 2018

	2016		2017		2018	
Tipo de Auditoría	Cantidad Auditorías	Cantidad de preguntas	Cantidad Auditorías	Cantidad de preguntas	Cantidad Auditorías	Cantidad de preguntas
PT pres	1.126	20.268	1.874	33.732	1.664	29.952
PT no pres	660	11.880	1.587	28.566	1.312	23.616
LOTO	473	5.203	912	10.032	1.061	11.671
CDE	343	3.430	778	7.780	859	8.590
LLCC	216	864	716	2.864	938	3.752
Cambio de Turno pres	151	1.963	485	6.305	654	8.502
Cambio de Turno no pres	185	1.665	584	5.256	692	6.228
Charla 15 min pres	294	1.764	698	4.188	747	4.482
Charla 15 min no pres	229	1.374	719	4.314	718	4.308
Operaciones en Campo pres	173	2.941	723	12.291	635	10.795
Operaciones en Campo no pres	105	1.155	461	5.071	58	638
AudiComp (Opi)	0	0	72	1.512	649	13.629
VCP	56	1.736	531	16.461	460	14.260
Housekeeping	0	0	187	22.440	247	29.640
Rondas Operativas pres	0	0	344	5.848	726	12.342
Rondas Operativas no pres	0	0	375	3.000	719	5.752
Consola	0	0	215	5.375	494	12.350
Operaciones No Rutinarias	0	0	46	460	822	8.220
Total	**4.011**	**54.243**	**11.307**	**175.495**	**13.455**	**208.727**

Tabla 1

4 Desarrollar Compromisos Totales

4.1 Principios

Definimos los Compromisos Totales como la dedicación y compromiso responsable y total de realizar todas las capacitaciones y entrenamientos que son necesarios para ser mejores operadores, dando confiabilidad a la operación del día a día y a la respuesta ante la emergencia.

Forman parte de los Compromisos Totales:

Refresco de las Expectativas de la División Producción

Cumplimiento del Plan anual de Simulacros de Emergencias Operativas

Conocimiento de las Matrices de Riesgo

Participación de las Prácticas de Lucha Contra Incendio

Asistencia perfecta al Curso de Operaciones Seguras

Participación de las charlas de difusión de lecciones aprendidas, boletines de seguridad, de hechos destacados, etc.

4.2 Aplicación

Cada una de las actividades mencionadas consta de un plan anual pre definido para ir cumpliendo mes a mes. Además de este plan, se incorporan según la necesidad, boletines de alertas de seguridad, boletines de actuaciones destacadas, repaso de procedimientos operativos, de sistemas de gestión, etc. Estas incorporaciones tienen que ver siempre con acciones concretas identificadas como una debilidad o foco de atención vinculado con la Disciplina Operativa.

Cada una de estas actividades queda reflejada en un acta con el nombre de los asistentes y los comentarios surgidos en la reunión. Se carga en un sistema informático que alimenta una base de datos que permite hacer seguimiento a las recomendaciones propuestas por operadores así como verificar que se cumpla el plan.

Los indicadores que permite obtener son principalmente dos. El primero que cada planta y guardia haya realizado la actividad, siendo el objetivo de 100%. El segundo tiene que ver con el porcentaje de personal alcanzado, siendo el objetivo anual de 80%. Este segundo indicador es muy importante ya que podría haber 100% de reuniones realizadas pero no haber alcanzado al total de la dotación por encontrarse algún operador ausente por enfermedad, vacaciones, etc. En este caso la supervisión es responsable cada mes de revisar el % de personal alcanzado y realizar un catch up, en especial en los Simulacros de Emergencia Operativa

5 Adquirir Responsabilidad Incondicional

5.1 Principios

Definimos la Responsabilidad Incondicional como el modo que tenemos dentro de la División Producción de enfrentarnos a los problemas del día a día, convirtiéndolos en desafíos que vamos a lograr resolver. Esto lo hacemos sin excusas y sin condiciones, dando siempre lo mejor de nosotros mismos hasta su resolución. Los principios de la Responsabilidad Incondicional que definidos son:

> No buscamos excusas para liberarnos de la culpa o la responsabilidad porque sabemos que las excusas deterioran la confianza y limitan nuestra capacidad para enfrentar desafíos y resolver problemas
>
> No adoptamos una posición de 'víctima' ante los problemas porque esto nos lleva a justificarnos y aceptar que la responsabilidad del error o desvío es de otro
>
> No queremos ser espectadores de los problemas, queremos ser los protagonistas de su resolución
>
> Siempre podemos hacer algo más para lograr que lo que queremos que suceda
>
> No decimos "hay que hacer..." o "podríamos ver...". Hablamos siempre en primera persona ("Yo") y decimos "... para evitar que esto vuelva a pasar YO voy a hacer...."
>
> Responsabilidad Incondicional es actuar siempre bajo los principios arriba mencionados de forma incondicional ante cada situación que nos toca enfrentar

5.2 Aplicación

A diferencia de la Disciplina Operativa y de los Compromisos Totales, la Responsabilidad Incondicional no tiene indicadores. Es el aspecto más *'blando'* del sistema de gestión.

Estos conceptos son repasados periódicamente por la gerencia y jefatura junto con la supervisión y a su vez estos lo repasan y discuten con sus operadores.

La forma de aplicarla consiste en conocer el concepto y desafiar los planteos que llegan al supervisor en forma de excusas o quejas improductivas por parte de los operadores, convertirlos en desafíos para estos.

La difusión y entendimiento del concepto, y el forzar a que todos hablasen en primera persona fue logrando un cambio que se notaba en el mayor involucramiento de los operadores. Esto era notado por el nivel supervisor que felizmente veía como sus operadores poco a poco comenzaban a llevar propuestas de solución a los problemas y no simplemente limitarse a transmitirlos.

6 Lograr la Autodisciplina

6.1 Principios

Todo el proceso descripto hasta ahora tuvo lugar durante un período de dos años. Paso a paso se fueron implementando cada una de las acciones detalladas, y poco a poco se fue notando un cambio en el nivel técnico y de compromiso del personal.

La Autodisciplina es el objetivo final de este sistema de gestión que pretende un cambio en la cultura y comportamiento de la organización y sus personas. Esto se logra transitando el camino de la seguridad desde la **disciplina** impuesta, la que paulatinamente evoluciona hacia el desarrollo de un **compromiso** total con todo aquello que nos permite aprender a ser mejores operadores y finalmente nos convierte en responsables de manera **incondicional** ante aquello que a cada uno nos toca enfrentar.

La combinación de estos tres conceptos es lo que dio nombre al sistema como ***'Disciplina Total Incondicional' (DTI)*** y su fin último es lograr la Autodisciplina del operador.

Con base en los tres pilares de la DTI se confeccionaron los "20 pasos de la Autodisciplina". Estos son:

Para todo y a todo momento:

1) Poseer actitud cuestionante positiva
2) Parar y preguntar cuando no esté seguro
3) Seguir siempre los procedimientos
4) Utilizar listas de verificación
5) Uso de EPP (elementos de protección personal)
6) Cumplir con las rondas del operador
7) Gestión de parámetros operativos, alarmas y dispositivos críticos
8) Mantener y hacer cumplir el Housekeeping

9) Comunicación efectiva

Al comenzar la jornada:

10) Cambio de turno

11) Charlas 15 minutos

Previo a cualquier tarea:

12) Repaso de procedimientos

13) Confección o repaso del Análisis de Tarea Segura (ATS)

14) Preparación de la tarea

15) Revisión del sitio de trabajo

En la entrega de la tarea:

16) Uso de Permisos de Trabajo

17) Aislamiento de energía. Correcta aplicación de LOTO (Lock out/Tag out)

Durante la tarea:

18) Análisis Previo de la Tarea (APT)

19) Observaciones para prevenir incidentes (OPIs)

Al finalizar la tarea:

20) Revisión del trabajo finalizado

Cada uno de estos pasos explican "por qué lo hago" y "cómo lo hago". En este orden y no al revés, ya que no es lo mismo decir "debes completar bien el permisos de trabajo para que no te lastimes" que decir "como no quiero que te lastimes tienes que completar bien el permiso de trabajo". En el segundo caso el mensaje es mejor recibido y el compromiso que genera es mayor.

Con estos 20 pasos se confeccionó una libreta de bolsillo que todo el personal de Producción debe llevar siempre consigo. Asimismo, la libreta de los supervisores incluye páginas adicionales ("Apéndice del Líder") que desarrollan conceptos de liderazgo enfocados en la seguridad y que constituyen una importante herramienta de ayuda para gestionar a los colaboradores a cargo con fuerte foco en la prevención de accidentes y/o incidentes. Los temas de este apéndice son: Disciplina Operativa y expectativas sobre el rol del Líder, Tres tipos de colaborares según su capacidad para identificar riesgos, Cuatro errores más comunes de un líder que pueden derivar en accidentes y/o incidentes, Refuerzos positivos y corrección de desvíos, Responsabilidades del líder como gestor de la prevención, Ciclo de la gestión segura del trabajo del líder, Roles del Líder, Delegación de la autoridad, Motivación, Comunicación, Construyendo Compromiso.

6.2 Aplicación

Para lograr la Autodisciplina la principal herramienta es la libreta. En primer lugar formó parte de los Compromisos Totales la difusión a todos los operadores del contenido de la libreta. Cada semana se difundían cuatro pasos, y al cabo de 5 semanas se repasaron los veinte. Entonces se lanzó la segunda etapa de la implementación de la libreta que se describe a continuación.

La libreta cuenta con un espacio donde cuatrimestralmente los operadores escriben en ella cuáles de los 20 pasos de la Autodisciplina se comprometen a aprender y dominar en el período. A su vez también tiene un espacio para que los supervisores o cualquier persona puedan preguntar por los 20 pasos y dejar registro de la calidad de la respuesta o grado de conocimiento.

De igual modo, la libreta de la supervisión cuenta también con un espacio para fijar compromisos cuatrimestrales vinculador al contenido del Apéndice del Líder que los supervisores se comprometen a aprender y aplicar.

7 Otras acciones

7.1 Un día en el Turno y Un día en otro turno

Jefes y gerentes de Área realizaron jornadas completas en turno rotativo junto a supervisores para detectar oportunidades de mejora en la gestión del día a día, así como realizar coaching sobre los supervisores, fundamentalmente relacionado con la gestión del personal. El plan armado contemplaba qué mandos medios de Producción según sus competencias más fuertes debían acompañar a qué supervisores según sus principales debilidades o puntos de mejora. El plan armado implicó que cada supervisor sería acompañado en total 3 veces en el año por un mando medio distinto en cada vez.

De un modo similar, los superintendentes de turno hicieron el acompañamiento de dos de sus compañeros con el objetivo de poder identificar oportunidades de mejora para copiar y compartir

7.2 Un día en HSE (Health, Safety and Environment)

Esta actividad consistió en programar recorridas de Housekeeping y Process Safety lideradas por el Departamento de Seguridad una vez a la semana por cada área productiva. La recorrida por planta es encabezada por el técnico de seguridad y participa un operario de Producción. Al final de la recorrida se realiza un informe de los hallazgos los cuales deben ser gestionados por el área, empoderando al operador de turno a poder él mismo gestionar la solución.

La finalidad de estas recorridas era:

1) Lograr el acercamiento y mejora de vínculo entre el personal de HSE y personal de Producción.

2) Capacitar al operario de turno al mismo tiempo que se hacía la recorrida de la planta.

Como parte de ser "coherente entre lo que pienso, digo, hago y demuestro" un representante del grupo de jefes y gerentes de la División Producción participan de la recorrida, mostrando su compromiso con la seguridad.

7.3 Tutelaje y coaching

A aquellos supervisores de turno más nuevos y con debilidades en sus competencias de liderazgo se les asignó un tutor. Tutelado y tutor un par de veces al mes salían juntos a la planta a realizar las tareas de supervisión o simplemente se sentaban a conversar sobre problemas existentes con la gente, con el gremio, con sus jefes, etc. y al final el tutor le daba feedback.

7.4 Refuerzo visual gráfico

Se desarrolló por parte de diseñadores gráficos cartelería con los conceptos de la Disciplina Operativa, Compromisos Totales, Responsabilidad Incondicional, Visión y Autodisciplina.

Se creó un logotipo de la DTI (Figura 3) el cual comenzó a estar presente en las planillas de las auditorías, informes, presentaciones, etc.

DTI DISCIPLINA TOTAL INCONDICIONAL

Figura 3

También se diseñaron cuadernos de notas tamaño A5 con el diseño de la DTI, incluyendo sus conceptos y los 20 pasos de la autodisciplina como separadores. Este cuaderno permite a los supervisores en cualquier charla con un operador remitirse rápidamente a los conceptos de la DTI y repasarlos.

7.5 Revisión de las Guías de Entrenamiento de todas las posiciones

En función de los preceptos determinados por la DTI se revisaron y actualizaron las guías de entrenamiento y exámenes de todas las posiciones de los operadores de Refinería.

8 Seguimiento

8.1 Informe mensual

Todos los meses se elabora un informe resumiendo las actividades de Disciplina Operativa y de Compromisos Totales. El mismo consta de más de 70 páginas

Por cada tipo de auditoría se representa gráficamente: cantidad total de auditorías, porcentaje con algún error, porcentaje que obtuvieron un puntaje menor a 90%, cantidad de auditorías realizadas por área y guardia, cantidad de auditorías con algún error por área y por guardia, cantidad de errores más comunes, cantidad de auditorías realizadas por cada guardia en cada sala de operadores.

El gerente de producción analiza cada gráfico del informe mensual y elabora un resumen comentando lo que más llama la atención en cada una de las hojas. Incluso para algunos puntos observados ya define una acción y responsable. Este análisis forma parte del mismo informe (última página) y se distribuye a todo el personal supervisor.

Como ejemplo, las observaciones que suele comentar son: que alguna guardia no cumplió con la cantidad objetivo, que alguna guardia no pasó por todas las salas de operadores a realizar algún tipo de auditorías, que un área (o planta) tiene más desvíos detectados que otra y que debe ser analizado, que alguna guardia no hizo todos los simulacros del mes, que el porcentaje de personal alcanzado por alguna capacitación es bajo y no llega al objetivo, etc.

A continuación se muestra como ejemplo algunas estadísticas y resultados de auditorías de Cambio de Turno Presencial. En cada gráfico se resalta en color rojo el principal punto que se desea destacar y comentar.

En Gráfico 1 se puede observar como el porcentaje de auditorías con puntaje menor a 90% fue mejorando mes a mes.

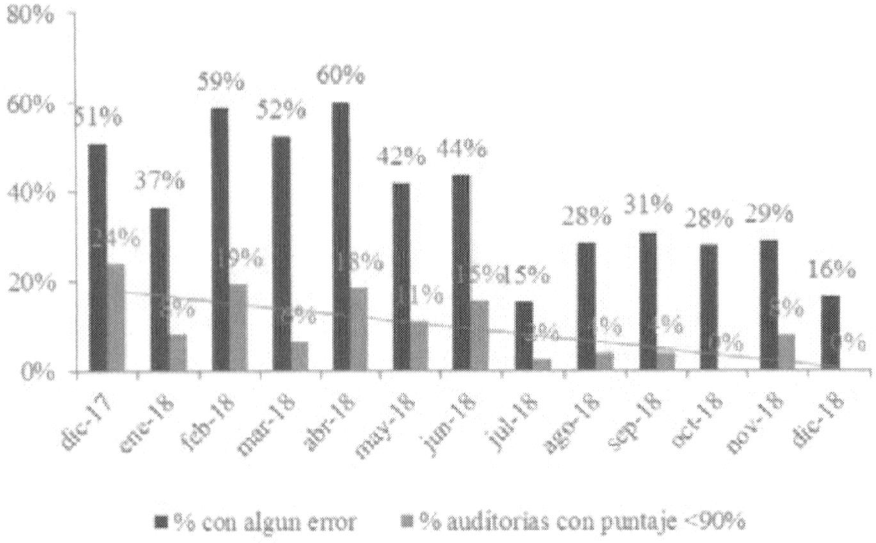

Gráfico 1

A continuación se grafican los desvíos más comunes detectados (Gráfico 2). Se observa como fue corrigiéndose la costumbre de no mencionar si hubo o no incidentes en Refinería.

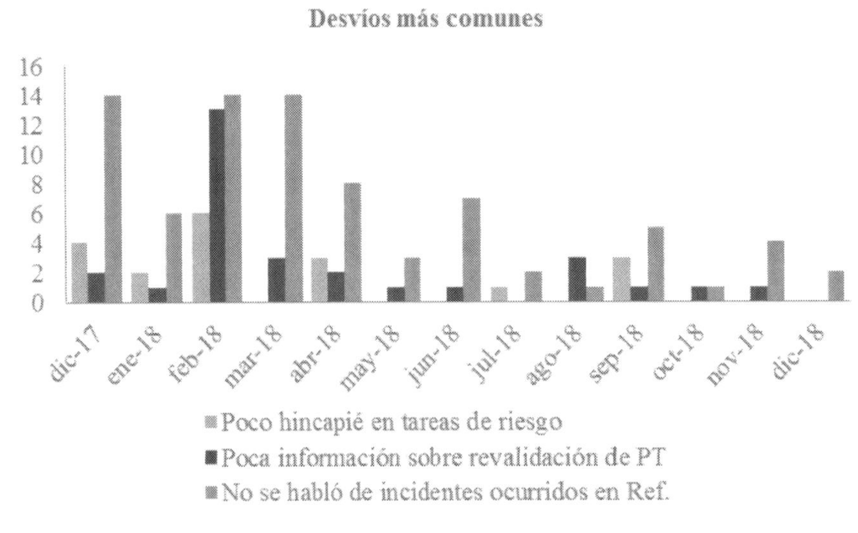

Gráfico 2

En el Gráfico 3 se observa como fue mejorando el hecho de que no se recordaban o mencionaban en el pasaje de guardia las condiciones no habituales de operación

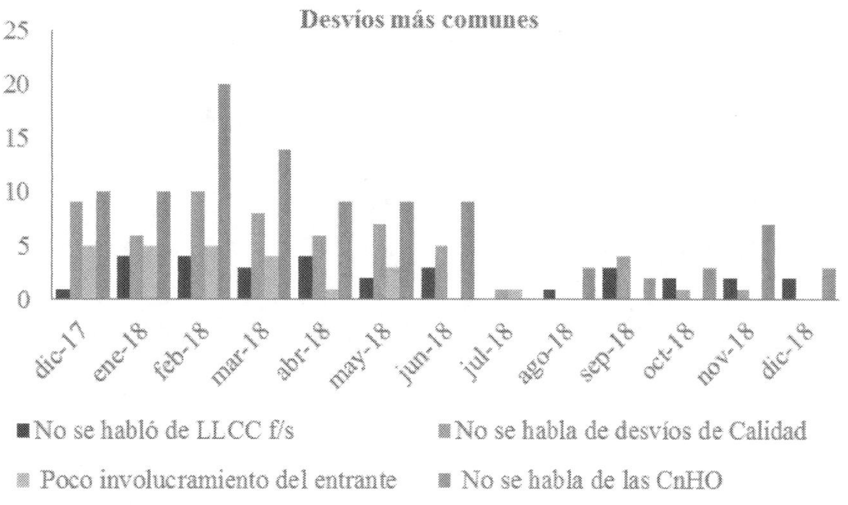

Gráfico 3

Observando el Gráfico 5 sirve para generar por ejemplo la pregunta acerca de por qué en el Área 2 y Área 3 se realizaron en diciembre la mitad de auditorías que las demás áreas.

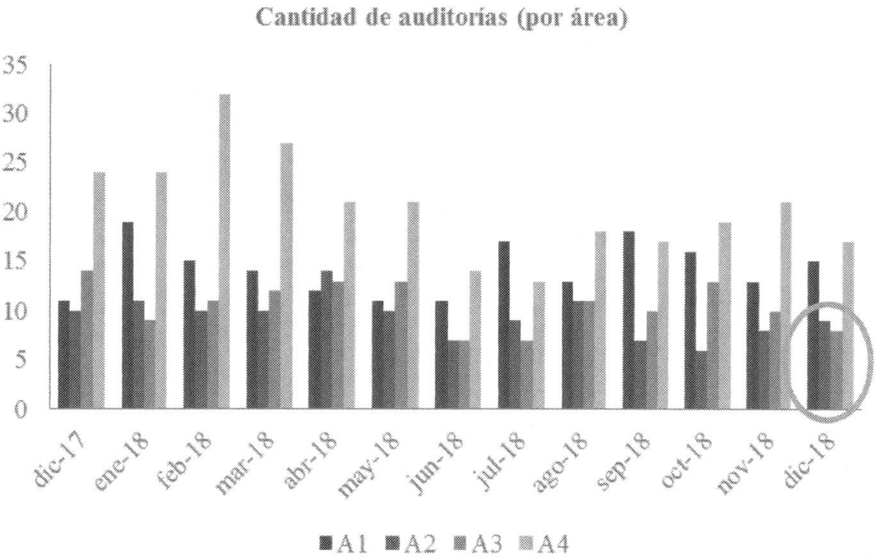

Gráfico 4

De un modo similar, de la lectura de los Gráficos 5 y 6 surgen las preguntas de por qué en diciembre en el Área 2 no se encontró ningún tipo de desvío mientras que en las demás áreas si hubo y por qué la Guardia A no encontró ningún tipo de desvío.

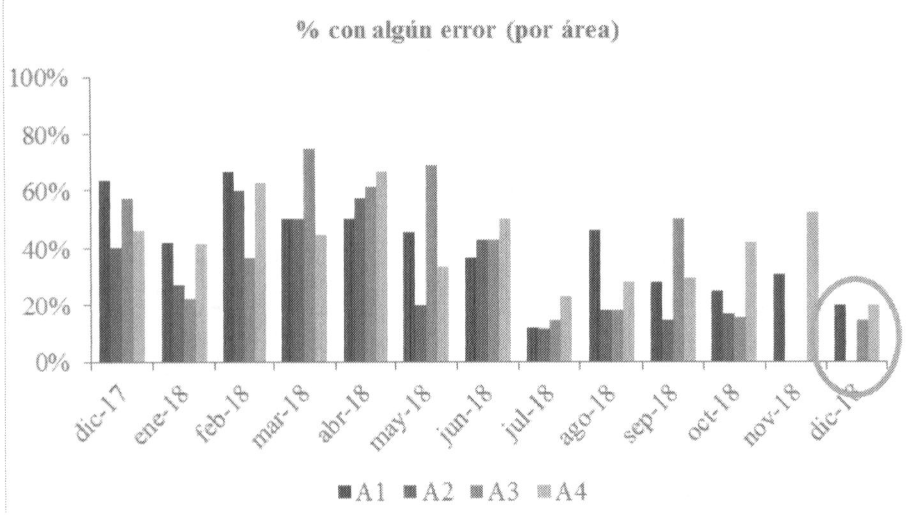

Gráfico 5

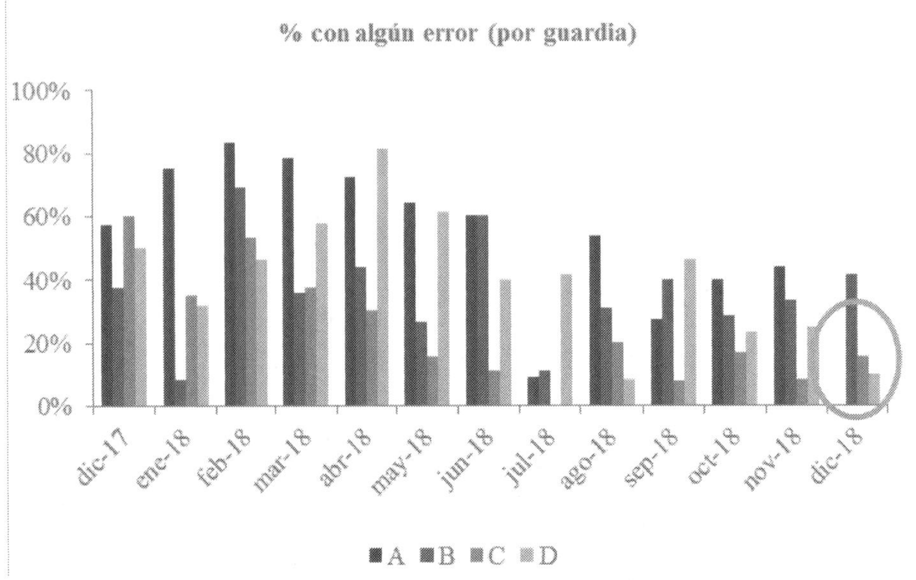

Gráfico 6

En el informe mensual también se hace seguimiento a la actividad de Compromisos Totales que también tiene sus indicadores como por ejemplo cumplimiento de los planes de simulacros de emergencias operativas y del personal alcanzado.

Además incluye un ranking (Top 5 y Bottom 5) donde se identifican aquellos supervisores que más y menos auditorías hicieron en el mes. Esto promueve una competencia entre ellos, tanto por estar en el "Top" como por no aparecer en el "Bottom". De un modo similar también se calcula un ranking por Guardia y otro por Planta generando el mismo efecto en los superintendentes y jefes de planta respectivamente.

8.2 Reuniones de seguimiento

Cada mes se realizan las siguientes reuniones

1) Por cada guardia se reúnen superintendente de turno con sus supervisores para analizar el informe elaborado por la gerencia

2) El gerente de producción se reúne con cada una de las 4 guardias (superintendente y supervisores) donde ésta le hace una presentación con su análisis de situación de la DTI

3) Finalmente, los supervisores difunden esa reunión con los operadores de su guardia.

Para lograr el contagio de la DTI en todos los gerentes y mandos medios de la División de Producción y que no solo los superintendentes tuvieran el ownership sobre ésta se decidió que el informe fuera elaborado cada mes por un gerente diferente y que alternativamente la reunión con el gerente de producción fuera presentada por Guardia (superintendente con un supervisor de cada planta) y por Planta (jefe de planta con todos los supervisores de su misma planta). De esta manera se logró mayor involucramiento de todos los mandos de la División.

Una vez establecido el estándar de presentación y que los planes iniciales fueran lanzados, se invitaron a las reuniones a operadores de consola y de campo con el objetivo de dar mayor visibilidad y entendimiento al trabajo que estaba realizando la supervisión de turno, enfatizando además que este sistema de gestión protege al operador, capacita, compromete y desafía a mantener a toda la División enfocada en la seguridad. De esta manera se logró demostrar a todos los colaboradores que la gestión de la DTI, estaba lejos de utilizarse de manera punitoria, y que en realidad ayudaba a mejorar las aptitudes.

9 Planes de acción

Se definieron guardias *'Champion'* que tienen a cargo entre cuatro y cinco tipos de auditorías de Disciplina Operativa que deben liderar para su mejora. No solo en su guardia sino en todas. Por ejemplo. La guardia A es la responsable de mejorar el Cambio de Turno y la Charla de 15 minutos, la guardia B aislamiento de energía mecánico y eléctrico, la C lazos críticos y alarmas, la D permisos de trabajo y rondas operativas.

Cada una analiza en profundidad estas auditorías y plantea planes de acción cross, estos planes en todo momento representan una acción proactiva, tendiente a corregir el desvío, extrapolándolo en ocasiones a todos aquellos sectores de la Refinería que por similitud de operaciones o escenarios pudieren incurrir en un desvío igual o equivalente.

En las reuniones con el gerente se realizaba una minuta con las acciones y compromisos surgidos, los cuales se volcaban a una base de datos para su seguimiento. En las reuniones del mes siguiente se repasan estos puntos para ver su cumplimiento

Como se mencionó en el punto anterior, luego de dos años comenzamos a incorporar a las reuniones de la gerencia con cada guardia a un operador de turno de cada planta con el fin de que vieran más de cerca como vivía la gerencia y mandos medios la DTI, y sobre todo trasmitirles que en ningún caso las auditorias y controles tenían como fin sancionarlos (nunca se usó para eso) y para poder escuchar de los propios operadores cuales eran sus principales preocupaciones y opinión sobre la utilidad o no de la DTI

10 Resultados

Gracias a esta gestión el indicador (Tier1+Tier2)/200.000 hh (Gráfico 7) se redujo un 75% en dos años. Pasamos de tener un incidente operativo Tier 1 vinculado a errores del operador cada tres meses a no tener por más de un año.

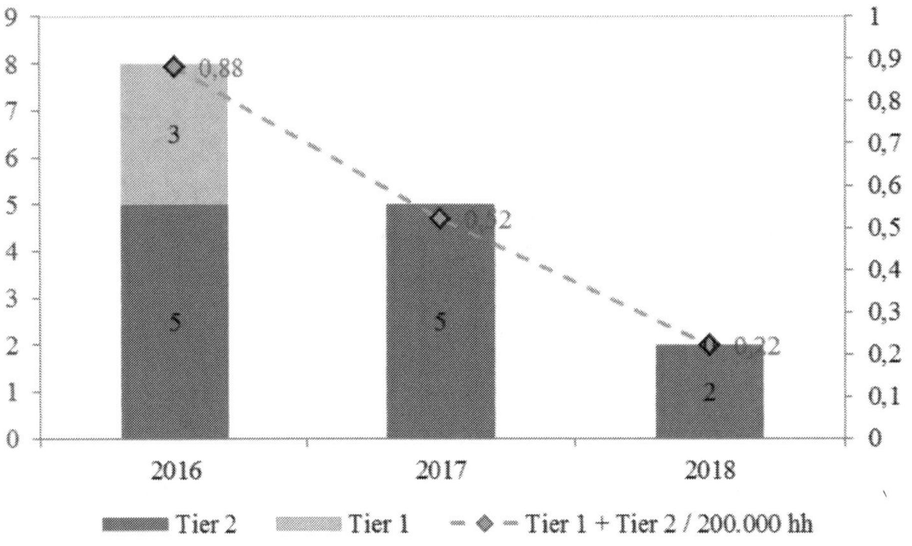

Gráfico 7

11 Conclusiones

Los anteriores resultados muestran un afianzamiento de la cultura de seguridad. Se pudo apreciar que los operadores pasaron en un principio de sentirse vigilados a entender que el objetivo de las auditorías era el de ayudar a registrar todo lo que se hacía y a detectar los desvíos más recurrentes o graves para poder gestionarlos.

Progresivamente los operadores empezaron a competir sanamente entre ellos, a querer ser la mejor guardia y a querer volverse referentes de temas específicos.

Los supervisores también cambiaron. En un principio consideraban que realizar las auditorías e ingresarlas al sistema era una carga administrativa, pero poco a poco vieron sus frutos a través del cambio en los comportamientos de los operadores quienes demostraban cada vez un mayor compromiso por hacer sus tareas con seguridad rigurosa.

12 Referencias

[1] American Petroleum Institute (API) Recommended Practice (RP) 754, Process Safety Performance Indicators for the Refining and Petrochemical Industries, 2nd Edition, April 2016

[2] *Métricas Predictivas y Retrospectivas para Seguridad de Procesos*. Center for Chemical Process Safety, American Institute of Chemical Engineers, Dic. 20, 2007.

546834 Seguridad y Medio Ambiente Como Valor! Un Camino Para La Toma De desiciÃ³nes Basada En Riesgo

Seguridad y Medio Ambiente como Valores

La búsqueda del sentido de la vulnerabilidad y toma de decisiones en función del Riesgo.

Juan Manuel Fernandez Becerra,

YPF S.A. - Regional Centro

jmfernandezb@ypf.com

Resumen

El Negocio Upstream de YPF cuenta con 4 regionales en Argentina;

Regional Norte; que incluye la provincia de Mendoza y norte de la provincia de Neuquén.

Regional No Convencional; que incluye parte de la provincia de Neuquén.

Regional Centro; que incluye parte del centro de la provincia de Neuquén y este de la Provincia de Río Negro.

La región Centro comienza con un programa para disminuir la cantidad de accidentes e incidentes ambientales en el año 2014, logrando excelentes resultados en su performance.

En el año 2018 este programa es Modificado para que incluya los 4 pilares de la seguridad de los procesos basada en Riesgo.

Introducción

El año 2013 lo finaliza la Región Centro con 61 accidentes con pérdidas de días, en una población de alrededor de 9.700 operarios.

Como consecuencia de esto, busca entre otras cosas, un nuevo desafío para reemplazar las capacitaciones tradicionales aspirando a que los grupos alcanzados incorporen a la Seguridad y Medioambiente como un VALOR.

A través de un cambio CULTURAL, con "**Talleres de Reflexión y Trabajo**", donde los participantes pudieran vivenciar y aplicar de manera inmediata en su trabajo y vida personal conocimientos, herramientas o procesos que lo ayudasen a desempeñarse de forma interpersonal logrando una continua mejora en sus resultados y en los de la organización.

Se formaron grupos denominados equipo motor donde su principal función es que participen voluntariamente en Talleres de Trabajo, donde se los forma y se les brinda competencias necesarias para que puedan certificar como "**Tutores en Cultura Preventiva de Seguridad y Medioambiente**". Para ello deben cumplir una serie de requisitos evaluativos y la presentación de un trabajo final que genere valor agregado a la organización. Los principales objetivos del equipo motor son:

v Inspirar e influir en su entorno más cercano para aumentar el nivel de conciencia en temas de seguridad de las personas y cuidados del medioambiente"

v "Generar efecto multiplicador de agentes de cambio en la cultura preventiva."

En la actualidad, este programa se está implementando en todas las regionales del Upstream de YPF.

Desarrollo

Cómo se mencionó en la introducción, el programa de Seguridad y Medio Ambiente como Valores surge como una medida para mejorar la performance de la Región Centro.

El resultado de la implementación de este programa impactó directamente en la cultura del personal de la Regional Centro. En el siguiente gráfico se muestra cómo disminuyeron los accidentes con pérdidas de días.

La creación del equipo motor surge en el año 2017 donde cada Gerente/Jefe eligió en sus equipos de trabajo, personas que venían participando activamente desde 2014 en los Talleres de Seguridad y Medioambiente como Valores y que en su desempeño diario mostraban estar **comprometidos espontáneamente** con la Seguridad y Medioambiente. El proceso re repite en 2018 con un 2do. Equipo Motor.

- *Identificar a través de un mayor acercamiento/conocimiento a las personas de nuestros equipos -circulo de influencia-, "agentes de cambio" (apóstoles) que promuevan e inspiren comportamientos seguros en búsqueda de la interdependencia para alcanzar mejores niveles en la prevención de accidentes.*

Previo a la creación de los equipos motor, en los inicios del programa se crean los Equipos Naturales con el objetivo de:

"Generar mayor confianza e interdependencia en equipos naturales, incluyendo personal propios y contratados para promover la cultura de Seguridad y Medioambiente como Valor, para reducir las pérdidas humanas, materiales y medioambientales para la sustentabilidad del negocio.

"Lograr relaciones de mayor confianza y entendimiento en nuestros circulo de influencia, de esta manera conocer en profundidad lo que piensa y siente la persona pudiendo así evitar accidentes generados por stress, cansancio, y problemas personales que alteran nuestra atención diaria"

En el año 2018 el programa se modifica asumiendo un modelo de seguridad basada en riesgo, tomando como partida los cuatro pilares RBPS.

Siendo los pilares:

v Compromiso.

v Entender los Peligros y Riesgos.

v Administración del Riesgo

v Aprender de la Experiencia.

El objetivo del programa es la toma de decisiones en función del riesgo y lograr sentido de la vulnerabilidad para todos integrantes de la Regional Centro.

Para cada pilar se han establecido elementos de trabajo con objetivos con objetivos individuales y grupales concretos.

Compromiso:

- Salidas a campo
- Comités CMASS en las operaciones
- Programa de "Cultura Compartida con Contratistas"
- Reconocimientos formales/Actividades con la familia
- Programa visita Reaseguradores
- Formación de líderes/Competencias en Seguridad de los Procesos
- Certificación de oficios

Entender Peligros y Riesgos:

- Cumplimiento de la Norma10065
- Auditorias y seguimiento acciones LPO (Licencia Para Operar)
- Auditorías Fichas de Riesgos
- Generar una herramienta de planificación de tareas en función del riesgo
- Normalización de los desvíos

Administración del Riesgo:

- Manejo del Cambio
- Plan reparaciones provisorias/Plan PSV
- By pass alarmas, enclavamientos
- Seguimiento de estudios de riesgo (acciones)
- Bow Tie en P&W y Pulling
- Plan de Inspección elementos críticos
- Control de vehículos/parque cerrado
- Implementación OMS
- Administración de la emergencia/Programa APELL

Aprender de la experiencia:

- Incorporar indicadores SP a los comités regionales
- Incorporación de nuevas herramientas LADS
- Difusión de lecciones aprendidas
- Presentación en comité regional de eventos
- Introducir el concepto de "Línea de Fuego"

Para cada uno de estos sub-elementos, se plantean objetivos personales y de equipo para todos los integrantes de la región centro.

Resultados.

El programa de Seguridad y Medio Ambiente como Valores representa para la Región Centro, el pilar fundamental para el cuidado de la Seguridad, la Salud, el Medio Ambiente y la Integridad de su Instalaciones.

Desde su implementación y en los 4 años de vida el resultado del programa ha sido totalmente exitoso llegando a disminuir en el año 2018 la totalidad de accidentes con pérdidas de días.

Con los nuevos objetivos planteados, se pretende dar al programa un nuevo enfoque de la seguridad basada en riesgo, que le permitirá que tenga un ciclo de 10 años, siendo un éxito para la empresa.

546289 Entendiendo Los Riesgos De Ciberseguridad y Su conexiÃ³n e Impacto Con La Seguridad De Procesos

Entendiendo los riesgos de Ciberseguridad y su conexión e impacto con la Seguridad de Procesos

Con el desarrollo del llamado internet de las cosas (IoT) , la conectividad que se tiene a todo nivel y específicamente entre las redes de control y las corporativas con el mundo exterior se han incrementado los riesgos por ciberseguridad y su impacto a nivel de seguridad de procesos por afectación en las redes de control u Operations Technology.

Se estima que actualmente hay 7.600 millones de smartphones conectados y para el año 2.020 se calcula que tendremos 200 mil millones de dispositivos incluyendo todo equipo de tecnología móvil, esto nos da una idea de que la conectividad será total, pero igualmente los riesgos a los que estamos y estaremos expuestos a todo nivel.

Desde el año 2.015 en el World Economic Forum se ha visto cómo los Ciberataques han pasado en nivel de riesgo de un quinto (5) lugar, a un tercer (3) lugar en el año 2.018, sólo superado por Desastre Naturales y en primer (1) lugar por eventos climáticos extremos.

Diariamente oímos hablar a los riesgos que estamos expuestos por ingreso a nuestras cuentas de ahorros, al robo de información de nuestros correos electrónicos , a la denominada ciberextorsion por acceso a nuestra información personal y sensible en nuestros computadores o celulares, pero nos hemos preguntado y somos conscientes de los riesgos a los que estamos expuestos en nuestras facilidades de operación de hidrocarburos, plantas nucleares, químicas por esta causa?, conocemos de eventos que hayan ocurrido de este tipo en alguna refinería, me atrevería a decir sin temor a equivocarme que es posible que a lo sumo hayamos escuchado de Stuxnet , un virus informático que atacó a los PLCs Siemens S-7 , los cuales controlaban las bombas centrífugas en una de las plantas de enriquecimiento de uranio que hacían parte del programa Iraní para el desarrolla de un arma nuclear, este ataque al parecer fue promovido por USA e Israel, la consecuencia un retraso de varios años en su programa debido a que sus bombas casi se destruyen sin que sus técnico se percataran de ello, las estaciones de operación mostraban un proceso dentro de parámetros normales , pero internamente sucedía todo lo contrario.

Aunque usted no lo crea los riesgos de Seguridad de Procesos originados por explotación de vulnerabilidades cibernéticas van en aumento, y han pasado de un tema de retos para accesar una gran compañía a objetivos tales como bloquear Sistemas Instrumentados de Seguridad para dejar desprotegida una refinería, seguramente para generar explosiones las cuales serían de una más alta probabilidad con sus sistemas de protección fuera de servicio.

El presente documento tiene como objetivos:

1. Explicar de una forma sencilla qué es Ciberseguridad y establecer la conexión que existe entre las vulnerabilidades de Ciberseguridad y que se pueden convertir en eventos iniciadores para eventos de seguridad de procesos e incluso cómo pueden estas vulnerabilidades inhibir incluso las capas de protección instrumentadas y componentes de los sistemas de control como las estaciones de operación, conduciendo a pérdida de capacidad operativa o entregando información errónea para la toma adecuada de decisiones.

Para explicar la conexión entre ciberseguridad y seguridad de proceso se empleará el diagrama del queso suizo , lo que permitirá identificar aquellos términos de ambas disciplinas que son diferentes pero que cumplen similares funciones

2. Concientizar sobre la importancia de un plan de manejo de la ciberseguridad mostrando qué tipos de motivaciones llevan a este tipo de ataques, los vectores que se emplean y el incremento de la complejidad tecnológica para el desarrollo de nuevos virus no sólo por "lobos solitarios", sino por grupos y estados que están desarrollando armas cibernéticas para atacar a sus enemigos, muchas veces en forma silenciosa y a relativo bajo costo . Muchos países tienen sus propios ejércitos cibernéticos, algunos de ellos como USA, Rusia, China, Corea del Norte, Israel , que están buscando atacar la denominada infraestructura crítica, dentro de la cual se encuentra la industria nuclear, de hidrocarburos, química, de alimentos, de generación energética, agua y sistema financiero.

Se hará una descripción cronológica de los diez (10) ataques cibernéticos más representativos a la industria de Oil and Gas.

3. Explicación del ciclo de vida de la ciberseguridad a la luz de la ISA 99 y las características principales de varias de la normas empleadas como la NIST 800-53, IEC 62443 e ISO 27002, de tal forma que no se preste para la confusión que tenemos muchos de nosotros hoy en día.

ACTUALIZACIÓN DE ESTÁNDARES DE PHA

Sergio Fernandez
Pan American Energy
Buenos Aires, Argentina
sfernandez@pan-energy.com

Matías A. Massello
Process Improvement Institute
La Plata, Argentina
mmassello@piii.com

Keywords: PHA, HAZOP, LOPA, Risk Management.

Prepared for Presentation at
American Institute of Chemical Engineers
2019 Spring Meeting and 15th Global Congress on Process Safety
New Orleans, LA
March 31 – April 3, 2019

ACTUALIZACIÓN DE ESTÁNDARES DE PHA

Autores

Sergio Fernandez – Safety & Risk Management Manager
Pan American Energy (PAE)
e-mail: sfernandez@pan-energy.com

Matías A. Massello – Process Safety Engineer
Process Improvement Institute, Inc. (PII)
e-mail: mmassello@piii.com

Colaboradores

William G. Bridges – President
Process Improvement Institute, Inc. (PII)
e-mail: wbridges@piii.com

Federico C. Caldora – Safety, Enviroment & Risk Management VP
Pan American Energy (PAE)
e-mail: fcaldora@pan-energy.com

Abstract

Durante la última década Latinoamérica ha dado pasos importantes hacia la implementación total de Seguridad de Procesos. Las compañías son más maduras y consientes de la necesidad de implementar un programa de PSM; a su vez, algunos países comenzaron a tener regulación gubernamental sobre PSM (como Perú, Chile y Colombia). Este proceso evolutivo presenta una buena oportunidad para aprender de lo que ha funcionado en otros lugares y para compartir nuestras experiencias con otros para crear un efecto sinérgico y avanzar con mayor rapidez hacia procesos más seguros.

Este paper se centra en el elemento de Identificación de Peligros. Los Análisis de Peligros de Procesos (PHA) es un elemento clave en la Seguridad de Procesos – es imposible de controlar una amenaza a las personas, medioambiente y/o negocio si dicha amenaza no ha sido identificada como tal. Este paper comparte los pasos tomados en Pan American Energy para mejorar sus estándares de PHA: el disparador para la revisión y actualización de los estándares, el análisis de brechas para comparar la situación actual versus las Mejores Prácticas, las decisiones de los Líderes para obtener la aprobación de todas las partes involucradas, y los resultados de las mejoras.

Introducción

La empresa

Pan American Energía es una compañía del sector de energía, en Petróleo y Gas con presencia en las cuencas del Golfo de San Jorge, Neuquina, Noroeste (NOA) y Golfo de México como operador; y en las cuencas Marina Austral y Tarija (Bolivia) como socio.

En refino y venta de combustibles y lubricantes, con la operación de una refinería en Campana (Argentina) y una red de estaciones de servicio en Argentina, Uruguay y Paraguay.

En Energía Renovables, con el proyecto del Parque Eólico de Garayalde en Argentina dentro del Programa Renovar, con PEM durante el 2018.

En números PAE puede describirse así:

- 18% de la participación en el mercado de hidrocarburos de Argentina.
- 147% de tasa de reposición de reservas durante los últimos 15 años.
- 22000 personas trabajan directa o indirectamente en la compañía.
- USD 15000 millones invertidos en exploración y producción de hidrocarburos entre 2001 y 2017.
- 15% de participación en comercialización de combustible en Argentina.
- 20% de incremento de producción de combustibles desde el 2012.
- USD 1500 millones de inversión en la expansión refinería de Campana.

Historia en análisis de riesgos

La gestión de identificación de peligros y evaluación de riesgos en la compañía data desde los años noventa. Se comenzó con la realización de HAZOPs (Hazard and Operability) en proyectos grandes o instalaciones de superficie importantes. En un principio, dichos estudios no se realizaron en forma sistemática o dentro de un sistema de gestión de riesgos, las mismas empresas que realizaban los proyectos proveían el servicio de análisis de peligros y los típicos de matrices de riesgo. No se contaba con el nivel de aceptabilidad del riesgo en formato de Matriz Riesgo que unificara lo criterios.

A partir del año 2004, aproximadamente, se incorporaron los análisis de peligros o PHAs con la modalidad HAZOP en forma sistemática a los proyectos que la compañía desarrollaba para las instalaciones de superficie y para instalaciones existentes con cambios. Los estudios HAZID (Hazard Identification) para nuevos proyectos con ingenierías no conocidas o en los inicios de un proyecto de envergadura para la identificación de peligros, de estudios o análisis específicos y la formulación de contramedidas en las etapas tempranas.

Paralelamente a esto, se trabajó en la implementación de una Matriz de Riesgo, y posteriormente se avanzó en el análisis de capas de protección mediante la incorporación de la metodología LOPA (Layer of Protection Analysis), donde se comenzó a analizar semicuantitativamente la frecuencia de ocurrencia de eventos iniciadores de alta severidad con el objetivo de verificar el nivel de protección de las capas de protección implementadas en relación al nivel de aceptabilidad del riesgo de la compañía.

A tales efectos, se han desarrollado estándares o guías que norman las metodologías a utilizar y que permiten conducir los análisis con el mismo nivel de resultados independientemente del facilitador que los conduzca.

Las actuales guías constan de tres principales análisis para los proyectos de nuevas instalaciones de superficie o para instalaciones sujetas a cambios o revamps y en operación:

- HAZID (Hazard Identification)
- HAZOP (Hazard Operability)
- LOPA (Layer of Protection Anlysis)

Auditoría a los estándares existentes

En 2013, PAE trabajo en forma conjunta con Process Improvement Institute (PII) para realizar una evaluación de los estándares existentes de PHA/HAZOP y LOPA y de su implementación.

Motivación de la auditoría

Si bien los documentos de trabajo de PAE contaban con un respaldo de historia importante, la mejora continua es parte de la identidad de la empresa, siendo esto lo que principalmente lleva esta dar curso a la auditoría externa.

Metodología de la auditoría

La auditoría se llevó a cabo en dos etapas:

1. Revisión de los estándares existentes y otros documentos relacionados (informes de HAZOP y LOPA, matriz de riesgos, etc.).
2. Participación del auditor en una sesión de HAZOP/LOPA como observador.

Hallazgos de la auditoría

De la evaluación realizada surgieron varios hallazgos/recomendaciones. A continuación se muestran algunas de ellas tanto para el estandar existente de HAZOP como para el de LOPA.

HAZOP

- **Falta de análisis de los modos de operación no rutinarios** (tanto en instalaciones existentes como en proyectos de nuevas instalaciones). Aproximadamente el 80% de los accidentes en la industria ocurren durante los modos de operación no continuos, por lo que es de vital importancia
- **Poca cantidad de PHAs a lo largo de un proyecto**. Generalmente, PAE solo realizaba HAZOP (PHA) una vez concluída la etapa de diseño, pero la Mejor Práctica es realizar entre 3 - 6 PHAs a lo largo del proyecto (dependiendo de la magnitud y complejidad del mismo). A su vez, es importante la inclusión del PHA de los Procedimientos una vez que estos hayan sido escritos.
- **El criterio para la selección de Nodos no seguía las Mejores Prácticas**. En la metodología HAZOP, durante el análisis del modo de operación continuo, los Nodos utilizados eran demasiado grandes. A priori incluir las cañerías de entrada y salida junto con el Nodo de un recipiente parecería ahorrar tiempo; pero la práctica demuestra que no es así.
- **La manera en que se conducían las reuniones no era eficiente**. En líneas generales, las sesiones de PHA no eran eficientes; agotando al equipo en discusiones poco relevantes y además, malgastando tiempo. Algunos ítems a mejorar relacionados a este punto son:
  - Utilizar un Asistente/Escriba (para PHAs de más de 4h)

- Utilizar las definiciones de IPLs (Capas Independientes de Protección) durante el PHA
- No utilizar la Matriz de Riesgos el 100% del tiempo
- No proyectar las tablas del PHA (sólo recapitular las Recomendaciones al final de la sesión)
- Analizar primero las Consecuencias y luego las Causas (si no tiene Consecuencias de interés, no es necesario invertir tiempo buscando Causas y Salvaguardas)

- **Falta de consideración formal de los Incidentes Previos.** Los estándares de PHA/HAZOP de PAE no requieren de una consideración formal de los Incidentes Previos. La Mejor Práctica de la industria es resumir los Incidentes Previos para cada área de proceso y tener esta lista disponible para el equipo de PHA.

LOPA

- **Demasiados escenarios son analizados mediante LOPA.** Uno de los principales problemas con la implementación de LOPA es que se utiliza para demasiados escenarios (cualquier escenario con consecuencias A, B o C). LOPA sólo debe ser utilizado cuando el equipo de PHA no se siente seguro de la evaluación de riesgo o cuando considera que el escenario es demasiado complejo para solo ser analizado cualitativamente.
- **Suma de varios escenarios con la misma consecuencia.** Las reglas del método LOPA fueron creadas para un escenario que consta de un único par Causa-Consecuencia. Sumar varios Eventos Iniciadores que llevan a la misma consecuencia aparenta dar mayor exactitud en la evaluación, pero debido a la gran incertidumbre de los datos de entrada esto es engañoso y aleja el foco del objetivo principal del método (Asegurar que hay IPLs suficientes para el escenario, y que los IEs e IPLs son mantenidos y validados).
- **Falta de datos de tasas de falla propios.** La clave para prevenir accidentes es la inspección y el mantenimiento de las IPLs. La prueba de esto es la medición, registro y analisis estadístico de las tasas de falla alcanzadas realmente en el sitio. Esto a su vez permitiría el uso de valores específicos (de PFD y IEF) en los LOPA.
- **Fallas en el criterio de Independencia de las IPL.** Varios de los escenarios realizados tenían IPLs que no cumplían con el requisito de Independencia de manera adecuada. Esto se debía principalmente a una mala interpretación de los límites de cada IPL y de cómo considerar las acciones humanas dentro de los escenarios en LOPA.
- **Realización de LOPA en equipo.** El LOPA debería ser realizado un por un único analista, con apoyo ocasional de otras personas para mayor información o clarificación sobre algún escenario en particular.

Reestructuración de los Estándares

Producto del proceso de auditorías, objetivos buscados y oportunidades de cambios identificados; se logró un sistema de gestión de riesgos que involucra no sólo nuevas prácticas técnicas sino que incluyen los procesos a implementar y la interacción necesaria con otras áreas de la compañía, tales como ingeniería conceptual, básica y de detalle, operación y mantenimiento.

La Figura 1 a continuación muestra el sistema de guías que se han desarrollado para implementar los análisis de peligros y/o evaluaciones de riesgos:

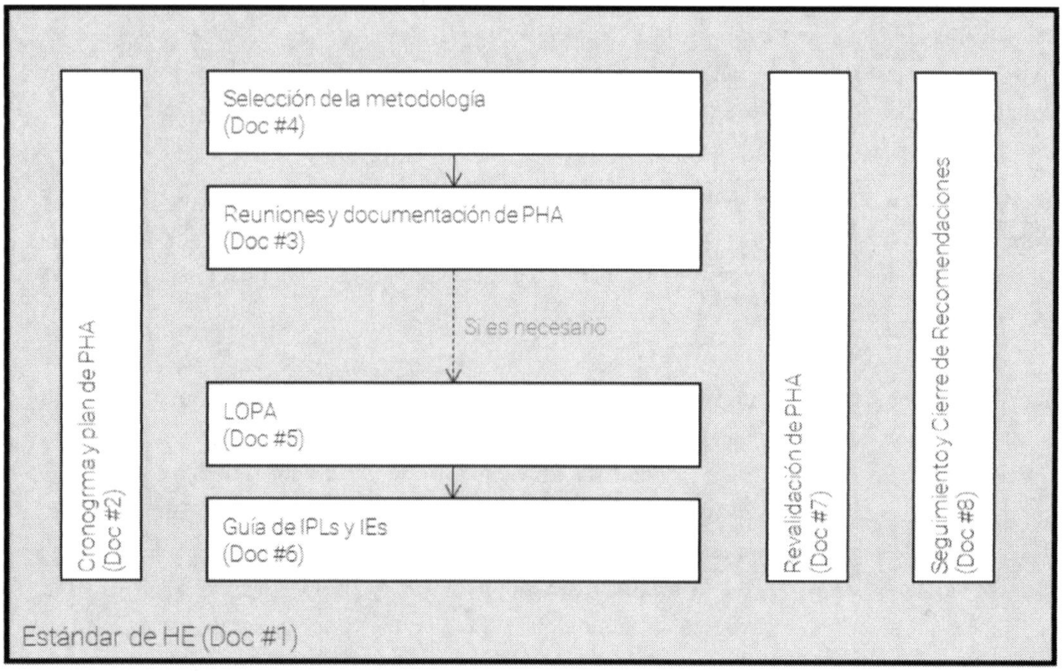

Figura 1. Sistema de Guías para PHAs

Las prácticas, en proceso de implementación en la actualidad, son ocho e incorporan mejoras en cuanto a la gestión de los procedimientos, planificación dentro del ciclo de vida del proyecto, discreción entre el tamaño de proyecto y tipo de PHA, entre otras.

A continuación una breve descripción de los temas tratados en los nuevos procedimientos:

1. **Estándar de Evaluación de Peligros:** Mantener plan de PHA para cada instalación, identificar el personal entrenado y calificado para conformar los equipos de PHA, realizar y documentar PHAs, documentar y comunicar los resultados de PHA, asegurar que las recomendaciones de PHA sean tratadas correctamente, auditar estos estándares periódicamente.
2. **Guía para el Ciclo de Vida de las evaluaciones de peligros:** Programar los PHA dentro de las diversas fases del proyecto, definir el alcance y el contenido y qué resultados espera obtenerse en cada fase, definir la información necesaria de seguridad de procesos durante el proyecto y una vez finalizado.
3. **Guía para sesiones y documentación de PHA:** identificar los recursos necesarios para la sesión de PHA enfocados en la identificación de peligros y los principales problemas operativos, evaluar qué consecuencias pudieran tener estos y que salvaguardas efectivos existen, proponer recomendaciones para mitigar los peligros y proveer actividades para gestionar el riesgo.
4. **Guía para seleccionar la metodología de los PHA:** describir los métodos aplicables para identificar los peligros y los principales problemas operativos, asignar la responsabilidad en la selección del método de PHA, proporcionar una guía para poder facilitar la selección del tipo de PHA en base a los objetivos buscados y la fase del proyecto.

5. **Guía para Análisis de Capas de Protección:** comprender los escenarios de PHA o investigación de accidentes, evaluar las salvaguardas para controlar el riesgo bajo un nivel aceptable, identificar la necesidad de una SIF y establecer su valor de SIL.
6. **Guía de Eventos Iniciadores y Capas Independientes de Protección:** compendio de valores estándar para utilizar en el estudio de LOPA cuando los datos específicos del sitio no estén disponibles.
7. **Guía para revalidar PHA:** descripción del proceso de revalidación de PHA, identificación de cuándo es necesario implementarla y de cómo gestionar los recursos necesarios.
8. **Guía para el seguimiento y cierre de recomendaciones y resoluciones:** gestión de los hallazgos o recomendaciones generadas a través de los elementos del sistema de gestión de seguridad de procesos.

No solo un trabajo aislado

Si bien los trabajos de Auditoría y adecuación de estándares son los ejes sobre los cuales se ha desarrollado este trabajo, es necesario mencionar que éstas no fueron actividades aisladas ya que se encuentran complementadas con el Proyecto de formación de Líderes/Facilitadores de PHA internos. Este proyecto tiene como fin tener mayor autonomía y agilizar las evaluaciones de riesgos, tanto de nuevos proyectos como de escenarios mas simples como un MOC. Se debe aclarar que dependiendo del alcance del PHA los requisitos de independencia del Líder se modifican, por ejemplo: para proyectos, el Líder debería ser independiente del proyecto y de los contratistas/proveedores; mientras que para una MOC, el Líder debería ser independiente de las unidades de procesos afectadas por el cambio.

El Proyecto de Formación de Líderes consta de dos etapas:

- Capacitación a Facilitadores (Etapa ya realizada)
- Coaching (Futuro)

Entrenamiento a Facilitadores

Con el objetivo de implementar la técnica en forma adecuada y asegurar que el conocimiento sobre los riesgos de proceso quede dentro de la compañía, se decidió asignar a un grupo de ingenieros de procesos y seguridad de procesos (10 en total) a realizar el entrenamiento de Liderazgo de PHA.

El entrenamiento como líderes de PHA tuvo como ejes tres temas principales:

- Capacitación general como Líderes de PHA/HAZOP, incluyendo formación en Factores Humanos.
- Entrenamiento en LOPA como complemento de PHA/HAZOP.
- Entrenamiento específico de PHA/HAZOPs para procesos de Manejos del Cambio.

Coaching

Si bien el entrenamiento brinda a los futuros Facilitadores el conocimiento necesario, se debe trabajar con mayor profundidad la competencia de las personas a la hora de llevar dichos conocimientos a la práctica; es por esto que se diseñará un programa de coaching para cada uno de ellos orientado a que luego de varias horas de práctica puedan comenzar a ejecutar los PHA de la forma adecuada y asegurando la calidad de los mismos.

Impactos en la Organización

Teniendo en cuenta que, como se dijo en párrafos anteriores, PAE solo realizaba en sus orígenes PHAs solo en una oportunidad a lo largo de todo el proyecto, los cambios implementados no solo impactan la metodología de las Evaluaciones de Riesgos, sino también la forma en que los proyectos de infraestructuras deben ser enfrentados.

Con el objetivo de alinear y comunicar a todas las áreas de la organización que tienen impacto directo sobre el proceso, se encuentran en proceso de despliegue las siguientes etapas:

- Ajuste conjunto de los procedimientos con todas las áreas involucradas.
- Mapeo conjunto de los procesos de PHA.
- Talleres de difusión de los nuevos procedimientos.

Esto nos permitirá minimizar los desvíos que podamos llegar a tener al momento del inicio de las actividades de implementación.

Conclusión

Mediante la implementación de un sistema de gestión de riesgos, PAE será capaz de identificar peligros latentes en la compañía que escapaban al análisis de los procedimientos pre-existentes.

El nuevo sistema de gestión enfoca sus energías a identificar peligros en los puntos más sensibles de las operaciones como los modos operativos no rutinarios y los factores humanos. Logrando un uso eficiente de los recursos de la compañía y potenciando la identificación de los peligros.

Finalmente, la sistematización del proceso y el entrenamiento de facilitadores internos, permitirá un seguimiento efectivo a través del ciclo de vida de los proyectos y trazabilidad de los estudios de PHA y sus actividades derivadas.

Referencias

CCPS/AIChE, Guidelines for Hazard Evaluation Procedures, 3rd Edition, Wiley, 2008.

CCPS/AIChE, Guidelines for Initiating Events and Independent Protection Layers in Layer of Protection Analysis, 1st ed., Wiley, 2015.

CCPS/AIChe, Guidelines for Risk Based Process Safety, 1st ed, Wiley-Interscience, 2007.

PRINCIPALES DEFICIENCIAS EN LOS PHA EN ARGENTINA (Y LA REGIÓN) Y CÓMO ABORDARLAS

Matías A. Massello
Process Improvement Institute
La Plata, Argentina
mmassello@piii.com

Prepared for Presentation at
American Institute of Chemical Engineers
2019 Spring Meeting and 15th Global Congress on Process Safety
New Orleans, LA
March 31 – April 3, 2019

PRINCIPALES DEFICIENCIAS EN LOS PHA EN ARGENTINA (Y LA REGIÓN) Y CÓMO ABORDARLAS

Autor principal

Matías A. Massello – Process Safety Engineer
Process Improvement Institute, Inc. (PII)
e-mail: mmassello@piii.com

Colaboradores

Diego Ferrucci Cura – Process Safety Engineer
e-mail: diegoferruccic@gmail.com

William G. Bridges – President
Process Improvement Institute, Inc. (PII)
e-mail: wbridges@piii.com

Keywords: *Identificación de peligros, Análisis de riesgos, Comprensión de peligros y riesgos, PHA*

Abstract

La identificación de peligros y situaciones que podrían causar daño de algún tipo es un aspecto fundamental para tomar decisiones tanto en la vida cotidiana como en la industria de procesos. En el ámbito de la industria de procesos, la manera de identificar dichos peligros es mediante los Análisis de Riesgos/Peligros de los Procesos (PHA, Process Hazard Analysis). Los Análisis de Riesgos/Peligros de los Procesos han estado presentes en la industria de procesos por más de medio siglo, y cobraron mayor importancia a partir de accidentes trágicos como los de Flixborough (1974), Seveso (1974), Bhopal (1984) y Phillips Petroleum Company (1989). Estos accidentes tuvieron una relevancia tal que fueron los disparadores de la regulación en Europa y Estados Unidos sobre Seguridad de Procesos (Directivas Seveso y US OSHA PSM Standard). A su vez, la industria ha ido desarrollando las Mejores Prácticas para la identificación de peligros, las cuales fueron condensadas por el CCPS en el libro *Guidelines for Hazard Evaluation Procedures.*

A pesar de que los PHA no son un tema nuevo, en Argentina siguen existiendo muchas deficiencias a la hora de realizarlos. Algunas de las deficiencias están relacionadas al alcance y contenido del PHA (que son las más importantes) y otras están relacionadas a la forma en la que se realizan los PHA, las cuales también impactan tanto en la calidad como en la eficiencia de los estudios. Este paper tiene el fin de describir cuáles son las principales deficiencias, por qué ocurren y cómo poder solucionarlas a fin de realizar una Identificación de peligros coherente con las Mejores Prácticas actuales. Este paper está basado en la experiencia del autor en más de 70 PHAs de diverso alcance y duración, en más de una docena de empresas diferentes en Argentina, combinado con la experiencia de Process Improvement Institute en alrededor de 8000 PHAs a nivel mundial.

Introducción

La identificación de peligros y situaciones que podrían causar daño de algún tipo es un aspecto fundamental para tomar decisiones tanto en la vida cotidiana como en la industria de procesos. Para poder estar preparados para contrarrestar los efectos indeseados de las situaciones peligrosas, es necesario tener dichas situaciones identificadas: Si no esperamos (o ignoramos) la ocurrencia de una situación peligrosa en particular, que se produzcan (o no) sus efectos indeseados dependerá de condiciones y factores que no están bajo nuestro control (o lo que a veces denominamos como "suerte"). En el ámbito de la industria de procesos, la manera de identificar dichos peligros es mediante los Análisis de Riesgos/Peligros de los Procesos (PHA, Process Hazard Analysis).

Los Análisis de Riesgos/Peligros de los Procesos han estado presentes en la industria de procesos por más de medio siglo, y cobraron mayor importancia a partir de accidentes trágicos como los de Flixborough (1974), Seveso (1974) y Bhopal (1984). Estos accidentes tuvieron una relevancia tal que fueron los disparadores de la regulación en Europa y Estados Unidos sobre Seguridad de Procesos (Directivas Seveso y US OSHA PSM Standard). A su vez, la industria ha ido desarrollando las Mejores Prácticas para la identificación de peligros, las cuales fueron condensadas por el CCPS en el libro *Guidelines for Hazard Evaluation Procedures*, cuya primer edición fue en 1985, y posteriormente actualizado y reeditado en 1992 y 2010.

A pesar de que los PHA no son un tema nuevo, en Argentina siguen existiendo muchas deficiencias a la hora de realizarlos. Este paper tiene el fin de describir cuáles son las principales deficiencias, por qué ocurren y cómo poder solucionarlas a fin de realizar una Identificación de peligros coherente con las Mejores Prácticas actuales.

Este paper está basado en decenas de PHAs de diverso alcance y duración (Tabla 1), distribuidos principalmente en el territorio argentino (Figura 1) combinados y comparados con la experiencia de Process Improvement Institute en ~8000 PHAs alrededor del mundo.

Tabla 1. PHAs considerados en este paper

Número de PHAs	75
Alcance	Plantas, Unidades, MOCs
Metodologías	HAZOP, What-if, Checklist y What-if+Checklist
Duración mínima	1 hora
Duración máxima	11 semanas
Lugares	22 (18 en Argentina, 3 en Perú y 1 en Bolivia)
Empresas	14
Sectores	Oil and gas, Refinería, Petroquímica, Química

Figura 1. Distribución de los PHAs

Las consecuencias de tener PHAs con deficiencias pueden ser muy significativas ya que es imposible gestionar una situación de la que no se tiene conocimiento de su posible ocurrencia. Es decir que puede haber escenarios cuyas consecuencias podrían ser catastróficas que no fueron identificados y no tener los controles o salvaguardas apropiados (bajo control). La Tabla 2 muestra las consecuencias y costos de 4 accidentes ocurridos en los últimos 15 años a modo de ejemplo. [1]

Tabla 2. Consecuencias y costos de accidentes históricos

Año	Accidente	Fatalidades	Costo (MM USD)
2005	Explosión BP Texas City Refinery (USA)	15 fatalidades y 180 heridos	1.500
2010	Deepwater Horizon (Golfo de México)	11 fatalidades y 17 heridos	21.000
2012	Exosión Chevron Refinery (USA)	~15.000 personas buscaron asistencia médica	2
2013	Explosión West Fertilizer Company (USA)	15 fatalidades y 260+ heridos	247

Principales deficiencias

Del análisis de los PHAs considerados para realizar este paper (Tabla 1), se pueden identificar sin hacer grandes esfuerzos algunas fallas, comunes a todos, de gran importancia. Otras fallas de menor importancia relativa también se encuentran prácticamente en todos los PHAs analizados, pero las posibles soluciones dependen de cada caso particular ya que las empresas abordan estas cuestiones de manera diferente en sus estándares de Análisis/Evaluación de Riesgos. Algunas de las deficiencias están relacionadas al alcance y contenido del PHA (que son las más importantes) y otras están relacionadas a la forma en la que se realizan los PHA, las cuales también impactan tanto en la calidad como en la eficiencia de los estudios.

Las dos principales deficiencias identificadas son:

- Falta de análisis de todos los modos de operación (incluyendo modos no rutinarios: Arranque, Parada, Parada de Emergencia, Mantenimiento en línea)
- Falta de análisis de los Mecanismos de Daños

En un segundo nivel de importancia se encuentran las fallas:

- Análisis pobre (o nulo) de Incidentes Previos
- Análisis pobre (o nulo) de los Factores Humanos.

En un tercer escalón, se identificaron las siguientes deficiencias que, si bien no son están relacionadas al contenido y alcance como las anteriores, impactan en la calidad y eficiencia de los PHA:

- Mala selección de la técnica de análisis
- No se siguen las Mejores Prácticas para la dinámica de las reuniones ni de documentación.

A continuación se describen en mayor detalle las dos principales deficiencias y de manera general las cuatro restantes.

Modos de operación no-rutinarios

Sin duda alguna, la principal deficiencia en los PHAs en Argentina es la falta de análisis de **todos** los modos de operación. Prácticamente la totalidad de los PHAs realizados se centran únicamente en el modo de operación continua (estado estacionario). Suelen incluirse algunos escenarios de otros modos de operación, pero depende de que algunos de los miembros del equipo los mencione ya que no se realiza una búsqueda sistemática para tratar de identificarlos.

Los modos de operación que actualmente **no son analizados** debidamente son:

- Arranque/Puesta en marcha
- Parada normal
- Parada de emergencia
- Mantenimiento en línea
- Otras tareas no estacionarias regidas por procedimientos

La pregunta que podría surgir es: *"Si estos modos de operación están presentes mucho menos tiempo que la operación normal, ¿qué tan grave puede ser no analizarlos en los PHA?"*. Para responder a esta pregunta y recalcar la importancia que tiene el análisis de todos los modos de operación, sólo basta con mirar las estadísticas de los accidentes ocurridos en la industria:

> *~80% de los accidentes investigados por el CSB fueron durante modos de operación no rutinarios o en procesos/tareas batch [2]*

Dicha estadística coincide con lo manifestado informalmente a Process Improvement Institute por miles de estudiantes (de diversas empresas) en los cursos de PSM, Investigación de Incidentes o PHA.

Es decir que, al no analizar adecuadamente dichos modos de operación durante los PHA, **no se identifican de manera apropiada los escenarios que causan el 80% de los accidentes!** Resulta evidente de por qué esta es la principal deficiencia actual.

Mecanismos de daños

Los Mecanismos de Daños (DM, *Damage mechanisms*) se refieren, de manera general, a las diferentes formas o modos de degradación que podrían sufrir los equipos de procesos o sus componentes. Algunos ejemplos de DM son:

- Shock térmico
- Corrosión bajo aislación (CUI) - *Causa del accidente de Chevron Richmond (2012)*
- Erosión
- Ataque de Hidrógeno a Alta Temperatura (HTHA) - *Causa del accidente en Tesoro, Anacortes (2010)*
- Fractura frágil

Desde la segunda edición del libro del CCPS *Guidelines for Hazard Evaluation Procedures* (1991) [3], se incluyó una buena descripción y ejemplos de cómo abarcar los Mecanismos de daños durante un PHA. Sin embargo, la mayoría de los PHA (no sólo los realizados en Argentina) no siguieron los ejemplos provistos en dicho libro y, en cambio, no abordaron (o lo hicieron de manera deficiente) los diferentes Mecanismos de Daños tales como corrosión, erosión, impactos externos, etc. La tercer edición del libro *Guidelines for Hazard Evaluation Procedures (2008)* [4] incorporó texto adicional enfatizando los ejemplos previos sobre cómo abordar los Mecanismos de Daños durante los PHA y en los informes de PHA. Además, se emitió la API 571 [5] que describe varios Mecanismos de Daños y cómo deben ser abordados por las compañías.

Muchas compañías a nivel mundial tomaron la iniciativa de incorporar las cuestiones relacionadas a los DMs en sus PHA, incluyendo las principales empresas químicas, petroquímicas y refinerías. Sin embargo, la gran mayoría no logra abordar de manera apropiada los Mecanismos de Daños durante sus

PHA/HAZOP. Como puede verse, esta deficiencia no existe sólo en Argentina; pero en Argentina el análisis es prácticamente nulo (con excepción de situaciones puntuales donde se discute el problema de corrosión).

El CSB hizo varias referencias a la necesidad de que los PHA/HAZOP abarquen todos los mecanismos de daños. El CSB reconoce esta carencia como la segunda deficiencia más importante en los PHAs actuales y lo ha explicado en varias presentaciones en el Congreso Mundial de Seguridad de Procesos (GCPS) durante los últimos años. Además, realizaron varios análisis profundos de algunos accidentes que se originaron en problemas con los Mecanismos de Daños, incluyendo el incendio en la refinería de Chevron en Richmond, California (2012) [6] y la ruptura catastrófica de un intercambiador de calor en la refinería de Tesoro Anacortes (2010) [7].

Otras

Como ya se mencionó con anterioridad, además de las dos principales deficiencias descriptas en los párrafos previos, existen otras deficiencias de menor importancia relativa pero que también tienen un gran impacto en la calidad y eficiencia de los PHA:

- Análisis pobre (o nulo) de los Incidentes Previos
- Análisis pobre (o nulo) de los Factores Humanos.
- Mala selección de la técnica de análisis
- No se siguen las Mejores Prácticas para la dinámica de las reuniones ni de documentación.

A continuación se presenta una descripción de cada una de estas deficiencias.

Incidentes Previos

Una buena fuente de información sobre los peligros de un proceso determinado son los informes de los Incidentes Previos, sin embargo, dichos informes raramente son analizados apropiadamente durante los PHA. Actualmente en Argentina, este análisis se realiza de manera muy poco sistemática y hasta casi "informal" durante los PHA ya que depende de que alguno de los miembros del equipo diga *"Esto ya nos pasó"* y describa el incidente según lo recuerde. Esta metodología tiene el problema de que depende prácticamente en un 100% de la memoria de los miembros del equipo (no es posible esperar que los participantes recuerden todos los incidentes, sus causas, consecuencias y qué acciones se tomaron de manera precisa).

Factores humanos

Trevor Kletz dice en el video del CSB del accidente de BP Texas:

> *"Por mucho tiempo la gente dijo que la mayoría de los accidentes eran debido a errores humanos y eso es cierto en cierto sentido, pero no es muy útil. Es como decir que las cosas se caen por culpa de la gravedad" [8].*

Argentina se encuentra en esa etapa. En la gran mayoría (por no decir todos) de los Análisis de riesgos existe al menos una causa documentada como "Error humano" sin ir más allá para tratar de identificar los posibles Factores Humanos qué contribuirían en llevar al trabajador a cometer un error: ¿El procedimiento es incorrecto?, ¿la identificación en campo es confusa?, ¿excesiva cantidad de horas de trabajo?, etc.

Errores humanos durante el diseño, construcción, instalación, operación, mantenimiento, inspección, gestión, etc., pueden ser considerados como la causa de casi todos los accidentes industriales. Los expertos generalmente dicen que aproximadamente el 85% de los accidentes son causados por errores humanos; aunque algunos dicen que es el 100%, exceptuando los desastres naturales. Sin embargo, simplemente atribuir estos accidentes a "Errores humanos" sin ahondar en las causas raíz implica que estos errores son inevitables, impredecibles e incontrolables; lo cual no es cierto.

Las personas cometen errores por varias razones, pero los expertos estiman que sólo el 10% de los accidentes debido a errores humanos ocurren por cuestiones personales (estado emocional, salud o desinterés). La mayoría de los errores cometidos por las personas en el ambiente laboral son por influencias externas tales como:

- Procedimientos deficientes
- Capacitación/Entrenamiento deficiente
- Supervisión inadecuada
- Interfaz Humano-Máquina deficiente
- Dotación de personal insuficiente

La Tabla 3 muestra diferentes categorías de Factores Humanos, sus niveles y cuánto influyen en hacer que sea más o menos probable un error humano. Estas causas de Errores humanos (Factores Humanos), que eventualmente resultan en otros errores humanos, están bajo el control de la gerencia. Al analizar escenarios relacionados a errores humanos durante el PHA, es importante la influencia de dichos Factores Humanos y hasta qué punto están controlados.

Tabla 3. Factores Humanos

Categoría	Cuestión/Nivel	Multiplicador
Tiempo disponible Solo para respuestas	Insuficiente	P(falla) = 100%
	Apenas suficiente (~2/3x Nominal)	10
	Nominal	1
	Tiempo extra (Entre 1x y 2x Nominal y > 20min)	0.1
	Tiempo extensivo (>2x nominal y > 20min)	0.01
Stress	Extremo	5
	Alto	2
	Nominal	1
Diseño y complejidad de la tarea	Muy complejo	5
	Complejo	2
	Nominal	1
	Diagnóstico obvio	0.1
Experiencia/ Entrenamiento	Bajo	10
	Nominal	1
	Alto	0.5
Procedimientos	No disponibles	20
	Incompleto	10
	Disponibles, pero pobres	5
	Nominal	0.5
	Orientados a diagnósticos/síntomas	1
Interfaz Humano-Máquina	Faltante/Confusa	20
	Pobre	10
	Nominal	1
	Buena	0.5
Aptitud para el trabajo	No apto (Alto nivel de fatiga, enfermedad, muy medicado, no físicamente capaz de trabajar)	20
	Aptitud degradada	5
	Nominal	1
Procesos y supervisión del trabajo	Pobre	2
	Nominal	1
	Buena	0.8
Ambiente de trabajo	Extremo	5
	Bueno	1
Comunicación	Sin comunicación o interferencia/daños en el sistema	10
	Sin reglas estándar de comunicación verbal	5
	Estándar bien implementado y practicado	1

Basado parcialmente en: Gertman, D.; Blackman, H.; Marble, J.; Byers, J. and Smith, C., *The SPAR-H Human Reliability Analysis Method*, NUREG/CR-6883, U.S. Nuclear Regulatory Commission, Office of Nuclear Regulatory Research, Washington, DC, August 2005. Otra iniciativa de US NRC, NUREG/CR-6903, *Human Event Repository and Analysis (HERA)*, 2007 (preparada por B. Hallbert, A. Whaley, R. Boring, P. McCabe, y. Chang), se basa en las categorías de SPAR-G para desarrollar una taxonomía para la recolección de datos de errores humanos en plantas nucleares. PII modificó la lista para contemplar datos y terminología genérica de la industria y para incorporar datos internos de PII.

Selección de la técnica

En Argentina, al igual que en otros lugares, el término "Análisis de riesgos" es prácticamente sinónimo de "HAZOP". Como consecuencia de esto, cuando se llega a la etapa de la Identificación de Peligros/Análisis de Riesgo de un proyecto o MOC se recurre directamente a la técnica HAZOP. Que esta asociación resulte ser acertada o no, dependerá de cada caso en particular pero, si se trata de un sistema o modificación simple, la técnica de HAZOP puede ser excesiva en cuanto a recursos requeridos (tiempo) o, peor aún, no ser la más indicada para identificar los peligros particulares del sistema en cuestión.

Como contrapartida, también se da la situación del uso excesivo de metodologías de PHA más simples, como por ejemplo What-if, en casos complejos donde se requiere mayor grado de rigurosidad para asegurar un análisis más exhaustivo.

Otro aspecto observado relacionado a esta cuestión, es el uso de una única técnica para todo el alcance del PHA. Esto genera una mala optimización de recursos (tiempo y personal). La mejor alternativa es seleccionar la técnica a utilizar dependiendo de la complejidad y tipo de peligros de cada sub-unidad a analizar, como se explica más adelante.

Reuniones ineficientes

La dinámica de las reuniones es, quizás, el aspecto más variable en los PHAs en Argentina ya que depende en gran parte del Líder del PHA y de los procedimientos o estándares de PHA de las diferentes empresas. Pero, en líneas generales, no se siguen las Mejores Prácticas para la dinámica de las reuniones ni de documentación con el fin de maximizar el *brainstorming* del Equipo de PHA (lo que resultaría en la identificación de una mayor cantidad de escenarios).

Sólo por nombrar algunas de las deficiencias:

- No se utiliza un Asistente/Escriba para PHAs de +4h.
- Se proyectan las tablas de PHA el 100% del tiempo.
- No se preparan los Nodos/Secciones y el software antes de la reunión.
- Se documenta de la forma *Causa-por-Causa* todo el PHA.
- Se utiliza la matriz de riesgos en todos los escenarios.

Seguir las Mejores Prácticas puede reducir hasta en un 30% el tiempo requerido para el PHA y, a su vez, permitiría la identificación de más escenarios. [9]

¿Por qué existen estas deficiencias?

Los motivos inmediatos que llevan a estas deficiencias en los PHAs podrían resumirse en dos:

- Los estándares de las compañías no son del todo completos y, por lo tanto, los PHAs basados en dichos estándares resultan incompletos.
- Falta de capacitación/conocimiento de quiénes están a cargo de los Estudios de PHA en las empresas.

Sin embargo, las razones subyacentes por las que existen estas deficiencias en los PHAs en Argentina exceden el ámbito de la identificación de peligros y están relacionadas a problemas con la implementación de la Gestión de Seguridad de Procesos (PSM) en general. Al no existir una sólida implementación de PSM, es lógico esperar que sus Elementos tengan deficiencias; en este caso particular "Identificación de peligros y Análisis de riesgos" (según los Elementos del programa *Risk Based Process Safety* del CCPS [10]).

Algunas de las trabas que impiden la implementación de Seguridad de Procesos en países que no tienen regulación sobre PSM (como el caso de Argentina) son:

- Es difícil despertar el interés sobre Seguridad de Procesos en países que:
  - No tienen regulación que requiera la implementación de un sistema de Seguridad de Procesos (como por ejemplo: OSHA PSM o UK HSE).
  - La vida humana no sea tan importante/valorada como en los países más desarrollados *- Este punto no es aplicable al caso de Argentina*
- Dado que no es requerido por el gobierno, algunas empresas argumentan que sus competidores tampoco lo están haciendo y por lo tanto no pueden implementar Seguridad de Procesos y seguir siendo competitivos, y que además "Los clientes no pagarán por Seguridad de Procesos".
- Debido a la falta de regulación, varias compañías no participan en organizaciones de la industria (tales como CCPS o API), ni realizan el seguimiento del desarrollo de estándares o prácticas recomendadas. Entonces, a menos que la empresa responsable del diseño o algún socio esté activamente involucrado, esta fuente de conocimiento clave de PSM permanece sin explotar.
- PSM generalmente es responsabilidad del área de "Seguridad" y no es parte de la disciplina operativa de la instalación/compañía.
- La Seguridad de Procesos (prevención de pérdidas de contención, incendios, explosiones, etc.) es más difícil de entender y explicar que la Seguridad Ocupacional (resbalones, golpes, caídas, etc.), especialmente en países que no tienen regulación específica de Seguridad de Procesos. Los beneficios de un buen programa/sistema de Seguridad de Procesos pueden no ser fáciles de calcular mediante un análisis Costo-Beneficio, y además, los beneficios son generalmente a largo plazo (previniendo grandes incidentes de seguridad de procesos).
- Algunos Gerentes Senior que sostienen *"Eso no ha pasado nunca"* o *"La compañía XYZ responsable de la tecnología no nos dijo que eso podría ser un problema"*.
- El nivel de conocimiento, habilidades y capacitación de los empleados no es suficiente para implementar Seguridad de Procesos. Existe poca oferta de personal apropiadamente capacitado para realizar estudios, escribir procedimientos y capacitar al personal. Y hay un porcentaje muy bajo dentro de la población de personal técnico con pericia en Seguridad de Procesos; por lo que las compañías que desean implementar PSM siguiendo las mejores prácticas deben importar trabajo (inicialmente mediante consultores).
- Otras

Para una descripción más detallada sobre la implementación de PSM en países sin regulación, referir al paper *"Obstáculos, beneficios y costos habituales de implementar completamente Seguridad de Procesos alrededor del mundo"* [11].

Cómo abordar las deficiencias

Modos de operación no-rutinarios

Varios analistas han tratado de identificar los peligros asociados a los modos de operación no rutinarios mediante el agregado de desviaciones genéricas tales como *"Desviaciones durante el arranque"* o *"Desviaciones durante el mantenimiento"* en cada Nodo/Sección. Desafortunadamente, este método sólo encuentra una fracción de los accidentes que podrían ocurrir durante estos modos de operación debido a que el equipo está más enfocado en el modo de operación continua y que la búsqueda de escenarios no es sistemática.

La evaluación de peligros de los modos de operación no rutinarios implica la **revisión de procedimientos** utilizando las técnicas de HAZOP, HAZOP simplificado o What-if para descubrir posibles escenarios de accidentes. El objetivo de esta revisión no es el de asegurar que los procedimientos son correctos y aceptables, sino es evaluar los escenarios de accidentes posibles si los procedimientos no se siguen. Incluso el mejor procedimiento puede no ser seguido por diversas razones.

> *El objetivo de la evaluación de peligros de los modos de operación no rutinarios (gobernados por procedimientos escritos) es el de asegurar que la organización tiene suficientes salvaguardas para el momento inevitable en que un paso sea realizado incorrectamente o sea omitido (ya sea inadvertidamente, por tomar "atajos" o por otras razones).*

Previo a cualquier análisis, la pregunta que debe hacerse es: ¿Son los procedimientos aptos para ser analizados? Si los procedimientos no son actualizados, correctos, claros y utilizados por los operadores, analizarlos representaría una pérdida de tiempo y, peor aún, la posibilidad de asumir que hay protecciones suficientes cuando no las hay o no identificar algunos escenarios. Si los procedimientos no son al menos un 90% exactos, el primer paso a realizar es el de desarrollar los procedimientos siguiendo las Mejores Prácticas[1] tan pronto como sea posible y luego realizar el PHA a estos nuevos procedimientos.

Es importante notar que para este análisis el rol del personal de Operaciones toma mayor relevancia. **Es necesaria la presencia dentro del Equipo de PHA de un Operador Senior y un Operador de campo**. Muchas veces sólo está presente un Supervisor/Superintendente que fue operador durante varios años pero que quizás no está al tanto de todas las actividades o "atajos" que toman los operadores en el campo actualmente.

A continuación se describen los tres métodos para analizar los modos de operación no rutinarios.

HAZOP

La variante del HAZOP orientada a procedimientos es la forma más antigua de HAZOP (originada en ICI por la década del 60). Los creadores optaron por subdividir los dos tipos de errores humanos (omisión y comisión) generando las Palabras Guía que se describen en la Tabla 4.

[1] Las Mejores Prácticas sobre escritura y formato de procedimientos pueden encontrarse en *"Mejores Prácticas para escribir procedimientos operativos y guías de solución de problemas"* [10]

Tabla 4. Palabras Guía de HAZOP para procedimientos

Tipo error	Palabra guía	Significado
Omisión	No	La intención específica de la instrucción no se realiza
	Parte de	Una parte de la intención de la instrucción no se realiza. Usualmente sólo aplica a tareas que implican 2 o más tareas prácticamente simultaneas (*"Abrir las válvulas A, B y C"*).
Comisión	Más	Incremento cualitativo. Se realiza demasiado de la intención de la instrucción (no aplica a tareas de Prender/Apagar o Abrir/Cerrar) o se hace demasiado rápido.
	Menos	Disminución cualitativa. Se realiza demasiado poco de la intención de la instrucción o se hace demasiado lento.
	Antes/Después	La instrucción se realiza fuera de la secuencia correcta
	Además	Se realiza otra acción además de la especificada en la instrucción (podría ser un atajo)
	Otro	Se selecciona, usa, lee, etc. el dispositivo incorrecto o se utiliza de otra manera. O se selecciona/agrega otro material.
-	Faltante	Un paso, instrucción o precaución está faltante en el procedimiento previo a este paso (Similar a *"Fuera de secuencia"*, pero el paso no está escrito).

Cada una de las Palabras Guía (Tabla 4) es aplicada a cada instrucción del procedimiento que implique una acción que modifique el estado del sistema. Es decir, todas las instrucciones del tipo Verificar/Asegurar/Chequear no son analizadas y son consideradas como "protecciones" para los otros pasos. En el caso de que el procedimiento no sigua las Mejores Prácticas [12] y cada instrucción no implique una única acción, el primer paso será subdividir las instrucciones en acciones individuales.

El desarrollo de la reunión es prácticamente idéntico a la de la revisión del modo continuo. Para cada desviación de la intención de la Instrucción/paso el equipo debe ir más allá de la causa obvia de *"Error operativo"* para identificar las Causas raíz asociadas a dicho error; como por ejemplo: *"No hay suficiente énfasis en este paso durante el entrenamiento"*, *"El Operador es responsable de realizar dos tareas simultáneamente"*, *"El etiquetado de las válvulas en campo es confuso"*, o *"La lectura del indicador es difícil de leer porque está borrosa"*. La Palabra Guía *"Faltante"* permite identificar causas como *"No hay un paso escrito o entrenamiento formal para solicitar el permiso de trabajo en caliente previo a este paso"* o *"No hay un paso escrito o entrenamiento formal para abrir la válvula de descarga antes de poner en marcha la bomba"*.

HAZOP Simplificado

Un enfoque simplificado del HAZOP (descripto en los párrafos anteriores) consta en utilizar dos Palabras (Frases) Guía en vez de las 7 Palabras Guía de la Tabla 4. Las dos Frases Guía que se utilizan se presentan en la tabla a continuación (Tabla 5).

Tabla 5 Frases Guía de HAZOP Simplificado para procedimientos

Frase guía	Significado
El paso/Instrucción no se realiza	El paso (o parte de él) no se realiza. Algunas posibles razones incluyen que el operador se olvidó de realizar el paso, no comprendió su importancia, o el procedimiento no incluía este paso.
El paso/instrucción se realiza de manera incorrecta	El Operador tiene la intención de realizar el paso (no omitirlo) sin embargo no lo realiza como es pretendido. Algunas posibles razones incluye

La primer Frase Guía está relacionado a las Palabras Guía de *"No"*, *"Parte de"* y *"Faltante"* de la Tabla 4, mientras que la segunda frase está relacionada a los errores de comisión abarcados en las Palabras Guía *"Más"*, *"Menos"*, *"Antes/Después"*, *"Además"* y *"Otro"* de la Tabla 4.

La aplicación y la dinámica de las reuniones de este método son equivalente a la variante de HAZOP con 7 Palabras Guía. Este enfoque ha demostrado ser útil en los casos que:

- El Procedimiento analizado no posea grandes peligros y/o no sea demasiado complejo.
- El Líder del PHA tenga mucha experiencia con el enfoque previo y pueda compensar las deficiencias de este enfoque debidas a la simplificación.

What-if

El método de What-if para analizar modos de operación basados en procedimientos es un *brainstorming* libre sin la ayuda (o restricción) de las palabras guía. El equipo de PHA lee todo el procedimiento y luego responde la pregunta: *"¿Qué errores podrían llevar a una consecuencia de interés?"*. El equipo lista estos errores y luego analiza las consecuencias, causas, protecciones existentes y realiza recomendaciones de ser necesario. Este método no se aplica paso por paso sino que se analiza la tarea completa (procedimiento) en conjunto.

Este método se describe en detalle en el libro del CCPS *Guideline for Hazard Evaluation procedures* [4].

Selección de la metodología a utilizar

Las tres metodologías anteriores fueron descriptas en orden de rigurosidad; y a mayor rigurosidad, mayor cantidad de tiempo requerido. De manera análoga a lo que ocurre con la selección de la técnica para el modo de operación continuo (que se describe más adelante) el criterio básico de selección es simple: Cuanto mayor sean los peligros asociados al procedimiento y cuanto mayor sea su complejidad; se requerirá mayor rigurosidad en la técnica de análisis.

La experiencia del Líder juega un papel fundamental en los PHA de modos de operación no rutinarios tanto en la selección de qué procedimientos a analizar como en la elección de la metodología de análisis para cada uno de ellos.

La Figura 2 muestra la distribución típica del uso de cada una de las técnicas en un PHA de una planta química o de una refinería. La gran mayoría de los procedimientos son lo suficientemente simples o no tienen grandes peligros asociados como para justificar el uso de las técnicas menos rigurosas: What-if y HAZOP simplificado. El método HAZOP (7-8 Palabras Guía) prácticamente no se utiliza ya que la mayoría de las tareas no requiere tanto nivel de escrutinio para identificar los posibles escenarios.

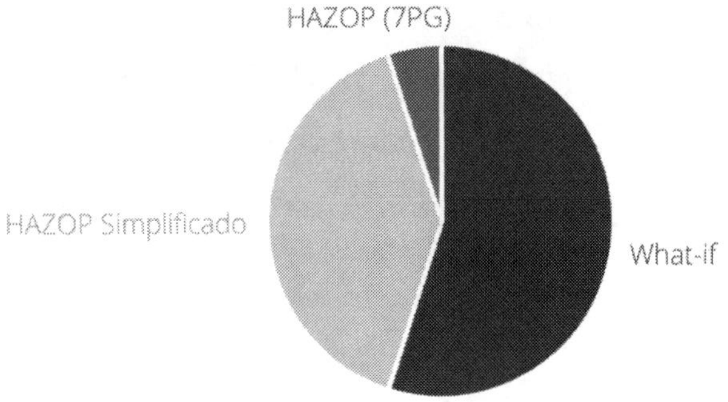

Figura 2. Distribución típica del uso de las técnicas de análisis de modos de operación no rutinarios

La Figura 3 muestra un diagrama de flujo resumiendo la metodología para realizar el análisis de los modos de operación no-rutinarios.

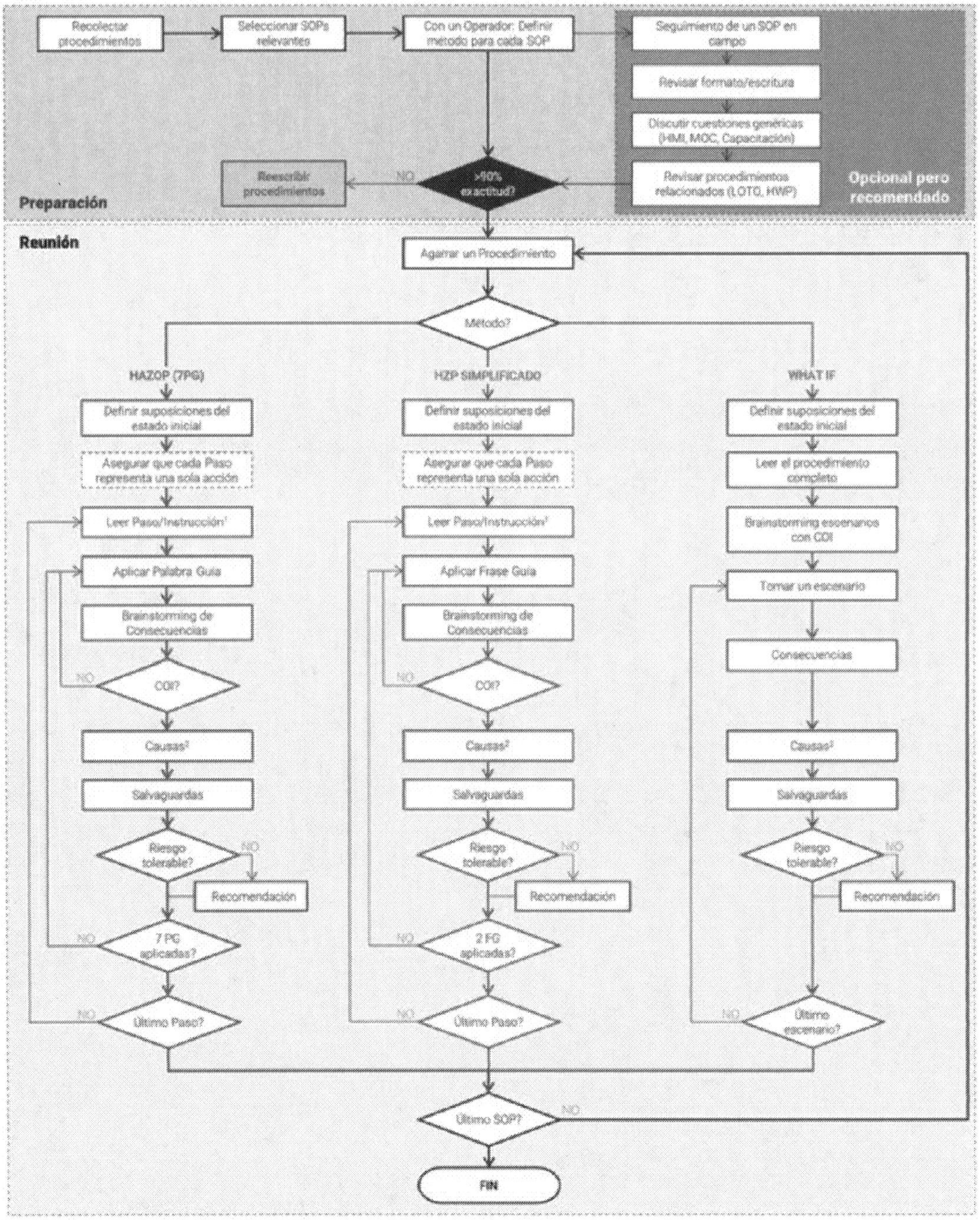

¹ Sólo analizar pasos que modifican el estado del sistema. Saltear pasos del tipo Asegurar/Chequear/Verificar
² Indagar más allá de la causa "Error humano" e identificar las cuestiones relacionadas a los Factores Humanos
COI: Consecuencia de Interés

Figura 3. Diagrama de flujo de un PHA de procedimientos

Mecanismos de daños

Existen diversas alternativas para abordar los Mecanismos de Daños (DM):

- Discutir los DM durante el PHA/HAZOP en la desviación "Pérdida de contención" – *Esta es la alternativa sugerida por PII*
- Registrar los DM que surgen durante la discusión de otras desviaciones del proceso (Presión, Temperatura, Concentración).
- Incluir DM como un nodo separado (para la técnica de HAZOP) – *Enfoque propuesto por Risk Management Professionals* [13].
- Realizar una mini-checklist de Integridad Mecánica en cada Nodo/Sección.
- Dividir los P&IDs (o PFDs) en diferentes servicios con el fin de realizar una discusión para una Revisión de Peligros de Mecanismos de Daños (DHMR) considerando:
  - Proceso
  - Química
  - Cambios de especificación
  - Caudales
  - Parámetros operativos

  Este enfoque es utilizado por algunas refinerías (fuera de Argentina) para cumplir con los requisitos de un DHMR recomendados en la API RP 571 [5]. *En base a la experiencia de PII, este enfoque tiene sus ventajas e identifica algunas cuestiones que el enfoque sugerido por PII no identifica; pero en líneas generales el enfoque sugerido por PII es mejor para identificar los escenarios claves relacionados a los DM y las salvaguardas asociadas porque la revisión es específica para cada Nodo/Sección del proceso.*

Como se mencionó más arriba, la opción recomendada por PII es de analizar los DM en cada Nodo/Sección (ya sea HAZOP o What-if). Esta alternativa tiene la ventaja de que el equipo puede identificar con mayor facilidad los cambios en los DM de Sección a Sección y además, puede identificar cuándo se requieren salvaguardas específicas tales como: válvulas de aislamiento de acción remota, diques de contención segregados o materiales de construcción específicos.

Metodología

Para cada Nodo/Sección analizado, el Líder junto al Equipo del PHA deben evaluar los posibles impactos de los DM en el Nodo/Sección en cuestión. Para ello, bajo la desviación *"Pérdida de contención"*, el equipo debe considerar las causas típicas de Pérdida de contención (incluyendo los posibles Mecanismos de Daños). La Tabla 6 lista algunas causas típicas de LOC.

Tabla 6. Causas típicas de Pérdida de Contención (Incluyen DMs)

Corrosión interna	Válvula de drenaje dejada abierta/con pase
Corrosión externa (Incluyendo CUI)	Válvula de venteo dejada abierta/con pase
Erosión	Apertura espuria de PSV
Fatiga	Impacto externo
Material de construcción inadecuado	Fuego externo
Mantenimiento inadecuado	Expansión térmica (cuando el proceso está bloqueado)
Rotura de línea de instrumentos	+ Otros DMs específicos según el proceso

Una vez identificadas las causas de LOC, el equipo deberá identificar las salvaguardas existentes (ya sean preventivas o mitigadoras). La Tabla 7 muestra una lista de salvaguardas típicas que podrían cumplir con los requisitos para ser consideradas IPL, y la Tabla 8 continua con ejemplos de salvaguardas contra LOC, pero que no son IPLs.

Tabla 7. Ejemplos de salvaguardas típicas de Pérdida de Contención (Posibles IPLs)

Válvulas de alivio	**Aislamiento**
Válvula de alivio	Válvulas de aislamiento remotas (auto/manual)
Disco de ruptura	Paro remoto de bombas, sopladores, etc.
Válvula de presión y vacío	**Contención y mitigación**
Válvula rompedora de vacío	Doble pared (tanques o cañerías)
Válvula de seguridad de expansión térmica	Proceso contenido en un edificio cerrado
Sistemas de supresión y extinción	Diques
Sistema de diluvio en tanques/recipientes	Drenajes
Alarma manual contra incendio	Aislamiento contra incendio
Extintores en Centro de Distribución de Potencia (PDC)	
Extintores en sala de control	

Tabla 8. Ejemplos de salvaguardas típicas de Pérdida de Contención (No IPLs)

Cañerías
Códigos aplicables de diseño
Soldaduras realizadas por soldadores calificados utilizando procedimientos calificados
Inspección periódica de espesor (ultrasonido, rayos X)
Mayor frecuencia de inspección en áreas críticas (zonas de mayor erosión/corrosión)
Inspección de fugas previo al re-arranque post-mantenimiento
Prueba hidráulica a equipos/componentes previo a su instalación
Bombas
Especificaciones escritas en los procedimientos de mantenimiento con el fin de garantizar que son mantenidas, reparadas y reemplazadas según los requisitos originales.
Válvulas
Programa de monitoreo de emisiones fugitivas (incluyendo válvulas y juntas)
Configuración de posición de "falla segura" ante perdida de aire, nitrógeno o electricidad.
Corrosión
Inspección periódica (cañerías, equipos, estructuras)
Pintura/revestimiento anti-corrosión
Identificación positiva de material (PMI) del 100% de los componentes y soldaduras en campo
Mantenimiento
Recorridas periódicas inspeccionando pérdidas y otras anormalidades
Sistema de mantenimiento preventivo
Inspección periódica de los mazos de tubos de intercambiadores
Instrumentación
Todos los controladores tienen indicación
Testeo y calibración de instrumentación crítica
Protección personal
EPP
Inspección y testeo de EPP
Disponibilidad de kit de primeros auxilios y personal de Emergencias médicas
Respuesta ante emergencia
Sistema de comunicación de emergencia remota
Brigadas de bomberos
Sistema de alarma en toda la planta
Plan de contingencia

Documentación

Hay dos estilos posibles de documentación para el análisis de los DM durante el PHA:

- Referenciar tablas genéricas/típicas de causas y salvaguardas
- Listar detalladamente los DM específicos y sus protecciones

Ambas alternativas son aceptables y dependen principalmente de la preferencia del Líder/Escriba.

La documentación mediante el uso de referencias a tablas de causas típicas (Tabla 6) y a salvaguardas globales (Tabla 7 y Tabla 8) ahorra la repetición de texto y tiempo de documentación pero requiere que el Líder (y el Escriba/Asistente) asegure que se discuten todos los ítems listados en las tablas referenciadas en cada Nodo para asegurar que son aplicables al nodo en cuestión. Una variante de este estilo es de eliminar algunas causas/salvaguardas de las tablas genéricas y listarlas solamente en los Nodos que sean aplicables (por ejemplo, si la fragilización por Hidrógeno solo es un problema en una sección en particular de la unidad; se elimina de la tabla de causas típicas y sólo se lista en los nodos aplicables). Un ejemplo de este estilo de documentación se muestra en la Tabla 9.

Tabla 9. Ejemplo A: Documentación de DM utilizando referencias a tablas genéricas

Ítem	Desviación	Causas	Consecuencias	Salvaguardas	Recomendaciones
9.10	Pérdida de contención	Corrosión acelerada Fuego externo Alta Presión (ver 9.7) Causas típicas de LOC (ver Tabla A.1)	Liberación de material tóxico. Posibles lesiones severas a varios trabajadores.	PSV-231A/B/C (1 de 3) @15psig – No IPL (ITPM inadecuado) Salvaguardas generales de LOC (ver Tabla A.2)	29. Considerar cambiar el cronograma de ITPM de las PSV. La frecuencia de inspección actual es demasiado baja de acuerdo a las Mejores Prácticas. Por ejemplo, la frecuencia de inspección actual de la PSV-8220 del Recibidor de Amoniaco (V-2005) es de 9 años; mientras que la recomendación típica es de 1-4 años para PSVs en servicios altamente tóxicos.

El otro método de documentación no utiliza las tablas típicas si no que crea una lista específica para cada Nodo (en la desviación de *Pérdida de Contención*). Este método es más fácil de leer, pero requiere un mayor esfuerzo de documentación. Un ejemplo de este estilo de documentación se muestra en la Tabla 10.

Tabla 10. Ejemplo B: Documentación de DM utilizando listado detallado

Ítem	Desviación	Causas	Consecuencias	Salvaguardas	Recomendaciones
15.10	Pérdida de contención	Alta Presión (ver 15.7) Corrosión ácida causada por Alta concentración de agua (ver 15.9) Corrosión/Erosión Fuego externo y/o incidencia directa de llama Falla de empaquetadura Mantenimiento inadecuado Material defectuoso Válvula de venteo dejada abierta/con pase Válvula de drenaje dejada abierta/con pase Alta temperatura ambiente Impacto externo (por ejemplo: otro vagón)	Liberación catastrófica de Cloro. Posibles fatalidades (trabajadores) y lesiones severas (público). Liberación continua de Cloro por una rotura de línea. Posible afectación moderada a los trabajadores. Liberación continua menor de Cloro por una fuga. Posible afectación moderada a los trabajadores.	Kit de reparación de cloro Descarrilador y la bandera de alerta para prevenir el impacto otro vagón Acceso limitado de vehículos a la zona Respuesta de Operaciones/Mantenimiento según se requiera, incluyendo aislamiento, si es necesario Monitoreo periódico de las válvulas de vagones durante la descarga por el Operador. Equipo de protección personal en el área Tapones instalados en todas las válvulas de cloro a la atmósfera cuando las válvulas no están en uso Válvula de seguridad en cada vagón Mantenimiento de vagones por el proveedor (según los requisitos del US DOT) Supervisión por video de la zona de descarga Durmientes de hormigón en ramal de ferrocarril Dique para evitar que cualquier combustible derramado alcance la zona del rack de descarga.	10. Considerar la instalación de un sistema de detección de cloro en el área de descarga y vaporización para ayudar a detectar fugas de cloro (especialmente en los probables puntos de liberación). 11. Verificar que se llevan a cabo las inspecciones de mantenimiento periódicas en conformidad con las recomendaciones del Instituto del Cloro. 12. Revisar el sistema de drenaje para el área de descarga, e identificar las áreas que podrían ser afectadas por una gran liberación de cloro. 13. Considerar la prohibición del uso de equipos pesados (por ejemplo, grúas) en el área de descarga y vaporización a menos que se promulguen precauciones especiales para evitar daños a los equipos. 5. Considerar proporcionar un sistema de diluvio de agua en la zona de descarga para ayudar a mitigar escapes de cloro del vagón.

Independiente del método utilizado para la documentación, el Líder (y el Escriba) deben asegurar que el equipo analiza rigurosamente todos los DMs y que se registran adecuadamente los resultados de dicho análisis.

Otras

Incidentes previos

El objetivo de los PHA es predecir qué podría salir mal, y los reportes de los Incidentes Previos contienen información real sobre secuencias de errores humanos y fallas de equipos. Realizar el análisis de los Incidentes Previos durante el PHA permitirá:

- Ayudar a identificar posibles causas o consecuencias de interés

- Evaluar las protecciones relacionadas existentes en el proceso
- Realizar recomendaciones sobre capas de protección adicionales necesarias para prevenir incidentes similares en el sistema.

Para esto es necesario listar todos los Incidentes Previos incluyendo Incidentes ocurridos en instalaciones similares y discutir aquellos de mayor relevancia para saber:

- ¿Qué causó los incidentes?
- ¿Qué acciones se tomaron en respuesta?
- ¿Fueron las acciones tomadas efectivas?¿Evitaron Incidentes similares?
- Si el accidente ocurrió en otra instalación (no en la que está siendo analizada), ¿funcionarían las mismas medidas en la instalación de interés?

La manera de documentar los Incidentes Previos en el PHA se muestra en la Figura 4.

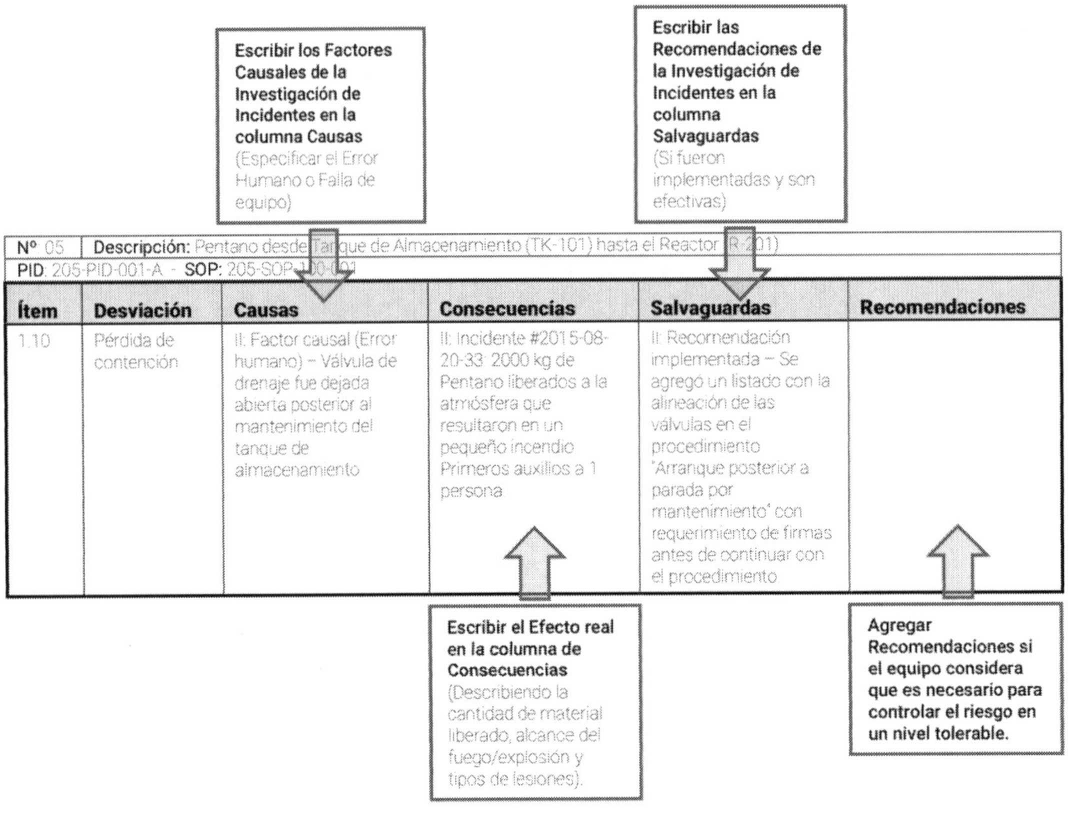

Figura 4. Ejemplo: Documentación de Incidentes Previos en un PHA

Factores humanos

A lo largo de este paper ya se ha mencionado en varias instancias cómo incorporar los Factores Humanos en un PHA: Cada vez que el equipo identifica un "Error humano" como causa de un escenario (ya sea HAZOP o What-if), el Líder debe preguntar *"Por qué?"* a fin de continuar con el proceso de *brainstorming* e identificar los factores que podrían llevar al humano a cometer el error.

Por ejemplo, durante el análisis de la desviación "Bajo Caudal" el equipo puede mencionar "Error operativo" dentro de las causas. Al ser consultados qué podría causar ese "Error operativo" el equipo menciona (1) El controlador de caudal es difícil de ajustar, (2) El indicador de caudal funciona incorrectamente dando un valor erróneo, o (3) Los operadores usualmente pisan sobre las cañerías adyacentes para maniobrar válvulas elevadas pudiendo romper una línea de suministro de aire al actuador de la válvula. Una vez identificadas las causas específicas, el equipo analizará las protecciones existentes y, eventualmente, realizará recomendaciones en caso de que dichas protecciones no sean consideradas como suficientes.

Además de identificar las causas raíz de los "errores humanos", es recomendable realizar un Checklist al final del PHA (al finalizar el análisis de todos los Nodos/Secciones) para complementar el análisis de los Factores Humanos realizado. Este checklist aborda cuestiones más generales sobre:

- Limpieza y ambiente general de trabajo
- Accesibilidad/Disponibilidad de controles y equipos
- Etiquetado
- Pantallas/Retroalimentación
- Carga de trabajo y factores de estrés
- Procedimientos
- Capacitación (empleados y contratistas)

La Tabla 11 presenta un extracto de un Checklist típico para abordar los Factores Humanos (por cuestiones de espacio sólo se muestran las primeras dos preguntas de cada tema).

Tabla 11. Checklist de Factores Humanos (Extracto)

Ítem	Tema	Preguntas	Respuestas	Salvaguardas	Recomendaciones
1.1	Limpieza y ambiente general de trabajo	Están las áreas de trabajo limpias en general? Se mantiene el ruido a un nivel tolerable?			
1.2	Accesibilidad de controles y equipos	Hay suficientes equipos de protección disponibles para su uso de rutina y de emergencia? Son capaces los trabajadores de realizar las tareas rutinarias y de emergencia de manera segura mientras usan el equipo de protección?			
1.3	Etiquetado	Están todos los equipos importante etiquetados de forma legible, precisa y sin ambigüedades? Se ha asignado responsabilidades para mantener y actualizar las etiquetas?			
1.4	Pantallas/ Retroalimentación	Son las pantallas adecuadamente visibles desde todas las posiciones de trabajo relevantes? Presentan pantallas separadas información similar de una manera consistente?			
1.5	Controles	Es el diseño de las consolas lógica, coherente y eficaz? Son los controles distinguibles, accesibles y fáciles de usar?			
1.6	Carga de trabajo y factores de estrés	Pueden operadores adicionales (por ejemplo, de otras áreas o fuera del sitio) ser llamados rápidamente para ayudar durante una emergencia? Es el nivel de dotación de personal adecuado para todos los modos de operación (normal, de emergencia, etc.)?			
1.7	Procedimientos	Creen los operadores que el formato y el lenguaje de los procedimientos son fáciles de seguir y entender? Son fieles/precisos los procedimientos (es decir, reflejan la forma en que realmente se lleva a cabo el trabajo)?			
1.8	Capacitación (empleados y contratistas)	Se realizan simulacros de emergencia periódicamente? Hay testigos que observan y critican los simulacros de emergencia?			

Selección de la técnica

La selección adecuada de la técnica de análisis es más un arte que una ciencia exacta, por lo que resulta difícil determinar a priori un conjunto de instrucciones para poder realizar dicha selección. La mejor alternativa es dejar que un Líder Experto en PHA determine la técnica en función de la complejidad del sistema, el tipo de peligros y su pericia, experiencia y criterio.

Sólo de manera genérica y a modo de orientación se podrían definir las siguientes "reglas" de selección:

- Utilizar **What-if** para peligros bajos/moderados con complejidad baja/moderada. La justificación es que habrá menor número de escenarios (desviaciones con consecuencias de interés) para los procesos de menor complejidad y consecuencias menos graves para peligros bajos/moderados. Incluso para procesos relacionados con productos químicos altamente peligrosos el Líder de PHA puede justificar el uso del What-If para procesos moderadamente simples, tales como la descarga de materias primas, almacenamiento de productos, sistemas de filtración, separadores, etc. Es poco probable que el equipo de PHA se pierda un escenario para estos procesos menos complejos, y la técnica estructurada de HAZOP no es necesaria para poder identificar la mayoría de los escenarios; lo que sería una pérdida de tiempo.
- Utilizar **HAZOP** para procesos de alta complejidad, con menos énfasis en la gravedad del peligro. La regla general es utilizar HAZOP para procesos que implican condiciones operativas extremas, reacciones químicas y operaciones unitarias complejas. Pero a veces las empresas optan por utilizar HAZOP en procesos poco complejos que implican materiales extremadamente peligrosos, por ejemplo, descarga de cloro (que es tóxico y puede tener un enorme impacto en la comunidad vecina).
- Utilizar **Checklist** como una técnica complementaria para ayudar al equipo a identificar los peligros que pueden haber pasado por alto. Una vez más, ya que cada proceso es único, una lista de verificación basada en peligros conocidos y el cumplimiento de códigos y normas seguramente no abarcará todos los peligros potenciales.
- Utilizar **FMEA** para sistemas electro-mecánicos, sistemas que exigen alto nivel de confiabilidad, y equipos muy costosos. Los ejemplos incluyen procesos con extrusoras, transportadoras, sistemas de distribución eléctrica, sistemas de agua contra incendios, sistemas de control por computadoras.

Reuniones ineficientes

Describir con detalle las Mejores Prácticas para la dinámica de las reuniones de PHA y su documentación excede el alcance de este paper. A modo de ejemplo, se mencionan a continuación algunas de las acciones a realizar para optimizar las reuniones:

- Utilizar una persona dedicada exclusivamente al rol de Asistente/Escriba en PHA que duren más de 4h.
- No proyectar las tablas de PHA durante la reunión – Solo hacerlo en algunos casos particulares (por ejemplo al utilizar la metodología de Checklist o recapitular las Recomendaciones).
- Preparar los Nodos/Secciones y el archivo del software de PHA antes de las reuniones
- Documentar mayoritariamente del modo *Desviación-por-desviación*. Utilizar *Causa-por-Causa* solo cuando el otro método sea muy confuso.
- Minimizar el uso de matrices de riesgo.

Se invita al lector a leer el paper *"Optimizing qualitative hazard evaluations for maximized brainstorming"* [9] de Process Improvement Institute en el cual se describen en detalle las Mejores Prácticas.

Conclusiones

La identificación de peligros juega un rol fundamental en la Gestión de Seguridad de Procesos:

> *Es imposible gestionar una situación de la que no se tiene conocimiento de su posible ocurrencia.*

Resulta evidente que enmendar las deficiencias descriptas en este paper es de gran importancia, ya que permitirá una mejor y más completa identificación de escenarios peligrosos; permitiendo así su gestión y control en niveles de riesgo tolerables.

En término de prioridades, las principales deficiencias a abordar son aquellas que están relacionadas al alcance de los PHA:

- Análisis de todos los modos de operación - No incluir dichos modos de operación en el análisis significa ignorar los modos de operación donde ocurren ~80% de los accidentes!
- Análisis de los Mecanismos de Daños
- Análisis de Incidentes Previos
- Análisis de los Factores Humanos.

En un segundo nivel, están las deficiencias relacionadas con la manera de realizar los PHA:

- Mala selección de la técnica de análisis
- No se siguen las Mejores Prácticas para la dinámica de las reuniones ni de documentación.

Enmendar las deficiencias descriptas en este paper requerirá de recursos, principalmente en capacitación (y la consecuente etapa de aprendizaje). De todas maneras, **el beneficio obtenido de realizar un PHA completo** (identificar situaciones peligrosas para su posterior gestión y así evitar accidentes) **es muchísimo mayor que el costo de realizarlos**.

Acrónimos

CCPS: *Center for Chemical Process Safety*

CSB: *US Chemical Safety Board*

CUI: Corrosión bajo aislación (*Corrosion Under Insulation*)

DM: Mecanismos de daños (*Damage Mechanisms*)

DMHR: Revisión de Peligros de Mecanismos de daños (*Damage Mechanisms Hazard Review*)

FMEA: Análisis de Modos de Falla y Efectos (*Failure Mode and Effects Analysis*)

HAZOP: Estudio de peligros y operatividad (*Hazard and Operability study*)

HF: Factores humanos (*Human Factors*)

HTHA: Ataque de Hidrógeno a alta temperatura (*High Temperature Hydrogen Attack*)

II: Investigación de Incidentes

IPL: Capa de protección independiente (*Independent Protection Layer*)

ITPM: Inspección, testeo y mantenimiento preventivo (*Inspection, Test and Preventive Maintenance*)

LOC: Pérdida de contención (*Loss Of Containment*)

MOC: Manejo del cambio (*Management Of Change*)

P&ID: Diagrama de cañerías e instrumentos (*Piping and Instrumentation Diagram*)

PFD: Diagrama de flujo de proceso (*Process Flow Diagram*)

PHA: Análisis de Peligros de Procesos (*Process Hazard Analysis*)

PMI: Inspección positiva de materiales (*Possitive Material Inspection*)

PSM: Gestión de Seguridad de Procesos (*Process Safety Management*)

RBPS: Seguridad de Procesos basada en riesgo (*Risk Based Process Safety*)

SOP: Procedimiento operativo estándar (*Standard Operating Procedure*)

Referencias

[1] U.S. Chemical Safety and Hazard Investigation Board, *Safety business case,* 2017.

[2] CSB, «Completed Investigations,» [En línea]. Available: www.csb.gov/investigations/completed-investigations.

[3] CCPS/AIChE, Guidelines for hazard evaluation procedures, 2nd ed., Wiley, 1992.

[4] CCPS/AIChE, Guidelines for hazard evaluation procedures, 3rd ed., Wiley, 2008.

[5] A. P. Institute, *API RP 571: Damage Mechanisms Affecting Fixed Equipment in the Refining Industry,* 2011.

[6] U.S. Chemical Safety and Hazard Investigation Board (CSB), «Report No. 2012-03-I-CA - Final investigation report: Chevron Richmond Refinery. Pipe rupture and fire.,» 2015.

[7] U.S. Chemical Safety and Hazard Investigation Board (CSB), «Report No: 2010-08-I-WA - Investigation Report: Catastrophic rupture of heat exchanger (seven fatalities).,» 2014.

[8] CSB, *CSB Safety Video: Anatomy of a Disaster,* 2008.

[9] W. G. Bridges y R. Tew, Tampa, Florida: 5th Global congress on process safety, 2009.

[10] CCPS/AIChE, Guidelines for Risk Based Process Safety, Wiley, 2007.

[11] W. G. Bridges, A. Dowell, J. Thomas y P. Casarez, *Common hurdles, benefits, and costs for fully implementing Process Safety worldwide,* San Antonio: 13th Global Congress on Process Safety, 2017.

[12] W. G. Bridges y R. Tew, *Best practices for writing operating procedures and trouble-shooting guides,* San Antonio, TX: 13th Global Congress on Process Safety, 2017.

[13] M. Nour, S. Maher y J. Schultz, *Effectively addressing new PSM/RMP Damage Mechanism Review requirements with an Integrated PHA (iPHA),* Austin, TX: 11th Global Congress on Process Safety, 2015.

Compromiso con la Seguridad de los Procesos

MORALES, Pablo Javier
YPF SA
Neuquén, Argentina
Pablo.j.morales@ypf.com

PISTAGNESI, Henry
YPF SA
Neuquén, Argentina
Henry.pistagnesi@ypf.com

XANDRI, Carlos
YPF SA
Neuquén, Argentina
Carlos.xandri@ypf.com

Prepared for Presentation at
American Institute of Chemical Engineers
2019 Spring Meeting and 15th Global Congress on Process Safety
New Orleans, LA
March 31 – April 3, 2019

Compromiso con la Seguridad de los Procesos

MORALES, Pablo Javier
YPF SA
Neuquén, Argentina
Pablo.j.morales@ypf.com

PISTAGNESI, Henry
YPF SA

XANDRI, Carlos
YPF SA

Keywords: Process Knowledge Management, Safe Work Practices, Culture in Process Safety

Abstract

Las técnicas de análisis de datos cualitativos de Encuestas aplicadas a investigaciones han aportado durante décadas valiosa información acerca de la estructura, dimensiones y significaciones respecto de los comportamientos sociales, políticos, de imagen, etc. Estos datos registrados pueden ser procesados mediante el tratamiento cuantitativo de lo cualitativo a través de la segmentación, la codificación, entre otras; de modo tal que el "dato textual" se reduzca a un tratamiento y análisis de datos numéricos.

En base a las conclusiones exhibidas en "The Report the BP U.S. Refineries independent safety review Panel" (2007) sobre el evento ocurrido en la Refinería de BP en Texas el 23 de marzo del 2005 en donde hubo 15 fatalidades y más de 170 personas heridas, es que la industria en general comienza a evaluar la Cultura en Seguridad de Procesos implantada en las personas que trabajan de manera directa e indirectamente en sus operaciones.

Basados en el concepto de benchmarking, el equipo de Ingeniería de Procesos y Control junto al equipo de operaciones de Plantas de Gas –ambos de la Regional Centro, Operaciones Oeste- toman como la mejor práctica existente para evaluar la Cultura en Seguridad de Procesos a la Encuesta realizada por International Survey Research (ISR) en el año 2006 y la ejecutan en las instalaciones de dicha Regional.

El compromiso de la Gerencia para con esta iniciativa ha permitido que casi la totalidad de las personas hayan mostrado interés en participar de esta actividad, lo que permitió tener mayores certezas de la Cultura en Seguridad de Procesos en estas instalaciones.

Como resultado de esta investigación, luego de la transformación de los datos cualitativos en cuantitativos, se llega a distintas conclusiones de sumo interés que se pretenden mostrar en este trabajo.

El equipo de trabajo entiende que fortalecer la Cultura en Seguridad de Procesos, como pilar principal del Sistema de Administración de la Seguridad de Procesos del CCPS, permitirá no sólo aportar en Conocimiento y Capacitación adecuada a los trabajadores que participan en las operaciones, sino que también conducirán a un ambiente de trabajo más seguro para las instalaciones de YPF SA.

1 Introducción

2.1 Presentación

En este trabajo de investigación se muestra cómo se evalúa la Cultura en Seguridad de Procesos instaurada en dos instalaciones de tratamiento de gas, del tipo LTS, pertenecientes a YPF S.A.

Para ello, y bajo el concepto de Benchmarking, se toma como la mejor práctica existente en la actualidad a la Encuesta realizada por el "Baker Panel" y el International Survey Research (ISR), que, como ya se dijo anteriormente, ayudara a determinar la Cultura en Seguridad de los Procesos de las Refinerías de BP en los Estados Unidos luego del trágico accidente del 2005. [1]

También con algunos ejemplos concretos, se ejemplifica como se traduce y adaptan las preguntas; y como se analizan las distintas respuestas que permite mostrar el gran potencial que tiene esta herramienta.

Finalmente, la Figura 1 expone parte del Proceso de Mejora Continua que el equipo de trabajo tiene como meta conseguir.

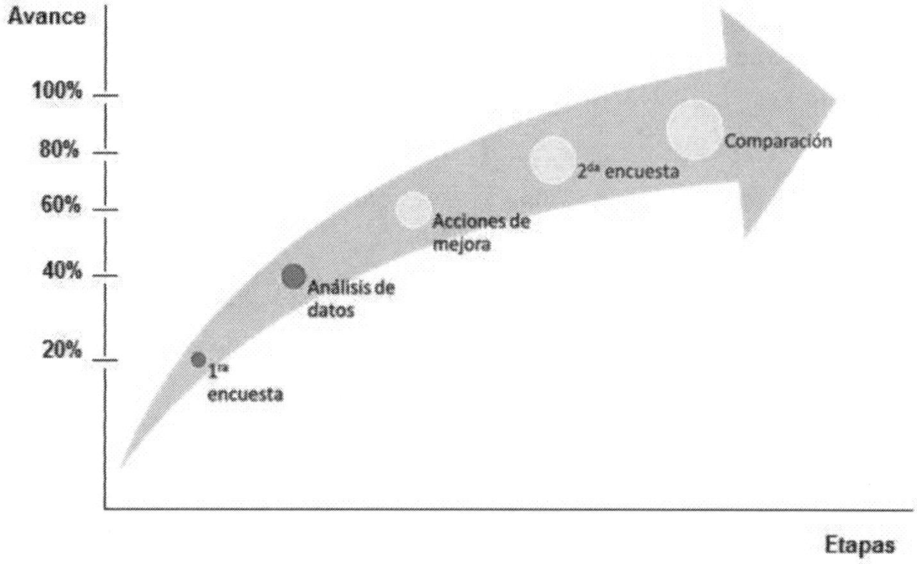

Figura 1. Proceso de Mejora Continua

Las dos primeras etapas se desarrollan a lo largo de todo el presente artículo: 1era encuesta y Análisis de datos.

3.1 Justificación

A largo del tiempo los avances tecnológicos han ocasionado que los procesos sean cada vez más complejos, involucrando así, nuevas materias primas, equipos y actores. Como consecuencia de ello surgen nuevos peligros y riesgos que, de no identificarse y controlarse, pueden ocasionar sucesos que afecten a las personas, al ambiente, a las finanzas y a la reputación de las empresas.

Los grandes accidentes industriales, aunque ocurren con poca frecuencia (baja probabilidad de ocurrencia), cuando se materializan, sus efectos son devastadores (alta consecuencia). (Morgan Torres, 2018) [2]

Desde que el Centro para la Seguridad de Procesos Químicos (CCPS) fue establecido en 1985 por el Instituto Americano de Ingenieros Químicos (AIChE), se fueron utilizando las recomendaciones dadas por esta institución para evitar o mitigar accidentes químicos catastróficos, por ejemplo: las métricas predictivas y retrospectivas, los estándares tales como API RP 754, entre otras. [3]

Sin embargo, a pesar de que existen políticas empresariales o gubernamentales para el cuidado del ambiente laboral, como los procedimientos de buenas prácticas o las métricas y estándares citados anteriormente, los accidentes, incidentes o cuasi accidentes continúan presentes en las operaciones industriales.

A pesar de tener capacitaciones y evaluaciones adecuadas, en las investigaciones de cada evento nos preguntamos a menudo: "¿Por qué se actuó de esa manera?", "¿Por qué se operó con una protección inhibida?", "¿Por qué se firmó un procedimiento donde se detallaban cuáles eran las tareas por realizar y al momento de ejecutarlas, se saltearon los pasos establecidos?". Evidentemente estas herramientas de formación no son suficientes.

La situación se torna hasta inquietante al oír estas preguntas que muchas veces no tienen respuestas, por lo que queremos saber hasta lo imposible: conocer cómo piensa o actúa la gente en momentos de presión o cuando no se la está observando.

El análisis de la Cultura en Seguridad de Procesos nos aproxima a ello, y aunque parezca complejo, nos permite actuar en las debilidades encontradas para que las operaciones en la industria puedan ser más confiables, seguras y sostenibles en el tiempo.

Como menciona Néstor Spósito en su trabajo "*Y a pesar de todo sigue ocurriendo*", **"la Seguridad de Procesos, la Prevención y la Planificación no es un mero ejercicio intelectual, es la diferencia entra la Vida y la Muerte"** [4]

2 Marco de Investigación

El trabajo de campo se llevó a cabo en las plantas de tratamiento de gas LTS 1 y LTS 2 de YPF S.A. ubicadas en el yacimiento Loma La Lata a 100 km de la ciudad de Neuquén, Provincia del Neuquén, República Argentina. Ver Figura 2.

Estas plantas reciben el gas de la separación primaria de yacimientos propios y de otras compañías. Para acondicionarlo a la especificación de transporte, ajustan el punto de rocío de hidrocarburo y agua mediante tecnología de separación por baja temperatura gracias a un sistema de enfriamiento mecánico (LTS: Low Temperature Separation).

El gas tratado es inyectado a gasoductos troncales para su transporte a centros de consumo o es enviado a plantas productoras de NGL[1] que se encuentran aguas abajo.

En las instalaciones trabajan un total de 21 operadores y 6 supervisores en turnos rotativos garantizando así la continua operación.

La LTS 1 tiene una capacidad de diseño para procesar de 21 MMm3/día. Consta de 5 (cinco) trenes de tratamiento; los 4 (cuatro) primeros construidos en el año 1988 y el último en el año 2000. La LTS 2, con capacidad de diseño para procesar de 9 MMm3/día fue construida en el año 1995.

Figura 2. A la izquierda la Planta LTS 1. A la derecha la Planta LTS 2

[1] NGL: Gas Natural Licuado (del inglés Natural Gas Liquids)

3 Metodología

La investigación se lleva a cabo en la Regional Centro, Operaciones Oeste de YPF S.A. en el 3er trimestre de 2018. Es de tipo exploratoria ya que se intenta conocer el nivel de Cultura en Seguridad de Procesos en las instalaciones LTS 1 y LTS 2, hasta el momento desconocido.

La unidad de análisis es el personal operativo. La construcción del objeto se determina por la agrupación de los operadores y supervisores que trabajan en las citadas instalaciones.

En cuanto a las propiedades, se toman preguntas de la encuesta decretadas por el equipo de trabajo como claves o bien, agrupaciones de ciertas preguntas que representan un interés en particular.

La muestra de datos se hace por medio de gráficos de barras y cuadros comparativos con las posibles respuestas del objeto de estudio en el vector vertical.

El procedimiento consta en encuestar personalmente a operadores y supervisores de la LTS 1 y LTS 2. Los investigadores asisten a cada instalación y como primera medida exponen al personal la intención de evaluar la Cultura en Seguridad de Procesos: se da la definición de estos conceptos, la importancia de ellos, se detalla la estructura de la encuesta y la modalidad de respuesta.

Por medio de una herramienta online se envía un link a cada encuestado dónde se encuentra el formulario y se asiste con las dudas o inquietudes que puedan surgir en el momento.

La recolección de información se hace online y luego se analizan los resultados. [5]

4 Encuesta

Respecto a lo conceptual, el diseño de la encuesta consta de preguntas cerradas con respuestas en Escala de Likert[2] y, al final de cada título, un espacio abierto para que el encuestado pueda expresar los comentarios que desee.

La encuesta posee preguntas que gramaticalmente son positivas o negativas. A pesar de que las respuestas sean del tipo “En desacuerdo/Totalmente en desacuerdo”, el porcentaje de ellas puede representar una tasa positiva para la Seguridad de Procesos.

Debido a las dificultades en el idioma y las diferencias operativas entre una refinería y una planta de gas, como así también en la cultura del habla y del trabajo, se realiza la traducción y adaptación

[2] Escala de Likert: es una herramienta de medición que a diferencia de preguntas dicotómicas con respuesta sí/no, permite medir actitudes y conocer el grado de conformidad del encuestado con cualquier afirmación que se le proponga.

de las preguntas de la encuesta original. A tales efectos es importante destacar que algunas de ellas se traducen de manera literal y otras, en cambio, se modifican.

En la Tabla 1 se listan algunos ejemplos de preguntas originales y los cambios incorporados a estas.

Tabla 1. Adaptación de encuesta

N°	Preguntas originales	Preguntas traducidas y adaptadas
1	I have received training on hazard identification, control and reporting in the last 12 months	He recibido capacitación formal sobre identificación y control de riesgos en los últimos 12 meses
2	I can report hazardous conditions without fear of negative consequences	Puedo informar condiciones peligrosas sin temor a consecuencias negativas por parte de YPF S.A.
3	I believe a culture exists at this refinery that encourages raising process safety concerns	Creo que existe una cultura en esta instalación que fomenta la preocupación por la seguridad en el proceso
4	Workers at this refinery feel pressured to work considerable overtime from a) Co-workers b) Supervisors c) Refinery management d) Their own sense of loyalty to their operating units	Los trabajadores de esta instalación se sienten presionados a trabajar considerables horas extras por parte de: a) Compañeros de trabajo b) Su Superior c) Por mayor remuneración d) Su propio interés en ver los trabajos finalizados correctamente
5	Persons with appropriate supervisory authority and expertise participate in hazardous process-related activities, such as startup	La supervisión o personal con experiencia apropiada participa en actividades críticas del proceso como la puesta en marcha inicial, paro, maniobras especiales
6	Process equipment is not regularly a) Tested b) Maintained	Los equipos del proceso (ej. equipos sometidos a presión) son: a) Testeados o probados regularmente b) Mantenidos regularmente
7	I feel that I can influence the process safety policies implemented at this refinery	Siento que puedo influir en la forma de llevar a cabo las acciones relacionadas a

		la seguridad de procesos en esta instalación
8	The process safety training that I have received allows me to recognize when a process should be shut down if safety critical interlocks, alarms or other process-safety devices fail or become unavailable during operation	La capacitación en seguridad de procesos que ha recibido le permite reconocer cuándo debe tomar medidas correctivas o detener una tarea si los enclavamientos críticos para la seguridad, las alarmas u otros dispositivos fallan o no están disponibles durante la operación

Por ejemplo, en la N° 6 el sentido gramatical cambia de negativo a positivo para un mejor entendimiento.

En otros casos se agregan elementos particulares o propios del sistema de trabajo, como por ejemplo en la adaptación de la N° 4 se añade "Mayor remuneración" como una posibilidad del origen de presión a trabajar considerables horas extras.

En igual sentido a lo expresado por "The Report the BP U.S. Refineries independent safety review Panel", **no hay normas reglamentarias ni datos de referencias externas con los que se puedan medir los resultados de la encuesta. No se puede concluir si las instalaciones tienen una cultura de seguridad más o menos positiva que otras instalaciones similares. Tampoco hay una encuesta previa para comparar si la Seguridad de Procesos está mejorando, empeorando o estabilizándose. Por lo que ningún porcentaje (%) de respuesta específico para ningún ítem de la encuesta puede considerarse como indicador.**

Sabiendo de esta limitación, se consideran los datos como información valiosa y los resultados de su análisis, una oportunidad para mejorar la Seguridad de los Procesos.

5 Análisis de Datos

2.5 Participación

Durante el lapso en el que se realizó la encuesta, participaron el 83% de los supervisores y el 71,4% de los operadores. El tiempo de respuesta fue de aproximadamente de 40 min por Encuestado.

3.5 Consideraciones

El equipo de trabajo define que para tasas positivas superiores a 70% y tasas negativas inferiores a 30%, la percepción en general es positiva y no requiere medidas correctivas inmediatas.

En cambio, si la tasa negativa es mayor a un 30% o si la tasa positiva es inferior al 70%, se debe prestar atención en los resultados obtenidos y proponer acciones de mejora.

En el caso de que el porcentaje de "Ni de acuerdo, ni en desacuerdo" tenga un valor mayor a un 25% se debe analizar el caso y plantear mejoras. Si el porcentaje es menor, no se considera necesario tomar acciones al respecto.

Para el análisis de datos, los ítems "De acuerdo/Totalmente de acuerdo" forman un solo grupo de respuesta positiva. De idéntica forma en "Desacuerdo/Totalmente en desacuerdo" forma otro grupo de resultados negativos.

4.5 Ejemplos de resultados dónde no se requieren inmediatas acciones de mejora

EJEMPLO 1:

El 90% de los encuestados (mayor al 70% fijado) cree que puede hacerlo sin sufrir alguna consecuencia negativa. Esto es positivo para la Seguridad de Procesos. Ver Figura 3.

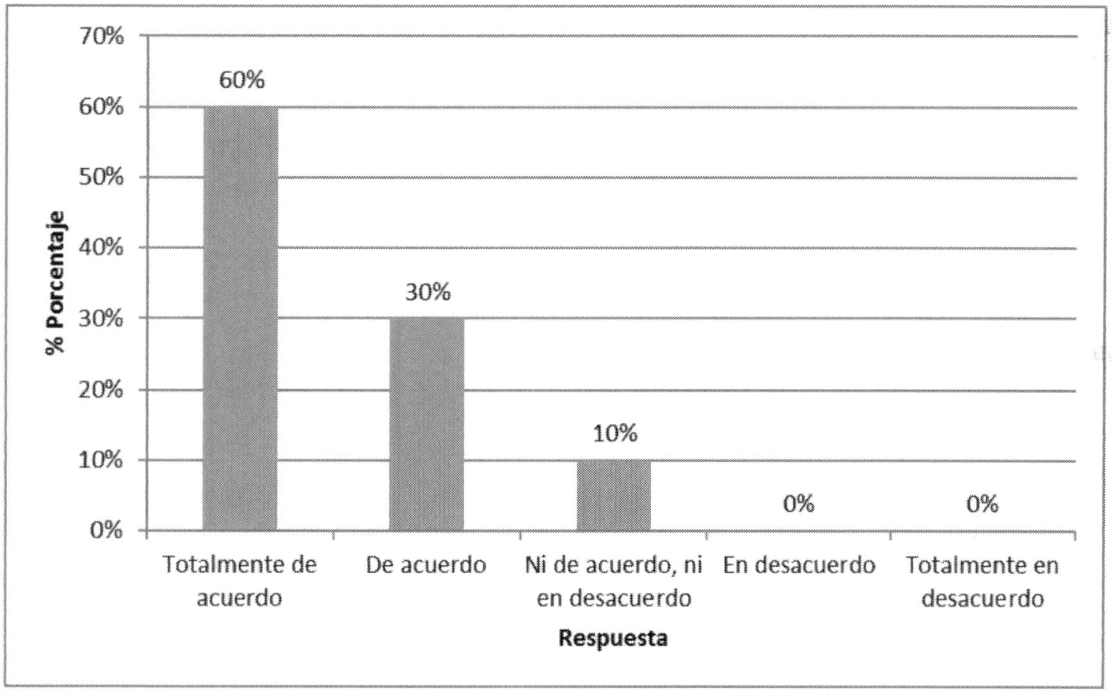

Figura 3. "Puedo informar condiciones peligrosas sin temor a consecuencias negativas por parte de YPF S.A."

EJEMPLO 2:

A pesar de que la respuesta es negativa (En desacuerdo/Totalmente en desacuerdo), para la Seguridad de Procesos esto es positivo. Al igual que en el caso anterior, el resultado del 95% positivo o bien, el 5% negativo restante, se encuentran fuera del límite fijado para una acción inmediata. Ver Figura 4.

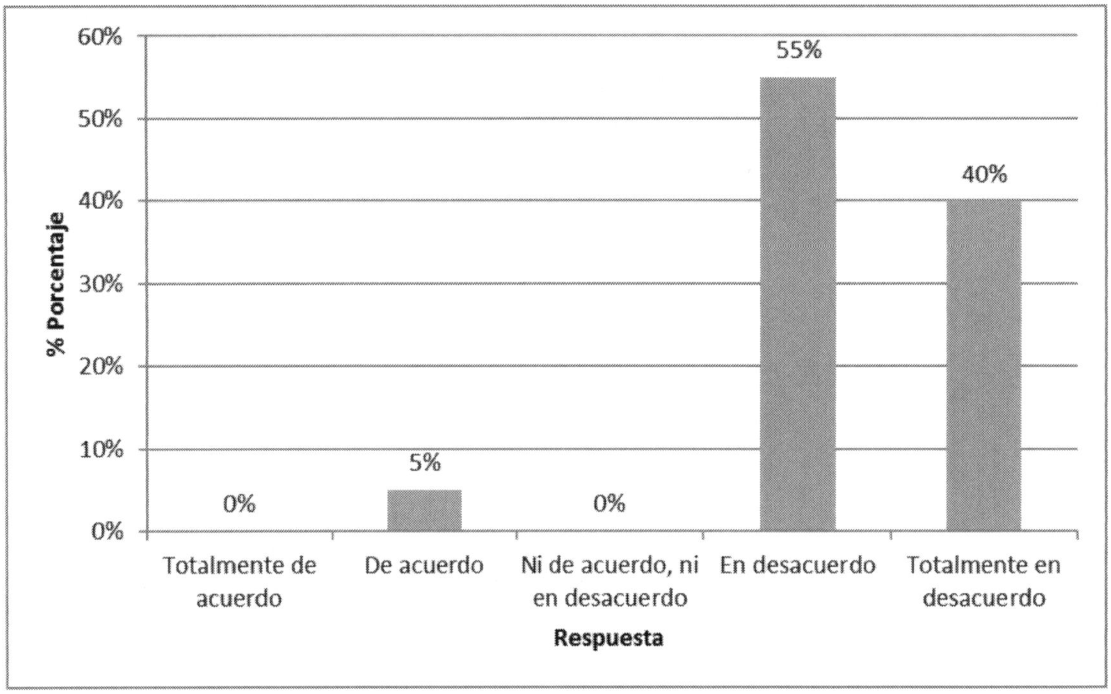

Figura 4. "Mi Supervisor a veces me pide que realice un trabajo inseguro"

EJEMPLO 3:

Para este ejemplo. se observa en la Tabla 2 los porcentajes obtenidos respecto a si los trabajadores se sienten presionados a trabajar considerables horas extras. Estos datos se ven gráficamente en Figura 5.

Tabla 2. "Los trabajadores de esta instalación se sienten presionados a trabajar considerables horas extras por parte de:"

Posibles respuestas	Compañeros de trabajo	Su superior	Por mayor remuneración	Su propio interés en ver los trabajos finalizados correctamente
Totalmente de acuerdo	0%	0%	10%	15%
De acuerdo	0%	0%	15%	5%
Ni de acuerdo, ni en desacuerdo	0%	0%	0%	15%
En desacuerdo	50%	50%	35%	30%

Totalmente en desacuerdo	50%	50%	40%	35%

Se puede concluir que no existe un causal que genere que el personal realice considerables horas extras.

Si bien la mayor remuneración es uno de los causales para ser considerables horas extras, el porcentaje de resultados no re quiere tomar acciones inmediatas, lo mismo que el propio interés que ver los trabajos finalizados correctamente.

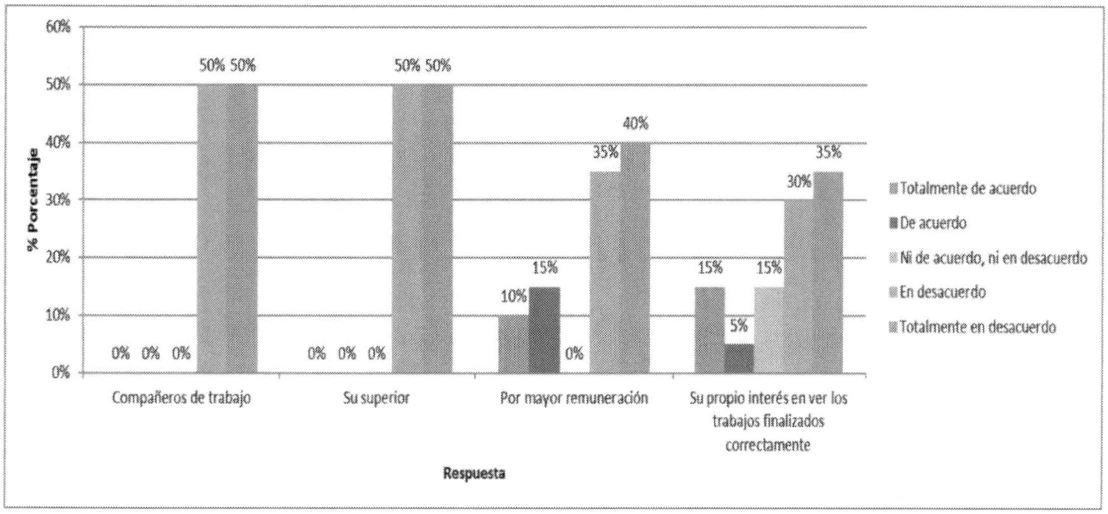

Figura 5. "Los trabajadores de esta instalación se sienten presionados a trabajar considerables horas extras por parte de:"

5.5 Ejemplos de resultados dónde si se requiere alguna acción de mejora

EJEMPLO 1:

Como se muestra en la Figura 6, el 60% de los encuestados dice que no ha recibido capacitación sobre identificación y control de riesgos en los últimos 12 meses.

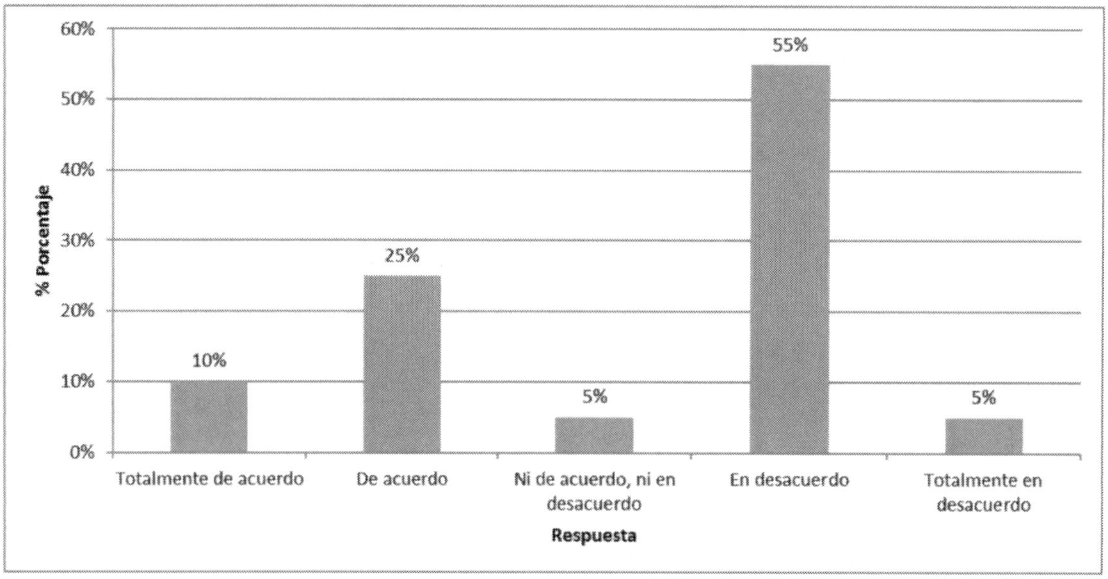

Figura 6. "He recibido capacitación formal sobre identificación y control de riesgos en los últimos 12 meses"

En la Tabla 3 se muestra el análisis particular de la población: el 40% de los supervisores sostiene no haber recibido capacitación formal en identificación y control de riesgos en los últimos 12 meses, en tanto que este porcentaje asciende a un 66,67% en caso de los operadores. En esta dirección se deberían tomar acciones inmediatas para capacitar al personal.

Tabla 3. "He recibido capacitación formal sobre identificación y control de riesgos en los últimos 12 meses"

Posibles respuestas	Operadores	Supervisores
Totalmente de acuerdo	6,67%	20%
De acuerdo	20%	40%
Ni de acuerdo, ni en desacuerdo	6,67%	0%

En desacuerdo	60%	40%
Totalmente en desacuerdo	6,67%	0%

EJEMPLO 2:

En otro resultado, el 32% de las personas que trabajan en las plantas de gas no tienen una respuesta definida respecto de si los problemas de Seguridad del Proceso sirven para tomar acciones correctivas e implementarlas en tiempo y forma. El grado de incertidumbre o de desconocimiento hace necesario que se tome algún tipo de acción. Ver Figura 7.

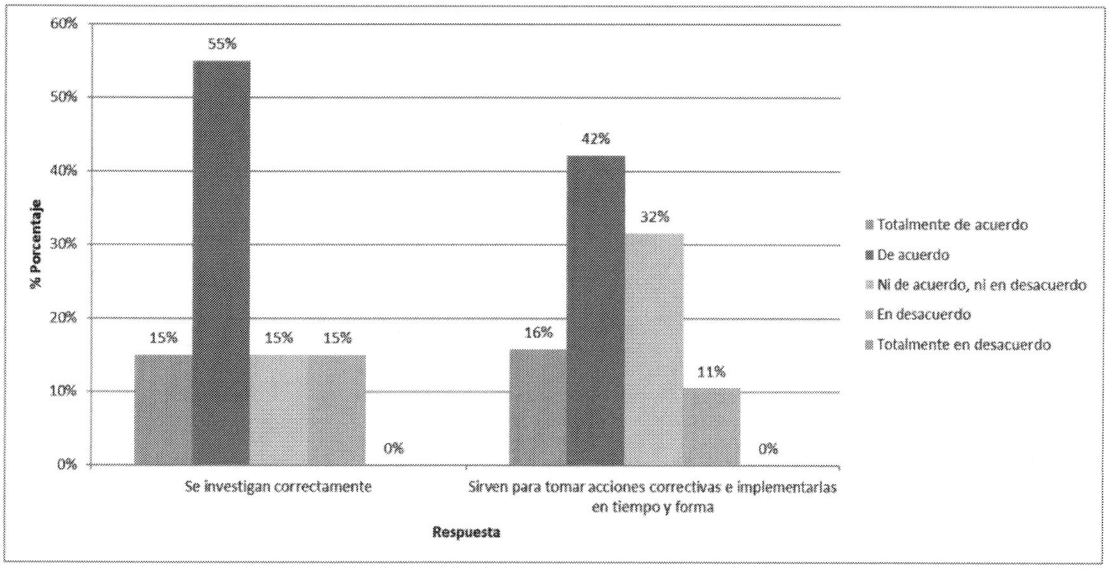

Figura 7. "Estoy seguro de que los problemas de seguridad del proceso: a) Se investigan correctamente; b) Sirven para tomar acciones correctivas e implementarlas en tiempo y forma"

EJEMPLO 3:

Respecto a que, si los procedimientos operativos escritos se revisan regularmente y se mantienen actualizados, los resultados de esta consulta se observan en la Tabla 4:

Tabla 4. "Los procedimientos operativos escritos"

Posibles respuestas	Revisan regularmente	Mantienen actualizados
Totalmente de acuerdo	5%	5%

De acuerdo	40%	25%
Ni de acuerdo, ni en desacuerdo	30%	45%
En desacuerdo	25%	25%
Totalmente en desacuerdo	0%	0%

En la Figura 8, se observa gráficamente que el grado de incertidumbre es tan elevado que requeriría tomar algún tipo de acción.

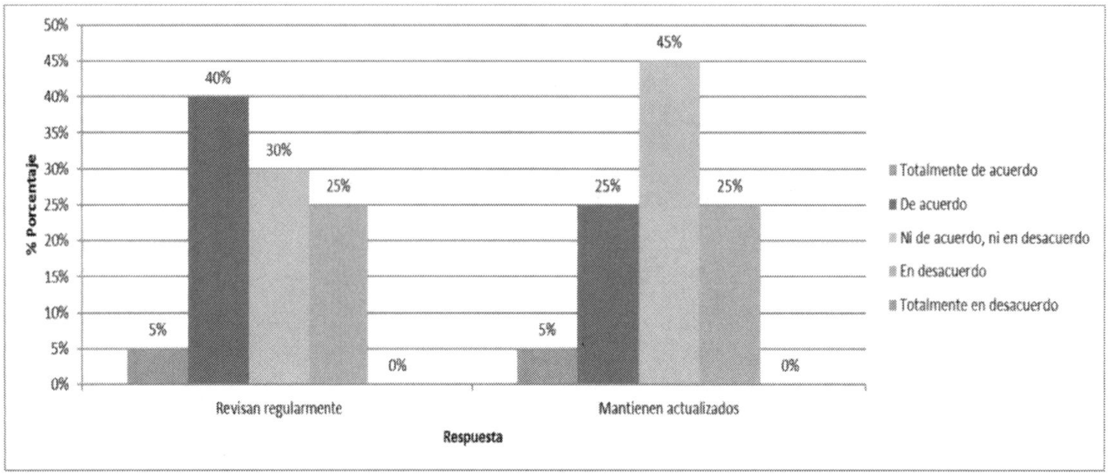

Figura 8. "Los procedimientos operativos escritos se: a) Revisan regularmente, b) Mantienen actualizados"

EJEMPLO 4:

En la Figura 9, se visualiza que el 60% de los encuestados considera que los trabajadores de la LTS 1 y LTS 2 no participa activamente en la revisión y evaluación de riesgo. Esto es un hecho relevante que requiere implementación de medidas correctivas. Además, que el 70% dice que no participan en investigación de incidentes y accidentes. Lo cual también requiere un estudio y posteriores acciones correctivas.

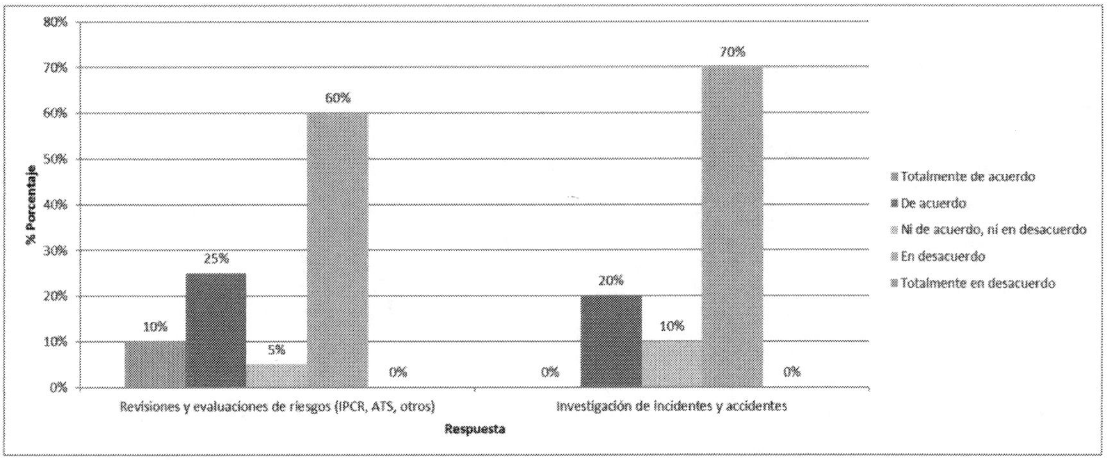

Figura 9. "Los trabajadores en todos los niveles de mi instalación participan activamente en a) Revisiones y evaluaciones de riesgos (IPCR[3], ATS[4], otros), b) Investigación de incidentes y accidentes"

EJEMPLO 5:

En la Figura 10, podemos realizar distintos análisis:

El 65% de los encuestados considera que el entrenamiento en seguridad de procesos a nuevos trabajadores no es suficiente.

El 35% considera que los trabajadores experimentados no reciben el entrenamiento adecuado para realizar sus tareas en forma segura.

El 85% piensa que el supervisor inmediato superior recibe el entrenamiento adecuado en seguridad de los procesos.

El 50% de los trabajadores advierte que el personal de empresas contratistas no recibe el entrenamiento necesario en Seguridad de Procesos.

Salvo los supervisores, el resto de los actores que intervienen en la operación no tiene entrenamiento suficiente en Seguridad de Procesos. Se deberán tomar acciones correctivas.

"¿Cree que las personas abajo indicadas reciben el entrenamiento de seguridad de proceso necesario para hacer su trabajo de forma segura?"

---

[3] IPCR: Identificación de Peligros y Control de Riesgos
[4] ATS: Análisis de Trabajo Seguro

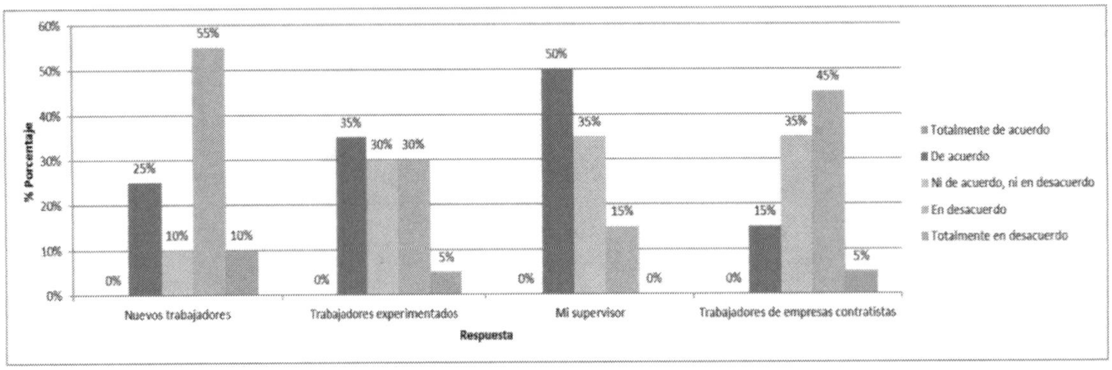

Figura 10. "¿Cree que las personas abajo indicadas reciben el entrenamiento de seguridad de proceso necesario para hacer su trabajo de forma segura?"

6 Conclusiones

Se observa que no hay base de datos que permita la comparación directa de la información obtenida, igualmente el diseño de la encuesta y el análisis de los resultados a posteriori han permitido tomar acciones de mejora inmediata en las áreas que requieren mayor atención.

Particularmente ha expuesto debilidades en temas como, por ejemplo, la frecuencia en la capacitación en seguridad de procesos y la participación en investigaciones de incidentes y accidentes. Además, nos ha permitido acercarnos a la realidad social de los diferentes actores de la planta. Realidades en ocasiones imperceptibles, dada la vorágine diaria en este tipo de industria.

La encuesta realizada por el "Baker Panel" y el International Survey Research (ISR), en nuestra adaptación, ha sido una herramienta efectiva al momento de analizar el nivel de Cultura en Seguridad de Procesos en las plantas LTS 1 y LTS 2.

Por tanto, la aplicación reiterada de la misma, su extensión en el resto de las instalaciones y en empresas contratistas que trabajan con esta compañía, será el desafío que permitirá continuar con el proceso de mejora continua.

7 Referencias

[1] J. A. Baker, F.L. Bowman, G. Erwin, S. Gorton, D. Hendershot, N. Leveson, S. Priest, I. Rosenthal, P. Tebo, D.A. Wiegmann y L.D. Wilson. *The report the BP U.S. Refineries Independent Safety Review Panel*, U.S., 2007.

[2] G, Morgan Torres, Seguridad Personal & de Procesos, Disponibe en https://revistaempresarial.com/salud/salud-ocupacional/seguridad-personal-procesos/, Accedido en Diciembre 2018.

[3] CCPS. *Process Safety Leading and Lagging Metrics.* Center for Chemical Process Safety, American Institute of Chemical Engineers, New York, NY, 2006.

[4] N, Spósito. *Y a Pesar de todo Siguen Ocurriendo, Prevención y Mitigación de Accidentes*, PPR Consulta, Buenos Aires, 2014.

[5] A, Marradi, N, Archenti y J.I. Piovani. *Metodología de las Ciencias Sociales*. Emecé Editores, Buenos Aires, 2007.

The Effects of Oxygen Enrichment

on the

Explosibility Properties of Flammable Gases and Vapors

Richard W. Prugh
DEKRA Process Safety
113 Campus Drive
Princeton, NJ 08540
www.dekra-process-safety.com
prughrw@aol.com

Prepared for Presentation at
American Institute of Chemical Engineers
2019 Spring Meeting and 15th Global Congress on Process Safety
New Orleans, LA
March 31 – April 3, 2019

The Effects of Oxygen Enrichment on the Explosibility Properties of Flammable Gases and Vapors

Richard W. Prugh
DEKRA Process Safety
113 Campus Drive
Princeton, NJ 08540

Keywords: Flammables, Oxygen, Explosion

Abstract

In the chemical industry, oxygen-containing chemicals are used in the production of partial-oxidation products, including many in the pharmaceutical industry. The vapor spaces in oxidation reactors can contain high concentrations of oxygen gas, in combination with flammable vapors. Thus, it is important to determine the explosibility properties of organic-compound vapors in atmospheres containing high oxygen concentrations.

1. Introduction

In the chemical industry, "nascent" oxygen is involved in the production of partial-oxidation products, including many in pharmaceutical operations. The oxygen is initially in the form of peroxides, such as hydrogen and sodium peroxides, and also as chlorates and perchlorates, periodates, nitrates, and dichromates [1]. The vapor space in oxidation reactors can contain high concentrations of oxygen gas, in combination with the flammable vapors of organic compounds. To evaluate the possible explosion hazards of such mixtures of fuel and oxidizer, it is important to determine the explosibility properties of organic compounds in atmospheres containing high oxygen concentrations.

The basic flammability/explosibility data for this study are from the following references:

Table 1. References for Flammability Data

REFERENCE	FLAMMABILITY/EXPLOSIBILITY PROPERTY					
	Flammable Limits	Auto-Ignition Temperature	Minimum Ignition Energy	Maximum Explosion Pressure	Deflagration Index Kg	Oxygen Conc. Limit
NFPA 49; NFPA 325 [2]	√	√				
NFPA 68 [3]				√	√	
NFPA 69 [4]						√
NFPA 77 [5]	√ *		√			
NFPA 497 [6]	√	√	√			
USBM 503 [7]	√ *					√
USBM 627 [8]	√ *	√ *				
Babrauskas [9]	√	√ *	√ *			
Britton [10]	√ *		√ *			√
Crowl and Louvar [11]	√ *	√	√	√	√	√
Yaws [12]	√	√				

* Limited data for oxygen

Several methods are presented in this paper for estimating explosibility properties for preliminary process design. However, it is prudent to confirm the fuel/oxygen relationships with tests where process safety depends on the reliable control of oxygen concentration.

2. Flammability/Explosibility Properties

The explosibility properties of combustible materials can be divided into Ignition Sensitivity and Explosion Severity [13, 14]. The Lower Flammable Limits, the Autoignition Temperatures, and the Minimum Ignition Energies would be important in evaluating the Ignition Sensitivity. The Maximum Explosion Pressure and the Deflagration Index – a measure of burning rate – would be important in evaluating the Explosion Severity. The following sections describe the relationships of these properties to oxygen concentration.

2.1 Lower Flammable Limit

Data from many sources [15, 16, 17, 18] indicate that there is no significant effect of oxygen concentration on the Lower Flammable Limit [LFL] from the 21 volume-percent oxygen in air to 100% oxygen. At oxygen concentrations greater than the LFL in air, the specific heat of oxygen is about the same as the specific heat of nitrogen; that is, excess oxygen behaves as a diluent at the LFL.

For comparison with the Minimum Explosible Concentrations of combustible dusts and for evaluating the hazards of mixtures of flammable gases/vapors of combustible dusts, the values of LFL can be converted to mass densities by:

$$C = [\,LFL / (100 - LFL)\,]\,[\,MW / 0.024\,] \quad \text{grams per cubic meter}$$

where 0.024 is the number of cubic meters per gram-molecular weight [at 25°C and one atmosphere pressure (14.7 psia)].

For example, the LFL for propane in air is 2.1 volume-percent and, for the molecular weight of 44.1, the Minimum Explosible Concentration would be given by:

$$C = [\,2.1 / (100 - 2.1)\,]\,[\,44.1 / 0.024\,] = 39 \text{ grams per cubic meter}$$

Since the LFL is not significantly affected by oxygen concentration, oxygen enrichment would not affect the ventilation and dilution or inerting precautions that should normally be taken for a flammable vapor or gas.

2.2 Autoignition Temperature

The Autoignition Temperature [AIT] of a fuel is a kinetic phenomenon; that is, the AIT is low for very-long exposures to a given temperature, and it is quite high for short exposures to a given temperature. This relationship can be shown to follow the Arrhenius "law" [19], as:

$$r = A\,e^{-Ea/RT} \quad \text{fraction per second}$$

where A is the pre-exponential factor (in "per second");
Ea is the activation energy (in calories per gram-mole);
R is a constant (1.987 calories per gram-mole per °K); and
T is the temperature (in °K).

If equal reaction rates are assumed for combustion in air and oxygen, at two different temperatures, a value for the activation energy can be obtained, from:

$$Ea = 1.987\,[\,\ln(100/21)\,] / [\,(1/T_{AIR}) - (1/T_{O2})\,] = 3.1 / [\,(1/T_{AIR}) - (1/T_{O2})\,] \quad \text{cal./gmol}$$

The Autoignition Temperatures for 90 chemical compounds in air and in 100% oxygen are available [20]. The average AIT in air was calculated to be about 394°C [about 667°K], and the average AIT in oxygen was about 361°C [about 634°K]. Thus, the differences in AIT for air and oxygen is quite small, at about 5% lower.

The calculated activation energy Ea, as based on the referenced data, would be about 40,000 calories per gram mole, which is a "reasonable" value for an oxidation/ combustion reaction [21].

Thus, the AIT for a given oxygen concentration can be estimated from:

$$AIT_{O2} = AIT_{AIR} / [1 + (1.987\ AIT_{AIR} / 40{,}000) (\ln (C_{O2}/21)] - 273 \qquad ^{o}C$$

Oxygen enrichment would not affect the precautions that should normally be taken for a flammable or combustible vapor, such as avoiding exposure to hot surfaces in ovens, in dryers, and on electrical equipment such as motors.

2.3 Minimum Ignition Energy

Previous investigators had found that the Minimum Ignition Energy [MIE] was inversely related to oxygen concentration [22]. The Minimum Ignition Energies for 13 chemical compounds in air and in 100% oxygen are available [23]. The average MIE in air [excluding hydrogen (very low) and trichloroethylene (very high)] was calculated to be about 0.30 millijoule, and the average MIE in oxygen was about 0.0027 millijoule. Thus, oxygen concentration has a very strong effect on MIE, with about a factor of 100 reduction for 100% oxygen. The exponent for oxygen concentration for the 13 chemical compounds [including hydrogen and trichloroethylene] is about 2.9.

Thus, the MIE for a given oxygen concentration can be estimated [+/- 50%] from:

$$MIE_{O2} = MIE_{AIR} / [(C_{O2} / 21)^{2.9} = 6{,}200\ MIE_{AIR} / C_{O2}^{\ 2.9} \qquad \text{millijoule}$$

As indicated above, oxygen enrichment would have a significant effect on the precautions [such as inerting] that should be taken for flammable gases and liquid against very weak electrostatic [and other] ignition sources, including corona discharges [24].

2.4 Maximum Explosion Pressure

The Maximum Explosion Pressure [Pmax] – for combustion in air or in an oxygen-enriched atmosphere – can be calculated rather accurately if the heat of formation of the fuel is known and the degree of dissociation of the carbon dioxide and steam combustion products can be determined. For combustion in an oxygen-enriched atmosphere, the degrees of dissociation are very high; thus, the Pmax in 100% oxygen could be expected to show a rather small increase in pressure as compared to combustion in air.

There is a direct relationship between flame temperature and explosion pressure, as governed by the Ideal Gas Law, as:

$$P_{Max} = 14.7\ (\ T_{FLAME} + 273\ )\ /\ (\ 25 + 273\ ) - 14.7 \qquad \text{pounds/sq.in. [psig]}$$

The available flame-temperature explosion-pressure data are as follows [25]:

Table 2. Flame Temperatures and Explosion Pressures for Fuel/Oxygen Mixtures

FUEL	FLAME TEMPERATURES (°C)			MAX. PRESSURES (barg)		RATIO (bara/ bara)	IDEAL-GAS-LAW RATIOS (°K/bara)	
	In Air	In Oxygen	Ratio [°K/°K]	In Air	In Oxygen		In Air	In Oxygen
Methane	1,950	2,705	1.37	7.1	10.4	1.41	274	261
Propane	1,988	2,822	1.37	7.9	10.2	1.26	254	276
Ethylene	2,088	2,902	1.34	8.0	10.7	1.30	262	271
Acetylene	2,262	3,069	1.32	10.6	11.2	1.05	219	274

Thus, approximate values for Pmax in mixtures of oxygen and in air would be given by:

$$P_{Max,O2} = [\ (\ T_{FLAME} + 273\ )\ /\ 261\ ] - 1 \qquad \text{barg}$$

Thus, if the flame temperature can be calculated, from heats of formation and specific heats – with consideration of dissociations – the maximum explosion pressure for a hydrocarbon can then be calculated. It is likely that this relationship would also apply to other hydrocarbons and carbohydrates. Also, as indicated in the preceding Table 2, the values of P_{MAX} in 100% oxygen are about 30% greater than the value in air [excluding acetylene].

The data in Table 2 above can be used to estimate the Maximum Explosion Pressure [P_{MAX}] for a given concentration of oxygen as:

$$P_{MAX,C} = 0.035\ C_{O2} + 0.92\ P_{MAX,AIR} \qquad \text{bara}$$

The following Table presents the results of calculations, based on the above relationship, for two typical values for P_{MAX} in air [26]:

Table 3. Maximum Explosion Pressures for Fuel/Oxygen Mixtures

$P_{MAX,AIR}$	MAXIMUM EXPLOSION PRESSURES IN FUEL/OXYGEN MIXTURES (barg)							
	30 %O_2	40 %O_2	50 %O_2	60 %O_2	70 %O_2	80 %O_2	90 %O_2	100 %O_2
7 barg	7.4	7.7	8.1	8.4	8.8	9.1	9.5	9.8
8 barg	8.3	8.6	9.0	9.3	9.7	10.0	10.4	10.7
9 barg	9.2	9.6	9.9	10.3	10.6	11.0	11.3	11.7
10 barg	10.1	10.5	10.8	11.2	11.5	11.9	12.2	12.6

Thus, oxygen enrichment would have a significant effect on the extent of damage that could result from ignition of confined flammable gases and vapors, and could affect the design of explosion-protection systems and devices, such as explosion vents.

2.5 Deflagration Index, Kg

The Deflagration Index [Kg for flammable gases and vapors] is a measure of the speed of combustion and, thus, is directly related to the speed of flame through a mixture of fuel and oxidant. In the absence of strong turbulence, the flame speed is related to the fundamental burning velocity [FBV] by the ratio of the densities of the unburned mixture and the density of the combustion gases, by [27]:

$$S_{FLAME} = FBV\,(d_{UNBURNED}/d_{BURNED}) \quad \text{centimeters per second}$$

Typically, the ratio of the flame speed to FBV is about 6.

Previous investigators had observed that there is a relationship between the FBV and the Minimum Ignition Energy [MIE], as given by [28]:

$$FBV = K/MIE^{0.8} \text{ centimeters per second}$$

Some data are available for both the FBV and the MIE, as indicated in the following Table [29] :

Table 4. The Relationship Between FBV And Oxygen Concentration

GAS/VAPOR	FBV, Air (cm./sec.)	Kg, Air (bar-m/s)	MIE, Air (millijoule)	FBV, O_2 (cm./sec.)	MIE, O_2 (millijoule)	Kg, O_2 (bar-m/s)	Kg Ratio O_2/Air
Hydrogen	312	638	0.016	1400	0.0012	3,000	4.5
Methane	45	55	0.21	450	0.0027	450	4.6
Ethane	47.5	103	0.24	375	0.0019	670	6.5
Propane	46	100	0.25	370	0.0021	720	7.2
Butane	-	92	0.25	-	0.009	-	-
Hexane	-	-	0.24	-	0.006	-	-
Heptane	42.5	-	0.24	-	-	-	-
Averages	-	-	-	-	-	-	5.7

As indicated in the above Table, the Deflagration Index – and the speed of combustion – in oxygen is almost six times greater in 100% oxygen than it is in air [at 21% oxygen]. The data in the preceding Table can be combined to yield two approximate expressions for the Deflagration Index in given concentrations in oxygen, as:

$$Kg_{,O2} = 0.04\ (K_{G,AIR})\ (C^{1.2}) \qquad \text{bar-meters/second}$$

$$Kg_{,O2} = 1.85\ C^{1.1} / MIE_{AIR}{}^{0.46} \qquad \text{bar-meters/second [excluding } H_2\text{]}$$

Thus, if the value of the Deflagration Index in air is known, or if the Minimum Ignition Energy in air is known, an approximate value for the Deflagration Index at a given oxygen concentration can be determined.

3. Combined Relationships

The explosibility/flammability properties of a given flammable gas or vapor can be compared with a standard or typical flammable gas, through the use of Ignition Sensitivity and Explosion Severity. These property combinations were first developed and utilized for the comparisons of the hazards of coal mines, starting in 1940, where combustible dusts presented an explosion hazard [30].

3.1 Ignition Sensitivity

The product of LFL [in terms of grams per cubic meter], AIT, and MIE form the basis for a calculation of Ignition Sensitivity. A high value of this product indicates a low likelihood of ignition by relatively-weak ignition sources, such as electrostatic discharges, friction, impact, and hot surfaces. For propane in a mixture with 40% oxygen, the product is only 39 x 445 x 0.035 = 600. For comparison, the similar product for typical combustible dusts is about 45,000.

To determine the Ignition Sensitivity of a combustible dust, the LFLxAITxMIE product would be divided into the corresponding product for Pittsburgh [bituminous] Coal Dust, at 2,000,000. Thus, the Ignition Sensitivity for typical combustible dusts is about 2,000,000/45,000 = 45. By comparison, the Ignition Sensitivity for propane in a 40% oxygen mixture would be about 2,000,000/600 = 3,300, indicating the much greater sensitivity of this propane mixture to ignition sources.

As another indication of the effects of oxygen enrichment, the Ignition Sensitivity for a stoichiometric [worst-case] mixture of propane and air would be about 450. Thus, the 40% oxygen mixture would be about 7 times more sensitive to ignition [3,300/450].

3.1 Explosion Severity

The product of Deflagration Index Kg and the Maximum Explosion Pressure Pmax [in terms of barg] forms the basis for a calculation of Explosion Severity. A low value of this product indicates a low likelihood of severe damage from explosion. For propane in a mixture with 40% oxygen, the product is about 275 x 9 = 2,500. For comparison, the similar product for typical combustible dusts is about 1,500.

For combustible dusts, the product is divided by the corresponding product for Pittsburgh [bituminous] Coal Dust, at 840. Thus, the Explosion Severity for typical combustible dusts is about 1,500/840 = 1.8. By comparison, the Ignition Sensitivity for propane in a 40% oxygen mixture would be about 2,500/840 = 3.0, indicating the greater likelihood of damage if a confined mixture of propane and 40% oxygen were to be ignited.

The effect of oxygen enrichment can be shown by the Explosion Severity for a stoichiometric [worst-case] mixture of propane and air, at about 0.9. Thus, the 40% oxygen mixture could be about 3 times more damaging than an air mixture [3/0.9].

3.1 Combination of Ignition Sensitivity and Explosion Severity

If the Ignition Sensitivities of fuel/oxidant mixtures are less than 0.2 AND the Explosion Severities are less than 0.5, then such mixtures "are generally not considered to be significant explosion hazards" [31].
For the mixture of propane and 40% oxygen – for which the properties are shown above – the combination of Ignition Sensitivity and Explosion Severity is about 5 orders of Magnitude [a factor of almost 100,000] above the "no significant explosion hazard" criterion. Thus, it is unlikely that a mixture of air with any flammable or combustible vapor could be considered to present "no significant explosion hazard".

3.1 Combination of Ignition Sensitivity and Explosion Severity as Explosibility Index

The product of Ignition Sensitivity and Explosion Severity can be considered to be an Explosibility Index [32 to 37]. This Index has been found to be useful in comparing the hazards of combustible dusts, but had not yet been applied to flammable gases and vapors.

4. Graphical Presentation of Flammability/Explosibility Properties

Figure 1 presents the flammability/explosibility properties of a mixture of flammable fuel, oxygen, and nitrogen, in terms of Ignition Sensitivity [IS], Explosion Severity [ES], and Explosibility Index [EI] at atmospheric pressure and at an initial temperature of 25°C, as compared to the similar explosibility properties of propane at room temperature and atmospheric pressure. The ratios are based on the available data and the results of the calculations that are described above. As could be expected, the explosion hazards increase as the stoichiometric line – where all carbon is oxidized to carbon dioxide, and all hydrogen is oxidized to steam – is approached.

5. Conclusions.

To evaluate the explosion hazards of mixtures of fuels with oxygen-enriched atmospheres, the available data were utilized to obtain methods for estimating the values of pertinent explosibility properties. The relationships are as follows:

Lower Flammable Limit	Equal to that of fuel/air mixtures	
Autoignition Temperature	$AIT_{O2}=AIT_{AIR}/[1+(2.0AIT_{AIR}/40{,}000)(\ln(C_{O2}/21)]-273$	°C
Minimum Ignition Energy	$MIE_{O2} = 6{,}200\ MIE_{AIR} / C_{O2}^{\ 2.9}$	millijoule
Max. Explosion Pressure	$P_{Max,O2} = [[0.035\ C_{O2}) + (0.92\ (P_{Max,AIR}+1))] - 1$	barg
Deflagration Index	$Kg_{,O2} = 0.04\ (K_{G,AIR})\ (C^{\ 1.2})$	bar-meters/second
	$Kg_{,O2} = 1.85\ C^{\ 1.1} / MIE_{AIR}^{\ 0.46}$	bar-meters/second

In addition, a method is presented for evaluating the Ignition Sensitivity and the Explosion Severity of fuel/oxygen mixtures, using the above relationships, together with a method for graphically showing the relative hazards of such mixtures.

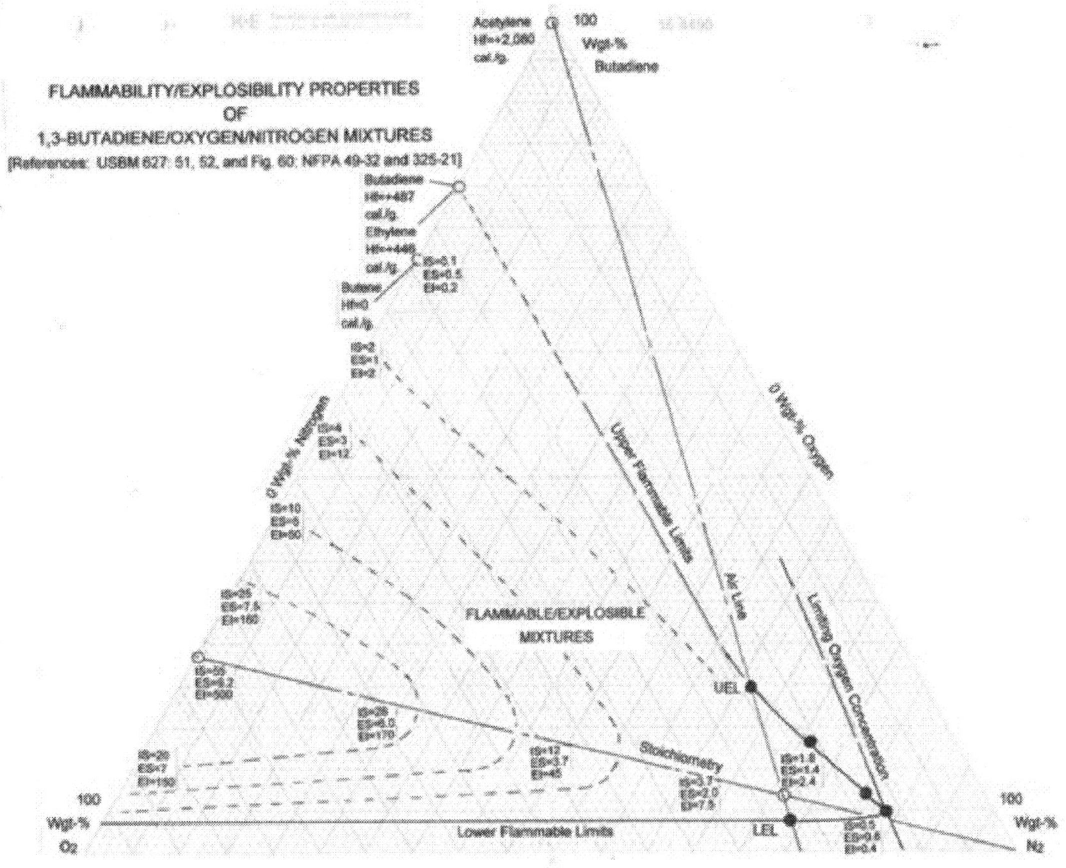

Figure 1 – Weight Concentrations of Gas Mixtures
with
Ignition Sensitivities, Explosion Severities, and Explosion Index
as compared to the properties of Propane

6. References.

1. Urben, P. G. [editor], "Bretherick's Handbook of Reactive Chemical Hazards", 6th Ed., Vol. 2, Page 287 (1999).

2. National Fire Protection Association, "Fire Protection Guide to Hazardous Materials", 13th Ed., including NFPA 49 and NFPA 325 (2002).

3. National Fire Protection Association, "Guide for Venting of Deflagrations", NFPA 68, Table D.1 (2002).

4. National Fire Protection Association, "Standard on Explosion Prevention Systems", NFPA 69, Table C.1(a) (2014).

5. National Fire Protection Association, "Recommended Practice on Static Electricity", NFPA 77, Table B.1 (2014).

6. National Fire Protection Association, "Recommended Practice for the Classification of Flammable Liquids, Gases, or Vapors and of Hazardous (Classified) Locations for Electrical Installations in Chemical Process Areas", NFPA 497, Table 4.4.2 (2017).

7. Coward, H. F., and G. W. Jones [U. S. Bureau of Mines], "Limits of Flammability of Gases and Vapors", Bulletin 503, Table 44 [and many "nose" Figures] (1952).

8. Zabetakis, M. G. [U. S. Bureau of Mines], "Flammability Characteristics of Combustible Gases and Vapors", Bulletin 627, Table A-1 [and many "nose" Figures] (1965).

9. Babrauskas, V., "Ignition Handbook", Table 1A (1992).

10. Britton, L. G., "Avoiding Static Ignition Hazards in Chemical Operations", Page 223 (1999).

11. Crowl, D. A., and J. F. Louvar, "Chemical Process Safety", 3rd Ed., Pages 251, 259, 261, 287, and 654 (2011).

12. Yaws, C. L., "Chemical Properties Handbook", Table 25-1 (1999).

13. National Fire Protection Association, "Recommended Practice for the Classification of Combustible Dusts and of Hazardous (Classified) Locations for Electrical Installations in Chemical Process Areas", NFPA 499, Paragraph A.5.2.3 (2017).

14. Frank, W. L., and S. A. Rodgers, "NFPA Guide to Combustible Dusts", Page 92 (2012).

15. Britton, L. G., "Avoiding Static Ignition Hazards in Chemical Operations", Page 223 (1999).

16. Babrauskas, V., "Ignition Handbook", Page 1023 (1992).

17. Zabetakis, M. G. [U. S. Bureau of Mines], "Flammability Characteristics of Combustible Gases and Vapors", Bulletin 627, Page 9 (1965).

18. Crowl, D. A., and J. F. Louvar, "Chemical Process Safety", 3rd Ed., Pages 263, 265 to 267, and 269 (2011).

19. Crowl, D. A., and J. F. Louvar, "Chemical Process Safety", 3rd Ed., Pages 398 and 400 (2011).

20. Babrauskas, V., "Ignition Handbook", Pages 1023 and 1024 to 1058 (1992).

21. Green, D. W., and M. Z. Southard, "Perry's Chemical Engineers' Handbook", 9th Ed., Page 19-24 (2019).

22. Babrauskas, V., "Ignition Handbook", Pages 75, 76, 203, and 625 (1992).

23. Babrauskas, V., "Ignition Handbook", Pages 1024 to 1055 (1992).

24. Babrauskas, V., "Ignition Handbook", Page 625 (1992).

25. Lewis, B., and G. von Elbe, "Combustion, Flames, and Explosions of Gases", 3rd Ed., Pages 717 and 719 (1987).

26. National Fire Protection Association, "Guide for Venting of Deflagrations", NFPA 68, Table D.1 (2002).

27. Babrauskas, V., “Ignition Handbook”, Page 35 (1992).

28. Babrauskas, V., “Ignition Handbook”, Page 74 (1992).

29. Zabetakis, M. G. [U. S. Bureau of Mines], “Flammability Characteristics of Combustible Gases and Vapors”, Bulletin 627, Page 43 and Figures 44 and 45 (1965).

30. Nagy, J., et al. [U. S. Bureau of Mines], “Explosibility of Carbonaceous Dusts”, Report of Investigations 6597, Page 2 (1965).

31. National Fire Protection Association, “Recommended Practice for the Classification of Combustible Dusts and of Hazardous (Classified) Locations for Electrical Installations in Chemical Process Areas”, NFPA 499, Paragraph A.5.2.3 (2017).

32. Dorsett, H. G., et al. [U. S. Bureau of Mines], Laboratory Equipment and Test Procedures for Evaluating Explosibility of Dusts”, Report of Investigations 5624, Page 2 (1960).

33. Jacobson, M., et al. [U. S. Bureau of Mines], “Explosibility of Dusts used in the Plastics Industry”, Report of Investigations 5971, Tables A-1, A-2, and A-3 (1962).

34. Jacobson, M., et al. [U. S. Bureau of Mines], “Explosibility of Metal Powders”, Report of Investigations 6516, Tables A-1, A-2, and A-3 (1964).

35. Nagy, J., et al. [U. S. Bureau of Mines], “Explosibility of Carbonaceous Dusts”, Report of Investigations 6597, Page 2 (1965).

36. Dorsett, H. G., and J. Nagy [U. S. Bureau of Mines], “Dust Explosibility of Chemicals, Drugs, Dyes, and Pesticides”, Report of Investigations 7132, Tables A-1, A-2, and A-3 (1968).

37. Grossel, S. S., “Safety Considerations in Conveying of Bulk Solids and Powders”, Page 62 in the “Journal of Loss Prevention in the Process Industries”, Butterworths, Vol. 1, No. 2 (April 1988).

7. Legal Disclaimer and Liability

(a) Limitation of Liability. The consulting services conducted by DEKRA Process Safety (the "Company") were performed using generally accepted guidelines, standards, and/or practices, which the Company considers reliable. Although the Company performed its consulting services pursuant to reliable and generally accepted practices in the industry, the Company does not guarantee or provide any representations or warranties with respect to Client's use, interpretation or application of the findings, conclusions, and/or suggestions of the consulting services provided by the Company. Moreover, the findings, conclusions, and the suggestions resulting from the consulting service are based upon certain assumptions, information, documents, and procedures provided by the Customer. AS SUCH, IN NO EVENT AND UNDER NO CIRCUMSTANCE SHALL THE COMPANY BE LIABLE FOR SPECIAL, INDIRECT, PUNITIVE OR CONSEQUENTIAL DAMAGES OF ANY NATURE WHATSOEVER, INCLUDING WITHOUT LIMITATION, ANY LOST REVENUE OR PROFITS OF THE CUSTOMER OR ITS CUSTOMERS, AGENTS AND DISTRIBUTORS, RESULTING FROM, ARISING OUT OF OR IN CONNECTION WITH, THE SERVICES PROVIDED BY THE COMPANY. The Customer agrees that the Company shall have no liability for damages, which may result from Client's use, interpretation or application of the consulting services provided by the Company.

(b) The Company's pricing of the consulting services provided does not contemplate that the Company shall have any liability resulting from its performance of the consulting services, except as otherwise set forth in the Quotation from the Company. Accordingly, the Customer shall indemnify and hold harmless the Company, its shareholders, directors, officers, employees and agents (the "Indemnified Parties") from and against any and all loss, cost, liability and expense, including reasonable attorney's fees and costs, which any of the Indemnified Parties may incur, sustain or be subject to, as a result of any claim, demand, action, investigation or proceeding arising out of or relating to either: (a) the consulting services provided by the Company; or (b) any material, equipment, specifications or safety information (or lack thereof) supplied to the Company (or which should have been supplied to the Company) by Customer and/or any failure of such materials, equipment, specifications and safety information to comply with any federal, state or local law or safety standard.

(c) For additional terms and conditions, which apply with respect to the provision of this report, see the Quotation provided by the Company and executed by Customer. If any terms set forth in the Quotation conflict with the terms set forth herein, the terms set forth herein shall apply.

Bridging Chemical Engineering (Risk Based Process Safety) to the Humanities – 7 Mega Elements' Approach (Protocol) of Accident Investigation in Light of Ethical Dimension, Architecture & Epistemology

Long(Steven) Zhang
No affiliation since Jul 2014 off-hiring from COP
11455 Bissonnet St Apt #514, Houston TX 77099
longz.houston@gmail.com

Prepared for Presentation at
American Institute of Chemical Engineers
2019 Spring Meeting and 15th Global Congress on Process Safety
New Orleans, LA
March 31 – April 3, 2019

Bridging Chemical Engineering (Risk Based Process safety) to the Humanities – 7 Mega Elements' Approach (Protocol) of Accident Investigation in Light of Ethical Dimension, Architecture & Epistemology

Long(Steven) Zhang
No affiliation since Jul 2014 off-hiring from COP
11455 Bissonnet St Apt #514, Houston TX 77099
longz.houston@gmail.com

Keywords: Investigation, Diagnosis, Humanities, *the Art of War, Narrative, Epistemology*

Abstract

In this paper, facing the challenges of Incident Investigation, actually whole CCPS 4th Pillar "Learn from Experience", chemical engineers' (process safety practitioners') brand new vocabulary (7 Mega Elements' narrative on failure diagnosis) would be presented in context of bridging chemical engineering (RBPS) to the Humanities, of which 3 core topics i.e. ethical dimension, architecture and epistemology would be explored and reflected upon so that a new era of RBPS diagnosis (Learn from Experience) predicted.

1 Introduction

In 2007 Center for Chemical Process Safety (CCPS) unprecedentedly presented to process industry Risk Based Process Safety (RBPS) 4 Pillars (and relevant 20 Elements), among which 4th Pillar "Learn from Experience" consists of 4 Elements: Incident Investigation, Measurement & Metrics, Auditing, and Management Review & Continuous Improvement[1].

In latest decades chemical engineers (process safety practitioners) have been facing some big challenges from accident/incident investigations, actually from whole CCPS 4th Pillar (Learn from Experience), for which there might have been many efforts, discussions, and even some 'solutions'. The author argues, however, the challenges comes not only from the needs of high quality and consistency of accident investigation, but, more fundamentally, from building up a brand new vocabulary (protocol) for chemical engineers to competently communicate of, and negotiate on, accident investigations or "why they occur and how ultimately to prevent".

Toward building a new vocabulary/protocol/narrative on accident investigation or, more broadly, on how Experience could be truly Learned (CCPS 4th Pillar), since 9th Global Congress, the author has had a series of efforts for exploration and reflections on the possible way forward, e.g. RBPS Core value (landlord for temple), Process Safety Cultural Communication (landlord's livelihood) & Egg model[2]; 4th Pillar as "espionage" not to enemies but ourselves[3][4]; 7 Mega Elements & 28 infra cultural activities (practices); RBPS Socio-Technical systems[5]; an ethical dimension of Vision 20/20[6] etc.. And in broader context of bridging chemical engineering (Risk Based Process Safety) to the Humanities, the author finally arrived at 7 Mega Elements' vocabulary/ protocol/ narrative/ "cosmology" on RBPS diagnosis for chemical engineers (process safety practitioners)[7].

2 7 Mega Elements' Protocol – Chemical Engineers' Vocabulary

Louis L. Bucciarelli claimed that he had taken seriously the "two cultures" challenge set by C. P. Snow and been working at bridging engineering to the Humanities for 35 years[8]; instrumental rationality does not suffice itself in engineering design, diagnosis and education[9]. And the author's experiences during the humble quest (through relevance of anthropology, intercultural communication, *the Art of War* etc.) for an ultimate solution to catastrophic (process safety) accidents (chemical engineering failure) seems not only resonating with Bucciarelli's bridging journey but a little bit unique, not only toward chemical engineering (Risk Based Process Safety) but also to engineering discipline at large, i.e., through benchmarking RBPS approach against Sun Tzu's *the Art of War,* 7 Mega Elements (vocabulary/ protocol/ narrative) not only were discovered for (chemical) engineers but function as 7 fundamental (chemical) engineering thoughts, which eventually sheds light on those entailed 28 practices of (chemical) engineering discipline.

Unlike Bucciarelli who tried to bridge philosophy and engineering, the author proceeded another way, i.e. first making relevance of anthropology & intercultural communication to chemical engineering (Risk Based Process Safety), then benchmarking Egg model against *the Art of War*, what the author calls 1st main ancient source for bridging (chemical) engineering to the Humanities, compared to 1st main modern one - Bucciarelli's *Engineering Philosophy*.

When guiding all visitors to some biggest lessons ("more original" two cultures) of ancient cultural traditions or civilizations of east and west, the author claimed that there should be an landlord (core value) to live within the temple building (CCPS 4 Pillars) and, furthermore, the landlord should have livelihood (Process Safety Cultural Communication (PSCC)) as well, and that should the landlord died the building seems still beautiful outwardly, however, with a purpose of tomb. In another words, when (chemical) engineers' vocabulary/ protocol/ narrative becomes deficient, or landlord and his livelihood are absent in the vocabulary, catastrophic (process safety) accidents are bound to happen, no matter how technically over competent relevant chemical engineers (process safety practitioners) might be, e.g., the lesson from the case of Bian Que–Duke Cai.

Through anthropology and intercultural communication, **RBPS Core Value** was defined in two ways, an unfolded one: *RBPS Core Value functions as the root of the tree of Process Safety*

Culture. Deficiencies are symptoms or signals for a dying or dead state of a core value, which means that the organization has strategically failed at the highest levels in managing process safety through a risk based approach – even if there has been no occurrence of process safety accidents till that moment. In another words, the RBPS temple is bound to collapse, sooner or later, due to the change or drift from the original design intent of the temple. And RBPS Core Value's simpler definition, through the concept of Power Distance in value studies: *there is no power, authority or governance on process safety except on behalf of process safety expertise (PSE).* (PSE means the divine essence of the highest strategic insights and vision of process safety management; the Tao for process safety culture; and the soul of *The Art of War* against catastrophic process safety accidents)

And process safety culture does not exist except through ***PSCC****, whose ultimate purpose is to convey the message of RBPS Core Value. Without nourishment from RBPS Core Value, the tree of Process Safety Culture or PSCC would get dry and dead.* So the temple building was eventually upgraded to Egg model, as showed within the scheme 1st. And these 2 new universes (or 2 Mega Elements) might be deemed what Bucciarelli called the rhetoric of engineering as the parcel of instrumental explanation[10], and probably 2 most fundamental (chemical) engineering thoughts as well.

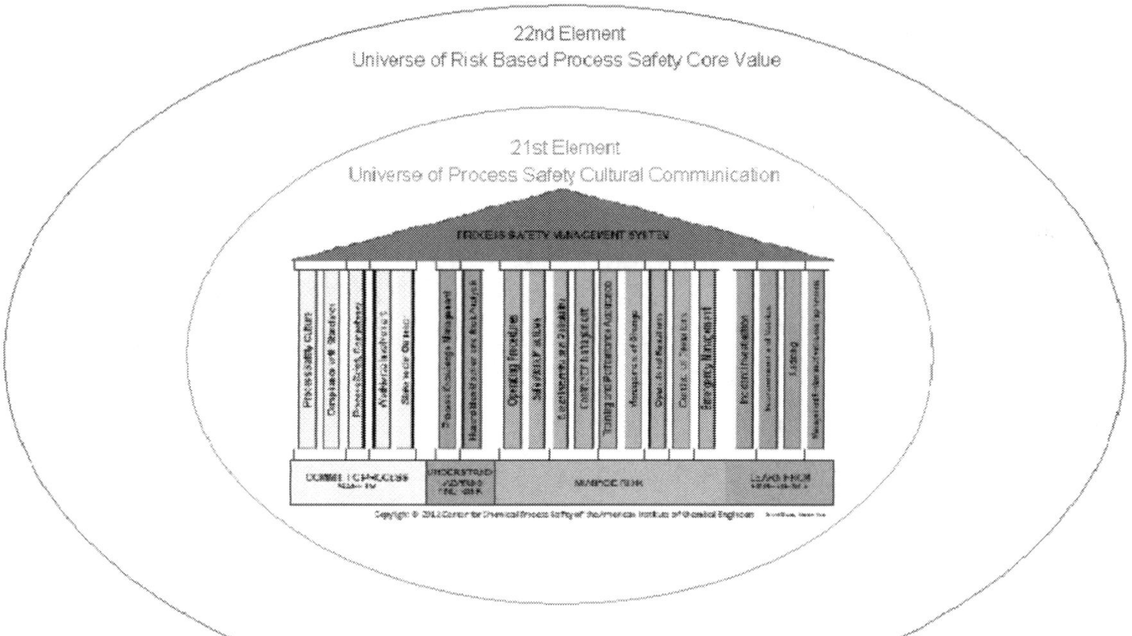

Egg Model (CCPS Temple + 2 universes) - Scheme 1st

The Temple (4 pillars) Figure Source: CCPS, Process Safety Boot Camp, contributed by David Guss, Nexen Inc. Copyright 2010 AIChE and used with permission

Then, in spite of the formidable strengths of most, actually 17, elements of CCPS 4 Pillars, except Process Safety Culture, Process Safety competency & Stakeholder Outreach, the outcome of benchmarking Egg model (4 Pillars + 2 Universes) against *The Art of War,* what the author calls a matured ancient vocabulary on engineering (military) diagnosis, not only shows that those 4 Pillars' (20 elements') narrative, even together with the new 2 Universes, would probably be not so architecturally organized at negotiation and communication on why process safety (chemical engineering) accidents (failure) occur & how to prevent, but makes the author further believe that Sun Tzu's vocabulary (6 Mega Elements, refer to Table 1st) might potentially become a narrative/ protocol paradigm for future (chemical) engineering diagnosis and *The Art of War* an main ancient source of bridging engineering to the Humanities.

For the fact that, however, we could not neglect the historical context of narrative/ protocol of *The Art of War*, the author, in light of modern globalization and global operations, fundamentally redefined Process safety Culture into Process Safety Cultural Communication (PSCC) as the 7th Mega Element, and upgraded another 3 ones i.e. Leadership, Weather/Climate & Espionage as below:

Leadership: a new chain of command, inherently from RBPS Core Value, rather than blindly from superiors;

Weather/Climate: RBPS Core Value enlightened outreach to the stakeholders & community, rather than only tactically maintaining seemingly good relationship with community; &

<table>
<tr><td colspan="2">RBPS Egg</td><td>The Art of War</td><td>Philosophy of Process Safety</td><td></td></tr>
<tr><td colspan="4">RBPS Core Value - Tao - RBPS Worldview</td><td>Worldview</td></tr>
<tr><td rowspan="4">Commit to Process Safety</td><td>Process Safety Competency</td><td rowspan="2">Leadership</td><td rowspan="2">People</td><td rowspan="8">Process Safety Cultural Communication (Process Safety Culture) RBPS Methodology</td></tr>
<tr><td>Workforce Involvement</td></tr>
<tr><td>Compliance with Standards</td><td rowspan="2">Weather/Climate</td><td rowspan="3">Plant</td></tr>
<tr><td>Stakeholder Outreach</td></tr>
<tr><td>Understand Hazards & Risk</td><td></td><td>Terrain</td></tr>
<tr><td>Manage Risk</td><td></td><td>Discipline</td><td>Procedure</td></tr>
<tr><td>Learn from Experience</td><td></td><td>Espionage</td><td>Engineering & Humanities</td></tr>
</table>

Table 1st

Espionage: 4th Pillar *Learn from Experience* (4 elements: *investigation, metrics, audit & review*) plays role "espionage", not to the enemies but to ourselves our organizational "national security"

And 7 Mega Elements might also be deemed as 7 (chemical) engineering thoughts.

3 Ethical Dimension (2 Universes) - What Chemical Engineers Count & How

While declared *there is no power, authority or governance on process safety except on behalf of Process Safety Expertise (PSE)* the author claimed that PSE shall eternally and exclusively chair at the highest strategic levels throughout all the processes of decision making and behavior determining, i.e. chemical engineers (process safety practitioners) need getting inspired from RBPS Core Value and persistently conveying the message of RBPS Core Value, both of which make up the ethical dimension (of 7 Mega elements' vocabulary/ protocol/ narrative), probably what Bucciarelli calls "best left sacred"[11], which ensures, or anthropologically designs, all relevant decision making and behavior determining correct. In this sense, at highest strategical levels, chemical engineers do not design at all but RBPS Core Value does.

From philosophical perspective, Bucciarelli claimed that the tensions at the interfaces between different design participants could not be solved by instrumental means alone [12]. And imitating American anthropologist Edward T. Hall's saying the author ever described process safety culture as a prison, unless we have RBPS Core Value as the key to unlock it and PSCC the brightness of the key. And Bucciarelli pointed out that *designing is a social process of negotiation, of iteration, of rectifying mis-steps, even misconceptions – a process rich in ambiguity and uncertainty* [13]. Similarly, the author in metaphor described PSCC as the photosynthesis of tree of process safety culture, the impulses from the heart (RBPS Core Value) and nourishment (highest designing principles) transportation from the root (RBPS Core Value). What does Bucciarelli mean by that 'social process'? What does he mean by "contribute to a sense of community among the citizenry" [14]? And what does the author mean by "photosynthesis" and "nourishment"? With hesitation & obligations, the author claims that the said social process, what the author puts PSCC, traditionally should be called Constitutional Democracy; and RBPS Core Value a nickname for Universal Value in chemical engineering.

As Universal Value & Constitutional Democracy in (chemical) engineering seem more subtle than those traditional ones the author here might call them Universal Value & Constitutional Democracy 2.0, what Bucciarelli described "old themes but in a new light" [15]. And this 2.0 might be acknowledged as two fundamental (chemical) engineering thoughts and the ethical dimension of (chemical) engineers' vocabulary/ protocol/ narrative on accident investigation and whole (chemical) engineering diagnosis (4th Pillars) as well. And probably only in this way could (chemical) engineers truly acquire what Bucciarelli called "new ways of perceiving and reading the world" [16], eventually leading the industries and communities into a new enlightened era of accident investigation and (chemical) engineering diagnosis. And the author might further states that, as the ethical dimension of (chemical) engineers' vocabulary (protocol), Universal Value & Constitutional Democracy 2.0 should function as the highest communication (designing) principles on accident investigation and (chemical) engineering diagnosis, and on accident prevention (foreknowledge) as well.

4 7 Mega Elements' Architecture – 28 Practices' Diagnosis Map

Inspired from Edward T. Hall's map of culture, the author developed a map of 28 infra cultural activities of Process Safety Culture (refer to 7 ones in the diagonal from upper left to lower right and 21 ones above), what Bucciarelli might like to call another, slightly different, "more complex vision of engineering thoughts and practices" [17]. And each infra cultural activity was defined as the critical statements & behaviors related to the interaction between Mega Element A and Mega Element B, e.g. infra activity 04 means: verbally & behaviorally communicate that the actual death of RBPS Core Value might be one of the root causes of accidents. And those 7 ones, extremely, are in fact of the critical verbal statements & behaviors related to the definition of the Mega Element (or engineering thought) concerned, e.g. infra activity 22: verbally and behaviorally embrace and reflect on Leadership - the engineering thought of new chain of command inherently from RBPS Core Value rather than blindly from superiors.

This matching between verbal statements & behaviors, or the inherent relationship between the message of RBPS Core Value and all relevant decision making & behavior determining, seems some resonance with Wittgenstein's recognition that rules and meanings (what the author calls ***tone***) in talk and texts (what the author calls vocabulary/ protocol/ narrative) were mixed with practice [18], or with what Bucciarelli put another way "the meaning of statements is embedded in practice" [19]. So these 28 infra cultural activities' map, an unfolded version of what & how chemical engineers (process safety practitioners) count, might accommodate as highest strategic encompassing diagnosis protocol (map).

	RBPS C.V. 0	P.S.C.C. 1	Leadership 2	Weather/C. 3	Espionage (Diagnosis) 4	Terrain 5	Discipline 6
RBPS Core Value 0	00	01	02	03	04	05	06
Process Safety Cultural Communication (PSCC) 1	10	11	12	13	14	15	16
Leadership 2	20	21	22 4 dimensions	23	24	25	26
Weather/ Climate 3	30	31	32	33	34	35	36
Espionage (Diagnosis) 4	40	41	42	43	44	45	46
Terrain 5	50	51	52	53	54	55	56
Discipline 6	60	61	62	63	64	65	66

28 Infra Cultural Activities' Diagnosis Map (Table 2nd)

As the Humanities e.g. philosophy, anthropology & intercultural communication, Sun Tzu's *The Art of War* etc. sheds more lights on the old themes (i.e. RBPS Core value, PSCC, Leadership and Weather/Climate), upon this Diagnosis Map, "another" complex vision of engineering thoughts and practices, the author claims that *the yellow grounded, or more exactly red number lettered area, are all about organizational "national security", and that the infra activity 44 (Espionage)'s 2nd rationality, next to the instrumental one, whose protocol area is painted green, SHALL be reactivated so that "espionage" to our organization would shift to a more proper name - "counter-intelligence", on accident investigation & prevention and on engineering diagnosis at large.*

Through maintaining the architecture/protocol of 7 Mega Elements could all these 22.5 infra activities of "national security" area be reliably visible, open and "problematic" over there, together with "traditional" green area (5.5 infra activities), to critique and diagnosis among board of directors, executive management, operation managers, strategy, legal, HR, chemical engineers (process safety practitioners), and even beyond. Pragmatically, from designing perspective, chemical engineers should pursue this more matured architecture/protocol/vocabulary, getting inspired from RBPS Core Value, persistently conveying its message and maintaining alive this complex vision of engineering thoughts & practices i.e. this 28 Practices (infra activities)' Diagnosis Map, otherwise the insightful topic "social construction of failure/ accident" [20] would turn out to be a truism. In another words, either RBPS Core Value highest strategically designs safety (and tranquility) through PSCC, or otherwise (chemical engineering/ process safety) failures/ accidents/ catastrophes are socially constructed.

5 Epistemology – Future (Chemical) Engineers' Knowledge Hierarchy

Walter Vincenti's understanding of *engineering knowledge,* as J. C. Pitt ever explained, *is task-specific and aims at the production of an artifact to serve a predetermined purpose* [21], of which, the author believes, CCPS 4 Pillars (20 Elements), as of (chemical) engineering/RBPS knowledge, might have been satisfactorily reflecting the main characters. Bucciarelli discovered, however, that engineering knowledge, of which he described *what engineers do, the reasons they do what they do, and their justification for believing the latter are good and true,* is *historically and culturally contingent, not fixed by the dictates of science and empirical fact alone.* [22]. The author might explain here, echoing the former statement "the landlord … even to buy a new house with new style of design to live within", that the (chemical) engineering knowledge we have here with CCPS 4 Pillars might not be the single unique design/house for the landlord [23], but the landscape (landlord, livelihood & house) of the contingencies, what Bucciarelli calls "a picture of both possibilities and limitations of engineering thinking" [24], is unique. In another words, both the landlord (RBPS Core Value), deeply rooted in most cultural traditions and civilizations, and his livelihood (PSCC), what the author likes to call Constitutional Democracy, should preside the landscape of the contingencies and chair and guide at the highest strategic levels the whole engineering design process, for surety including on accident investigation and diagnosis, so that the objects of design, which "traditionally" have been confined to artifacts e.g. a mega petrochemical project/ operations, might also need to include all processes of decision making and behavior determining (refer to Table 2nd – 28 Infra Activities' Diagnosis Map).

The author might state that *a complete (chemical) engineering design process, including on accident investigation, diagnosis & prevention, shall consist of 3 components: RBPS Core Value (Universal Value), PSCC (Constitutional Democracy) and the application of (chemical) engineering knowledge (e.g. one with CCPS 4 Pillars).* And at same time, we recognize that *future (chemical) engineers need a new epistemology - a new hierarchy of knowledge which consists of Universal Value Knowledge, Constitutional Democratic Knowledge & the relevant (chemical) engineering knowledge.* Edward T. Hall had an earth-breaking saying: *culture is communication and communication is culture* [25], from which the author got inspired and developed the concept of PSCC. When (chemical) engineers verbally (of text) and behaviorally (of practice) abandon the ethical dimension of 7 Mega Elements' vocabulary/ protocol/ narrative, they lose the whole process safety culture or, we could say, both their knowledge hierarchy is deficient, akin to a genocide, and their engineering design process seriously flawed, like building concentration camps.

6 Conclusion

In context of bridging chemical engineering to the Humanities, on accident investigation, diagnosis and prevention, and on engineering design process at large, the author presents 7 Mega Elements' Vocabulary/ Protocol/ Narrative, 28 Infra Cultural Activities' Diagnosis Map and a newly galvanized old Hierarchy of Knowledge, so that some outlooks might be made: a) as one of the largest communities of engineering professions, AIChE would pursue leading roles at dialogues between engineering & the Humanities; b) some small groups of chemical philosopher/anthropologist-engineers would emerge; c) an enlightened era of (chemical/process safety) accident investigation, diagnosis & learning, and engineering design process at large, would arrive; & d) engineering education's future would get brighter than ever.

7 References

[1] Center for Chemical Process Safety, *Guidelines for Risk Based Process Safety*, John Wiley & Sons, Inc., 2007

[2] Zhang, L., An Upgraded Definition of Risk Based Process Safety Core Value and Its Impact in Light of Anthropology & Intercultural Communication on the Future Evolution of Process Safety Culture & Management, 2013 AIChE Spring Meeting and 9th Global Congress on Process Safety, San Antonio, April 2013

[3] Sun-Tzu and Ames, Roger T., Sun-tzu: *The Art of Warfare*, Ballatime Books, 1993

[4] Zhang, L., Benchmarking Risk Based Process Safety Egg (RBPS Temple+2 Universes) against The Art of War, 1st CCPS Asia-Pacific Conference on Process Safety, Qingdao, Sept 2013

[5] Zhang, L., People Means Leadership, NOT Simply mean People – 4 New Dimensions of Process Safety Competency, 2014 AIChE Spring Meeting and 10th Global Congress on Process Safety, New Orleans, April 2014

[6] Zhang, L., How Committed to Process Safety is Committed Enough? – Doing Process Safety Cultural Communication (PSCC) as the Inevitable Ethical Dimension for Vision 20/20, 2016 AIChE Spring Meeting and 12th Global Congress on Process Safety, Houston, April 2016

[7] Zhang, L. and Wasel, W., Corporate Memory Loss Ongoing – Except CCPS 4th Pillar (Learn from Experience) as Espionage, 2018 AIChE Spring Meeting and 14th Global Congress on Process Safety, Orlando, April 2018

[8] Bucciarelli, L. L., *Engineering Philosophy*, Delft: DUP Satellite, 2003, p.2

[9] Ibid. p.99

[10] Ibid. p.6

[11] Ibid. p.4

[12] Ibid. p.4-5

[13] Ibid. p.7

[14] Ibid. p.5

[15] Ibid. p.2

[16] Ibid. p.7

[17] Ibid. p.101

[18] Ibid. p.6

[19] Ibid. p.6

[20] Ibid. p. 101

[21] Pitt, J. C., "What Engineers Know", *Techne* 5:3, Spring, 2001, p.25

[22] Bucciarelli, *Engineering Philosophy*, p.99

[23] Pitt, "What Engineers Know", p.26

[24] Bucciarelli, *Engineering Philosophy*, p.99

[25] Hall, E. T., *The Silent Language*, Anchor Books, 1990, p.186

548354 Applying PSM in Vaca Muerta Shale Oil & Gas Operations: The Experience

How to develop, implement and sustain Process Safety in greenfield shale oil and gas projects? This has been a challenge to many new ventures.

Although conventional oil operations have been in place for decades, the exploration, drilling and fracking operations in unconventional shale fields are relatively new in the World. Most of the projects have kicked off less than eight years ago and there are some aspects of the operations that are unique to unconventional fields.

The objective of this paper is to share with the Process Safety global community, how a fit for purpose Process Safety program was developed and is being implemented in the Upstream Unconventional shale oil and gas YPF & Chevron joint venture in Vaca Muerta / Argentina. The project started in 2013, has currently 14 D&C rigs, 5 fracking sets, several separation and treatment production facilities, 3,500 employees and is expected to grow even more in the next 5 years. The paper describes how the project team developed the idea, planned the steps, engaged all stakeholders and selected the key elements of the Process Safety Management System that would add more value in each stage of the project, including drilling, fracking and oil and gas production. The paper also explains the key challenges, such as the difficulties faced to instill Operational Discipline and other key cultural and technical aspects related to Contractor field Operation crews. The journey to implement Process Safety in the Unconventional shale oil & gas operations is very dynamic and is in progress, however, there are already several positive aspects as well as pitfalls to take away as a lesson learned.

548596 A Survey of Process Safety Workers' Attitudes, Behaviors, and Outcomes: Moving Beyond Human Error and Toward a Systems Model

In an effort to move beyond the "human error" explanation for safety incidents, we explored issues with procedures from the worker's point of view by using an anonymous, holistic survey developed from interviews with then currently employed operators. The current sample ($N = 174$) included individuals employed in the process safety industry (MKO/Advanced Procedures Initiative) and were primarily from the Oil & Gas and Chemical industry. The survey was deployed over the course of approximately four weeks. Twelve distinct constructs emerged from the survey (e.g., perceptions of procedure quality, procedure deviation, attitudes toward the procedure change process, etc.). Results indicated that perceptions of procedure quality was the focal variable in all analyses including positive relationships with attitudes toward the procedure change process and negative relationships with procedure deviations, and safety incidents and near-misses. We conclude that these elements are important factors to consider when evaluating and developing procedure systems and provide additional information beyond what the industry considered paramount.

546680 A Method of the Evacuation Pathway Optimization Under Toxic Gas Releases in Chemical Plants

Toxic gases release in chemical plant is a key factor contributing to major accidents and may cause tremendous human life and economic losses. A scientific planning evacuation pathway is an important guarantee of reducing the losses of accidents. Many studies regarding evacuation planning in chemical plant have been carried out by simulating the accident scenarios. However, at present, a systemic method of evacuation pathway optimization has not been developed, and the effectiveness of emergency preparedness has been compromised. To address this problem, an approach based on the CFD modelling of hazardous gas release consequences to plan evacuation pathways and appropriate shelters is proposed in this paper. The methods stochastic programming and dose-response model are mainly employed in this paper. The toxic gas concentration can be converted into poisoning fatality probabilities by dose-response model. Then, stochastic programming method is used to select all the optimal pathways from the evacuation point to the safe area. Setting the minimal fatality probability as the optimization goal, the optimized evacuation pathways under all the leakage scenarios are finally determined through solving the stochastic programming model. As for places with limited evacuation opportunities, an approach of setting up a safe room is proposed in this paper and relevant design guidance are put forward in this paper considering the detailed situation of the chemical plant. A case study of the optimization of evacuation pathways and safe room setup in a salt chemical plant that contains 88 leakage scenarios of chlorine leakage is investigated. With the proposed methodology, the optimized evacuation pathway and the appropriate safe room can provide scientific references for decision makers and evacuees on reducing losses when toxic gases release in chemical plants.

Four Major Gaps That Are Preventing Most Companies Worldwide from Achieving Excellent Process Safety Performance

William G. Bridges
Process Improvement Institute Inc. (PII)
1321 Waterside Lane, Knoxville, TN 37922 USA
wbridges@piii.com

Jeff Thomas
Process Improvement Institute Inc. (PII)
jthomas@piii.com

Prepared for Presentation at
American Institute of Chemical Engineers
2019 Spring Meeting and 15th Global Congress on Process Safety
New Orleans, LA
March 31 – April 3, 2019

FOUR MAJOR GAPS THAT ARE PREVENTING MOST COMPANIES WORLDWIDE FROM ACHIEVING EXCELLENT PROCESS SAFETY PERFORMANCE

William G. Bridges
Process Improvement Institute Inc. (PII)
wbridges@piii.com

Jeff Thomas
Process Improvement Institute Inc. (PII)
jthomas@piii.com

Keywords: process safety, PSM, human factors, near misses, PHA, HAZOP, procedures

Abstract

Process safety requires implementing many management systems, specific engineered features, and operating and maintenance practices effectively. Most companies believe they have done just that, and yet major accidents continue to occur. Why is that? What is missing?

This paper looks at the statistics of major accidents, combined with results from audits and assessments from more than 50 chemical, petrochemical, oil/gas, and related processing companies world-wide. The paper illustrates the four major gaps that are common to the companies/sites that keep having major accidents, compared to those companies/sites that do not have such accidents:

- Accurate and clear operating and maintenance work instructions (procedures)
- PHA of all modes of operation, especially startup, shutdown, abnormal operations, and online maintenance modes
- Getting Near Misses reported and investigated
- Addressing human factors that are missing from most management systems

The paper also provides guidance for closing each of these gaps.

Introduction

"Process safety management (PSM) is the application of management principles to the identification, understanding, and control of process hazards to prevent process-related incidents"[1]. PSM entails development and implementation of programs or systems to ensure that the practices and equipment used in hazardous processes are adequate and are maintained appropriately. The primary categories of programs or systems have come to be called elements of PSM. However, the basic elements of PSM have been defined by many groups in a number of ways. Table 1 lists the elements of PSM systems from various industry and government groups. Many of the elements with different names have essentially the same meaning. For instance, "maintenance and inspection of facilities," together with some aspects of "personnel" practices, both under American Chemistry Council's (ACC's, formerly CMA's) Process Safety Code of Responsible Care™[2] are essentially the same as the single element, "mechanical integrity," under 29 CFR 1910.1190)[3].

Table 1. Comparison of PSM systems

OSHA 29 CFR 1910.119 EPA 40 CFR 68	AIChE/CCPS Risk-Based Process Safety (RBPS) Standard	Responsible Care© Process Safety Code
Management System Employee Participation Process Safety Information Process Hazard Analysis* Operating Procedures* Training* Contractors Pre-Startup Safety Review Mechanical Integrity Hot Work Permit Management of Change Incident Investigation Emergency Planning and Response Compliance Audits Trade Secrets	**Commitment to Process Safety** Process Safety Culture* Compliance with Standards Process Safety Competency* Workforce Involvement* Stakeholder Outreach **Understand Hazards and Evaluate Risk** Process Knowledge Management Hazard Identification and Risk Analysis **Manage Risk** Operating Procedures* Training and Performance* Safe Work Practices * Asset Integrity and Reliability Contractor Management Management of Change Operational Readiness* Conduct of Operations* Emergency Management **Learn from Experience** Incident Investigation Measures and Metrics Auditing Management Review and Continuous Improvement	**Management Leadership** Commitment* Accountability* Performance Measurement Incident Investigation Information Sharing CAER Integration **Technology** Design Documentation Process Hazard Information Process Hazard Analysis Management of Change Facilities Siting **Codes and Standards** Safety Reviews Maintenance and Inspection Multiple Safeguards* Emergency Management **Personnel** Job Skills* Safe Work Practices* Initial Training* Employee Proficiency* Fitness for Duty* Contractors

* Contains some of the 6 elements of a Human Factors element, but all missing elements not covered completely. Mostly covered in CoC/OD element

Note that the newest definition of process safety is CCPS's *Risk Based Process Safety (RBPS)*[4] replaces their earlier process safety definition. In the older definition from CCPS [1], there was an element on Human Factors, which brought strong focus to this necessary element; in RBPS, the human factors sub-elements are now spread across 6 different elements, though some human factors were inadvertently weakened in the transition to *RBPS*.

Although nearly the entire industry agrees that implementing PSM is the right thing to do, interpreting and converting the PSM requirements into practices is unique to each company, and even unique to each plant site. Not only can the requirements be interpreted differently for each site based on local needs, but each site also starts from a different point when they begin to implement a system that is consistent with this regulation.

This paper looks at the statistics of major accidents, combined with results from audits and assessments from more than 50 chemical, petrochemical, oil/gas, and related processing companies world-wide. The paper illustrates the four major gaps that are common to the companies/sites that keep having major accidents, compared to those companies/sites that do not have such accidents:

- Accurate and clear operating and maintenance work instructions (procedures)
- PHA of all modes of operation, especially startup, shutdown, abnormal operations, and online maintenance modes
- Getting Near Misses reported and investigated
- Addressing human factors that are missing from most management systems

The paper also provides guidance for closing each of these gaps.

1. Procedures Need to be More Accurate and Clear

Operating and maintenance procedures have always been crucial to the safety, quality, and productivity of process systems. With the advent of new safety and quality standards such as OSHA's process safety management (PSM) regulation, OSHA's personal protective equipment (PPE) regulation, EPA's risk management program (RMP) regulation, and the ISO 9000 quality standard, many companies are facing the daunting task of developing or upgrading their procedures to satisfy varied and sometimes complex and conflicting requirements.

Although there are overlapping characteristics that the various regulations and standards share, each approach the procedure-writing process from a somewhat different perspective (e.g., quality or safety, protection of workers or protection of the public/environment), and the required level of detail for documenting the procedures differs greatly between these regulations and standards. For instance, the ISO 9000 quality standard advocates documenting procedures and work instructions that impact quality and suggests keeping the procedures as simple as possible, while the OSHA PSM regulation requires detailed procedures that address: all operating modes (startup, shutdown, etc.), operating limits, consequences of deviations, means to avoid hazards, safety and health considerations, and safety systems and their functions. How can companies develop and maintain procedures that ensure productivity and simultaneously satisfy the different regulations and standards? The key is to remember the ultimate goal of the regulations and standards: to reduce human errors that can impact quality, productivity, and/or safety.

Procedure-related errors are errors that occur because some characteristic of the procedure caused task performance to fail. **This is currently the most critical human factor at most sites since 90% of accidents have at least one root cause related to mistakes within procedures**. Reducing these procedure deficiencies can reduce human error rates by a factor of 2 to 20 [5].

1.1. Identified problems

Deficient Procedures are the most prevalent problem in process industries since procedures have not traditionally been developed from the perspective of optimizing human factors; instead procedures have been traditionally developed to meet a compliance requirement. Examples of procedure deficiencies (inaccuracies) include:

- Incorrect/incomplete/nonexistent (most procedures we have audited have been only 70-75% accurate – the inaccuracies include missing critical steps, steps as written are not what needs to be done, or the steps are out of sequence) [5]
- No/misplaced/incorrect information in warnings (for example, a warning should **never contain** the action to take; it should instead **emphasize** the action to take) and warnings should be located BEFORE the step applicable to the warning
- Poor format and presentation rules

1.2. Objective

The flowchart (Figure 1) nd the procedure quality checklist (Table 2) describes the basic sequence of steps for writing effective procedures. Note that some basic rules are contained in the flowchart, such as "operators write operating procedures," "maintenance craft-persons write maintenance procedures," and "Lab technicians write lab procedures." Procedures are NOT written by engineers or department superintendents. This is a critical to ensure procedures are written in the user's common language and are written at the right level of detail.

Figure 1. Best Approach to developing written procedures, once the unit has decided which procedures are needed (Extracted from PII Course, Writing Effective Operating and Maintenance Procedures © PII 2003-2017) [6]

The procedure content controls the human error related to "procedure-based" error – if content is wrong, the trainee is very likely to learn the step wrong (trainer uses the procedure as a basis for training or the trainee refers to the procedure step that is wrong when executing the procedure).

Even if the content is correct, not following page formatting best practices can still increase human error rates. If the content of the step is wrong it does not matter if the presentation is clear. If the content is perfect, you can further reduce error rates by 50% to 70% by presenting the information properly.

Rules for developing content[7]

- The procedures should be written at the level of someone who has just completed the basic training for that task. Do not write the procedure for someone just hired or for the 10+ veteran.
- Clearly identify ahead of time what activities need procedures to help ensure error rates are controlled low enough, and identify which activities are considered common "skills" of all of the staff.

  Example: Starting a pump is typically considered a "skill." This means a procedure step can simply be "Start the benzene recycle pump (P-119)." The procedure does not need to explain how to start the pump. But, starting a pump is a skill that nonetheless must be learned; it requires doing several sub-steps related to positioning of intake and discharge valves, how the pump is throttled at startup, checking local pressure gauge, etc. So, a training module is needed for starting pumps of various types, and the operator needs to learn this necessary skill. Once the skill is learned, it will be applied to so many various cases and so often, that a procedure is no longer needed as a "refresher" on how to "start the pump."

- The first draft of the procedure must be walked-down in the field by another user. Simply reading through the procedure does not catch enough of the mistakes.
- The revised draft should be reviewed in the field by a technical staff person, such as a process or operations engineer (for operating procedures).
- The procedure needs to be checked to ensure it follow the page format and step writing rules described in the next section. This ensures the accurate steps are clearly presented.

The target is to reach an accuracy of 95% or better[7] (so, no more than one wrong or missing step out of 20 steps). The following have been observed in the field by PII staff by direct data collection at more than 110 sites/plants around the world, based on walk-down review of several to dozens of procedures at each unit/plant:

- At 95% accuracy or better and when the same procedures follow 80% of the rules for procedure clarity (presented next), then most users will follow the written procedures and will try to keep the procedures up-to-date
- At about 85% accuracy or less, about half of the users stop using the procedures
- At about 75% accuracy or less, less than 10% of the users will refer to the procedure or will try to keep it up-to-date. **So, the written procedures are not very effective.**
- Unfortunately, the typical operating procedure walked down by PII staff (accompanied by senior operators, a process engineer, and a shift supervisor) is about 75% accurate (so one step in four is missing or wrong). Usually these inaccurate procedures also follow less than half of the best practices for procedure format, presented next [7].

Procedure Clarity

The importance of the accuracy of the procedure steps were discussed above. The clarity of the procedure steps (how they are written and how the page is formatted) is also important. Following best practices for step and page format reduces human errors by a multiplying factor of 3 to 5. [5], [7]

The best practice rules for writing and validating procedures have been published for many years *(see Bridges & Williams, 1997; Madden & Bridges, 2016, 2017)* [7],[8],[9]. These rules include best practices for formatting of the pages and steps. These have been gradually improved over the past decades and now are incorporated into PII's training materials.[6]

Table 2. Procedure Quality Checklist (courtesy PII, copyright 2003-2017)[6]

#	Issue	Response
	Procedure Content Checklist	
1	Is the procedure drafted by a future user of the written procedure? (Engineers should not author procedures to be used by operators or maintenance staff.)	
2	Is the procedure validated by a walk-down in the field by another future user of the procedures?	
3	Is the procedure reviewed and commented on by technical staff (engineers or vendors)?	
4	Is the procedure checked versus the Page and Step format rules below?	
5	Is a hazard review of step-by-step procedures performed to make sure there are sufficient safeguards (IPLs) against the errors that will occur eventually (when a step is skipped or performed wrong)?	
6	Is the content measured using "newly trained operators" to judge the % of steps that are missing, steps that are confusing or wrong, and steps that are out-of-sequence? (A score of 95% accuracy of content is good; 98% should be the targeted average.)	
	Page Format Checklist	
1	Is the title of the procedure the largest item on the page?	
2	Is the procedure title clear and consistent with other titles, and does it uniquely describes the topic?	
3	Are the document control features the smallest items on the page?	
4	Are temporary procedures clearly identified?	
5	Is white space used effectively? • Is there one blank line between each step? • Does indentation help the user keep their place? • Are the margins large enough to reduce page congestion?	
6	Is type size is 12 pt font or larger?	
7	Is mixed case used for words of steps, with ALL CAPS used only for special cases (such as IF, THEN, AUTO, and WARNING)?	
8	Is the step number very simple (single level of number used)? Only an integer?	
9	Have sections or information not necessary to performing the steps been moved to the back or to another part of the manual or training guide?	
10	Are section titles bold or larger than the text font? Do sections have clear endings?	
11	Is the decision on electronic presentation versus hard copy correct? Are electronic linkages to procedures clear and accurate and easy to use? If paper is chosen, is the color of the paper appropriate?	
12	Is the overall page format (such as Outline format or Two-Column [T-Bar] format) appropriate to the use of the procedure?	
13	Are play script features added for tasks that must be coordinated between two or more users? • Play script is normally used when there are two or more hand-offs of responsibility for steps.	
14	Are rules followed for formatting of Warnings, Cautions, and Notes? (See annotated rules, such as Warnings are for worker safety and Warnings must clearly standout from rest of page.)	
	Step Writing Checklist	
1	Is each step written as a command?	
2	Is the proper level of detail used throughout? This is judged based on: • *Who* will use the procedures • Same level of detail used in similar procedure steps	
3	On average, is there only one implied action per instruction? Best practice is to average 1.2.	
4	Does the procedure indicate when sequence is important? • If sequence matters, each step should be numbered (with an integer or letter)	

	• If sequence does not matter, bullet lists should be used	
5	Are only common words used? Apply "education" level test (5 grade reading level is best)	
6	Do all acronyms, abbreviations, and jargon aid understanding? • Develop a list of such terms for use in procedures *and* communication. • Use terms that users use (within reason)	
7	Is each step specific enough? No room left to guess/interpret: • The meaning of a word or phrase (Check vs. Make sure) • The intent of a step or series of steps • A desired quantity or value • To what equipment the step applies	
8	Is the procedure free of steps that require in-your-head calculations? • Values expressed as ranges rather than targets with error bands • Conversion tables, worksheets, or graphs provided where needed	
9	Are graphics to the user's advantage? • No explanatory paragraphs or lengthy instructions that could be replaced by a picture • No impressive graphics that provide no real advantage	
10	Are references to the user's advantage? • No lengthy explanations or instructions that could be replaced by branching to a reference • No references to a procedure that references still another • No gaps or overlaps between this procedure and a referenced document • If branching, must branch to a procedure, not to a specific step in a procedure	
11	Are rules followed for writing warnings, cautions, and conditional steps? • No more than 2 per page • No actions within a warning or caution (actions must always be numbered steps) • Warnings and Cautions contain descriptions of potential consequences	

For a complete description of best practices, refer to PII's paper *"Best practices for writing operating procedures and trouble-shooting guides"*. [7]

2. PHAs (Process Hazards Analyses) Need to Close Two Major Gaps

Process Hazards is a broad term. The relationship between many components define process hazards. For a PHA team, Process Safety Information (information on the chemicals, technology and equipment) and operating procedures are necessary to identify specific components of and analyze a hazard scenario. These components include: process deviations (parametric and procedural), their causes, the process safety consequence and various safeguards for detecting deviations and causes, preventing causes, detecting and mitigating consequences and intervening safeguards. A thorough review of incident investigations also offers an opportunity to further analyze process hazards.

2.1. Identified problems

Over the last twenty years the CSB has brought needed attention, through its investigations of industry incidences, to two significant gaps in many current PHAs; where the teams did not identify and analyze:

- Process hazards unique to non-routine modes of operations resulting in catastrophic consequences, and
- Process hazards related to Damage Mechanisms.

Though the intent of the US OSHA PSM regulation has always been for PHA teams to analyze the hazards for all modes of operation, OSHA did not give specific requirements for addressing damage mechanisms and only recently have they increased their enforcement attention on PHA of non-routine modes of operation.

Industry has responded by modifying the *Guidelines for Hazard Evaluation Procedures*, 3rd Edition, 2008, CCPS/AIChE[10] to improve the coverage of these two main issues. API in turn issues guidance on addressing damage mechanisms[11]. Industry best practices now exist for addressing both of these weaknesses. Now, with proper approaches, a team with process safety competencies and process knowledge can thoroughly address the issues during a PHA.

Another less significant gap noted by regulators and by the US CSB has been an incomplete consideration, within the PHA, of previous incidents.

This section of the paper lays out the case from the US CSB and US OSHA on these PHA gaps:

- Poor or no coverage of hazards during non-routine modes of operations
- Poor or no coverage of damage mechanisms

2.2. Objectives

2.2.1. Hazard Evaluation of Non-Routine Modes of Operation

Non Routine modes of operation include start-up, shut-down, online maintenance and other abnormal operations. The hazard evaluation of non-routine modes of operation involves reviewing procedures using a HAZOP, simplified HAZOP, or What-if analysis to uncover potential accident scenarios associated with non-routine operations, for continuous or batch operations. Human error is more likely and more critical during non-routine operations. By analyzing procedural steps where human error is more likely, and where human error or component failure could lead to a consequence of interest, risk can be reduced. The hazard evaluation team's objective is to evaluate the risk associated with skipping steps and performing steps wrong.

The purpose of a hazard evaluation of non-routine modes of operation (governed by written procedures) is to make sure an organization has enough safeguards for the inevitable instance when a step is either performed wrong or skipped (inadvertently or due to shortcutting or other reasons).

Industry has found that a HAZOP or what-if analysis, structured to address procedures, can be used effectively for finding the great majority of accident scenarios that can occur during non-routine modes of operation[10],[12],[13],[14]. Experience shows that reviews of non-routine procedures have revealed many more hazards than merely trying to address these modes of operation during the P&ID driven hazard evaluations.

General guidelines for analyzing non-routine modes of operation or batch processes

- Define the assumptions about the system's initial status. "What is assumed to be the starting conditions when the user of the procedure begins with Step 1?"
- Define the complete design intention for each step. "Is the step actually 3 or 5 actions instead of one action? If so, what are the individual actions to accomplish this task?"
- Don't analyze safeguard steps that start with ensure, check, verify, inspect, etc., or where the consequence of skip is "loss of one level of safeguard/protection against…" There is no reason to analyze these steps since they will show up as safeguards of deviations of other steps. This approach is similar to not analyzing a PSV during a HAZOP of continuous mode (i.e., during a parametric deviation analysis); instead the PSV is shown as a safeguard against loss of containment.
- Together with an operator before the review, identify the sections of the procedures that warrant use of:
  - 7-8 Guide Words (extremely large consequences can happen if deviations occur)
  - 2 Guide Words (the system is complex, mistakes are costly, or several consequences could occur)

   - On others, use What-If (no guide words or guide phrases; for use on simpler or lower hazard systems)
- Decompose each written step into a sequence of actions (verbs)
- Apply guide words directly to the intentions of each action

The following preparation steps may also be needed prior to the procedure PHA:
- Walk through procedure in the plant with one or more operators to see the work situation and verify the accuracy of the written procedure.
- Determine if the procedure follows the best practices for "presentation" of the content; the best practices will limit the probability of human error
- Discuss generic issues related to operating procedures, such as:
   - staffing (normal and temporary)
   - human-machine interface
   - worker training, certification, etc.
   - management of change
   - policy enforcement
- Review other related procedures such as lock out/tag out and hot work

- **If the procedures are NOT >90% accurate, then redo procedures before the PHA.**

2.2.2. Damage mechanisms

Recently, more and more regulators, including the US CSB, are making analysis of Damage Mechanisms (DMs) a PHA requirement. This requires that potential process damage mechanisms and their potential consequences of interest be identified. Damage mechanisms can be broken down into three main categories. These categories and examples of each are shown in the following table (Table 3).

Table 3. Types of Damage Mechanisms [15]

Category	Examples
Mechanical	• Mechanical loading failure • Mechanical fatigue • Buckling • Cracking • Embrittlement • Ductile fracture • Brittle fracture

Chemical	• Corrosion
	  - Uniform
	  - Localized
	  - Pitting

Physical	• Thermal failures
	  - Creep
	  - Thermal fatigue
	  - Transformation

The team may use aids such as piping specs, established literature and standards, any applicable MOC documents, etc. to help identify DMs. The PHA leader has several options of how to address damage mechanisms, but PII preferred approach is to discuss during HAZOP under "Loss of Containment" deviation in each node/section.

Damage mechanisms can (and sometimes ***should***) be reviewed prior to the PHA. For example, all MOCs should consider damage mechanisms and their potential consequences of interest before the change is approved. The MOC should also require updates to process safety information which might be relevant to protect against DMs.

Coverage of Damage Mechanism (DM) for each major section [14]

As mentioned above, this is PII's preferred method for thoroughly covering all DM within a PHA/HAZOP of an entire unit. The benefits are that the team can more easily catch changes in DM from section to section and they can more easily identify when unique safeguards, such as Remote Isolation Valves, segregated containments (dike), and unique materials of construction are needed. To facilitate this, a list of generic causes of loss of containment and a generic list of safeguards against loss of containment are covered in the Loss of Containment (LOC) deviation of each section.

For each node (each line, each vessel, each column, etc.), the PHA team should discuss and document each damage mechanism listed in Table 3 (as a cause of loss of containment), consequence of the failure if the damage mechanism occurs, and the safeguards in place to prevent the damage mechanism, detect the mechanism before failure, prevent the release, detect and response to the release, and mitigate or contain the release.

The documentation style varies between PHAs. There are two different styles that have been acceptable in the past.

- One style (Example A, Table 4) uses a reference to a summary table of typical causes (this summary table is not shown but is similar to Table 5 of this paper) instead of listing each individual damage mechanism. In Example A, the same approach is used for safeguards. This approach saves redundant text and some time, but it requires the leader and scribe to be diligent to cover everything in the Typical tables that were referenced in the Cause and Safeguard columns. A modified approach is to relegate some generic causes and safeguards to Typical tables for reference and then to carefully list the specific cause and/or safeguards of interest for LOC for each specific node.

Table 4. Example A – Referencing Generic Tables for Typical Causes and Typical Safeguards [14]

Dev	Deviation	Causes	Consequences	Existing Safeguards	Recommendations
9.10	Loss of containment	Accelerate corrosion External fire High pressure (linked from 9.7) Typical causes of loss of containment (see Table A.1)	Release to atmosphere leading to potential injury of workers and/or community	Pressure Relief Valve's 231A, B, C on Reactor and no valves in line to reactor Most lines / connections are welded construction; only a few flanges Drills conducted each year on evacuation, rescue and isolation Emergency Response personnel are trained at SABIC FTC Generic safeguards protecting against or mitigating process material releases (see Table A.2)	Safety 7. Consider changing the ITPM schedule for managing most PSVs whose inspections are too infrequent based on industry standards and best practice. For instance, the current inspection frequency for PSV-8220 on the Ammonia Receiver (I-2005G) is 9 years, whereas, consensus codes typically recommend testing/inspection every 1-4 years for PSVs in highly toxic services.

- Example B (**Table 5**) does not use or reference a generic or Typical list of causes (damage mechanism) or Typical safeguards, but instead develops a specific listing for the LOC deviation of each node. This style has proven easier to justify to regulators and other outside reviewers but takes more documentation effort.

Table 5. Example B – Specific Listing of Causes and Safeguards [14]

Dev	Deviation	Causes	Consequences	Existing Safeguards	Recommendations
1.9	Loss of containment	Corrosion/erosion External fire and/or flame impingement Gasket, packing, or seal failure Improper maintenance Material defect Operator failing to close or inadvertently opening a valve to the atmosphere (e.g., a valve at a hose connection) Railcar inadvertently derailed Valve leaking to the atmosphere High pressure (linked from 1.5) Acid corrosion caused by high concentration of water (linked from 1.8) High ambient temperature External impact (such as from a mini-engine or another railcar)	Catastrophic release of chlorine from a ruptured railcar Steady release of chlorine from a ruptured connection Steady release of chlorine from a leaky connection High pressure caused by thermal expansion of liquid chlorine if railcar is also over-full	Chlorine repair kit Derailer and warning flag to prevent impact by a mini-engine or another railcar Limited vehicular access to area Maintenance/operator response as required, including isolation if needed Operator periodically monitoring the railcar valves while unloading Personal protective equipment in the area Plugs installed in all chlorine valves to the atmosphere when the valves are not in use Relief valve on each railcar for mitigating releases caused by overpressure Supplier maintenance of railcars (per strictly enforced US DOT requirements) Video monitoring of the unloading area Concrete crossties on rail spur Dike preventing any combustibles spilled nearby from reaching the unloading rack area Concrete railroad ties in chlorine unloading area to prevent fires near railcar	10. Consider installing a chlorine detection system in the unloading and vaporizing area to help detect chlorine releases (especially at likely release points) 11. Verify that periodic maintenance and inspections are being performed in accordance with Chlorine Institute recommendations 12. Review the drainage system for the unloading are, and identify the areas that may be affected by a large chlorine release 13. Consider prohibiting the use of heavy equipment (e.g., cranes) in the unloading and vaporizing area unless special precautions to prevent equipment damage are enacted 33. Consider providing a high pressure alarm for each vaporizer 40. Consider providing a water deluge system in the unloading area to help mitigate chlorine releases from the railcar

Regardless of method, the team leader and scribe must ensure the team rigorously discusses all of the damage mechanisms in Table 5 and adequately documents the results of the damage mechanism review in the PHA/HAZOP analysis tables.

For a complete description of PHA best practices, refer to PII's paper *"Recipe for a Complete Process Hazard Analysis"*. [14]

3. Near Misses Need to be Reported and Enough of Them Investigated

We must learn from accidents and near misses to prevent recurrence. The first step in the learning process is investigation to determine the causes and underlying reasons why accidents and near misses occur. A thorough investigation of root causes will identify the management system weaknesses. Learning which management system weaknesses are leading to near misses and accidents is one of the highest value activities in which a company can invest, and learning from near misses is much cheaper than learning from accidents. Many chemical companies have implemented process safety management systems, and now they are beginning to focus on getting near misses reported and on root cause analysis. This is a very exciting trend. Unfortunately, the chemical industry gets very few near misses reported (the chemical industry is certainly not the only industry with this problem).

To understand more about near misses and getting them reported, it is best to first review the basic definitions.

> *An incident is either an Accident or a Near Miss.*
>
> *An Accident is a sequence of unplanned events and conditions that result in harm to people, environment, process, product or image.*
>
> *A Near Miss is an unplanned sequence of events that could have caused harm if conditions were different or is allowed to progress, but did not in this instance.*

Given a consistent understanding of the definition of a near miss, it is possible to estimate how many near misses should be reported for every accident. Studies in several industries indicate that there are between 50 and 100 near misses for every accident. Also, data indicates that there are perhaps 100 erroneous acts or conditions for every near miss. This gives a total population of roughly 10,000 errors for every accident. **Figure 2** illustrates the relationships between accidents, near misses and non-incidents.

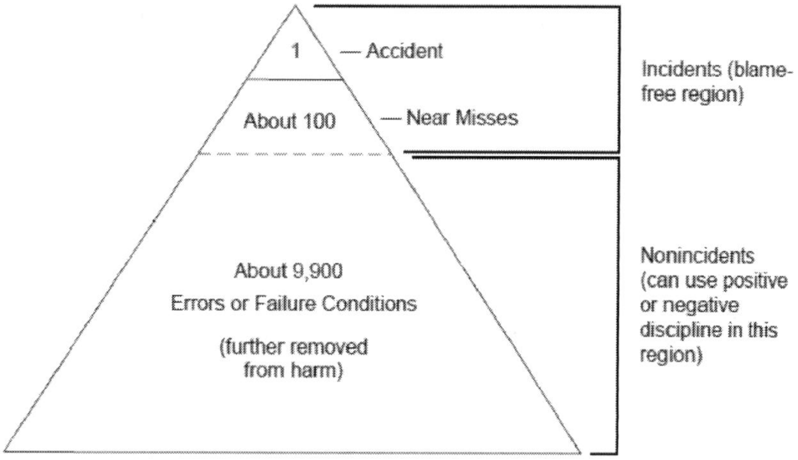

Figure 2. Relationship between Errors and Potential or Actual Impact [16]

Investigating near misses is critical to preventing accidents, because near misses share the causes and root causes of accidents; they are one or two barriers away from the loss/accident. We are very likely preventing many apparently unrelated accidents when we prevent the ones that are obviously related.

3.1. Identified problems

During the past 15 years, we have asked more than 5,000 students of our process safety management (PSM) courses and 3,500 students from our investigator leadership training courses how many near misses they get reported for every accident. The students represented about 400 companies, predominantly in chemical-related process industries (chemical, polymer/plastic, petrochemical, refining, oil and gas exploration, pharmaceutical, pulp and paper, etc.). The answers are quite disturbing. More than 95% said that their ratio of near misses reported to accidents reported falls in the range of 0 to 20. Less than 5% of the individuals said the ratio was greater than 5, and less than 2% said the ratio was higher than 10. Students noted that fear of disciplinary action, lack of management commitment, and lack of understanding of the difference between a near miss and a non-incident were the main reasons why near misses do not get reported. [16]

In conducting about 150 PSM audits and 7000 process hazard analyses (PHAs) during the past 22 years, we have found ratios from 0 to 105. For the first half of the 1990s, more than 90% of the facilities we talked to had ratios in the range of 0 to 0.5, and more than 95% had ratios in the range of 0 to 1. In the last half of the 1990s, more than 90% of the facilities had ratios in the 0 to1 range, and more than 95% had ratios in the 0 to 2 range. In addition, only a few (less than 2% of the facilities) had a ratio higher than 5. By 2005, many facilities had ratios in the 20 to 100 range and each of these had seen great gains (more on this in future articles), but the average across all companies we talked to increased only slightly. By 2011, the ratio of near misses to accidents (any loss event) was about 2 for companies we deal with. But by the end of 2011, many major companies had ratios well above 50! [16]

Our auditors and PHA leaders commented that the primary reasons for lack of reporting were about the same as those found in the survey. The barriers to getting near misses reported, discussed in detail later, are:

- Fear of disciplinary action.
- Fear of teasing by peers (embarrassment).
- Lack of understanding of what constitutes a near miss versus a non-incident.
- Lack of management commitment and lack of follow-through once a near miss is reported.
- An apparently high level of effort is required to report and to investigate near misses compared to low return on this investment.
- There is No Way to investigate the thousands of near misses per month or year!
- Disincentives for reporting near misses (e.g., reporting near misses hurts the department's safety performance).
- Not knowing which accident investigation system to use (or confusing reporting system).
- Company discourages near-miss reporting due to fear of legal liability if these are misused by outsiders.

The good news is that near-miss reporting appears to be improving, and the industry appears to recognize most of the barriers to near-miss reporting. The bad news is that the ratio is still very poor and improvement appears slow!

The number of companies that participated in the informal surveys suggests that the results are a statistically significant representation of the chemical industry.

A formal (written) survey was developed and e-mailed, faxed, and/or mailed to more than 100 companies. Of these, more than 12 replied originally in 1997. Since then, we have reviewed data in detail from more than a dozen major companies. These 25+ companies or affiliates provided data from more than 400 facilities, including more than 150,000 employees in manufacturing and operations. The data were from the chemical industry, polymer industry, refineries, drug/pharmaceutical companies, pulp and paper mills, petrochemical companies, and oil exploration/production.

Table 6 summarizes the results of both the informal and formal surveys.

Table 6. Summary of survey results [16]

Survey Items	Informal surveys		Formal Survey (or actual organization data)
	Classroom polls	**Found during Audits and PHAs**	
Current Near-miss Reporting Ratio Range (0-5 years prior)	1 to 105	2 to 20	0 to 105
Previous Near-miss Reporting Ratio Range (5-10 years prior)		1 to 5	NA
Current Near-miss Reporting Ratio Average (0-5 years prior)	2-3	3	5
Previous Near-miss Reporting Ratio Average (5-10 years prior)		1	NA
Goal for Near-miss Reporting Ratio (average)	20+	NA	20
Theoretical Upper Limit on Near-miss Reporting Ratio (average)	About 100	NA	100
Number of Participating Companies	400	300+	25+
Number of Participating Facilities	8500+	500+	400+

3.2. Objective

A company should strive to reach a ratio of 50-100 and investigate about 20 near misses per accident. This will provide a statistically significant sample of all incidents (and all important errors) and provide a company with sufficient feedback on which management system weaknesses are causing the errors and component failures. Various companies with different cultures have achieved high ratios with great return on investment.

Table 7 shows a brief description of possible solutions to overcome each one of the barriers identified in the survey.

Table 7. Summary of Survey Results[16]

Barriers	Solutions
Fear of disciplinary action	Implement a policy to NOT punish individuals when their errors lead to accidents and Near Misses.
Fear of teasing by peers (embarrassment)	Ensure that all employees understand the importance of near-miss reporting; demonstrate, through feedback of lessons learned.
Lack of understanding: Near miss vs Non-incident	Develop a list of "in-context" examples that illustrate what you consider to be Near Misses and what you consider to be non-incident
Lack of management commitment and lack of follow-through on reported near misses	Hold management accountable for achieving a Near-Miss reporting ratio
Apparently high level of effort is required to report/investigate Near Misses	Ensure that the data are entered in a database and queried regularly. Share the results with employees so they can see the value of the reported near misses
There is no way to investigate the thousands of Near Misses per month	Let front-line foremen or supervisors decide if a Near Miss or accident needs to be investigated
Disincentives for reporting Near Misses	Ensure that goals and incentives are not tied to lower incident rates (since this discourages reporting), but instead provide incentives for high Near-Miss reporting ratios
Not knowing which accident investigation system to use	Have ONE incident reporting system with ONE approach
Company discourages Near Miss reporting due to fear of legal liability	Involve legal on major Near Misses and accidents to ensure the results are protected as much as possible under attorney/client privilege.

For a complete description of the possible solutions for each barrier, refer to PII's paper *"Gains from Getting Near Misses Reported"*. [16]

4. Key Human Factors are Not being addressed

All accidents (or nearly all, if you consider that there are some natural phenomena that we either cannot guard against or choose not to guard against) **result from human error**. This is because humans govern and accomplish all of the activities necessary to control the risk of accidents. Humans influence other humans in the process – not only do humans cause accidents (unintentionally) by making errors directly related to the process itself, but they also cause errors by creating deficiencies in the design and the implementation of management systems (i.e., we make errors in authorities, accountabilities, procedures, feedback, proof documents and continual improvement provisions). Ultimately, these management systems govern the human error rate by directly causing or indirectly influencing the process. The process-related activities where errors have the most influence include:

- Designing a process
- Engineering a process
- Specifying components

- Receiving and installing equipment
- Commissioning
- Operating a process
- Maintaining, inspecting and repairing a process
- Troubleshooting and shutting down the process
- Managing process, procedure, materials, facility and personnel changes

The 10 human factors categories to be controlled by companies are:

- Available Time (includes staffing Issues) – for responses only
- Stress/Stressors (includes staffing issues)
- Complexity & task design
- Experience/Training
- Procedures
- Human-Machine Interface (includes tools)
- Fitness for Duty
- Work processes & supervision
- Work Environment
- Communication

4.1. Identified problems

Recent major accidents have highlighted the need for increased focus on human factors. The U.S. Chemical Safety Board (CSB) cited (US CSB Video, 2006) human factor deficiencies as one of the main contributors of the catastrophic accident at the BP Texas City Refinery in March 2005[17]. The human factor deficiencies included lack of control of worker fatigue, poor human-system-interface design, poor communication by radio/phone, out-of-date and inaccurate operating procedures, and poor (no) communication between workers at shift handover. The CSB cited similar issues from many other accidents[18] and has urged industry and the U.S. OSHA (the regulator) to pay much more attention to human factors. As a result, the recent U.S. OSHA National Emphasis Program for Refineries included human factors as one of the 12 core elements.

Implementing human factor engineering and policies to prevent accidents is not a new concept. Nearly all (or all, from a more complete perspective) of the causes and root causes of major accidents in the past 30 years have been the result of poor control of human factors. This has been cited in many root causes analysis reports and papers concerning these major accidents.

Process Safety Management (PSM) systems based on OSHA's PSM standard are likely lacking the fundamental human factor elements and implementation guides that, if applied across the applicable PSM elements, would work together to reduce human error. The *Risk Based Process Safety* guideline from the CCPS/AIChE (2007) does contain these human factor requirements.

Many site and company staff do not know there are standards for the control of human factors. However, at last count there were more than 300 non-governmental standards (such as from ANSI, ISO, IEEE, etc.) for the control of human factors, and there are more than 100 government regulations and standards and guidelines. Some of the better government regulations cover aviation, marine operations (shipping) and aerospace. An organization should devote resources to finding and implementing such best practices.

The following is a description of each of the human factors mentioned above and specific ways to control each. Our data from review of more than 2000 incidents reveal that:

- 90% of accidents had at least one root cause related to deficiencies in written work instructions.
- 70% of accidents had at least one root cause related to miscommunication between workers or between workers and supervisors.

- 40% of accidents had at least one root cause related to excessive fatigue of the worker. [5]

4.2. Objective

Not all errors can be prevented. Since the beginning of time, humans have tried to control error rate with more or less success. Human errors have been measured for hundreds of years. Psychologists have studied why humans make mistakes and have gradually put a science around human error probability. Today, the best models for control of human error in the workplace are generally agreed to be related to control of human factors; in turn, these have grown out of what was previously denoted performance shaping factors. What can be done to control each Human factor is discussed in greater detail next.

For a complete description of each human factor, its importance and how to control them; refer to PII's paper *"Human factors and their optimization"* [5].

Time available for the task

Task design is directly linked with this category. Process design, procedure development and worker scheduling need to take into consideration not only the time it takes to complete each task but also the time it takes to manage usual as well as potentially unusual or unexpected situations. True; we do not create norms for exceptions, but exceptions sometimes are more usual than realized.

Stress/Stressors

A well-designed, organized and managed work process helps maintain and promote individual health and well-being, but stresses will arise through no fault of the organization. Supervisors should be trained to recognize stress in workers and, in severe cases, reassign them and help them get assistance to deal with the stress and in other cases, reduce their job pressures for a day by shifting assignments. A wellness program may also be necessary to help cope with stress. Simple approaches, such as training workers to recognize and deal with stress, can help most individuals. Approaches to deal with stress include positive self-talk and practicing relaxation techniques (i.e., breathing control).

Complexity & task design

If roles and responsibilities for accomplishing tasks are not clearly defined, then there will be a risk of serious errors. It is also important that employees and engineers/designers understand the characteristics of the work elements involved, how each element passes information to the other, how each person involved communicates, and that each person has guidelines to adhere to during task design and procedure development.

For optimal performance, task designers often must integrate human and automated equipment in their task analysis.

It is crucial for task designers also to have an understanding of the appropriate allocation of roles and responsibilities to the various participants in a task. Understanding the characteristics and limitations of all of the humans and automated systems involved in the task (particularly critical tasks) will allow for establishing additional controls where needed.

The task analysis process includes *Task definition* which is the initial operational description of the operator tasks required for execution of a given system function; *Performance analysis* of all tasks with respect to human performance requirements; *Error and workload analysis* to be included in the design process, particularly when the likelihood of unacceptably operator error is high; and *Evaluation* since evaluating the design of a new or complex task is important to ensure that the task can be effectively performed as designed. For very high consequence scenarios involved complex human interaction, it is likely prudent to invest in a human reliability analysis (HRA).

Experience/Training

Training to master the Knowledge, Skills, and Abilities (KSAs) required for a job may be obtained from a variety of sources. Companies provide the training or it is obtained from other sources (e.g., trade schools, universities, contractor organizations, etc.) to reach the skills and knowledge established in job descriptions. There are several factors that may result in personnel not mastering the required KSAs for a job. These include course design and delivery methods, course completion, practical skills demonstrations or simulation, and training frequency. Course design begins when the learning objectives have been identified. The design process consists of determining the delivery methods (simulator, on-the-job training, etc.), number of hours required to cover the materials, instructor qualifications, etc. Although some methods, materials and instructors may be more effective or efficient than others, the important issue is that the course content is complete and addresses all of the relevant KSAs, so that the learning objectives are met and the KSAs mastered.

Another determinant of KSA mastery is course completion. Although this factor appears obvious, there are often competing demands on personnel that may pull them out of training at times. As a result, they may miss the instruction related to specific KSAs. Testing may not identify the KSA deficiency because it is impossible to test mastery of all KSAs. Sampling techniques are used to generate examinations. If attendance and participation are not controlled, some personnel may miss training on specific KSAs and testing may not identify the deficiencies before an error is committed.

Another factor affecting KSA mastery is forgetting. An individual's ability to perform a task will degrade over time unless the relevant KSAs are refreshed. Proficiency training (i.e., refresher) will be required for some tasks to maintain the level of mastery that was demonstrated following initial training. One function of training programs is to identify those tasks that require proficiency training.

If certain tasks are performed frequently, proficiency training may be unnecessary. By performing a task, personnel practice the task and obtain feedback on where they have weaknesses. Task performance refreshes the KSAs and successful task performance verifies that proficiency has been maintained. Furthermore, since humans learn by comparison to similar activities, as a worker practices one task, they are learning about similar tasks. Care must be taken here to avoid the practice drifting away from what the procedure requires. Also, if new operators are trained by experienced operators, you must ensure that the new employee is trained according to the procedure, not whatever practice is actually being used. (Of course, ideally, the practice and procedure will be identical.)

Operating and maintenance procedures

Many procedures do not follow best practices for controlling human error, and so the written procedure actually "contributes" to increased error rates. Additionally, many organizations do not have guides on how to troubleshoot (what to do when process deviations occur). The best practice rules for writing and validating procedures have been published for many years. [6], [7], [8], [9]

Procedures have not traditionally been developed from the perspective of optimizing human factors; instead, procedures have been traditionally developed to meet a compliance requirement to have written procedures. Examples of procedure deficiencies (inaccuracies) include:

- Incorrect/incomplete/nonexistent (most procedures we have audited have been only 70-85% accurate – the inaccuracies include missing critical steps, steps as written are not what needs to be done, or the steps are out of sequence)
- No/misplaced warnings or misuse of warnings (for example, a warning should never **state** the action to take; it should instead **emphasize** the action to take an the statement of the action should be in a numbered step)
- Poor format and presentation rules

Human-machine interface

Design for operability refers to designing the Human-System Interface (HSI) to be consistent with the abilities and limitations of the personnel who will be operating it. Weaknesses in the design processes can result in an HSI that is not well suited to the tasks that personnel must perform to ensure plant safety, resulting in increased workload, decreased performance by personnel, and an increased likelihood of errors.

Design for maintainability refers to designing the HSI and associated plant equipment to ensure that personnel are able to perform necessary maintenance activities efficiently. Weaknesses in the design process can result in systems that impose excessive demands on personnel for maintenance and, therefore, are prone to maintenance errors or problems with reliability and availability. *If it is hard to reach, workers will make more errors, including the error of deciding it is not worth it.*

Design for flexibility refers to the way that changes, such as upgrades to the HSI, are planned and put into use. A new HSI component may require the user to perform functions and tasks in new ways. Skills that the user developed for managing workload when using the former design, such as ways for scanning information or executing control actions may no longer be compatible with the new design. The new HSIs must also be compatible with the remaining HSIs so that operators can use them together with limited possibilities for human error. Also, HSI modifications may not be installed or put into service all at one time, causing the user to adapt to temporary configurations that are different from both the original and final configurations. Weaknesses in HSI implementation can increase operator workload and the likelihood of errors.

Fitness for duty

Implement a company fitness-for-duty program with a primary responsibility for detecting and preventing impaired personnel from performing tasks that may affect productivity and safety. Medical evaluations of personnel, behavioral observation programs, employee assistance programs, and drug and alcohol testing are used to detect impairment. Weaknesses in this program may allow impaired personnel to have access to vital areas in a plant where they could commit errors.

Overtime Policies and Practices – Most companies establish limits for work hours to reduce on-the-job fatigue and the potential consequences for poor task performance. Routine authorization for work hours in excess of those recommended may result in fatigued workers. Furthermore, a practice of excluding training or meetings that occur outside of an individual's normal work schedule from work-hour limitations will also contribute to fatigue. But in many cases these have weak enforcement, especially during shutdowns when the company is under stress to re-start soonest.

Shift Scheduling – Shift scheduling may also affect the likelihood that personnel will show performance decrements due to fatigue. A change in the assigned shift or a rotating shift schedule will disrupt circadian rhythms and may increase the likelihood of errors. So, a company must choose the proper shift rotation to allow adjustments to sleep patterns.

Work processes & Supervision

Written plant policies and procedures are meaningful only when they are enforced; otherwise, worker practices are the policy. Once discrepancies are tolerated, individual workers have to use their own judgment to decide what tasks are necessary and/or acceptable. Supervisors and managers should vigorously enforce plant policies/procedures and ensure that plant policies/procedures and practices are revised as necessary to be consistent. To enforce well, a supervisor must (1) lead by example, (2) watch workers often to ensure they are following company policies and procedures, and (3) determine the correct course of action (including recommending changes to policy and procedures) when workers are found deviating from procedures.

Work environment

Programmatic causes of task environment errors are typically found in the company's processes for designing human-system interfaces or in managing maintenance activities. Other programs may also be implicated. Common programmatic causes of task environment errors include:

Industrial Hygiene and Radiation Protection – These programs are responsible for ensuring that task environments have been evaluated to identify hazards and that needed controls are implemented to minimize exposures. Weaknesses in these programs may result in personnel working in task environments that are conducive to errors.

Work Planning and Control – Weaknesses in the work planning and control system may allow work to be planned without consideration of adverse environmental conditions and performed without the necessary compensatory measures. For example, communication devices may not be provided in noisy environments to support task performance. For tasks that involve unusual physical positions or cramped workspace, additional time to complete the task may not be scheduled. Rest breaks for hot and cold environments may not be planned into the work, or additional temporary lighting may not be provided if the work site is not adequately lighted.

Procedures – Weaknesses in the company's procedure development process may result in the design of procedures that are inappropriate for the conditions in which they will be used. For example, procedures that may be used at night, outside and in the rain should be laminated and the type size should be larger to ensure the procedure can be read. Procedures that will be used in vibration conditions may also require larger type size than procedures read in the stationary environment of the control room.

Human Factors Engineering – Weaknesses in the human factors engineering program may result in the installation of new equipment or systems without consideration of task environment characteristics. For example, the impact of control room lighting on the visibility of digital displays or effects of vibration on the legibility of dials or gauges at local control stations should be considered before installation.

Communication

Successful communication requires several steps. The sender first develops the intention to communicate either verbally or in writing. The sender then composes a message that presents the meaning as clearly as possible. The receiver must pay attention to the message and then interpret its meaning. If the communication is successful, the receiver interprets the message consistently with the sender's intended meaning. Table 8 summarizes the rules for successful verbal communication.

The similarity of the meanings given to the message by the sender and receiver can be verified through feedback. An example of feedback verification in verbal communication is when the receiver "repeats back" the message and the sender either agrees with the receiver's repeat back or corrects it. Verification feedback serves an important error-checking function in the communication process. It also allows supervisory oversight of communications to catch errors before they have consequences.

A sender and receiver must both be active for communication to be effective. The sender and receiver share responsibility for ensuring successful communication. However, when companies analyze the causes of events, errors in sending messages are more often identified than errors in receiving. The reasons for the difference are unclear. A company's investigation should consider sending and receiving errors and corrective actions should address both to be effective.

Shift handover is another critical aspect of communication. The status of equipment, repairs that occurred, problems faced, upcoming chores, etc., are all key information the oncoming worker will need to know. One good guide on shift turnover is U.S. DOE 1038-93. [19]

Table 8. Rules for effective verbal (face-to-face, radio, phone) Communication (© PII, 2012)

#	Communication rules
1	Senders MUST require repeat back by the receiver; ask questions to make sure of understanding.
2	Receivers of messages MUST repeat a message they receive.
3	Communicate one task at a time.
4	Sender & receiver must use the standard name and component number (and building name and number) and/or for related reference material (drawing, procedure name/number) in each message.
5	Sender must require receiver to give feedback of foreseeable conflicts. Sender (if message is to subordinate) should follow-up in the field if possible, to make sure message was understood.
6	Receiver must request clarification anytime they believe the message is confusing.
7	Do NOT give verbal instructions to workers who do not have the demonstrated skills to correctly understand and perform the task assignment (unless you plan to supervise/coach the task yourself).
8	Communicate to the most senior member on the work crew.
9	Know your audience & change message accordingly to compensate for lack of knowledge (when appropriate; see other related rules).
10	Use units (and use approved units, such as SI) for process parameters.
11	Always count from left to right when giving instructions related to multiple choices of equipment.
12	Use approved jargon only.
13	Use the words **check**, **make sure**, and **actions** consistently (see procedure writing rules).
14	When possible, write out the task (use a special procedure or work order) rather than communicating verbally. Use formalized templates for consistency and make sure the receiver is provided all information (such as work orders and drawings) necessary for the task.
15	Perform pre-job meeting with work crew (including with contractors or construction company).
16	Perform walkthrough at the location where the work will be completed.
17	Have a backup communication method. Don't rely on one mode of communication. In emergency situations have hand signals as backup to loss of verbal communication.
18	Talk to the operator or technician who is doing the work (no delegation of work); but see Rule 8.
19	Receiver (worker or group) must report back when work is complete.
20	If confusion exists in implementing a task, the shift supervisor (and higher, if supervisor is unavailable) must be contacted to make sure he/she understands the problem.
21	Do NOT use a PA (public address system) for process instructions, since this method of communication does not allow repeat-back.
22	Use Open questions and non-confrontational questioning methods when requesting clarification.
23	When communicating remotely (by phone or radio), if the message is not understood on the second attempt at clarification the Sender must find the Receiver and communicate face-to-face.
24	For large or complicated jobs, in addition to repeat back, the sender must ask the receiver for an assessment of the pre-job briefing to ensure the workers are not confused.
25	Workers and supervisors must keep a shift log to aid in turnover between shifts. Workers and supervisors and support staff (if necessary) must have a minimum or 15 minutes overlap with a relieving shift. *See shift turnover standard for more details.*

Conclusion

Closing the gaps in these 4 areas has a greater than 100:1 payback!

Acronyms and Abbreviations

ACC - American Chemistry Council
AIChE – American Institute of Chemical Engineers
API – American Petroleum Institute
CCPS – Center for Chemical Process Safety (of AIChE)
CMA - Chemical Manufacturers Association
CSB – Chemical Safety Board
DM – Damage Mechanisms
FMEA – Failure Mode and Effects Analysis
GCPS -- Global Congress of Process safety
HAZOP – Hazard and Operability; as in HAZOP Analysis or HAZOP Study
HRA – Human Reliability Analysis
HSI – Human-System Interface
KSA – Knowledge, skills and abilities
LOC – Loss of containment
LOPA – Layer of Protection Analysis
MOC – Management of Change
OSHA - US Occupational and Health Administration
PHA – Process Hazard Analysis
PSM – Process Safety Management
RBPS – Risk Based Process Safety
SOP – Standard Operating Procedure

References

[1] *Guidelines for Technical Management of Chemical Process Safety,* CCPS/AIChE, 1985.

[2] "The Cost and Benefits of Process Safety Management: Industry Survey Results," W. Bridges, *Process Safety Progress*, AIChE, January 1994.

[3] "Necessity of Performing Hazard Evaluations (PHAs) of Non-normal Modes of Operation (Startup, Shutdown, & Online Maintenance)", W. Bridges and Mike Marshall (US OSHA), *18th Annual International Symposium, Mary Kay-O-Connor Process Safety Center, College Station, TX, October 2015.*

[4] *Guidelines for Risk Based Process Safety*, CCPS/AIChE, 2007.

[5] "Human factors and their optimization" W. Bridges and G. Collazo-Ramos, *8th Global Congress on Process Safety*, Houston, AIChE, April 2012.

[6] PII Course 5, *Writing effective operating and maintenance procedures*, © 2003-current

[7] "Best practices for writing operating procedures and trouble-shooting guides", W. Bridges and R. Tew, *13th Global Congress on Process Safety*, San Antonio, AIChE, March 2017.

[8] "Effective Procedure Writing – A Two-Part Series (Part 1: Reducing Risk of Worker Errors; Part 2: Addressing Process Safety Management Requirements for Operating Procedures)," *Chemical Engineering Progress*, 1997, W. Bridges with T. R. Williams (both of JBFA).

[9] "Best Practices for Writing Operating Procedures and Trouble-Shooting Guides", L. Madden and W. Bridges, *12th Global Congress on Process Safety*, Houston, TX, AIChE, March 2016.

[10] *Guidelines for Hazard Evaluation Procedures*, 3rd Edition, 2008, CCPS/AIChE.

[11] "API RP 571 - Damage Mechanisms Affecting Fixed Equipment in the Refining Industry", API, 2nd edition, 2011.

[12] Bridges, W.G., et. al., "Addressing Human Error During Process Hazard Analyses," *Chemical Engineering Progress*, May 1994.

[13] "Necessity of Performing Hazard Evaluations (PHAs) of Non-normal Modes of Operation (Startup, Shutdown, & Online Maintenance)", W. Bridges and Mike Marshall (US OSHA), *18th Annual International Symposium, Mary Kay-O-Connor Process Safety Center, College Station, TX, October 2015 and 12th Global Congress on Process Safety, April, 2016.*

[14] "Recipe for a Complete Process Hazard Analysis – Especially Addressing the Key Demands from US CSB", R. Tew and W. Bridges, *13th Global Congress on Process Safety*, San Antonio, AIChE, March 2017.

[15] PII Course 8, *PHA/HAZOP Leadership training*, © 2003-current.

[16] "Gains from Getting Near Misses Reported", W. Bridges, *8th Global Congress on Process Safety*, Houston, AIChE, April 2012.

[17] US Chemical Safety and Hazard Investigation Board (CSB), *Anatomy of a Disaster: Explosion at BP Texas City Refinery,* published 2006.

[18] US Chemical Safety and Hazard Investigation Board (CSB), *"Pesticide Chemical Runaway Reaction Pressure Vessel Explosion – Bayer CropScience, LP",* published 2008.

[19] "DOE-STD-1038: Guide to Good Practices for Operations Turnover", US DOE, 1998.

546668 Recycle Line from NGL Spheres to Residual Gas Header

Abstract:

HNGL plant is received 1.6 BSCFD of Haradh gas flows into the DGA Gas treating units where CO2 is removed and the produced sweet gas is directed to the NGL recovery units .Hawiyah sales gas is pretreated at HGP for CO2 removal ; hence the produced 2.4 BSCFD flows directly to the NGL units where it is mixed with the Haradh gas. The mixed gas then is processed in the NGL units producing C2+ NGL and sales gas stream. The NGL is pumped to storage spheres and subsequently shipped to NGL piping network through Shedgum Gas Plant. The Residue Gas flows to the sales gas compression area which increase the pressure allowing routing back the gas to the sales gas grid.

The purpose of this project is to install vent line from the two NGL Spheres to Sales Gas (Residue Gas) common header at Hawiyah NGL Plant. The aim of this project is to minimize the continues flaring from the NGL spheres , safe plant resources and protect the environment

Sphere PVs were venting the NGL vapor to the cryogenic burn pit when the pressure of the sphere exceeds 350 psig. In order to lower sphere pressure without losing NGL production, newly installed PVs are set at lower pressure set point of 305 PSIG to recycle the spheres vapors to NGL trains residue gas line, downstream of the brake-compressors..

During summer months July and August (if associated with maximum plant production), spheres are venting about 4.0 MMSCFD of Methane. Based on Methane price of 4.4 $/MMBTU then total saving is 1.05 MM $/year.

547776 Practical Notes on Ragagep Compliance

Process Safety Management (PSM) and Risk Management Plan (RMP) regulations have long required facilities to ensure that they comply with Recognized and Generally Accepted Good Engineering Practice (RAGAGEP). In recent years, authorities having jurisdiction have placed increased emphasis on RAGAGEP compliance. Specifically, emphasis has been placed on documentation and compliance with the industry codes and standards to which a facility is designed and maintained. This poses a unique challenge to older facilities, which have been designed to codes and standards long out of date and especially to owners and operators that do not have access to engineering documents used in the original design of their facility. For facilities using outdated codes, OSHA requires that the operating company document that "equipment is designed, maintained, inspected, tested, and operating in a safe manner."

Over time, the gradual evolution of industry codes and standards may cause facilities to fall out of compliance with the RAGAGEP requirements of the regulations. This paper provides example cases based on authors' experience with various owners and operators of older facilities. Practical methods are discussed for achieving and maintaining RAGAGEP compliance. It includes examples of comparing systems designed to outdated codes and standards against the latest codes, to ensure that they are able to maintain compliance with RAGAGEP.

548482 Propane Safety: Investigation Findings and Lessons Learned in the 2014 Philidelphia FOOD Truck Explosion

On July 1, 2014, a propane cylinder catastrophically failed and exploded on the back of a food truck in Philadelphia, PA. The explosion led to a sudden release of pressurized propane vapor and superheated propane liquid, which rapidly evaporated resulting in a large white cloud that entered and engulfed the rear of the food truck. This cloud ultimately found an ignition source within the food truck and resulted in a significant fireball burning both inside and outside of the food truck. The explosion and ensuing fire fatally injured the food truck owner and her daughter, and also caused injuries to 11 other individuals. This paper will first present the results of our investigation into the cause and origin of the catastrophic failure and explosion event, which included: (1) analyzing video footage of the explosion and ensuing fire; (2) applying propane phase diagrams to determine the expansion and resulting pressures within the cylinder prior to failure; and (3) applying blast techniques to determine the post catastrophic blast and ignition during the event. These techniques included advanced computational fluid dynamic modeling.

The paper will also discuss safety critical design features and processes in the propane industry that must be in place to ensure the safe use and filling of propane cylinders, which include: cylinder requalification; having a fixed maximum liquid level gauge; having a pressure relief valve; and filling cylinders by weight or volume.

548391 Using the Risk Analysis Screening Tool (RAST) in Process Design to Manage Risk

The Risk Analysis Screening Tool (RAST), available through the Center for Chemical Process Safety (CCPS), courtesy of the Dow Chemical Company can enable users to identify hazards and risk scenarios to better manage process safety risks. RAST is a spreadsheet based tool that uses simplified algorithms to identify and evaluate risk scenarios of a unit operation. The risk scenario identification is based on generic HAZOP deviations applied to the unit operation. An engineer can use RAST to identify risks to proposed process designs or plant modifications in the design phase of a project, before P&IDs are developed and a full process hazard analysis study is done. This can save time in process design by weeding out higher risk process alternatives as the process flow diagram is being developed.

The utility of RAST is demonstrated by application to actual case studies. RAST is applied to the runaway reaction at T2 Laboratories in 2007 and to the chlorine release at DPC Enterprises in 2003 showing how the hazardous events and consequences could have been analyzed during the design phase of the projects.

19 AIChE Spring Meeting
15TH GLOBAL CONGRESS ON PROCESS SAFETY
NEW ORLEANS • 2019

Drug Product vs. Process Safety: Integrating Both into the Design Process from Concept Development to Start-up

Jeffrey Heil, P.E.
Lead Process Engineer
DPS Group Global
959 Concord Street, Suite 100
Framingham, MA 01701
Jeffrey.Heil@dpsgroupglobal.com

Prepared for Presentation at
American Institute of Chemical Engineers
2019 Spring Meeting and 15th Global Congress on Process Safety
New Orleans, LA
March 31 – April 3, 2019

Drug Product vs. Process Safety: Integrating Both into the Design Process from Concept Development to Start-up

Jeffrey Heil, P.E.
Lead Process Engineer
DPS Group Global
959 Concord Street, Suite 100
Framingham, MA 01701
Jeffrey.Heil@dpsgroupglobal.com

Keywords: Process Safety, Process Design, Biotechnology, Inherently Safe Design

Abstract

Process safety has been around for over 350 years but may easily be overlooked or treated as an afterthought in the design process, often leaving hazard reviews and safety valve considerations to the end of detailed design. In the chemical industry, emphasis on safety is generally in respect to process safety, but in the growing biotechnology industry the emphasis is often shifted to product and patient safety. Due to this subtle yet significant difference in focus, it is often challenging for the design team on these projects to secure the necessary project resources to ensure both robust product and process safety in the facility being designed, often on an accelerated schedule. Leveraging experience from working in the oil and petrochemical industry for over 10 years, an in-depth understanding of the project design cycle, and by asking some simple questions along the way; process safety in design can be assured for even the most challenging projects, without a significant impact on project resources.

1 Introduction

1.1 Brief History

The first safety valve, developed in 1681 by Frenchmen Denis Papin, protected the first ever pressure cooker, one of which was used to make meat jelly for England's King Charles II. [1] Papin first invented the steam pressure cooker in 1676 to make the jelly and other food products but then realized no one would be willing to operate the cooker given the risks. Having identified the hazard, he set out to invent a way to mitigate that hazard while ensuring the King got his jellied meat pies.

Following development of the first weighted lever style safety valve invented by Papin, little progress was made until the early 1800's when locomotive steam engines gained notoriety both as a superior means of transportation and a hazardous machine often operated beyond its limits in the name of more power and speed. Safety valves were revised throughout the early 1800's to make these locomotives safer and formally modernized in 1856 with the invention of the first spring actuated pressure relief device that could not be modified to a different pressure, while in service [2].

Modern process safety, as we can still see it in existence today, dates to the early 1900's with the development of modern oil refineries and all the hazards that they include. With these refineries and their complex chemical processes came a wealth of experimentation, safety system design, and industry supported safety initiatives. The goal was the same as it is today, ensuring safety to the employees, the environment, and the facility. The problem with many of these early safety initiatives is that different companies had different ideas of what safe meant and relied on historical data to predict the future. The notion that "it has been running like this for 20 years and will continue for another 20 years" lead to a raising number of deadly process safety incidents, particularly throughout the 1970's to 1980's (i.e. Bhopal, India Methyl Isocyanate release and Piper Alpha explosion in the North Sea.). Plants were operating to the status quo, not improving their practices new technology became available. In parallel, urban sprawl forced industrial plant operation closer to populated areas, increasing risk to the community. The OSHA process safety management (PSM) initiative was introduced into law in the United States to address this very problem [3]. The PSM standard was performance-based meaning that as new technology was developed and safer systems could be designed, those safer designs had to be implemented; i.e., the status quo was no longer good enough. The PSM standard was ratified in 1992: hence, even though the oil and petrochemical industries in the United States have been around for about 100 years, they have only been operating under this standard for 26 years.

Contrast this with today's biotechnology companies. Gentech, one of the earliest, received FDA approval for the insulin drug Humulun in 1982. Like any new technology, investment and growth in biotechnology was slow to start, not taking off until early 2000's. There are many metrics for tracking growth of the Biotechnology industry such as venture capitalist investment and the number of companies in operation, but the high number of mergers and acquisitions over the past 30 years makes such comparisons incomplete. The Nasdaq Biotechnology Index (NBI) was established in 1993 with a baseline index value of 200 and provides a good metric for tracking industry growth over time. After holding steady in the low 200-300s throughout the 1990's, the NBI surged to 1,300 in 2000 as the Biotechnology bubble was forming. The index has steadily risen to a peak of 4,003 on 01-July-2015 and is hovering currently around 3,500 (Q1 2019), exemplifying the continued growth of the industry over the past 20 years. [4]

Pharmaceutical companies (e.g., Aspirin, invented by Bayer in 1899) have been around almost as long the oil companies have and are sometimes confused with biotech. Biotechnology companies are adolescent at best, having typical company life of less than 20 years with little process safety oversight beyond building code requirements as they do not meet the hazard requirements of the OSHA PSM Standard. Not meeting the hazard requirements of the PSM standard does not exempt these companies from a responsibility to ensure a safe working environment. If anything, their proximity to high density population areas and minimal operating experience places them at a higher risk with respect to process safety, but in many cases the operators and designers have only worked in these types of facilities and may not fully appreciate the safety concerns present. Coming from a background where process safety is paramount and the product quality, while important, takes second place, the author has made it his personal mission to increase awareness throughout the Biotechnology industry one project at a time.

1.2 Process vs. Product Safety

In the chemical industry, emphasis on safety is always in respect to process safety; ensuring that the public, employees, environment, and facility (in that order) are sufficiently protected from inherent hazards of a process. "Safety" refers to containment of the process and protection of personnel from harm. If a hazardous material cannot escape the process; it cannot be breathed in, released to the environment, set on fire, or result in damage to equipment that may cause injury to nearby personnel. The previously described process safety management standard formalized the set of tools used by these industries to ensure current best practices are used when it comes to process safety in hazardous environments. The goal of the PSM standard and implications of it being "performance based" is summarized as follows:

> *"... Recognizing the dangers associated with highly hazardous chemicals, Congress mandated (by the Clean Air Act Amendments of 1990) that OSHA set requirements for the management of highly hazardous substances to prevent and mitigate hazards associated with catastrophic releases of flammable, explosive, reactive, and toxic chemicals that may endanger workers. As a result, OSHA developed the Process Safety Management (PSM) standard (issued in 1992), which covers the manufacturing of explosives and processes involving threshold quantities of flammable liquids and flammable gasses (10,000 lbs.), as well as 137 listed highly hazardous chemicals. The PSM standard (29 CFR 1910.119) requires employers to implement safety programs that identify, evaluate, and control these hazards. Unlike some of OSHA's standards, which prescribe precisely what employers must do to comply, PSM is a "performance-based" standard-it outlines key features of safety programs for controlling highly hazardous chemicals, and employers have the flexibility to tailor their safety programs to the unique conditions at their facilities..." [4]*

The biotechnology industry on the other hand often has a more focused emphasis on product safety. Product safety ensures that all current good manufacturing practices (cGMP) are being followed with the goal of protecting the end-user or patient from hazards of a contaminated biological drug product. Product safety was important for the pharmaceutical industry because the medicines being produced were intended for use in the human body and would chemically alter the bodies function. Product cross contamination could result in serious health side effects, especially with highly potent or specialized substances. Biotechnology processes work by growing viruses and manipulating this viral material to deliver highly specialized medicines to the patient. Poor manufacturing conditions could allow unintended bacterial growth alongside the desired product, leaving the future patient with a potentially deadly virus or bacterial infection instead of a cure. The US Food and Drug Administration (FDA) passed the first cGMP regulations in 1969 [6], making them older and with many respects better understood in the industry than the PSM standard is in the oil and petrochemical industry. The practiced engineer will note several similarities in the quoted statements describing the PSM standard and the cGMP standards.

> *"The CGMP requirements were established to be flexible in order to allow each manufacturer to decide individually how to best implement the necessary controls by using scientifically sound design, processing methods, and testing procedures. The flexibility in these regulations allows companies to use modern technologies and innovative approaches to achieve higher quality through continual improvement. Accordingly, the "C" in CGMP stands for "current," requiring companies to use technologies and systems that are up-to-date in order to comply with the regulations. Systems and equipment that may have been "top-of-the-line" to prevent contamination, mix-ups, and errors 10 or 20 years ago may be less than adequate by today's standards." [7]*

Product safety requires process containment too, but in many cases the consequences are less likely to cause bodily harm so long as product quality control procedures ensure quarantine of the affected drug product. Note that this statement applies to the biotechnology industry, the consequences of a loss of containment in the pharmaceutical industry can be as deadly as in a chemical plant, depending on the substance released. Fifty years ago, when the cGMP's were adopted, pharmaceutical operators already knew about process safety and adopted it as readily as oil refineries and industrial chemical plants did. Twenty years ago, biotechnology was in its infancy and there were few large-scale biological processes in operation. Ten years ago, most biological processes ran at atmospheric pressure using water-based chemistry. The risk associated with potential process safety events was minimal in comparison. Today, many biotech processes use advanced solvent-based chemistries needing large volumes of hazardous flammable solvents, a key element in the severity of an incident when a loss of containment occurs. Add to this unique process chemistries such as dichloroacetic acid, methyl chloride, hydrochloric acid, and sodium hydroxide; all of which can easily corrode stainless steel, a common metal alloy used in biological processing, and unique hazards become a reality even at small scale.

2 The Design Process

Designing processes to produce goods takes time, time that the future producer of those goods doesn't always like to part with because "time is money". Each month spent designing or building the facility to manufacturer a product is a month of revenue missed or often in biotech, a month of trials that could be running. A typical design process includes several independent stages or phases. An ideal project would separate each of these phases with well-defined checks creating structure, an opportunity for detailed reviews, and adding time to the project. Initial concept development, feasibility, and design of large-scale chemical or industrial facilities often involves acres of real estate, spans years, and costs millions of dollars. By contrast small-scale bioprocessing facilities, often less than 40,000 ft^2, are fully designed in terms of weeks or months and leverage parallel work paths where possible.

The impact of such a short design schedule is that time for staggered activities, separated by well-defined stage-gates, is often sacrificed in the name of the project schedule. It is possible to crash a project schedule, completing the design early, but doing so limits the amount of design work that can be completed based on previously completed design elements. If a project schedule allots for 6-8 weeks between establishing the basis of design for a facility and documents that are to be issued for vendor bid, scheduling one week for a hazard analysis and another week to implement the resulting design changes (25% of the schedule) is a challenge. By the time the hazard analysis is complete, many aspects of the design have been progressed and resulting changes will result in unbudgeted rework. The same is true for the design of key safety systems such as relief devices or process interlocks that require a complete system design to be finalized. The engineer can anticipate credible overpressure scenarios and safe operating limits for a process but cannot complete these design elements until piping to and from the relief device is complete and the hazard analysis is approved, accepting the level of protection assigned to each process step. It is for these reasons that it is not uncommon to see relief devices for a process being sized and planned for as one of the final project tasks. Explaining to an architect that a storage tank must be moved to meet a flammable liquid set-back identified during a hazard review or that another roof penetration is required for a relief device tail pipe two weeks before the completion of a design are just two examples of real-world challenges that project teams may be faced with.

Understanding the different design phases or stage-gates, the tasks that occur during them, and their impact on the ultimate facility design is important if one is to create a plan for addressing process safety on such a short schedule. Table 1 summarizes one example of a design process, key stage-gates, and design elements to be considered in each phase. Note that jumping a phase and for example combining schematic design and design development (the most common violation of the design process) into one phase can confuse the separation of tasks between these phases and eliminate checks important to the design process.

Table 1 – Key Aspects of the Design Process

Phase	Process Design	Process Safety
Feasibility Study	• Identify product to be manufactured and major unit operations • Establish block flow diagram of process • Identify facility requirements • Confirm business case for proposed facility	• Identify hazardous materials used in the process • Identify hazardous operating conditions (i.e. high pressure) • Establish acceptable risk criteria for operation of the facility

Phase	Process Design	Process Safety
Concept Design	• Breakdown individual processing steps to the equipment level • Develop process flow diagrams for each major process step • Develop preliminary heat and material balance • Develop equipment list and identify long lead equipment (to prioritize design development tasks) • Detailed layout of process equipment • Identify utility requirements and size utility systems • Size and layout major process headers	• Develop a comprehensive hazardous and flammable materials inventory / determine if additional design precautions are required • Identify potential material compatibility concerns and material specifications for each process step • Complete a code review in collaboration with the full design team to ensure all building, fire, and design codes are being followed • Ensure facility citing requirements are met in the layout
Design Development	• Develop initial P&IDs for the process • Identify and document key control loops and automation requirements • Finalize design of all process equipment (size, pressure rating, operating conditions) • Complete detailed sizing and specification of major process headers and utility lines • Finalize process heat and material balance • Prepare equipment and piping specifications and datasheets	• Confirm inherently safe design where possible (pressure profile versus equipment design) • Identify safe operating limits for each unit operation, add control points as required • Review P&IDs for relief devices and confirm their discharge locations • Complete scenario analysis and preliminary sizing for all relief devices. Based on preliminary sizing, define relief device inlet/outlet piping • Confirm piping specifications versus services • Confirm vessel sizing versus hazardous material inventory and update as required • Complete a preliminary hazard analysis on the full process at the end of the design development stage

Phase	Process Design	Process Safety
Detail Design	• Incorporate all recommendations from the design development preliminary hazard analysis • Finalize all equipment and line sizing • Fully specify all process components including equipment, lines, valves and instruments	• Ensure that all actions from the preliminary hazard analysis are incorporated into the design • Review and finalize relief device sizing • Complete sizing and design of effluent handling system for relief device discharges • Double check the set points of key process variables versus the operation's safe operating limits, previously defined. • Review final equipment specifications against material compatibility matrix and equipment specifications

3 Safety Throughout the Design Process

Process safety should be considered in the earliest stages of design. One principle of process safety design is that inherently safe design, creating a system that physically cannot fail, is the only way to prevent an incident from occurring. The easiest example of an inherently safe system is designing a pressure vessel to withstand maximum system pressure, even if all process controls were to fail. While an easy goal on paper, inherently safe design is not always possible and often never fully achievable. A vessel can be designed to the proper pressure, but an unforeseen corrosion issue could reduce the design pressure making the vessel susceptible to premature failure. In anticipation of this and to protect against it, engineers design-in multiple layers of protection to a process starting with inherently safe design at the concept level and ending with proper emergency response procedures developed during commissioning and simulated during plant operation. Continuing with the vessel example, minimizing the quantity of material stored and having an inventory management plan to support smaller on-hand inventories is another way to reduce hazards associated with the process. Additional design principles or layers of protection used are summarized in Figure 1. Different layers of protection against a process safety event are established during concept development and built-up throughout the design process.

Figure 1. Typical Layers of Protection Applied to a Design

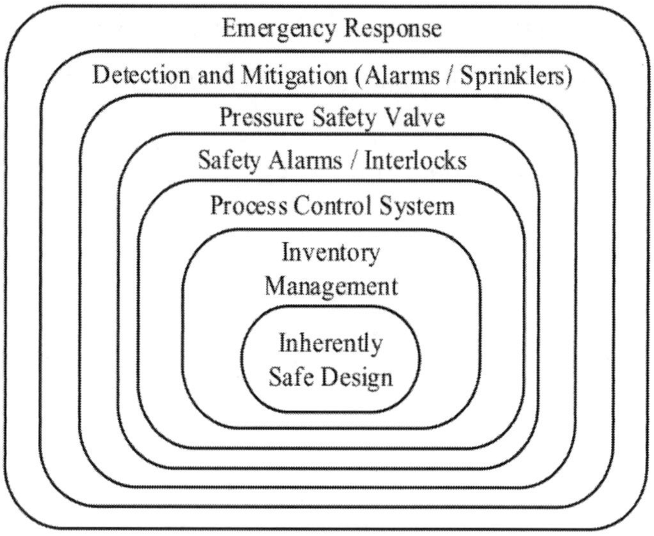

3.1 Concept Design

Concept design is one of the earliest stages of the design process. During concept design types of process equipment are first selected, their location mapped out, and consideration given to how materials will be heated up, cooled down, transferred, and stored when not in use. By asking a few questions as these fundamental design decisions are made, the design predisposes its self to being inherently safer.

- What is the worst-case operating condition, can the system be mechanically designed to handle those conditions?
- Where will hazardous materials be present and can that equipment be installed away from high concentrations of personnel?
- What hazards do the materials pose (toxic, flammable, corrosive)?
- How much of a potentially hazardous raw material must be on-hand? Weekly vs. monthly delivery may reduce stockpiles by 75%.
- Is there a safer way to produce the desired product?
- Is there a safer way to complete this unit operation?

3.2 Detail Design

Concept design develops enough detail to prove that the processing goal is obtainable and business case can be made. Concepts developed such as system location, connectivity, and required unit operations are further developed into a constructible design that can ultimately be operated to produce a finished product. In detailed design, detailed questions are needed as more layers of protection are added. These questions are to be asked informally throughout design and will be formalized in a process hazard analysis (PHA), completed during the detailed design phase.

- Where are the areas of high pressure, temperature, and hazardous material inventories; can they be reduced?
- Are there any specific code requirements or industry best practices for this system?
- What are the expected operating parameters, how can they be exceeded, identified ahead of time, and mitigated?
- Are all systems protected by a mechanical pressure relief device (if needed), do those devices discharge to safe locations?
- How and where could a loss of containment occur, how can it be quickly identified (leak or fire detection)?
- How will the consequences of an event be mitigated once started; how can the event be stopped (process shutdown or fire protection)?
- How can the public, employees, and environment be protected from the consequences of an event if it were to occur (emergency management)?

3.3 Start-Up

Following detailed design and construction, a process must be started, commissioned, qualified, and transitioned to manufacturing. Throughout this process, design assumptions must be confirmed, formal procedures verified, and changes to the design noted so that their impact on the process can be determined. Process safety considerations during start-up must include the full team from engineering to operations and maintenance.

- How does the as-built design compare to the final design, what are the impact of any changes?
- Were changes made to the operating envelope for any of the systems (pressure, temperature, level, concentration)?
- Has a pre-start-up safety review (PSSR) and as-built PHA been completed?
- Do operating procedures match the actual operation of the process?
- Have emergency drills been completed, are there lessons learned from these?

Addressing process safety throughout the design process, versus at the end of design when final design reviews are completed; minimizes late stage design changes, results in a better designed facility, and ensures safe facility operation for years to come.

4 Conclusion

Process safety is a well-established element in the design of oil and petrochemical facilities, taking a front row seat in the design process and being an area of focus once operations begin. Product safety holds a similar role and has been common place for a similar period in pharmaceutical and recently bioprocessing facilities. While the gap between these two elements of design has been closing in recent years as more pharmaceutical companies enter the bioprocessing arena and bioprocessing operations grow, a considerable gap remains. Based on the authors experience designing processes and facilities in both of areas, a few causes for the gaps are easily identifiable:

- Historically, bioprocessing facilities have not contained hazardous processes. Areas of pressure were limited to gas delivery systems (i.e. nitrogen and compressed air) and storage tanks contained aqueous based chemistries (i.e. bioreactors) minimizing the hazards with a loss of containment of these products.
- The shortened design timeline for bioprocess facilities challenges the ability to incorporate process safety into the design at distinct stage gates. The schedule does not allow for a discrete process safety design effort while other design work is on hold.
- There is a general lack of knowledge regarding safety system design in the biotechnology industry. This goes back to the opening statement that in many cases engineers responsible for these systems don't know what they don't know.

Product and patient safety are well defined requirements for biomanufacturing companies and are considered in the earliest stages of the drug discovery process. Applying similar principles to process safety throughout the design lifecycle, and not at distinct stage gates, will ensure a smooth transition from process development to design, validation, and ultimately commercial manufacturing.

5 References

[1] F.M. Editors of Encyclopedia Britannica. "Denis Papin." *Encyclopedia Britannica*, Encyclopedia Britannica, Inc., 18 Aug. 2018, www.britannica.com/biography/Denis-Papin.

[2] Ramsbottom, J. "On an Improved Safety-Valve." *Process Institute of Mechanical Engineers*, vol. 7, 1856, pp. 37–47.

[3] US Department of Labor Final Rule, "Final Rule on Process Safety Management of Highly Hazardous Chemicals; Explosives and Blasting Agents", 1910.109, *Fed Register #: 57.6356*, 24 Fen. 1992.

[4] NASDAQ Biotechnology Index, NBI, https://finance.yahoo.com/quote/%5Enbi/.

[5] *Process Safety Management Introduction*. Occupational Safety and Health Administration, www.osha.gov/dsg/psm/index.html.

[6] US Food and Drug Administration, *Good Manufacturing Practices (GMPs) for the 21st Century - Food Processing*, "GMPs Section One: Current Food Good Manufacturing Practices", 09 Aug. 2004, www.fda.gov/food/guidanceregulation/cgmp/ucm110907.htm.

[7] US Food and Drug Administration, "*Facts About the Current Good Manufacturing Practices (CGMPs)*", 25 Jun. 2018, www.fda.gov/food/guidanceregulation/cgmp/ucm110907.htm.

546706 LOPA and Sil Risk Analysis As Critical Tools for PSM Culture and Accidents Prevention

Industrial facilities that operate with dangerous products, whether chronic or acute, are increasingly common to be present and part of society at large, often allowing an aggregate of increasingly rich and economically beneficial professions and specialties to the community, be it directly or indirectly. Accidents in these types of units, especially those operating with hazardous products of acute consequence (those that have the capacity to cause fatalities when released to the environment), such as flammable, toxic, explosive products, due to the possibility of occurrence of undesirable events and the behavior of the effects resulting from releases of these types of product, could cause serious damage to the people and / or the installations that are in its vicinity. The various types of accident studies are known as EAR (Risk Analysis Studies), some of which are used more widely and others are specific to specific response needs.

According to BERNSTEIN (1997), the word "risk" derives from the ancient Italian risicare, which means "to dare". While DUARTE (2002) describes that the meaning of the word risk is an expression of thought that has a lot to do with the social, scientific and technological evolution of man. As BILLINTON and ALLAN (1992) note, society has great difficulty distinguishing between a hazard, which can be prioritized in terms of its severity, but does not take into account its probability, and risk, which considers not only the dangerous event, but also their probability of occurrence. KAPLAN (1997) explains that when a person asks "What is the risk?", 3 (three) questions are actually being questioned: What can happen? How can this happen? And if this does happen, what are the consequences? The answers define the scenario, frequency and consequence.

Within the agencies that tend to manage corporate risk, it is noticeable that the risk associated with a particular accident scenario is calculated according to its frequency of occurrence and its severity. Frequency is related to the temporal chance of occurrence of the initiating event, which leads to the release of dangerous product, such as LPP (Liquefied Petroleum Gas) emanation by ball rupture through overpressure of this equipment or, as a second example, the intrinsic rupture of a stretch of line. The consequence, also called severity, considering personal safety, is the capacity that the physical effects, due to the release of dangerous product, associated to the analyzed initiating event, can cause fatality (s), related to the potential of damage related to each effect analysis. The risk, qualitatively or quantitatively observed, and defined in a simplified way, is evaluated as the multiplication of frequency of occurrence and severity. The total social risk related to a particular facility, which can be defined as the risk that a community outside the facility is subject to by its presence in its vicinity, can be calculated as the sum of the risks of all the initiating events arising of the same. In order to do this, for each accident scenario, not only the generating causes, known as Initiating Events, but also the likelihood of ignition, the probability of affecting people, and several other parameters are considered, being one of them very important and permissible of analyzes, which are the failures

safeguards. These, if they act effectively when demanded, can prevent the accident from occurring, since the Initiating Event happens.

Hazardous product risk studies, such as qualitative (Hazard Operational Preliminary Analysis (APPP), Hazard Operability (HAZOP, among others) or those known as quantitative analyzes (Vulnerability Analysis and Quantitative Risk Analysis per se), in addition to being increasingly requested by regulatory agencies, have often been seen as an internal requirement of companies, representing one of the pillars of industrial safety management, in order to keep their risks at low or acceptable levels. However, these analyzes are usually performed in stages

until the final definition of the risk, comparison with acceptability criteria, and later mitigation, if necessary, each of these stages having a minimum time required for its execution, which may make the analysis complete rather time-consuming. Situations similar to these are detrimental to Risk Management when, in fact, what is desired is not a complete risk analysis, but only the achievement of a specific response, such as if my identified risk scenario is effective and properly protected.

The extent of possible damage related to each identified accident is delimited by the intensity of the physical effect causing the damage, and the relationship between the intensity of the physical effect and the corresponding damage is established through the vulnerability models.

On the other hand, semiquantitative risk analyzes such The extent of possible damage related to each identified accident is delimited by the intensity of the physical effect causing the damage, and the relationship between the intensity of the physical effect and the corresponding damage is established through the vulnerability models.

On the other hand, semiquantitative risk analyzes such as LOPA (Layer of Protection Analysis) and SIL (Safety Integrity Level), even though they are less demanded by chemical and petrochemical companies than qualitative or quantitative steps can have beneficial and effective results in terms of knowledge of all the characteristics of the analysis scenarios and of their protection barriers and, in a practical way, much faster to be developed. In addition, they can provide results that do not make up the analysis range of qualitative and / or quantitative studies in depth, such as knowledge and prior review of safeguards, many of which are known as Independent Protection Layers (CIP or IPL - Independent Protection Layers).

Therefore, with the anticipation of an increase in the number of future installations operating with hazardous products and the current installations that already do so, the semi-quantitative studies of LOPA represent an enormous value for the observance of the existing protections in each scenario of process accidents, also serving as the basis for the determination of new Independent Protection Layers or even the reduction of their number, tending to simplify the steps of determining the frequency of the scenario., even though they are less demanded by chemical and petrochemical companies than qualitative or quantitative steps can have beneficial and effective results in terms of knowledge of all the characteristics of the analysis scenarios and of their protection barriers and, in a practical way, much faster to be developed. In addition, they can provide results that do not make up the analysis range of qualitative and / or quantitative studies in depth, such as knowledge and prior review of safeguards, many of which are known as Independent Protection Layers (CIP or IPL - Independent Protection Layers).

Therefore, with the anticipation of an increase in the number of future installations operating with hazardous products and the current installations that already do so, the semi-quantitative studies of LOPA represent an enormous value for the observance of the existing protections in each scenario of process accidents, also serving as the basis for the determination of new Independent Protection Layers or even the reduction of their number, tending to simplify the steps of determining the frequency of the scenario.

546375 Controlling Static Discharge Using Antistatic Additives

Controlling Static Discharge using Antistatic Additives.

Abstract

The presentation will discuss controlling static discharge in non-fuel, non-conductive/low-flash point fluids. The scope of the presentation will concentrate on industrial producers, distributors, consumers of these fluids. Furthermore,furthermore, the presentation covers several industries where these fluids are most commonly found. Common Static-Accumulating Flammable Liquids That May Form Ignitable Vapor-Air Mixtures:

- VM&P naphtha
- Cyclohexane
- n-Heptane
- Benzene
- Toluene
- n-Hexane
- Xylene
- Ethyl benzene
- Styrene

Static discharge has caused catastrophic explosions and fires (U.S. Chemical Safety Board, Static Spark Ignites Explosion Inside Flammable Liquid Storage Tank No. 2007-06-I-KS).

Issues

- Nonconductive flammable liquids can accumulate static electricity during transfer and storage.
- Static sparks can readily ignite flammable vapor-air mixtures inside storage tanks.
- Material Safety Data Sheets (MSDSs) often do not adequately communicate hazard data and precautions.

Key findings:

- Safety publications omit antistat options. (see Process Safety Beacon, August, 2016)
- US affiliates of multinational corporations have different guidelines for antistatic use than their parent companies.
    - NFPA®77 and the American Coatings Association's, *Generation and Control of Static Electricity in Coatings Operation* has guidelines that are different than EU regulations ATEX- Appareils destinés à être utilisés en ATmosphères EXplosibles and ESIG's *Flamability: Safe Working With Industrial Solvents.*
    - The presentation will highlight these differences and how they contribute to the disparity in antistatic additive applications.
- Normal Bonding and Grounding May Not Be Enough! Companies that handle, transfer, and store flammable liquids should contact manufacturers to determine if these liquids can accumulate dangerous levels of static electricity, and if they can form explosive vapor-air mixtures inside storage tanks. If so, *extra precautions*—beyond normal bonding and grounding—may be necessary. Antistatic Additive is an extra precaution.

Recommendations: Knowledge members can use

- Who adds the Antistat?
- How much Antistat is added?
- How is the Antistat added?

546677 Led Your Process Safety Culture, Lead Beyond Hierarchies

The Oil & Gas industry is trying to advance at the next crossroads: "How to achieve that leaders without skills and without training in the management of culture, transform their own culture?"

Our research presents three key ideas that leaders from all over the world

can apply immediately to lead risk management with effectiveness:

frequency, consistency, meaning.

When specific leadership competencies to manage risk have not yet been developed,

Leaders can use this three concepts of easy adoption.

Increasing frequency of interactions with people will allow us to generate safer operations. In full attention economy and contexts of increasing complexity, for risk management, more communication is better communication. But it is not enough to improve the frequency of interactions. That frequency must be accompanied by consistency. The challenge is to achieve possible consistency between what we say and what We do. Leading by example is one of the central keys of risk management. The lack of consistency provides symbolic permissions to normalize deviations. We must maintain a high frequency of face-to-face interactions, but in addition, we must be consistent at every opportunity. And finally, to give meaning to our collaborators. It is necessary to ensure the understanding of risks. Each collaborator need to build his or her own sense of vulnerability and Operational Discipline.

We need workers who operate according to procedures and not adjusted to what the bosses say. This is what meaning allows.

548192 The Importance of Using Color Scale in Process Safety

Everyone knows the importance of color. From physics to fashion going through the automotive industry she will always be present. There is no world without color. Often, they come up on the scale form or not, but the variation in the scenarios are always important. It is so on the traffic sign and the ph tape. From the Diamond of Hommel to across the street. Colors, colors and colors. In addition to not necessarily needing ididomas, such as color of the sky and molten metal have a variety of shades. The main idea of the work is to choose a suitable color to be used in communication within a work area, choosing which and how many elements can evaluated suggest a color scale from the chosen one and present its contribution in the process safety.

550372 Flammable Gas Release Modeling for Real-Time Analysis Using Adversarial Variational Bayes

In this study, we have suggested a method to calculate high-accuracy damage range from flammable material in petrochemical complex. There are many flammable and toxic substances in chemical processes. Computational Fluid Dynamics(CFD) is one of the best way to calculate and simulate scenarios of release range of various fluids or gases. However, it is not used to calculate damage range, because of a lot of time to simulate one scenario. Thus, lots of researchers have enthusiastically studied the surrogate model or reduced order model to get the results of release range in real-time [1-3]. Recently, deep learning based generative model has receive attention because of its superior performance of non-linear manifold estimation [4]. Herein, we propose surrogate model based on adversarial varational Bayes (AVB) [5] for real-time analysis of probability of death. The proposed surrogate model can extract the latent space of high-dimensional flammable and toxic substances dispersion contours. Furthermore, it can generate the estimated contour image from variables of interest much faster and more precise than CFD data memorization with linear interpolation because it does not memorize the data but train the non-linear manifold regarding exact posterior distribution on given observation which basis is predefined in standard variational autoencoder (VAE) [4, 6].

The acrylonitrile was studied about risk assessment for health damage like cancer hazard.[7] But the component has flammable and toxicity, so we should calculate range of damage. The main scenario in this study is the acrylonitrile released from underground model, and the result is shown different damage following hole size, mass flow rate, wind velocity, and wind direction. The superior performance of the proposed model was exemplified by comparing with other surrogate models.

[1] Chen, T., Hadinoto, K., Yan, W., & Ma, Y. (2011). Efficient meta-modelling of complex process simulations with time–space-dependent outputs. *Computers & chemical engineering*, *35*(3), 502-509.

[2] Wang, K., Chen, T., Kwa, S. T., Ma, Y., & Lau, R. (2014). Meta-modelling for fast analysis of CFD-simulated vapour cloud dispersion processes. *Computers & Chemical Engineering*, *69*, 89-97.

[3] Long, K. J., Zajaczkowski, F. J., Haupt, S. E., & Peltier, L. J. (2009). Modeling a Hypothetical Chlorine Release on a College Campus. *JCP*, *4*(9), 881-890.

[4] Na, J., Jeon, K., & Lee, W. B. (2018). Toxic gas release modeling for real-time analysis using variational autoencoder with convolutional neural networks. *Chemical Engineering Science*, *181*, 68-78.

[5] Mescheder, L., Nowozin, S., & Geiger, A. (2017, August). Adversarial Variational Bayes: Unifying Variational Autoencoders and Generative Adversarial Networks. In *International Conference on Machine Learning (ICML)* (pp. 2391-2400). PMLR.

[6] Kingma, D. P., & Welling, M. (2013). Auto-encoding variational bayes. *arXiv preprint arXiv:1312.6114*.

[7] Haber, L. T., & Patterson, J. (2005). Report of an independent peer review of an acrylonitrile risk assessment. Hunam & Experimental Toxicology, 24, 487-527.

546615 Modeling and Analysis of Flammable Gas Dispersion and Deflagration from Offshore Platform Blowout

The operational safety of offshore platform is challenged by a series of adverse factors including harsh environmental conditions and process accidents. Blowout is a type of major accident that may result in catastrophic consequences. This paper uses a computational fluid dynamics (CFD) model to simulate the blowout accident on an offshore drilling platform. The simulation includes the dispersion of flammable gas and the resulting deflagration based on dispersion results. Through the modeling, the blowout gas dispersion processes under different wind speeds and directions are analyzed, and a deflagration scenario due to the ignition of flammable gas is predicted in which the impacts of jet fire, overpressure and heat radiation are assessed. The simulations indicate that wind speed and direction have a great impact on the dispersion behavior of blowout gas. During the deflagration, the maximum overpressure is 280 KPa, the jet fire mainly distributes in the derrick, and the maximum radiation flux of thermal radiation field is 2300 W/m^2. Essentially, the outcomes from simulations could help to conduct a prior risk assessment and make an emergency response plan. Additionally, it could give a reference for the inherent safety design of offshore platform to mitigate the unexpected consequences due to blowout accidents.

548682 Application of Inherently Safer Design Tool (*i*-SDT) for Hazards Assessment and Safer Operation

Inherently ***S***afer ***D***esign ***T***ool (*i*-SDT) is a newly developed semi-quantitative metrics for hazards identification and characterization[1]. It is built on property platform, probabilistic analysis of historical accidents information and characteristic equations of safety parameters (i.e., flammability, toxicity, explosiveness, reactivity, etc.) which are developed in-house to assess potential hazards under various operating conditions. Designers can identify dangerous components and vulnerable equipment of the investigated industrial process and can take necessary corrective actions by applying the inherent safety principles[2] (e.g., minimization, substitution, moderation and simplification). The final output of this *i*-SDT tool is a cluster safety parameter score (C_{SP}) which provides insights regarding the development of safe design and operational parameters to manage the hazards using a very limited amount of process information and safety incidents information.

In this work, the application of the proposed *i*-SDT tool is explained for assessing different alternative routes of an industrial process supply chain. The proposed *i*-SDT safety metric identifies the chemical components with highest risk and the major accident-prone operational units. The detail analysis using *i*-SDT tools shows how different operating conditions can affect the safety performance of the given process. The ultimate object is to find the safer route while considering the techno-economic and environmental performance. It provides Pareto curve to compare alternative routes and gives the freedom to the end-users choosing their safer operating route/technology and process conditions. Therefore, safety can be considered as an important selection criteria for choosing best alternative from the beginning.

Final cluster safety parameter scores (C_{SP}) obtained from the analysis enables designers to compare between available technologies and provides information on various hazardous situations. It can also provide in depth knowledge regarding the cause behind the hazardous behavior. The proposed *i*-SDT metrics allows designer to do the safety analysis simultaneously while doing the initial process synthesis which will reduce the future modification cost of any add-in safety features. This work shows how the unique features of *i*-SDT tool can be applied in a real life scenario while doing inherent safer design during early stage of design.

1. Kazi, M.–K, Eljack F. T., & AlNouss, A. Application of a property-based inherent safety quantification framework for integrating risk assessment into process safety life cycle. *AIChE Spring Meeting, Orlando, Florida, USA, 2018.* (156c), p. 2260.

2. Kletz, T. A., & Amyotte, P. (2010). Process Plants: A Handbook for Inherently Safer Design, Second Edition: Taylor & Francis.

546631 An Approach to Understanding Worker Interactions with Procedures Under Normal Working Conditions

Incidents in high-risk industrial settings are often attributed to the way in which procedures are used. Despite the identification of procedures misuse, little research has been conducted on how workers have interacted with procedures in normal conditions. To understand these interactions a behavior assessment technique has been developed which can describe how a worker interacts with a procedure to complete tasks in everyday work situations. interactions under normal conditions. Using the concept of work-as-imagined/work-as-done, the techniques describes the interactions at a macro whole-of-task level as well as an individual step level. At the macros level, assessment considers if the task was completed correctly and safely (Tier 0). At the micro step-level, assessment considers if the step was completed as it was imagined (Tier 1) or if the step was done in an alternate way (Tier 2). For the alternate approaches, assessment considers if the worker struggled or required assistance, completed the steps out of order or with iterations or if the step was not completed as intended.

Two studies, one involving industrial tasks Second Life® (Study 1), the other at Shells Robert, LA training facility (Study 2), were used to test and develop the behavior assessment technique. In Study 1, the behavior of 13 participants from offshore drilling conducting tasks in either day or night conditions was assessed. In Study 2 the behavior assessment technique was used to test if experience accounted for behavior difference in 25 participants from high-risk industries. Results demonstrate the success of applying the assessment techniques through identifying how the behavior of the participants changes according to their experience and familiarity with the tasks the were performing. Further analysis identified recommendations for design and work flow changes of both written procedures and specific equipment. In conclusion, the initial results suggest the assessment technique is an important contribution to the methods use for understanding early interventions to improve safety in the process industry.

PROCESS HAZARD ANALYSIS (PHA) PREPARATION: A ROLE BASED OPTIMIZATION

Susan S. Frizzell, PE
Adley Services, LLC
Lafayette, LA
Susan@AdleyServices.com

Biography:

Susan Frizzell, PE is the President of Adley Services. Shortly after high school, Susan began her career alongside her father on a drilling rig. Upon graduating from the University of Louisiana with a BS in Mechanical Engineering, she obtained her MBA and began working for a firm performing engineering and process safety reviews on refineries and other land facilities in the New Orleans area. After Susan relocated to Lafayette, LA, she began engineering offshore facilities, complimented with process hazard analysis facilitation. During her almost 20-year career, she has worked upstream, midstream, downstream, and all points in between. She is a licensed professional engineer and an executive member of various organizations.

Abstract

A PHA meeting consumes large resources and can place a strain on departmental manpower. Teams are often faced with the want to be more efficient in PHA conflicting with the need to be thorough and accurate. The key to solving this paradox is efficient preparation. By examining the lessons learned from past PHA's, a role by role analysis of best practices with regards to preparation will allow a PHA team to reduce the costly time spent in the meeting and yet still stay true to the methods and deliver a great, safe result.

When it comes to an optimal PHA, preparation is everything. For example, in a 6-member team PHA, 60 minutes of misused time equates to 6 hours of misused manhours. Whether it is managerial, operations, engineering, or even facilitation, there are lessons learned from previous PHA's that when shared, will not only make the PHA more efficient and the facility safer, but the industry as whole better. The goal being that those reading this be able to take back to their facility these practical preparation tips to share with their present and future PHA team members so that the resources of the facility can be better spent.

1. PHA Preparation Optimized

To fully optimize the Process Hazard Analysis (PHA) preparation, the team needs a clear vision of the end product: "To identify, evaluate, and control the hazards" [1, 2] of an operation or process to minimize the likelihood of occurrence and the severity of consequences of release of a dangerous substance [3]. With that in mind, the team can focus time and energy preparing for the PHA on factors within their control to meet that goal. An efficient, cost-effective PHA begins with including the different roles in the preparation. The PHA planning team should be comprised of management, engineering, operations, the scribe, and the facilitator.

An active commitment to reduce risk begins by assigning tasks to each role prior to the PHA. From the start, management should provide support and direction for the PHA team. They set the tone for obligation and importance. Second, engineering involvement in the preparation ensures accurate and complete process safety information (PSI) is available and accessible. Without correct information, the hazards may not be fully identified. Next, operations should walkdown the facility and identify areas of concern. They have the most intimate knowledge of the workings of the process, which is extremely valuable to the PHA when properly communicated to the team for hazard recognition. The scribe should have the technical skills to ensure equipment is functional and to properly and efficiently document the team's findings. Lastly, the facilitator needs to coach the team before and during the PHA to ensure the methodology is followed.

Successful PHA preparation can improve risk identification and limit cost through appropriate management direction, current and detailed documentation, and clear expectations of the timeline leading up to the PHA. We identify ways that companies can assist in preparation in the months leading up to a PHA and the benefits that preparation has on hazard recognition, team morale, and, most importantly, the bottom line.

2. Management – *"Set the Tone"*

It is important for management to set the tone of the PHA. More specifically, management needs to set the tone before, during, and after. In a PHA, the attitude, the posture, and the dedication of the team is usually a direct reflection of management's attitude, posture, and dedication. In our experience of both facilitating and participating in PHAs, we've seen attitudes of management in a variety of flavors. The following are observations we've made in the past regarding a successful PHA and one not as successful.

The corporate culture and support infrastructure set by management begins well before the first day of the PHA. A typical PHA is done every 5 years [1]. Given that schedule, it is imperative that management starts out on the right foot and begin planning well in advance. A management component that is very serious and very organized will typically induce an attitude of seriousness and organization to the team members. For example, a management team that waits until the last moment and rushes preparation will typically have PHA members that delay and do little preparation, if any at all, for the PHA. PHAs are very expensive and very time consuming. Lack of adequate and efficient preparation will be very damaging to an efficient use of PHA budget and resources. When management in advance sets the tone that a PHA needs to be done correctly and efficiently, the PHA team members tend to take preparation seriously. Just as it is well documented that an effective safety culture of a company starts at the top with management, effective preparation for a PHA also starts at the top with management [4]. The management teams we've seen in the past that do this most effectively set the tone of preparing and doing a PHA as part of your job, not as additional work on top of your daily duties. In other words, management sets the tone that PHAs are in your job description and that we are all PHA team members. For example, a PHA is not additional work given to engineers, it is in an engineer's job description.

The other things we have observed that effective management does by setting the tone is setting up a support team for the PHA team members. These PHA team members will have to transition out of their daily duties, be present at the PHA, and transition back into their regular duties. Often, their project managers and/or functional managers may give some resistance to that. It is imperative for upper management to set the tone with the managers of the team members that they need to have a plan for them to transition into the PHA and transition out of it with regards to their regular duties. Most of these facilities run around the clock all year long and the work never stops. Therefore, it is imperative for management to have a plan for these employees to be present at the PHA while it is going on. A PHA which is constantly stopped for employees dealing with issues outside of the PHA is an unacceptable waste of resources. For example, if a PHA consists of ten team members and one team member must stop the PHA to deal with outside problems unrelated to the PHA, the nine other team members are wasting company resources. When management sets this tone early on that we are dealing with real human lives and significant company resources, the employees and their managers tend to put greater weight on a PHA. And sometimes, the employees just need the help of upper management to persuade lower and mid-level management that their presence, uninterrupted, is necessary to conduct a PHA.

During the PHA, management's role is typically only at the very beginning. It is very effective during the opening remarks of a PHA for a member of management to also make an opening remark. During this address, management will set the tone on several things. It is not uncommon for management to begin opening remarks with a thank you for the team members for their hard work in preparing and committing of their time. Some members of management stress employees' safety; others may stress the importance of safety to the community or environment. Opening remarks is a great time for management to state known worst-case consequences initiated within the limits of the PHA scope or define the assumptions regarding severity. This guides the team towards consistent company-wide risk rankings and avoids the team paranoia over risk ranking too high or too low. It's not management's position to force upon the team a subjective belief; it is management's position to reaffirm the team uses best judgement and not be hesitant of ranking the risks as they believe they should be. Finally, management can give the PHA team members feedback from previous PHAs as to the wording, substance, and ranking of the recommendations to assist with review and completion post-PHA.

Once the PHA begins, management's role is very limited. We've seen effective managers from time to time pop-in, review recommendations, and discuss with team members their thoughts on such. Effective managers will pick a recommendation, look at the supporting data, and ask specific questions of how the team arrived at its conclusion. This serves the purpose of correcting any doubts that the PHA team may be having and serves the purpose of reaffirming to the PHA team what they are doing is correct. Management can also review recommendations with only the facilitator at specific progress intervals of the PHA. The facilitator can then bring suggestions back to the team without the direct watchful eye of management. This minimizes the intimidation factor but gives the team the opportunity to clarify wording or details.

We find in PHAs that the PHA team members want to be very cooperative with management. That is why it's imperative for management to not only set the expectations and tone, but also provide a supportive means and network for a high-quality PHA to be conducted. By management providing a support team and stressing the importance of a PHA during the preparation phase, the PHA team members are motivated to attend the PHA prepared with their full attention. It goes back to the old saying, garbage in, garbage out. Anyone who has sat in a PHA will attest to the opinion that PHAs are grueling and a marathon of being attentive. The PHA team members tend to be very technically sound and understand the PHA process and goal. The biggest hurdle is usually morale and minimizing outside distractions. A great opening statement by management and periodic check-ins to keep morale and encourage the team members goes a long way. Nothing lifts the moral of battle wary troops like a visit to the battlefield by the general. After the PHA is over, feedback and encouragement go a long way in instilling confidence for the PHA team members that will carry over to their next PHA. Although hardly any of this is required by the statutes, these are observations we have noticed with regards to effective management participation in PHAs.

3. Engineering – "The Devil is in the Details"

As an engineer, it is understood that your job is to ask the questions that have not been answered, review the material that has not been presented, and observe that which is not obvious. PHAs are no different. One of the duties of the engineer during a PHA is to gather all the process safety information (PSI). However, we've observed the more effective engineers will take this list, check how current it is, and review the design pressures and temperatures, such as pump discharge pressures. This review helps determine if the PSI is accurate. It also serves the dual purpose of observing any abnormalities that will need to be discussed during the PHA. Looking at this before a PHA is very effective when the engineer's eyes are fresh, and his mind is clear. Once the PHA is ongoing, this task becomes a little more complicated with lower morale and PHA fatigue. This holds true to the majority of the engineer's tasks.

Drawings, drawings, drawings. One of the largest time and money wasters we observe in PHAs is showing up to a PHA with either outdated drawings or needing drawings that are missing. It is imperative for the effective engineer to not only review the drawings to make sure they are up to date and accurately reflect what is present in the field, but it is equally important to review which drawings have been gathered and submitted to ensure that there are no surprises during the PHA. Anyone who has sat in a PHA and heard the words "these are not the most up to date drawings" or "where are all the other drawings at", will attest that it is going to be a long day. The PHA facilitator, the scribe, and pretty much everyone else in the room relies on the drawings being both accurate and complete in their preparation. If the drawings submitted for the PHA are either outdated, incomplete, or just wrong; it will have a ripple effect for the rest of the team both before and during the PHA. If the reader of this paper takes nothing away but one thing, please note that accurate and complete drawings are imperative for an effective PHA team.

PHA team members include those who are present in the room full time and other team members who may be called upon as needed. Effective engineers anticipate ahead of time what possible other resources will be needed during the PHA. For example, engineers for adjacent facilities need to be made aware of an upcoming PHA and given a heads-up as to be ready to answer questions regarding upset conditions and their effects. Likewise, engineers for skid packages or other equipment requiring expertise of vendor engineers need to also be made aware of an upcoming PHA and request their availability if called upon. Similarly, engineers of other disciplines may have better knowledge of failure mechanisms or a specific design review already conducted which is beneficial to the PHA team. It is not uncommon in a PHA to have a question that needs to be answered by a vendor engineer, subject matter experts, or an adjacent facility engineer. Examples of information needed include possible flow rates or pressure to/from receiving/feeding units, adequacy of safe guards in packages provided by vendors, or electrical area classifications. Oftentimes, a simple phone call to these engineers asking if they would be available by phone during the PHA is enough. Other times, effective engineers will require that other personnel be available to come in as requested. PHAs involve a lot of moving parts with each part having its own calendar or schedule. Giving the other possibly necessary personnel a heads up will ensure that should their presence become necessary, the PHA team will not have to recess until they are available.

Effective engineers will typically review MOCs and make phone calls or e-mails to update their current status. The adage that change is constant is very applicable to MOCs. MOCs may have been implemented, postponed, in progress, or completed. It is important to update the status of an MOC. One that has been canceled may require different attention than an MOC that is in progress or completed.

Effective engineers provide the technical expertise of the specific process under review [1]. Before a PHA, effective engineers will take a closer look to evaluate operating trends and upset conditions looking for a pattern that is not readably apparent. Engineers should discuss the facility and how it runs day-to-day with operations prior to the PHA. They should review procedures to ensure they are current and understood. It can be an awkward moment during a PHA when the engineer says, "we've never had a problem with that", and the operator rebuts with "we have a problem with that every day". Engineers know that engineered controls are the strongest defense against hazards. They look for ways to solve problems without relying on administrative controls.

Finally, effective engineers are present during a PHA. This does not mean they are sitting in a chair, it means they are focused, engaged, and completely dedicated to the PHA while they are in that room. Engineering is a life of dealing with problems. To be an effective PHA team member, it is important to make sure the engineer has support to deal with problems while he/she is in the PHA. Effective engineers can have a clear mind because they know that if normal, everyday problems occur, someone else is taking care of them. An engineer that is constantly worrying about what great surprises are awaiting them when they check their email at the end of the day, are not present during a PHA.

4. Operations – "The Ambassador of the field"

The operator is the ambassador of the conditions in the field. The operator, not engineering, management, or the facilitator, is the eyes and ears of the field. As an ambassador, it is important that the operator speak to his constituents (i.e. opposite crews), learn how to effectively communicate, and be present for the meeting of the PHA team. Effective operators speak to their coworkers and get a history of problems, concerns, or other issues they may be having that might not be well documented. Effective operators also walk around and observe conditions of the facility and document such. Documentation being key. Effective operators also review previous incidents not only in their facility, but other similar facilities or neighboring facilities to bring up any concerns that may not be readily apparent when reviewing documentation during a PHA. A great ambassador becomes the eyes and ears of his constituents by listening to what the other operators have to say and physically observing the condition of the facility.

Other things effective operators will do is review operating procedures and note deficiencies. Sometimes operating procedures are written in a way that no one understands them making them ineffective as a safeguard. Other times, there is no written procedure, forcing operations to react to conditions "the way we always do it." If operators are not following the procedures because they are vague, ambiguous, confusing, or out of date, it is up to the operator during the PHA to bring that to the attention of the team members. Frankly, effective operators understand that management and engineers assume that everything they write is perfect, clear, and easy to follow. It is up to the operator to explain why that may not always be the case. Some of the most effective operators we have encountered in a PHA not only describe what the problems are but are very specific about why these are problems.

The next characteristic of the operator as an ambassador for the field conditions is communication. For an operator to communicate with a PHA team, they must speak the same language. Effective operators know that to communicate with a PHA team, they must become a master at reading P&IDs. Being able to understand how the elements of a facility and its attributes are represented on a P&ID are extremely important for an operator. Effective operators can look at a P&ID with little or no guidance and interpret it to see if it is an accurate reflection of what is going on in the field. When the engineer asks a question regarding the P&ID, effective operators are on the same page. If the reader wanted to take away one important thing from this section on operators, please take away that an operator being a master of a P&ID is extremely important to be an effective operator in a PHA.

5. Scribe – "The Show Must Go On"

In American football, a great offensive lineman only receives attention when things do not go as planned. The same can be said about a scribe in a PHA. An effective scribe will give the outward appearance of very little responsibility or having very few complicated tasks. Nothing can be further from the truth. With the goal of quickly and accurately recording information [4], effective scribes understand that a lot of thought and preparation must go into making sure that nothing goes wrong, and if something does go wrong, the PHA suffers minimal interruption. Also, effective scribes are great at being very efficient in the tasks they complete.

At the end of the day, a scribe's responsibility is to ensure that the audio/video equipment works as planned. Although some facilities may be tasked with providing projectors and so forth, the scribe cannot scribe unless all this equipment works. Therefore, an effective scribe will oversee the equipment being used to make sure the parties who are responsible for providing it are communicating and that all the items are compatible. The scribe will ensure that both the scribe's computer and the display have the proper connectors and cables needed. An effective scribe not only brings the connectors and cables needed, but also bring the connectors and cables that were not anticipated as being necessary. An example is the conference room being used for the PHA has HDMI and ethernet and Wi-Fi...or at least that what was told to the PHA facilitator. In the past, we have shown up to conference rooms to find that the HDMI port is in fact a VGA port, or that the HDMI cable is three feet long and the scribe's position is ten feet away. Very effective scribes keep with them a toolbox of supplies such as VGA to HDMI converters, extra batteries for mice and Bluetooth keyboards, and extra cables. Sometimes, the right connectors and cables are present but are malfunctioning. The truly great, effective scribes seamlessly work through these problems and the show continues without interruption. Along those lines, a great practice for scribes to have is to not only back up the PHA file regularly, but also have a written procedure that has been reviewed regarding the backup. We always joke that we PHA the PHA. In other words, what are the risks for file corruption or data loss. Which risk are we willing to accept, and which risk are we going to mitigate? Knock on wood, so far, it's been very effective.

A lesser problem, but a problem none the less, is having to wait on a scribe to catch up. Although not as catastrophic as the PHA having to be stopped completely while technical issues are worked out, a scribe who is not able to keep up with the PHA team stifles the free flow of thought and can leave the door open for other interruptions such as taking phone calls and checking email. Very effective scribes will have experience and practice with the software, know the shortcut keys, and have a vocabulary list prepared for technical terms and names. A typical vocabulary list will be something as simple as a word document open where the scribe has already checked the spelling of names and technical equipment and as the scribe goes along, they can copy and paste the names and equipment labels very quickly without having to correct for spelling. A truly exceptional scribe will have already created templates with standard verbiage for safeguards, recommendations, and comments. For example, when the team decides to make a recommendation, the scribe can copy and paste it into the software and the team can edit the few words that need to be changed to modify the recommendation for the specific application. What this does is it allows the team to make a recommendation and move on to the next node without getting bogged down in the weeds of language choice. We find it is always easier to give a group of people a starting point for a recommendation as opposed to generating the language of a recommendation by consensus of the group. Along those lines, effective scribes review the drawings, MOCs, and other information to become familiar with the vocabulary in the drawings and have an inventory of what PSI they have and what node it corresponds to. Therefore, if a team member erroneously refers to a drawing that does not exist or refers to a drawing or MOC by incorrect name, the scribe will be quick to point out that the item does not exist in the inventory.

Effective scribes will also begin opening vital documents for a specific node as soon as the review of that node begins. Therefore, when the team would like to reference a specific document such as a drawing, MOC, or incident report, the document will be readily available. A common, effective means of doing this is having a second monitor only seen by the scribe, in addition to the screen shared with the projector. The scribe can work on his/her monitor opening documents or correcting language. As the team is ready to view specific information, he/she can drag the window over to the extended screen that is being displayed overhead. This allows the scribe to work without being micromanaged by the team and minimizes the distraction of the team from the scribe flipping through windows or searching for files. Again, the overall theme for an effective scribe is to keep the free flow of thoughts and the team moving. Anytime the team must pause or take a break because of the scribe performing its duties, the team's momentum and productivity can be lost.

Finally, truly effective scribes are great at covering the intangibles. This can include filling all the supplies or doing things to the environment in the room to lift morale. For example, effective scribes will typically have a box of random office supplies, whether it be pens, hi-liters, or post-it notes. At the beginning of each day and after each break, he/she should refill the table or the workspace of the PHA team with extras. So, if a team member would require a specific color pen, they would simply grab it and not have to pause the PHA. This gets back to the overall theme that an effective scribe keeps the evaluation moving forward without interruption. Another intangible that we often see effective scribes will do is they will place motivational posters around the room, such as large check-off list of nodes to visually see progress, to hang inspirational quotes, and so forth. PHAs are really an endurance test of morale and motivation and these little intangibles that the scribes do are of very limited cost, but of great benefit. In all honesty, we often do not realize how great and effective our scribe is until we do a PHA with an ineffective scribe. When you do a PHA with an ineffective scribe, you feel like you're stuck in the mud and going nowhere fast. Although realistically, the delay may add just a few minutes a day, with regards to morale and motivation, it feels like an eternity.

6. Facilitator – "What's Your Plan, Coach?"

Like a good coach, an effective facilitator enables great people to do great things. Effective facilitators setup the team members to be successful and, during the PHA, guide them to produce a high-quality product. Although a facilitator's duty is to ensure the methods are adhered to and the regulations concerning a PHA are complied with [1], a great PHA facilitator is so much more.

A great PHA facilitator, like a coach, ensures that the team is prepared come game day. Effective PHA facilitators will do such things as set deadlines for collection and development of information and suggest further actions to better prepare for game day. Not only do they set the deadlines, an effective facilitator will check in with the team members prior to those deadlines to ensure that progress is proceeding as planned and that the deadline will likely be met. Effective facilitators are great at helping team members overcome obstacles and hurdles in collecting this information. A team being well prepared on game day when they take their chairs and listen to management's opening remarks is no accident. Also, during the preparation, an effective PHA facilitator will ask the hard questions of the team members, such as: Do you have a transition plan for someone to fulfill your role during the PHA? Or, are there any allergies, physical, or dietary restrictions a team member may have so that these can be addressed well ahead of the PHA? To be an effective PHA facilitator, it really comes down to having the team prepare for the work they anticipate doing and addressing as many possible obstacles as one can prior to the PHA commencing. The litmus test for a PHA facilitator is simple - once the PHA commences, no surprises.

In addition to ensuring the rest of the team is preparing for the PHA, the effective facilitator puts in just as much effort as the rest of the team into preparing the PHA file, determining the node segregation, and sharpening his skills as a technical guide. On the day of the PHA, the PHA file should be prepopulated with node descriptions, drawings, common safeguards, etc. The facilitator should have reviewed the node list and noded drawings with other member(s) of the team to ensure the segregations make sense and all parts of the process are included. The effective facilitator should also review recent changes to regulations or process conditions that might affect the hazard identification. He should gain a basic understanding of the process under review along with the potential hazards specific to the facility. Effective facilitators have a list of questions to pose to the team during the appropriate node to ensure all hazards have been addressed.

Again, following the theme of the PHA facilitator being a coach, once the PHA commences, it is game day. It is the PHA facilitator's job to adhere to the methods and regulations regarding PHA facilitation. However, great, effective PHA facilitators understand that the true challenge in a PHA is morale, motivation, and focus. Effective PHA facilitators will typically start off the PHA with a "rules of the road" presentation. They do not assume that everyone in a room regularly participates in PHAs and is familiar with the process. They review the PHA process, the end goal of risk mitigation, the timeline, the daily schedule, acceptable and unacceptable behaviors, and expectations of participation. An effective PHA facilitator will go over these rules of the road and get each team member to commit to their role, to fully participate, and to strive towards the goal. It creates a sense of responsibility and obligation among the members. Once the PHA commences, the effective PHA facilitator will have a sense of when progress is slowing down and when to step in to move the team along. The effective PHA facilitator knows that the team must be given opportunity to discuss causes and consequences, but also has goals and objectives that need to be met within the time allotted.

As the team leader, the facilitator should plan on how to react to differences among team members. The great PHA facilitators have a way of getting conflicts resolved without upsetting or demeaning anyone in the room. Typically, in an argument, somebody must be wrong and somebody must be right. However, the great PHA facilitators have a way of allowing both parties to express their beliefs, guiding them to understand the beliefs of others, and concluding with a common solution. A great way to kill the free flow of ideas in a PHA and stifle productivity is to tell someone, "you're wrong, please be quiet". Great PHA facilitators know how to do that in a way that allows the person to maintain dignity and continue to perform in a PHA.

Another attribute of great PHA facilitators is that they are systematic and thorough. They have a plan of attack for each node. This usually includes the node description and limits review, a review of the appropriate drawings, written list of deviations to encourage the team, and review of MOCs, previous recommendations, and incidents related to the node. The effective facilitator also has a plan in place to limit idle time when critical information necessary to properly review a node is found to be missing. The action is documented as a follow-up for a specific team member to complete prior to the next day allowing the PHA to move forward. Any outstanding to-dos are documented and presented to the team members at the end of each day prior to dismissal. The PHA facilitator is responsible for meticulously documenting the team's findings, which cannot be done without adequate preparation of a game plan.

Finally, great PHA facilitators understand they have an obligation to future PHA facilitator to send their team off with a sense of accomplishment and pride in PHA participation. If the team members of the PHA leave with a bad taste in their mouth or feel discouraged, their participation in the next PHA may not be to the level it could be. Great PHA facilitators compliment their team members and make sure they leave the PHA on a positive note. Therefore, when the team members approach their next PHA, the new facilitator will not be dealing with baggage leftover from the previous PHA. It is a courtesy that great facilitators give to and expect from other facilitators. PHA facilitators are often asked by team members during a PHA, "how can you do this every day?" The response we often use is, "what can we do better to make you volunteer for the next PHA?" Soliciting feedback from team members besides a multiple-choice survey is imperative to be an effective facilitator. The feedback needs to be of substance that can be acted upon and improved upon. We truly love doing PHAs and at the end of every PHA, we express that to the team and solicit their feedback to what we can do better to make them truly love doing it too.

7. Summary

In conclusion, when it comes to an optimal PHA, preparation is everything. Each team member plays an important and specific role. Management sets the tone with their level of commitment to preparation, resources, and action. Engineering ensures the quality of information available to the team and contributes a deep understanding of the process design. Operations represents the field conditions, the procedures, and the most intimate knowledge of the process. The scribe provides accurate documentation through technical skill, proficiency with hardware and software, and team morale support. The facilitator leads the team to identify, evaluate, and control the hazards utilizing a systematic approach to preparation. It has been proven that a small amount of time spent preparing for each role will result in improved hazard identification, higher team morale, and reduced PHA cost.

8. References

1. Occupational Safety and Health Standards, Toxic and Hazardous Substances, Hazard Communication, Available at https://www.osha.gov/laws-regs/regulations/standardnumber/1910/1910.1200 Accessed on December 12, 2018.
2. Bureau of Safety and Environmental Enforcement, Department of the Interior, Offshore, Oil and Gas and Sulphur Operations in the Outer Continental Shelf, Available at https://www.ecfr.gov/cgi-bin/text-idx?tpl=/ecfrbrowse/Title30/30cfr250_main_02.tpl Accessed on December 12, 2018.
3. American Petroleum Institute Recommended Practice 750: Management of Process Hazards, First Edition, American Petroleum Institute, Washington D.C., 1995.
4. CCPS. *Guidelines for Hazard Evaluation Procedures*. Center for Chemical Process Safety, American Institute of Chemical Engineers, New York, NY, 2008.

551204 Building Process Safety Intelligence: What Should Every Graduating Chemical Engineer Know?

Process safety intelligence is derived from derived from a vast corpus of sources - including textbooks, handbooks, journals, conferences, incident reports, safety datasheets, and available materials and thermophysical property data. Dedicated decision-support tools enabling practitioners to harness this intelligence in their daily work, such as Knovel are an important component of chemical engineering practice today. However, many students don't realize the breadth and depth of technical knowledge needed to support process design and safety – they tend to overrely on non-specialist search, and miss pedigreed information available on specialized knowledge platforms, such as Knovel.

The undergraduate engineering space offers an excellent opportunity to plant the foundational information seeking behaviors in future process safety professionals and leaders. Knovel focuses on applied a preview of the types of insight they will be forced to leverage in industry careers, without a direct cost to students and faculty.

Common Process Safety Management Gaps Based on Occupational Safety and Health Administration Process Safety Management of Highly Hazardous Chemicals Violation Trends

Janet Persechino
Engineering Planning and Management, Inc.
959 Concord Street
Framingham, MA 01701
jap@epm-inc.com

Ryan Welch
Engineering Planning and Management, Inc.
959 Concord Street
Framingham, MA 01701
rdw@epm-inc.com

Prepared for Presentation at
American Institute of Chemical Engineers
2019 Spring Meeting and 15th Global Congress on Process Safety
New Orleans, LA
March 31-April 3, 2019

Janet Persechino
Engineering Planning and Management, Inc.
959 Concord Street
Framingham, MA 01701
jap@epm-inc.com
Keywords: OSHA, violations, regulations, process safety

Abstract
On July 17, 1990 the U.S. Occupational Safety and Health Administration (OSHA) published in the Federal Register the proposed standard *Process Safety Management (PSM) of Highly Hazardous Chemicals (HHCs)*. Approximately four months later it was enacted into law via Clean Air Act Amendments (CAAA). These amendments require that the U.S. Secretary of Labor, in coordination with the U.S. Administrator of the Environmental Protection Agency, promulgate, pursuant to the Occupational Safety and Health Act of 1970, a chemical process safety standard to prevent accidental releases of chemicals that could pose a threat to employees [1]. The full text of the final standard was issued in the Federal Register on February 24, 1992.

The OSHA standard is listed under Title 29 (Department of Labor) of the Code of Federal Regulations (CFR) as 1910.119. This standard leans towards a performance-based guidance document that outlines methods and programs to decrease the likelihood of flammable, toxic, reactive and explosive chemical releases. The intent is to reduce the risk to employees, visitors and contractors that may enter or work at chemical processing facilities.

Facilities are subject to the PSM regulation if they hold a threshold quantity of any HHC that is listed in Appendix A of the standard, with some exceptions. The violations that OSHA issues for the 1910.119 standard will cite the specific element and its relevant sub-section. Published data allows for review of the violations, and based on that information an indication of common PSM program gaps can be assessed.

A ten-year span of OSHA PSM violation data was examined (2007-2017). The data shows the top four OSHA cited PSM sections (often referred to as elements) are (highest to lowest):

1) Process Safety Information
2) Mechanical Integrity
3) Operating Procedures
4) Process Hazard Analysis

Each of these elements have known solutions, and specific corrective actions can be easily implemented given the lessons learned, the current level of expertise that exists in the process safety industry and the cost-effective software tools that are now available.

Regulatory compliance is not the only goal of adequate PSM programs, nor is it the goal of this review. The purpose of reviewing the OSHA violation information is not meant to help facilities avoid OSHA citations, but to uncover gaps found by OSHA and bring awareness and attention to some common issues, hopefully leading to focused solutions for safer, more reliable facilities.

1.0 Introduction

There are approximately 24,000 sites that are subject to the OSHA PSM standard nationwide [2]. Facilities are subject to the standard if they process any of the 130+ chemicals listed in Appendix A (of the standard) in excess of the associated threshold quantities, also listed in Appendix A, on-site in one location. They are also subject to the standard if they process flammable liquids and/or gases in quantities equal to or greater than 10,000 pounds. OSHA defines "on site in one location" as areas under the control of an employer, or affiliated employers, that have control over any group of vessels that are interconnected, or in separate vessels that are located in close enough proximity that a HHC could be involved in a catastrophic release.

Exceptions to the applicability include [3]:

- ✓ Retail facilities
- ✓ Oil/gas well drilling or servicing operations
- ✓ Normally unoccupied remote facilities; the standard defines "remote" as a facility that is operated, maintained or serviced by employees who visit the facility periodically to check its operation, no permanent employees are stationed at the facility, and is geographically remote from all other buildings and persons.
- ✓ Hydrocarbon fuel used solely for workplace consumption as a fuel if such fuels are not part of a process containing another HHC covered by the standard
- ✓ Flammable liquids with a flashpoint below 100°F stored in atmospheric storage tanks or transferred, which are kept below their normal boiling point without benefit of chilling or refrigeration

The above information outlines specific threshold quantities that trigger, and exclude, the PSM regulatory requirement. It is strongly recommended that some form of PSM be practiced at plants that may not be subject to PSM, especially those that fall into the following categories:

- Plants that store quantities of HHCs close to the threshold quantities listed in Appendix A of the standard
- Plants that utilize hazardous chemicals not listed in Appendix A of the standard
- Processes exempt from the standard, for example, those listed as exceptions in the standard, plants that have runaway reaction potential, etc.

An overview of "PSM Lite", and related fire code requirements, can be found in these conference proceedings in the paper titled *Fire Code Chemical Process Safety Requirements*. Implementation of an abbreviated form of PSM should be evaluated on a case-by-case basis which will take into account various factors that would clarify the actual risk and the residual risk that would remain after programs are incorporated into the operations.

2.0 Violation and Corrective Action Evaluation

The PSM elements are listed in the 1910.119 standard in the following order/sections:

(c) Employee Participation
(d) Process Safety Information
(e) Process Hazard Analysis
(f) Operating Procedures
(g) Training
(h) Contractors
(i) Pre-startup Safety Review
(j) Mechanical Integrity
(k) Hot Work Permit
(l) Management of Change
(m) Incident Investigation
(n) Emergency Planning and Response
(o) Compliance Audits
(p) Trade Secrets

OSHA has been administering violations for the 1910.119 standard since 1992. For purposes of relevancy, the last 10 years of violation data was examined (2007-2017). These violations were issued when OSHA completed an inspection and/or accident investigation visit to a particular facility.

Each one of the elements listed above has subsections pertaining to the requirements for that element. For example, when OSHA administers a PSM violation during a visit it is listed as: 1910119 E03 I. The nomenclature correlates to a violation for section (e) Process Hazard Analysis (PHA), part 3(i), or (3) the process hazard analysis shall address (i) the hazards of the process. This particular example indicates that the PHA did not properly address any or all of the hazards of the process. The OSHA inspection information also includes other data such as penalty dollar amounts, dates and type of inspection. The information provided in this summary reviews a few segments of this data from years 2007-2017 [4].

3.0 Violation Data Interpretation

All OSHA PSM violations were tallied over the 10-year period (2007-2017), this included:

- ✓ All citation classifications types (serious, repeat, willful etc.)
- ✓ All types of inspections/visits (accident, planned etc.)
- ✓ All types of facilities

This review does not distinguish whether violations were given during the same visit or during separate visits.

The data shows that there are 4 elements, or 4 deficiencies, that clearly stand out as the most prevalent (see Figure 1):

Most Commonly Cited PSM Elements (highest to lowest):

1) Process Safety Information (PSI)
2) Mechanical Integrity (MI)
3) Operating Procedures (OP)
4) Process Hazard Analysis (PHA)

Total Monetary Amount (highest to lowest):

1) Mechanical Integrity
2) Process Safety Information
3) Operating Procedures
4) Process Hazard Analysis

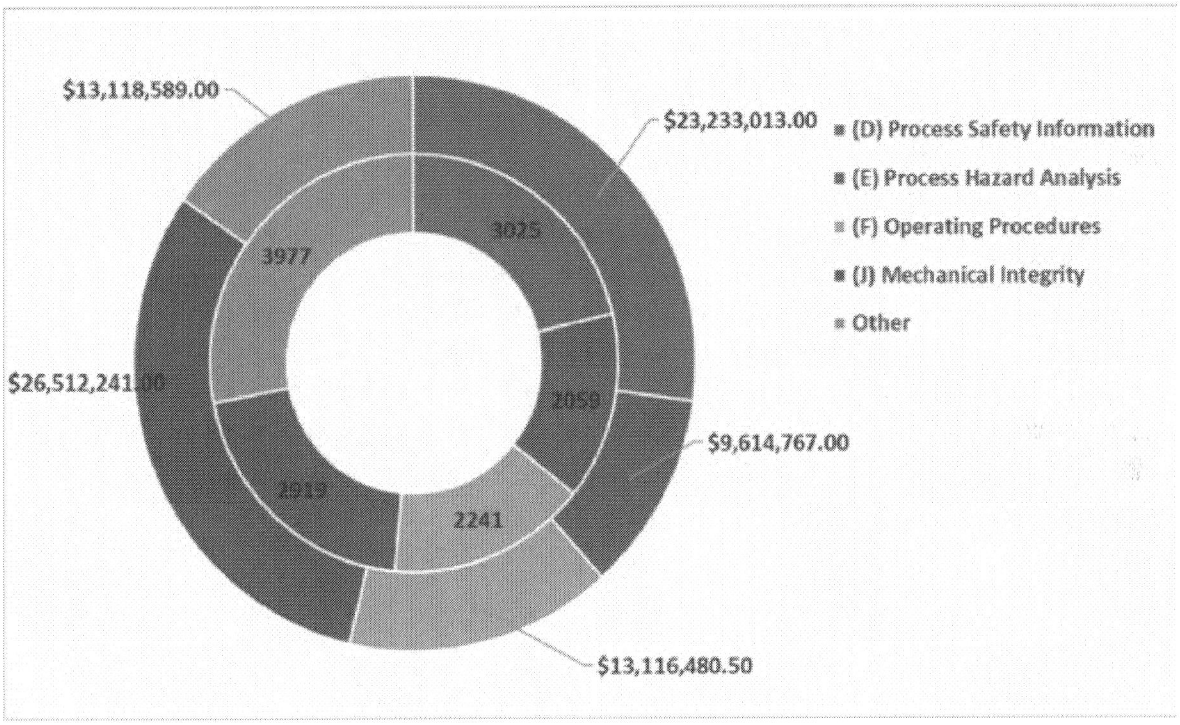

Figure 1: Total Number of OSHA PSM Violations and Total Citation Dollar Amounts by Element: 2007-2017

The top two elements (PSI and MI) have close to the same number of violations and the same goes for the third and fourth most common violations, giving the impression that the struggle to execute PSI and MI is about equal and OP and PHA both come in as close second.

It is alarming that PSI is the most cited element. Having correct and adequate PSI is the foundation for implementing, not only all PSM elements correctly, but any and all chemical safety precautions. PSI is crucial in completing an adequate PHA, the analysis that examines the hazards of the process and ensures that the proper safeguards are in place. The lack of PSI could lead to process hazards being missed, or incorrectly evaluated, during a PHA. Incomplete PSI can also lead to having inappropriate MI programs since some of these programs can be based on the hazards of the process, for example, more frequent inspections than the manufacturer's recommended practices based on the characteristics of the material being processed.

MI is the second most cited element but has the largest monetary amount, this could point to more instances of MI being noted within a single violation. OSHA will increase a fine for a single violation based on the number of "instances", or how many times they see a violation for the same element during the same visit. This review did not take into account the number of instances per violation.

MI can be difficult to implement, but new software solutions and training programs exist to help with proper execution. There are MI software products on the market that include capabilities to capture PSI, PHA and OPs, addressing all of the top-cited elements within one program.

It is helpful to utilize PSM experts for implementation and execution of PSM programs, as they are aware of the depth of information that is needed to meet the intent of the standard and for overall safety. They can often recommend products depending on your particular operation. Individuals dedicated to PSM will have access to resources that might not be available to others who are not embedded in the PSM industry. For instance, access to fee-based databases with previous incident data, one item that must be documented as part of a PHA. The expenditure for bringing in this type of expertise can provide a beneficial return on the investment when examining the average cost per violation and average number of violations per visit (Figures 2 and 3).

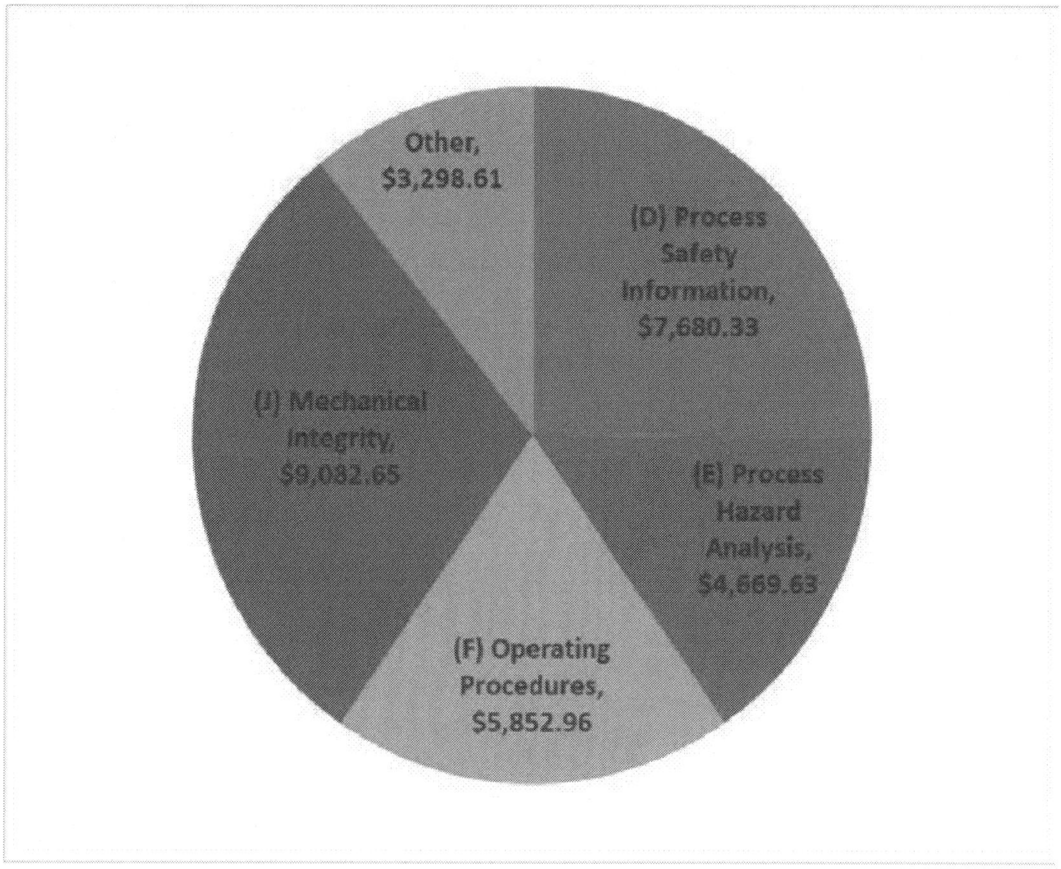

Figure 2: Average Cost per Violation by Element 2007-2017

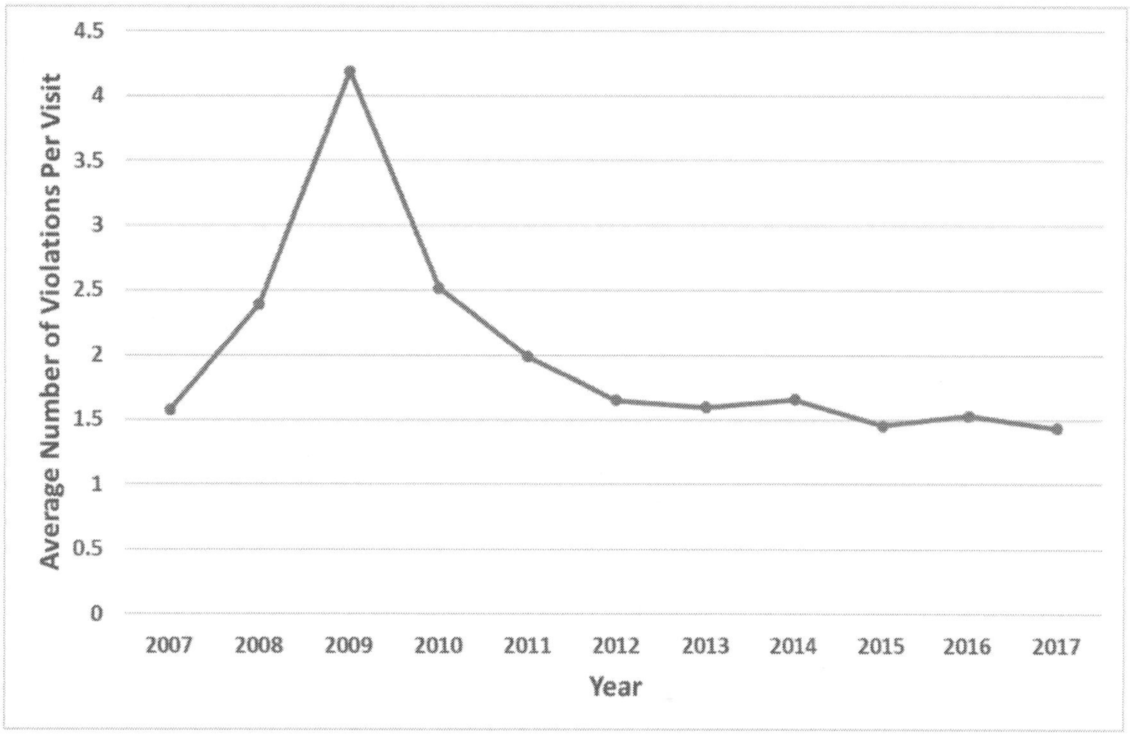

Figure 3: Average Number of Violations per Visit 2007-2017

(To note about Figure 3: the spike in the number of violations in 2009 could be due to the National Emphasis Program (NEP) in which OSHA extended to include a pilot on PSM-covered facilities from the 2007 NEP which only included refineries. The NEP went on from 2011 to only include chemical facilities and not petroleum refineries)

The citation costs are only one aspect to consider when reviewing the return on the investment. Other benefits that can be quickly realized once the programs have been established include but may not be limited to:

- ✓ Improvement in operating efficiencies
- ✓ Increased employee morale
- ✓ Lower turnover rates
- ✓ Safer working conditions
- ✓ Uptime gains
- ✓ Insurance benefits
- ✓ More consistent product quality
- ✓ Decreased likelihood of an incident

A deeper dive into the subsections of the most cited elements yielded the results in Figure 4.

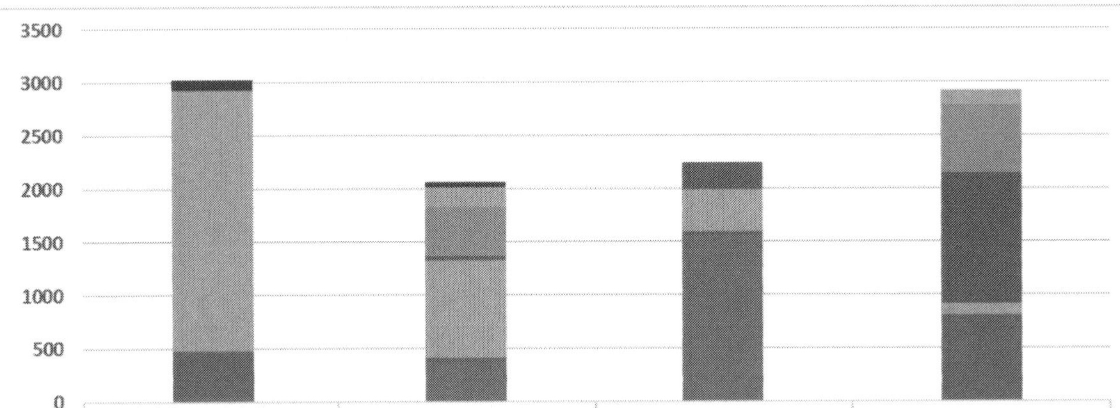

	(D) Process Safety Information	(E) Process Hazard Analysis	(F) Operating Procedures	(J) Mechanical Integrity
■ Blank	101			
■ 07		48		
■ 06		186		135
■ 05		463		640
■ 04		37	255	1235
■ 03	2452	918	391	105
■ 02	418	26	23	798
■ 01	54	381	1572	6

Figure 4: Sub-section Cited in the Top 4 OSHA Most Cited PSM Elements 2007-2017

Figure 4 shows that the biggest gaps over the last 10 years include the following sub-sections of these elements:

1. Process Safety Information (d)
    - Equipment Information (section 3)
2. Mechanical Integrity (j)
    - Written Maintenance Procedures (section 4)
    - Inspection and Testing (section 2)
    - Correcting Equipment Deficiencies (section 5)
3. Process Hazard Analysis (e)
    - Inadequate coverage of the hazards of the process (section 3)
4. Operating Procedures (f)
    - Develop written operating procedures (section 1)

There is a common theme among these four (4) subsection violations. While PSI was the most cited element, the most cited PSI sub-section is Equipment Information, which ties into the MI portion of PSM. Based on this, deficiencies in equipment related information and MI ranks very high on the list. MI related programs help to ensure that chemicals remain inside the vessels, pipes, etc. A review of the OSHA data shows that the industry may be struggling with proper execution of this critical piece of process safety.

As noted above, MI programs are complex and can be difficult to establish, but there are current solutions on the market to help correct these deficiencies, such as economical software solutions that improve compliance by organizing maintenance tasks. The right solution can alert staff to overdue maintenance, and store the maintenance history for each particular piece of equipment. Some tools can also track inventory of spare parts and link Management of Change documents related to the process and/or the equipment. Consultants versed in the regulations and MI techniques can help establish appropriate programs and find solutions useful to your specific operation.

Development of operating procedures for both process and maintenance activities has a high number of citations in the MI and OP elements. It is essential that these procedures are comprehensive, accessible, updated and easy to interpret. Mistakes in operating steps, or failure to recognize abnormal conditions, can quickly lead to catastrophic releases.

There are various writing and auditing techniques that can be used to ensure that these procedures include the correct information and address all stages of the process (start-up, normal conditions, abnormal conditions and shut down). Software tools currently on the market can also assist in organizing this critical information. This approach, while time consuming up-front, is time well spent in providing a safer and more efficient workplace.

Once authored, it is easier to audit the documents for accuracy and audit the operators using a behavior-based system to determine that procedures are being followed correctly.

Vast benefits can be gained using this approach, and many PSM consultants offer services to help author these documents and provide suggestions for document control and review processes.

4.0 Resources

Despite the intent for OSHA PSM to decrease the likelihood and severity of accidents, the OSHA data is showing that PSM gaps still exist in the most fundamental PSM elements. These gaps present an unacceptable risk. This is evidenced by the accidents that continue to occur. The various incident investigations performed by the U.S. Chemical Safety and Hazard Investigation Board often attribute inadequate PSM programs as the root cause.

There are many resources available for assistance with PSM implementation and improvement for plants of all types and sizes. There has been an increase in consulting organizations providing guidance and software solutions for PSM. The list of resources readily available to even the smallest of plants are much more prevalent than they were 10 years ago.

OSHA has published documents with guidelines for PSM for storage facilities and for small businesses. They also have a PSM resource page: https://www.osha.gov/SLTC/processsafetymanagement/. The Center for Chemical Process Safety has many publications, classes and tools to help with all aspects of PSM. Some of these resources are free, or provided at reasonable cost, and CCPS/AIChE (American Institute of Chemical Engineers) membership increases those benefits.

Small businesses often find it difficult to execute PSM. In 2016-2017, in almost the same number of inspections, OSHA administered 57 PSM related violations to companies with under 100 employees and only 18 violations to companies with over 100 employees. OSHA also points to a study that estimates "*employers with 1-25 employees are 47 times more likely to have a release and 17 times more likely to suffer an injury, per employee, than facilities with 1500 or more employees.*" [5]

There are a few PSM consulting firms that target small to medium companies, the companies that cannot dedicate resources to PSM, nor have the expertise in-house to meet the intent of the standard. Many of these consulting firms not only focus on OSHA compliance but becoming a partner to help embed PSM concepts into the plant's overall safety culture. Bringing in these third parties is worth the cost because they are embedded in the PSM realm and can bring along expertise from other industries, the latest information and years of experience.

There are cost effective software solutions, some aimed at small businesses from a price point and feature standpoint. These solutions can improve PSM while improving the bottom line through improved efficiency, production uptime gains, insurance benefits and violation and incident avoidance. The smaller chemical processing companies are yet to realize that these solutions exist at very reasonable prices, and the small upfront investments have very beneficial returns that easily justify the expenditure.

It is ideal to use the same consulting firm for various locations to standardize PSM solutions and audits across multiple sites that may have different risk. This type of PSM implementation helps to normalize the oversight and tracking to give a better picture of risk at the corporate level.

5.0 Conclusion

The research objective of this paper was to spot the potential for common gaps and issues in implementing PSM across the United States at all types of chemical processing industries using OSHA violation data, and to use that information to identify opportunities for solutions and preventative/corrective actions. The OSHA violation picture is only one piece of the gap analysis puzzle, but it is one place that collects enough data to increase awareness of PSM weaknesses nationwide regardless of the type of operation. This information can aid in creating education/awareness programs and campaigns aimed at improving PSM execution.

The top four cited violations are PSI, MI, PHA and OPs. Solutions to the deficiencies revealed by the OSHA data are well-known and PSM experts are aware of the many practical and feasible methods that exist to close these deficiency gaps. The return on investment on items outside of the safety aspect are usually realized soon after the solutions are put in place, including: more efficient and consistent work procedures (which can lead to improved product quality), increased up-time and insurance benefits to name only a few.

A review of the OSHA data can help quantify the costs that can be expected from OSHA PSM violations and citations. However, other impacts of violations and/or PSM incidents that are not so easily quantified include the business, legal, liability, company reputation and the impact that the injuries have on families. Sometimes plants are not able to rebuild in the same location or regain the customer share they may have lost due to an incident. Additionally, the OSHA violations are public record and can be accessed by lawyers, local authorities having jurisdiction, insurance companies, activists, etc. Companies may also find it difficult to obtain insurance due to OSHA violation or incident history. While these impacts are difficult to quantify, they must be considered in the overall cost to the business.

PSM implementation is a continuous improvement journey with new lessons learned, new technologies and new resources being created every year. Using the plethora of tools available can help close the deficiency gap.

References

[1] U.S Department of Labor OSHA 3132 Publication *Process Safety Management* 2000
[2] Washington State Department of Labor & Industries July 2010
[3] U.S Department of Labor OSHA 3133 *Publication Process Safety Management Guidelines for Compliance*, 1994
[4] OSHA Violation Data 2007-2017
[5] OSHA *Process Safety Management for Small Businesses* 2017

546725 Safety Performance Evaluation: A Benchmark of Practices across Various High-Risk Systems

The concept of performance has invaded all management spheres and imposed itself as a strong driver of decision making. Following the funding work of Kaplan and Norton (1996), performance appears as a multidimensional concept which evaluation and use require adapted tools and well calibrated decision processes.

Safety performance does not escape these requirements as it englobes various dimensions (workplace safety, process safety, environmental safety), can be approached through various evaluation paradigms (auto evaluation, external assessment through audits and KPIs) and may be used by several stakeholders (risk owner, regional and national authorities, public perceptions). Indeed, different combinations of these three dimensions allow for a large variety of performance governance paradigms carrying various understandings and practices with regard to the way performance should be assessed and used.

By comparing safety performance management in three different risk systems (process, railway and health) in Europe, this works aims at providing safety management actors with a description of the richness and variety of safety performance understandings and practices. Insights gained from this work should enhance safety managers' abilities to grasp the strengths and weaknesses of their approaches and explore the possibilities of a beneficial benchmarking.

Lessons learned from analyzing an explosion accident at Shanghai Secco chemical plant

Hanhai Dong
China University of Petroleum (East China)
No. 66, West Changjiang Road, Huangdao District, Qingdao, China
donghanhaiupc@126.com

Yi Liu
China University of Petroleum (East China)
No. 66, West Changjiang Road, Huangdao District, Qingdao, China
liuyi@upc.edu.cn

Dongfeng Zhao
China University of Petroleum (East China)
No. 66, West Changjiang Road, Huangdao District, Qingdao, China
zhaodf@vip.sina.com

Jian Chen
China University of Petroleum (East China)
No. 66, West Changjiang Road, Huangdao District, Qingdao, China
836771528@qq.com

Yanchun Wang
China University of Petroleum (East China)
No. 66, West Changjiang Road, Huangdao District, Qingdao, China
Wychun25@163.com

Wei Wang
China University of Petroleum (East China)
No. 66, West Changjiang Road, Huangdao District, Qingdao, China
15621097585@163.com

Prepared for Presentation at
American Institute of Chemical Engineers
2019 Spring Meeting and 15th Global Congress on Process Safety
New Orleans, LA
March 31 – April 3, 2019

Lessons learned from analyzing an explosion accident at Shanghai Secco chemical plant

Hanhai Dong
China University of Petroleum (East China)
No. 66, West Changjiang Road, Huangdao District, Qingdao, China
donghanhaiupc@126.com

Yi Liu
China University of Petroleum (East China)

Dongfeng Zhao
China University of Petroleum (East China)

Jian Chen
China University of Petroleum (East China)

Yanchun Wang
China University of Petroleum (East China)

Wei Wang
China University of Petroleum (East China)

Keywords: event sequence diagram; Multilayer Bow-tie Analysis; FLACS

Abstract

On May 12, 2018, six workers from Shanghai Aijinke Engineering Construction Services Co., Ltd. died on the spot of an explosion when they were maintaining a benzene tank in the tank farm at Shanghai Secco Petrochemical Co., Ltd. The accident was characterized as a disastrous accident caused by hot work in a confined space, which resulted in massive casualties and a direct economic loss of 11.66 million yuan. This paper aims to analyze the accident occurrence process in a chronological order firstly by using the event sequence diagram (ESD). Multilayer Bow-tie Analysis was adopted to explore and classify the causes of the explosion of the benzene tank, along with the corresponding consequences, and the preventive measures. The analyses show that the primary inducements were the unsafe behavior of on-site workers, hazardous conditions of in-tank substances, risky factors in the construction environment, and deficiencies in management. Casualties, fire and explosion, equipment damage, and property losses were all included as consequences. The FLACS software was used to simulate the variation of explosion pressure and explosion temperature varying with time when benzene vapor was ignited at the tank bottom during the accident, and a flow field diagram for the explosion was achieved correspondingly. The results obtained are of direct significance for investigation of similar accidents in the future and are instructive to equipment safety evaluation as well. Based on the qualitative and quantitative analyses, we emphasize three considerable reasons leading to the accident, which includes inapposite flammable gases detection points before the hot work, failing to test the flammable gases during the hot work, and deficiency of the Management of Change (MOC). Finally, the paper concludes with a summary of lessons learned from the accident as well as suggestions for accident prevention and safety improvement.

1 Introduction

The term confined space refers to closed or semi-closed facilities and places with limited entrance and exit, and poor ventilation, which may lack oxygen or contain flammable, explosive, toxic, or harmful substances. The main causes of accidents in a confined space include fire, explosion, poisoning, asphyxiation, and physical injury [1]. According to the statistics of the Ministry of Emergency Management of the People's Republic of China, from January 2012 to April 2018, a total of 64 accidents involving three or more deaths in a confined space occurred in China, resulting in 225 deaths and 82 injuries [2]. Among them, on April 8, 2014, three people were killed and two injured by an explosion during the repair and welding of the desulfurization tank at a factory in Wuhai, China. On March 18, 2015, the operators of a chemical company in Binzhou, China, suffered an explosion accident during an operation in a confined space, resulting in four deaths and two injuries. Hot work in a confined space is the common cause of the two accidents above.

Due to the poor ventilation, a large amount of flammable gas may be accumulated in a confined space. When construction workers perform hot work inside, they will face potential safety risks and problems, such as difficulty in escaping, delayed personnel rescue, and inconvenient communication. Hot work in a confined space can easily lead to fire, explosion, and personal injury accidents, which threaten not only the personal health of operators but also the property of enterprises.

Researchers have done a lot of risk analysis and exploration for the particular operating environment in a confined space. Selman et al. [3] analyzed the mortality rate of confined space operations in similarly industrialized countries from 0.05 to 0.08 per 100,000 workers and pointed out that the main risks are toxic atmospheric hazards and physical hazards. Selman et al. [4] divided the emergency rescue of confined space accidents into three levels: individual self-rescue, non-entry rescue, and progressive rescue. Botti et al. [5] applied the Internet of Things (IoT) technology and proposed a holistic approach to analyzing the procedure of confined space operations. Meanwhile, many studies adopted experimental and numerical simulation methods to observe the evolution of characteristic parameters during the diffusion and detonation of flammable gas in a confined space. The results obtained are helpful to the quantitative analysis of confined space accidents. Bang et al. [6] established a theoretical model to predict the pressure rise caused by the deflagration in a confined space. He et al. [7] presented an LES turbulence model to simulate the diffusion process of LPG in the underground confined space.

This paper mainly analyzes a disastrous accident caused by hot work in a confined space at Shanghai Secco Petrochemical Co., Ltd (Secco). On May 12, 2018, the workers from Shanghai Aijinke Engineering Construction Services Co., Ltd. (Aijinke) triggered the explosion of flammable gas when they were performing inspection and maintenance inside a benzene tank. The serious accident resulted in the immediate death of six on-site workers and direct economic loss of 11.66 million yuan. Through a series of qualitative and quantitative analysis methods, we reveal the root causes of the accident, simulate the development process of the explosion, and propose corresponding rectification suggestions. Here we want to emphasize that the reason why we write this paper is not to pursue responsibility and punishment but to reflect on how to avoid the recurrence of similar accidents.

2 The occurrence of the explosion accident

On the morning of May 12, 2018, Aijinke workers arrived at the utility tank farm of Secco to prepare for maintaining TK-75-0201 benzene tank (0201 benzene tank). Before the start of the operation, a Secco field operator used a hand-held gas detector to measure oxygen and flammable gas at the manhole outside the tank. The on-site supervisor of Aijinke, the on-site supervisor of Secco, and the on-duty monitor of Secco utility tank farm signed the work ticket successively without verifying the measurement data and checking the personal protective equipment (PPE) and explosion-proof tools of the on-site workers according to the requirements specified in the operating permit.

At 1:15 p.m., six workers of Aijinke continued to dismantle the floating pontoon in the tank and two workers worked outside for monitoring. At the scene, another Secco field operator was in charge of regular oxygen and flammable gas measurement. At 3:25 p.m., a sudden explosion occurred in the 0201 benzene tank.

After the explosion, Secco immediately launched their emergency response plan and reported the accident information to the response center of Shanghai Chemical Industry Zone Management Committee (SCIZMC), the fire brigade, the medical center, and the production operations monitoring center of Sinopec. At 3:50 p.m., the fire brigade put out the fire by spraying water. At 5:50 p.m., the on-site rescue work ended and a total of six Aijinke workers in the tank died on the spot. Figure 1a shows the scene of the 0201 benzene tank after the explosion.

Figure 1. The 0201 benzene tank and floating pontoons

2.1 Description of the timeline

The occurrence of accidents is a complex system problem, and it gives useful information on how accidents arise in practice [8]. In March 2018, Secco found that the breather valve of the 0201 benzene tank leaked slightly, which caused the VOC concentration to exceed the standard. After inspection, it was judged that the floating pontoon leaked. In order to find the root causes, we describe in detail the fundamental behaviors of the primary responsible party and the contractor prior to the explosion.

- On April 16, Secco established a maintenance order and a work list in the SAP system.
- On April 19, Secco began to empty the 0201 benzene tank, performing four days of steam dilution and eight days of nitrogen replacement.
- On April 24, Secco informed Aijinke to formulate maintenance plan.
- On April 26, Secco carried out the Job Hazard Analysis (JHA) around the maintenance work.
- From May 1 to 7, Secco opened the manhole of the tank for natural ventilation and inspection, and found that more than half of the floating pontoons had effusion.
- On May 8, the relevant departments of Secco held a special meeting and decided to change the original plan to replace all the floating pontoons.
- On May 9, Secco asked Aijinke workers to enter the tank and destructively dismantle all floating pontoons at the regular OSBL equipment maintenance meeting.
- From May 9 to 11, Aijinke workers entered the tank and dismantled the floating pontoons. A total of 38 floating pontoons were dismantled, as shown in Figure 1b.

2.2 Analysis by event sequence diagram

Event Sequence Diagram (ESD) is an accident scenario-oriented simulation method, which treats events as the basic research object, and takes into account the domino effect and risk transfer effect [9]. ESD is also an intuitive reflection of the accident scenario. It provides a more literal representation of the system state, indicating the behavior of crucial process variables, and even operator and hardware state changes [10].

In order to further illustrate and point out the critical links of the accident, this paper makes an intuitive description of events by using ESD. As shown in Figure 2, the basic events of accidents are divided into initial events, judge events, action state events, and termination events. At the same time, the path indicated by the dashed line represents the potential events that may occur in this accident, and helps to make a comprehensive and logical analysis of the whole process. The graphic symbols in the ESD are explained in the Appendix.

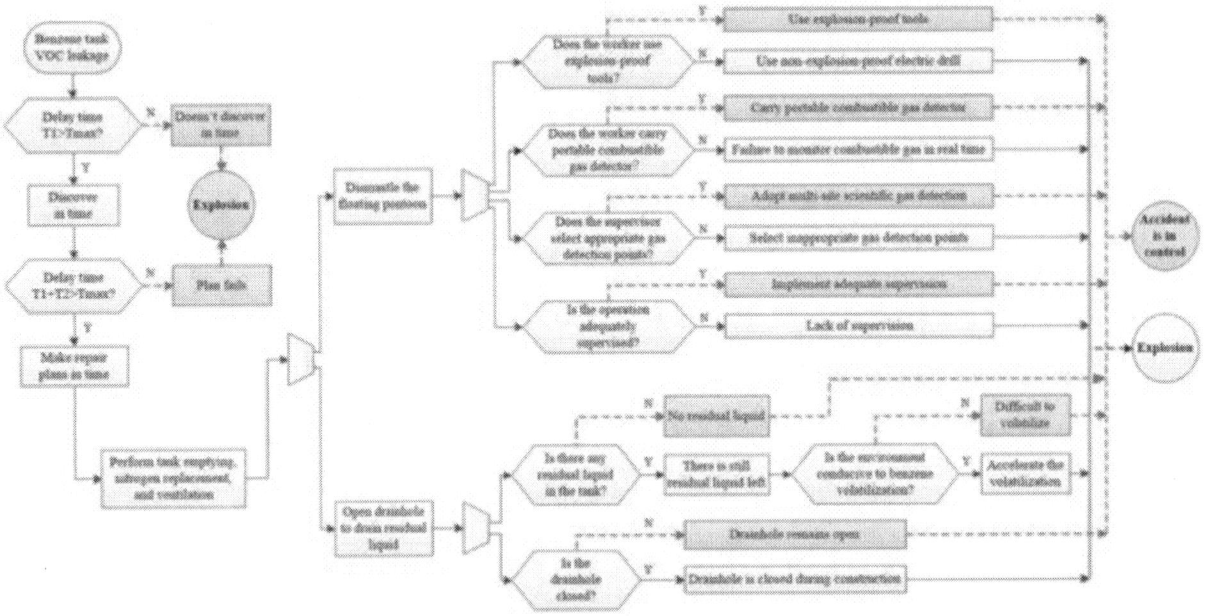

Figure 2. The ESD of the accident

As can be seen from the ESD, there are nine pivotal judgment nodes on the occurrence of the event chain, which can determine the direction of the incident. Among them, whether to discover leakage in time and whether to make a repair plan in time are the nodes at the time aspect. When the detection instrument and worker cannot find leakage and formulate the response plan in the maximum allowable delay time, they will directly lead to the explosion. Enterprises need to enhance equipment reliability and inspection frequency, so as to reduce emergency response time. Whether there are residues in the tank and whether environmental conditions are conducive to benzene volatilization are two nodes at the environmental aspect, which are also the leading causes of accidents. They require the workers to make a detailed hazard identification of the construction environment, and improve scientific detection. The critical nodes regarding human factors include whether the drainhole is closed, whether the on-site workers use explosion-proof tools, whether they carry portable combustible gas detector, whether they select appropriate detection points, and whether the supervisors take adequate monitoring [11]. Strengthening safety training before the operation, improving individual risk identification ability, perfecting risk control measures, and enhancing safety supervision level are essential means to avoid accidents caused by human factors.

All of these factors, organizational and operational, contributed to the accident, with a hierarchy of casualties among them. Through the comprehensive description, ESD can not only find out the nodes where the problem lies, but also guide the prevention of similar accidents. Therefore, enterprises should pay attention to the basic events expressed by nine judgment nodes and take appropriate interventions to block the development path of dangerous accidents.

3 Multilayer bow-tie analysis

3.1 The process of MBTA

Among many techniques available, bow-tie analysis is prevalent because it organically combines the two sophisticated risk analysis methods of Event Tree Analysis (ETA) and Fault Tree Analysis (FTA) [12], systematically summarizes the causes and consequences of accidents, and further integrates the prevention and mitigation of accidents or risks [13]. Bow-tie analysis has been proven to be a reliable and efficient tool, mainly used for process safety analysis, risk management, and accident risk analysis. It helps to understand which significant event combinations will lead to the top events, and which security barrier failures would escalate the top event to a certain consequence.

Traditional bow-tie analysis mainly includes top event, causes, consequences, preventive barriers, and mitigation barriers, which shows a lot of information. However, the single-level structure lacks logic and is not conducive to people's deep understanding. Thus, this paper establishes a Multilayer Bow-tie Analysis (MBTA), which resolves the top-event-related ETA and FTA individually into hierarchical structures, to deeply analyze the root causes, specify the severe consequences, and propose the corresponding safety barriers and scientific disposal methods. MBTA uses graphics to visually display the relationship between different levels and avoid repeated analysis effectively. Figure 3 shows the MBTA model, in which the code explanations are listed in Table 1.

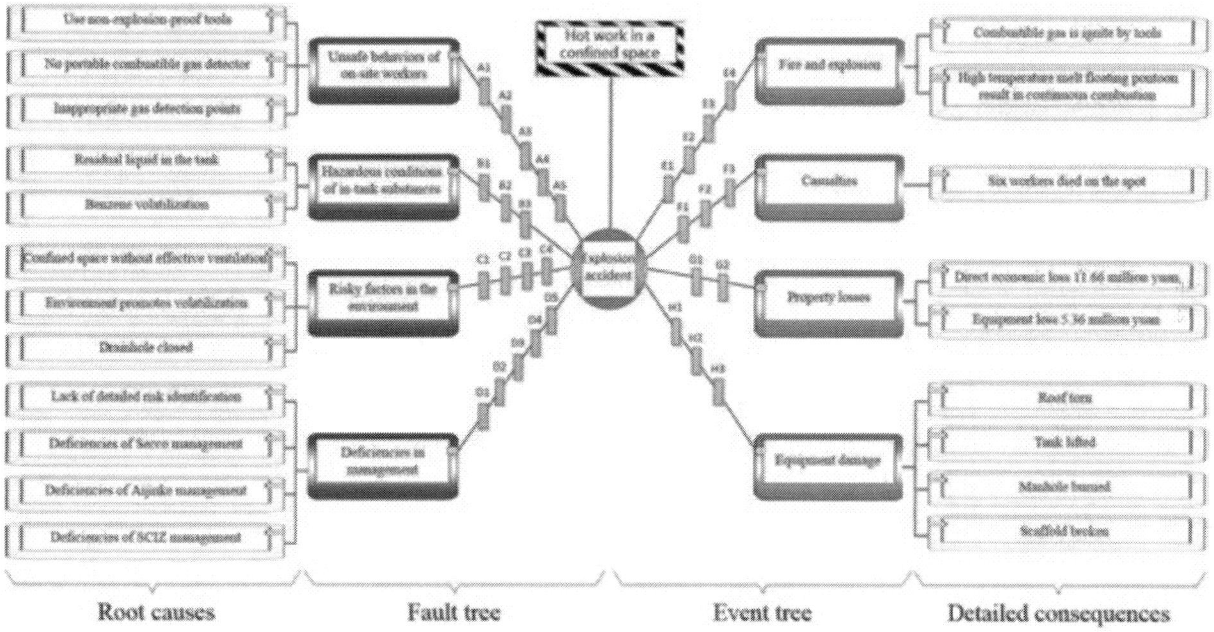

Figure 3. The MBTA model

Table 1. The explanations of MBTA

Code	Safety barrier	Code	Safety barrier
A1	Use explosion-proof tools	D4	Implement the main responsibility
A2	Carry portable combustible gas detector	D5	Improve contractor management
A3	Strengthen operation training	E1	Call the fire brigade in time

A4	Enhance supervision and assessment	E2	Spray water to extinguish fire
A5	Set up warning signs	E3	Fire isolation
B1	Remove the residual liquid	E4	Launch the emergency plan
B2	Hazardous material isolation	F1	Call for medical treatment
B3	Combustible gas detection in the tank	F2	Search and rescue the wounded
C1	Forced ventilation	F3	Launch the emergency plan
C2	Open drainhole	G1	Insurance claim
C3	Continuous detection during operation	G2	Launch the emergency plan
C4	Take environmental risk assessment	H1	Dangerous equipment isolation
D1	Compliance with rules and regulations	H2	Equipment repair
D2	Increase the intensity of punishment	H3	Equipment replacement
D3	Strengthen process safety management		

In the above MBTA model, we adopt the analytic hierarchy process (AHP) to extract the common characteristics of 12 causes of accidents and classify the causes into four categories: unsafe behaviors of on-site workers, hazardous conditions of in-tank substances, risk factors in the construction environment, and deficiencies in management. Moreover, we could conclude the direct and indirect reasons for the accident.

3.1.1 Direct reasons

Benzene is a colorless and transparent liquid with a strong aromatic odor. It has a boiling point of 80.1 ℃, a saturated vapor pressure of 10 kPa at 20 ℃, a flash point of -11 ℃, an explosion limit range of 1.2%-8.0%, and minimum ignition energy of only 0.20 mJ. Benzene is a highly flammable substance. The explosive mixture of benzene vapor and air can burn and explode when exposed to ignition source or high temperature.

At the construction site, the temperature is 24.7 ℃, and the pressure is 100.68 kPa. Under these conditions, benzene is easy to volatilize and form a gas explosion environment in a confined space. During the process of dismantling, residual benzene in the floating pontoons leaked onto the bottom of the tank and was not cleaned up in time. Because there was no effective ventilation in the tank, benzene vapor mixed with air to form an explosive environment and benzene concentration reached the explosion limit. The non-explosion-proof tools used by the operators produced ignition energy, which brought about the deflagration of the flammable gas. The high temperatures generated by the combustion melt the other aluminum alloy pontoons, resulting in sustained combustion for a short period.

3.1.2 Indirect reasons

Aijinke failed to abide by the relevant safety production rules and regulations. Managers did not fulfill their duties of safety management and supervision. Aijinke failed to give safety technical information to workers following the operating rules before the construction. Without implementing and supervising PPE, carrying portable gas detector, and using explosion-proof tools, operators were arranged to work in a confined space. When significant changes occurred in the job conditions, Aijinke did not promptly request workers to stop working and implement the Management of Change (MOC).

Secco had some omissions in safety risk management, professional management, and special operation management. In addition, Secco did not follow the safety production rules and regulations. On-site inspectors failed to carry out gas detection in accordance with specifications. After the change of operation, the manager did not ask the contractor to stop the process in time and revise the construction plan according to the regulations. Secco failed to report the maintenance plan to SCIZMC as required.

SCIZMC failed to implement the safety management system fully and did not discover that Secco performed maintenance operations without permission.

3.2 Lessons learned from MBTA

This paper puts forward a series of preventive measures, such as strengthening operation training, removing residual benzene in the tank, improving risk assessment of the environment, and enhancing process safety management. It is noticed that the environment in a confined space is harsh and there are many potential unsafe factors. Furthermore, the contractor's maintenance work cycle spans the statutory weekend. Based on the two reasons above, the enterprise should upgrade the level of hot work and adopt a broader system of operation approval and management. On-site workers need to stop working immediately when safety problems are found in the construction process and report to their superiors by relevant regulations.

Through refining the consequences of accidents, we summarized that detailed safety emergency plans and energy isolation methods are universally applicable in various measures to control and mitigate the impacts of accidents. They require Secco and Aijinke to formulate a comprehensive emergency response plan before, including fire extinguishing, personnel search and rescue, equipment maintenance, property damage remediation and so on. When an accident occurs, workers have to stay away from dangerous substances and places immediately.

As with many disasters, what ended up in a sequence of equipment damage and human errors started mostly as a management problem. Many unsafe practices were tolerated and even encouraged. The risk management of confined space operations should be divided into ten steps according to relevant regulations, namely training, pre-operation planning, job approval, hazardous energy isolation, ventilation, gas detection, personal protection and tool safety, warning signs, monitoring, and emergency response, so as to strengthen process safety management comprehensively. It is necessary to enhance the issuance of management of work tickets for hazardous operations such as hot work and confined space operations. Enterprises must strictly inspect illegal operations and supervise safety precaution measures to ensure that the operation process is safe and controllable. For changes occurring in the construction process, enterprises ought to strictly implement the MOC, analyze the potential risk and possible accidents, formulate corresponding safety measures, and carry out safety education for all operators.

MBTA provided a detailed analysis of the accident and established the corresponding safety barriers. However, there are still some certain subjective problems, and its content should also be flexible. Therefore, we encourage employees, accident investigators, case learners, and other relevant personnel to improve and develop the content of MBTA continually, thereby effectively expanding the application scope and effectiveness of MBTA.

4 Quantitative simulation of the explosion

The analysis of the accident should include not only qualitative analysis but also quantitative analysis [14]. We use FLACS software to simulate and analyze the occurrence of the benzene explosion. According to the relevant information, the 0201 benzene tank is an internal floating roof tank with a volume of 10,000 m^3. Its diameter is 30 m and height is 19 m. The design temperature range and pressure range are -10 – 65 ℃ and -0.49 –1.96 kPa, respectively. The tank adopts box-type aluminum alloy internal floating pontoons, and a manhole with a diameter of 0.6 m is set at the position of 0.75 m above the ground. Figure 4 shows the 0201 benzene tank model.

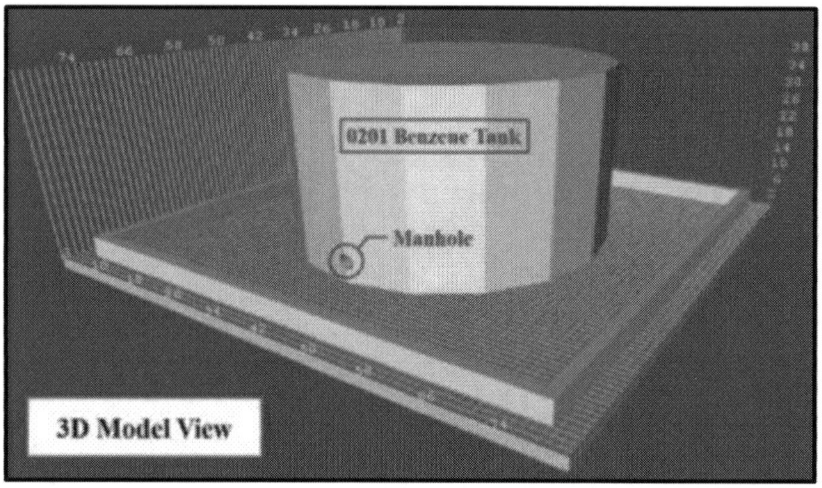

Figure 4. The 0201 benzene tank model

The simulated environmental conditions are fixed at the temperature of 24.7 ℃ and the pressure of 100.68 kPa, which are the same as the construction environment. We make reasonable assumptions: (a) The residual benzene in the tank evaporates continuously into the confined space and forms a uniform mixture of flammable gas. When the volatilization concentration of benzene reaches the lower explosion limit (LEL), the ignition source that persists during the construction operation detonates the flammable gas and results in an explosion. (b) Six on-site workers dismantled the float pontoons at the bottom of the tank, so the ignition source is set at the height of 0.75 meters from the geometric plane center, as shown in Figure 5 point I. There are several detection points of pressure and other characteristic parameters in the model of benzene tank, as shown in Figure 5, points M1-M5. (c) The benzene tank is windless, and the flammable gas accords with the real gas state equation. (d) The gas explosion process satisfies a series of fundamental equations, such as the continuity equation, momentum equation, and energy equation.

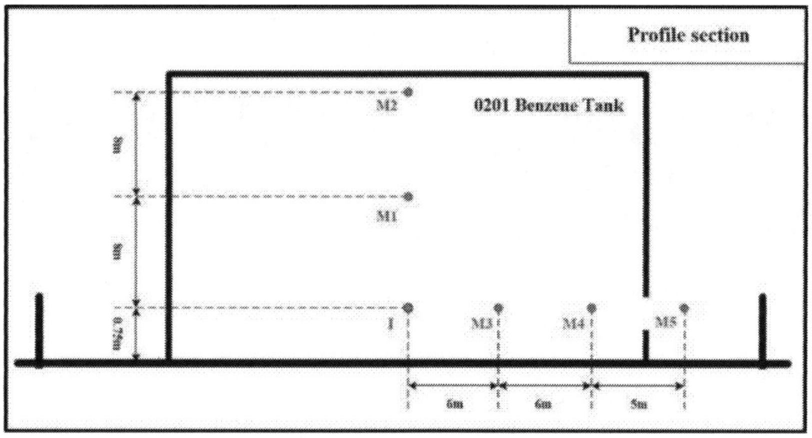

Figure 5. The profile section of the model

We selected the maximum profile section of the tank as the research object and intercepted the temperature field and pressure field at four different moments in the explosion process, as shown in Figure 6 and 7. The simulation results intuitively describe the development process of the explosion. When the flammable gas in the tank was ignited, the flame diffused around in the form of a spherical wave and reached the bottom of the tank at 0.35 s, as shown in Figure 6a and 7a. Since the radial distance of the tank is less than the vertical axial distance, the flame propagated horizontally to the wall, as shown in Figure 6b and 7b. At 1.52 s, the flame reached the manhole and sprayed out to the outside of the tank in the form of jet flow. During this period, the pressure and temperature increased continuously, which are the leading causes of the manhole being burned, as shown in Figure 8a. When the flame propagated to the top of the tank at 2.11 s, the uncombusted gas was recirculated by the viscous effect and blocking effect of the wall, which further caused the recirculation vortex. The recirculating vortex enlarged the contact area of the flame front, resulting in flame acceleration and pressure increase, as shown in Figure 6c and 7c. Figure 6d and 7d illustrate the flammable gas burned out at 3.24 s. Simultaneously, the explosion temperature and pressure reached the peak value contributing to the tearing of the roof and the lifting of the tank, as shown in Figure 8b and 7c.

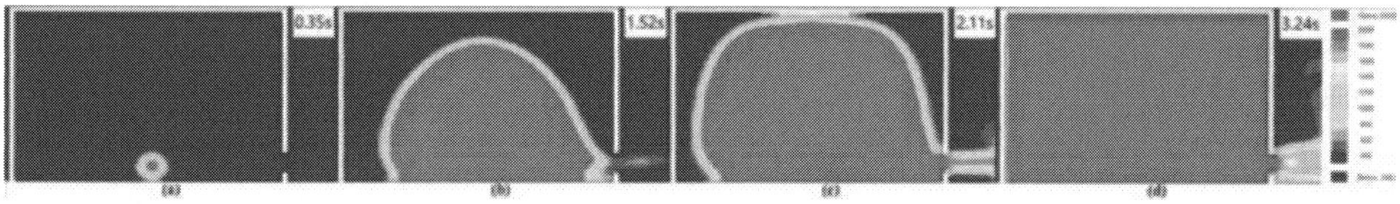

Figure 6. Temperature simulation of the explosion

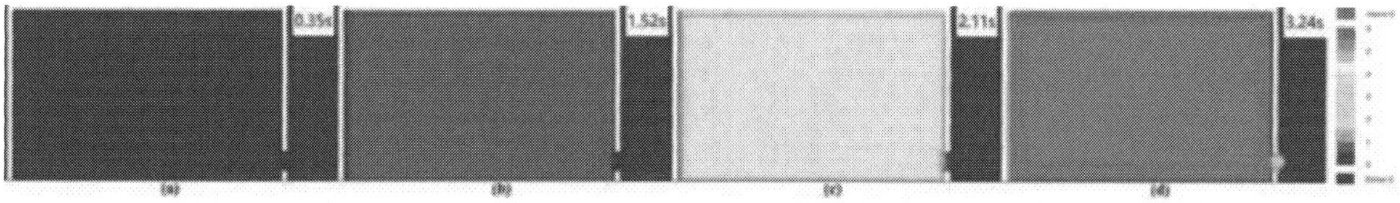

Figure 7. Pressure simulation of the explosion

Figure 8. The results of the accident

Figure 9 reveals the change of temperature and pressure recorded by the detection points M1-M5 during the explosion. Table 2 describes the data of the detection points, including maximum and minimum explosion pressures, maximum and minimum explosion temperatures, peak data arrival times and so on. Combining the data of detection points and the pressure field diagram in Figure 7, it is shown that the pressure in a confined space will be uniform immediately for a certain time and further proves the accuracy of the analysis of the flame propagation process. In particular, the external detection point M5 was affected by the propagation of explosion flame in the tank at 1.6 s, resulting in a peak explosion overpressure of 0.031 bar and a peak temperature of 1895 ℃. Subsequently, due to the pressure rise in the tank, the pressure at the detection point M5 dropped to negative pressure until the flammable gas burned out.

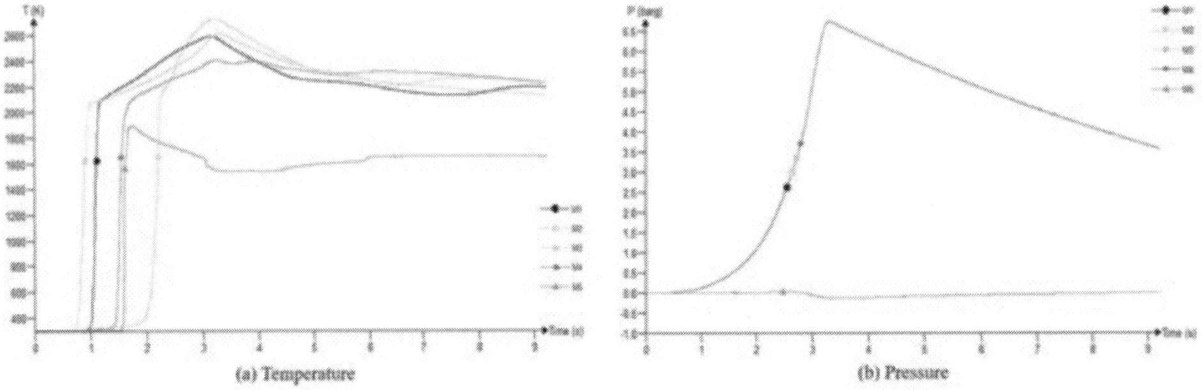

Figure 9. Parameters change of the detection points

Table 2. Data from the detection points

No.	Temperature				Pressure			
	Tmax/K	Time/s	Tmin/K	Time/s	Pmax/bar	Time/s	Pmin/bar	Time/s
M1	2589.4	3.11	298.1	0.04	6.75	3.24	0.00	0.04
M2	2732.2	3.24	298.1	0.04	6.76	3.24	0.00	0,04
M3	2609.3	3.28	298.1	0.00	6.76	3.24	0.00	0.00
M4	2306.2	3.25	298.1	0.00	6.75	3.24	0.00	0.00

M5	1895.5	1.68	298.0	0.69	0.05	1.65	-0.15	3.29

Although there is only a small amount of residual benzene, the benzene volatilization forms a flammable gas with the air inside the tank, which generates a substantial destructive power when ignited. The simulation results show that the flame propagation velocity of the flammable gas increases constantly, with the peak value reaching up to 41 m/s. Since the flame is propagated in the form of subsonic velocity, it can be judged that there are mainly the phenomena of deflagration occurred in the process of explosion [15]. Soloviova-Sokolova et al. [16] calculated the flame burning rate of the benzene vapor was 0.6 m/s at 298 K, 0.1 MPa and 0.6 eq., which clearly illustrated that the peak propagation velocity of the benzene vapor in a confined space would be several times higher. Meanwhile, the explosion pressure in a confined space is usually greater than that in an open area due to lack of heat loss and volume expansion.

When workers are engaged in hot work in a confined space, they should carry out comprehensive risk identification, select several representative flammable gas detection points, and detect the flammable gas in real time. Workers ought to stop work immediately and withdraw in an orderly manner once the flammable gas exceeds the standard. Through the analysis of the explosion process, the flame propagates to the wall in a very short time. Thermal radiation and explosive shock wave are the primary causes of casualties in the tank. Owing to the special working environment in a confined space, the phenomenon of mass casualties and injuries often occurs under the effect of high temperature and high pressure, which needs people's sufficient attention.

5 The subsequent event

The explosion accident of Secco has aroused widespread concern, and 22 responsible persons have also been punished. But unfortunately, lessons from the accident did not prevent a subsequent similar event. On November 26, 2018, three contractor workers of the Ningbo Engineering Company (NEC) and one worker of Secco inspected the discharge tank of cracking gas compressor in the olefin separation unit. Without ascertaining the concentration of oxygen in the tank, a contractor worker of NEC entered the tank for construction and fainted unexpectedly. A Secco worker immediately went into the tank to rescue him, but he also fell into the tank. The accident resulted in the death of two workers who had been rescued after asphyxia.

It is the second confined space operation accident in the same company within six months. The main causes of the accident are still the deficiency of contractor management and risk identification in a confined space. While expressing sadness, the public should profoundly reflect on why accidents continue to occur, what is the role of punishment and accountability, and how relevant regulations and reform measures can be truly implemented. Drawing lessons from the previous accidents is necessary. Enterprises ought to raise the safety awareness of their employees and use strict supervision to ensure safe production of chemical plants.

6 Conclusions

This paper provides an introduction of an accident caused by hot work in a confined space, together with the detailed ESD-based description of nine pivotal judgment nodes leading to this disastrous accident. It concludes that enterprises should take suitable intervention measures to block the pathway of the accident. The MBTA, covering the causes and consequences of the accident, offers proactive and reactive measures based on an intuitive analysis. Meanwhile, it emphasizes the importance in selection of appropriate detection points, real-time detection of in-tank operation, and implementation of MOC. Also, we apply the FLACS software into quantitative simulation of the explosion process in the tank. The results contribute to the guidance of accident investigation and safety measures selection. Finally, it suggests that enterprises adopt technical remedies and better safety management practices, since they are the essential basics for effective implementation of risk control measures.

Appendix

Symbol	Annotation
	Initial event: Beginning event of the ESD.
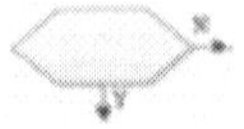	Judge event: The event has two possible outcomes that determine whether subsequent events occur.
	Action state event: Describe the action state of the event.
	Termination event: Ending event of the ESD.
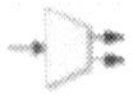	Output AND gate: Input event lead to multiple output events occurring.

7 References

[1] D.Burlet-Vienney, Y.Chinniah, A.Bahloul and B.Roberge, Design and application of a 5 step risk assessment tool for confined space entries, Safety science 80 (2015), 144-155.

[2] The Office of the State Council Security Committee Notified 8 Cases of Blindly Rescuing in a Comfined Space Leading to Casualties Enlargement, Available at http://www.chinasafety.gov.cn/default.htm Accessed on June 8, 2018.

[3] J.Selman, J.Spickett, J.Jansz and B.Mullins, Safety science 109 (2018), 333-343.

[4] J.Selman, J.Spickett, J.Jansz and B.Mullins, Safety science 113 (2019), 78-90.

[5] L.Botti, V.Duraccio,M.G.Gnoni and C.Mora, A framework for preventing and managing risks in confined spaces through IOT technologies, ESREL (2015), 3209-3217.

[6] B.Bang, C.Ahn, J.Lee, Y.Kim, M.Lee, B.Horn, D.Malik, K.Thomas, S.James, A.Yarin and S.Yoon, Theoretical, numerical, and experimental investigation of pressure rise due to deflagration in confined spaces, International Journal of Thermal Sciences 120 (2017), 469-480.

[7] B.He, X.Jiang, G.Yang and J.Xu, A numerical simulation study on the formation and dispersion of flammable vapor cloud in underground confined space. Process Safety and Environmental Protection 107 (2017), 1-11.

[8] I.M.Shaluf, F.Ahmadun and A.M.Said, Fire incident at a refinery in West Malaysia: the causes and lessons learned, Journal of Loss prevention in the Process Industries 16.4 (2003), 297-303.

[9] S.Swaminathan and C.Smidts, The event sequence diagram framework for dynamic probabilistic risk assessment, Reliability Engineering & System Safety 63.1 (1999), 73-90.

[10] J.Zhou, G.Reniers and N.Khakzad, Application of event sequence diagram to evaluate emergency response actions during fire-induced domino effects, Reliability Engineering & System Safety 150 (2016), 202-209.

[11] E.Zarei, M.Yazdi, R.Abbassi and F.Khan, A hybrid model for human factor analysis in process accidents: FBN-HFACS, Journal of Loss Prevention in the Process Industries 57 (2019), 142-155.

[12] N.Khakzad, F.Khan, P.Amyotte, Dynamic risk analysis using bow-tie approach, Reliability Engineering & System Safety 104 (2012), 36-44.

[13] M.B.Mulcahy, C.Boylan, S.Sigmann and R.Stuart, Using bowtie methodology to support laboratory hazard identification, risk management, and incident analysis, Journal of Chemical Health and Safety 24.3 (2017), 14-20.

[14] C.Zhu, J.Zhu, L.Wang and M.S.Mannan, Lessons learned from analyzing a VCE accident at a chemical plant, Journal of Loss Prevention in the Process Industries 50 (2017), 397-402.

[15] I.C.Tolias and A.G.Venetsanos, An improved CFD model for vented deflagration simulations–Analysis of a medium-scale hydrogen experiment, International Journal of Hydrogen Energy 43.52 (2018), 23568-23584.

[16] J.V.Soloviov-Sokolova, V.A.Alekseev, S.S.Matveev, I.V.Chechet, S.G.Matveev and A.A.Konnov, Fuel 175 (2016), 302-309.

Strengthening Process Safety Culture throughout Learning from Each Other´s Vulnerabilities

Luciana Poli SIlva
PETROBRAS
Av. Republica do Chile, 65, 4th floor, aisle 402.
lucianapoli@petrobras.com.br

Prepared for Presentation at
American Institute of Chemical Engineers
2019 Spring Meeting and 15th Global Congress on Process Safety
New Orleans, LA
March 31 – April 3, 2019

Strengthening Process Safety Culture throughout Learning from Each Other´s Vulnerabilities

Luciana Poli Silva
PETROBRAS
Av. República do Chile, 65, 4th floor, aisle 402.
lucianapoli@petrobras.com.br

Keywords: andragogy, dialogue, leadership, safety leadership culture, trust, vulnerabilities.

Abstract

A fundamental approach to any program that targets the implementation of a process safety culture is to work within reality, going beyond procedures and compliance. To facilitate the emergence of the unspoken is essential to stimulate the dialogue among leaders that brings about their vulnerabilities and opens new perspectives to address them effectively. This paper presents a pilot program designed to increase the Process Safety Leadership Culture in a Brazilian Energy Company. Based on a benchmark made within the industry and in andragogic principles, the program created a trust ambiance among peers and inspired leaders to recognize and share their vulnerabilities, as well as to take actions towards the strengthen the role of being a safety culture leader.

1 Introduction

A lot has been said and written about Leadership Process Safety Culture, but there is a gap between what the compliance procedures dictate and what managers really do. When it comes to the Brazilian Energy Industry, the pressure leaders are under in order to achieve the results they are expected to can be considered significant. How to balance the production challenges with a safety culture that must be among the core values of the operation? Are these factors compatible?

The pilot program entitled '*Strengthening Process Safety Culture*' dealt with reality, using andragogic approaches to bring about the complex factors forty managers face in their routine, such as the existence of several levels separating top managers from the operational team, privileging safety over production goals, and being an effective representative of the Company´s policies. All these factors can jeopardize the safety culture from becoming stronger every day.

This paper describes the implementation of a methodology designed to create a safe environment, in order to build trust among participants and at the same time, make them reflect and dialogue about how they can evolve from compliance to commitment.

In the coming sections, we will present the central concepts used to develop the program – Culture, Safety Culture and Leadership - exploring its implication on an actual work environment.

In the sequence, we will talk about what it takes to make the unspoken, unwritten rules emerge. From there, how we stimulated the increase in the self-conscious level as well as in the collective one, building a matrix entitled 'Personnel Influence on Safety' to diagnose how one really sees himself concerning his leadership in safety culture.

We will then present the method used to address the issues managers pointed out, in order to transform the discussions into actions to strengthen the safety process leadership culture.

In the conclusion, we will present our findings and discuss possible future approaches to amplify the safety leadership culture further.

2 Safety Culture and Leadership: definitions and implications

There are many possible definitions for culture. Taylor (2015) says culture has to do with the behaviour patterns that are encouraged, discouraged or tolerated over time in an organization. Schein (2015) says that culture is "the total sum of everything an organization has learned in its history in dealing with the external problems – which would be goals, strategy, how we do things – and how it organizes itself internally."

Since we humans are gregarious social beings, we are naturally wired to fit in with our community. We use our perceptions to read what the tribe accepts and we adapt by behaving accordingly, because we want to fit in and be rewarded. Therefore, we reinforce the existing norms. This process is perpetuated over time; established behavior influences the behavior of new employees and so on.

Most cultural sources are non-verbal. Taylor (2015) says it is a myth to believe that culture reflects the values statements that are declared in the company's annual reports. Most commonly, the statement will be the organization´s intents, not the culture as it actually is.

If we want to change culture, it is necessary to have a clear direction and a focused, effort-driven leadership to lead people towards the goal culture. Culture has to do with what is encouraged and allowed. For Taylor (2015), three major sources send cultural messages: behaviours - of other people, especially important, recognized ones; symbols - artefacts, events and decisions that are meaningful to the community; and systems - mechanisms that manage people and tasks.

In this sense, culture has to do with the manifestation of what is valued in the organization.

According to Lafraia (2014), culture shapes reality. Cultural lenses structure the perception of the people involved in the organization.

When it comes to safety culture, its meaning may vary from one organization to another, or even in different departments of the same company. It is perfectly possible to have diverse lenses that translate the meaning and the importance of safety practices in diverse ways.

OGP, in its 452 Report (2013), says that a positive Safety Culture has to do with a culture in which safety is a core value for everyone who works in the organization, and it is pursued in an everyday basis. In order to build a Safety Culture, there are factors that must be considered, such as "leadership commitment to safety; employee involvement and motivation; responsibilities and accountability; production and bottom line pressures versus quality issues; and actions, or lack of action, to correct unsafe behaviours and unsafe conditions"; among others.

In order to build a consistent safety culture, leaders must educate their teams by demonstrating, by their practices, that some policies are fundamental. Safety culture should use this lens, rather than a command and control opaque one. Leaders must educate people to understand why safety is a core value that concerns the most precious gift: one´s life.

Safety culture is not about only following the rules, but making the team understand that the rules serve a greater principle. This approach can influence values and beliefs and change the individual and team´s behavior. The challenge leaders face is to conciliate the educational role with the policeman role, guaranteeing that compliance is required and everyone in the organization, starting by himself, values that as an unnegotiable principle.

In order to develop a strong safety culture, some elements must be considered (OGP, 2013):

Elements of a Strong Safety Culture

- an informed culture: the company has quality data concerning safety and uses it well;
- a reporting culture: people feel they can report safety concerns without worrying about being punished;
- a learning culture: the organisation actually learns from its unsafe conditions and changes them, banning any equipment, practice or whatever takes place that hazards safety;
- a flexible culture: the company changes its command chain if necessary to adapt;
- a just culture: people understand what is acceptable and unacceptable. What is not tolerated is dealt fairly.

For Peter Drucker (Cohen, 2018), leadership can be defined as the lifting of a man's vision to higher sights. Throughout leadership, people perform in higher standards, they are inspired and evolve in multiple levels.

When it comes to safety leadership, we can quote Ram Charan (2007) when he says that "a leader who does not produce other leaders is not a great leader." A safety leader should act in a way to enhance in everybody´s minds and actions a safety leadership commitment.

In order to make it possible, it is crucial that the leader himself is aware of his beliefs and constantly monitoring his behaviours. It takes time and a lot of effort to build a strong safety culture. Despite that, a safety culture is fragile and easily threatened by wrong messages. (OGP, 2013)

In the next section we explore the method used to address self-consciousness about safety.

3 Creating a trust environment that favors Safety Process Leadership Culture

According to Malcon Knowles, who develop the paradigm of andragogy, adults have a learning style characterized, among other factors, by changing one´s concepts according to new experiences one lives; the pursuit for autonomy; increasing one´s own motivational level to match the social role assumed; and acquiring knowledge to immediately solve practical issues (Muresan, 2013).

Considering this practical approach, we began the program to strengthen safety leadership culture by making each participant reflect on its daily practices concerning management, as well as the safety implications involved in routine decisions.

To start this reflection, we did a benchmark and decided to invite a senior top manager from another company in the Energy Industry to give a testimony. He made an honest statement about his team perception of his management practices. Concerning safety process, he mentioned that he realized that, in order to influence people, he must first change himself. He then shared with an amazed group of participants some practical examples of safety interventions he changed in his auditing procedures, for example, in order to be coherent with the message he wanted to pass forward.

The testimony had an impressive impact on the managers. A remarkable phrase that was repeated by managers was: "*Simplify. Be the Powepoint. Less Presentations, More You.*"

After the testimony, there was a presentation on the invited company's safety philosophy, about its journey into safety leadership. The practical steps mentioned were followed by videos with managers from different parts of the world talking about the way they see safety culture benefits.

With that background environment set, we forwarded to a self-perception diagnosis. The goal was to stimulate the increase in the individual conscious level as well as in the collective one. We used a matrix entitled 'Personnel Influence on Safety' that was obtained in the benchmark process with the above mentioned company.

It aimed to diagnose how one really sees himself concerning its leadership in safety culture. It involved two dimensions: the personal influence towards oneself; and the personal influence concerning other people´s safety.

The figure below illustrates the matrix:

Figure 1. Personal Influence on Safety Matrix

We used the Self-perception as a tool to promote reflection and consciousness. We asked the participants to fill in the matrix in two different papers: one in his scrapbook (explained below) and another one on a plain sheet to hand over to the program facilitator. All the matrices were compiled and the result is shown in figure 2. As indicated, most participants initially evaluated themselves as influencing more themselves than others.

The coordination staff evaluated the matrix result as requiring attention from the participants. The staff observed the fact that some managers attending the program were obese, others worked extra hours, and most of them had reported having trouble balancing personal and professional life. Therefore, to promote deeper self-reflection, the consultant who was facilitating the program posed some questions to the participants regarding personal care. The remarks made were an invitation to honestly look at oneself and feel if the level of personal influence corresponds to what one would recommend to a loved person. Quoting Brené Brown (2015), the consultant said "Treat yourself like someone you love. Do you do it on a regular basis? Do you put yourself first?" Those remarks made some participants change their self-evaluation.

The intention was to call leaders ´attention to the fact that genuine influence on safety starts with their own safety, and goes beyond process or operational safety. To be legit and truly be a safety leader, one should take good care of oneself.

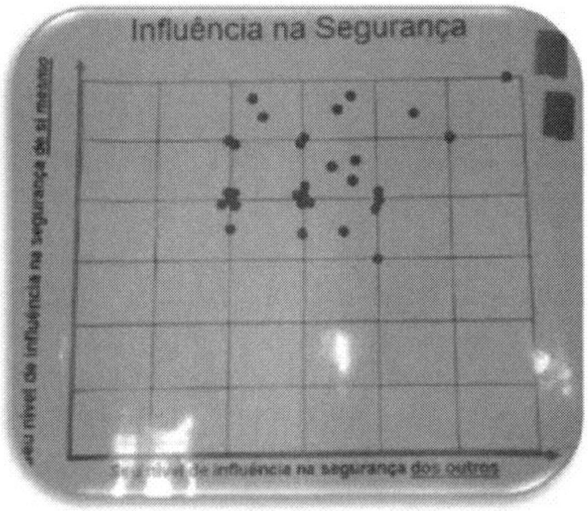

Figure 2. Compiled Personal Influence on Safety Matrix with the results from all the participants

We also developed a scrapbook that was given to each participant as a job-aid material to be used both during the program and after. The content of the scrapbook followed the program plan and was an invitation to self-analysis. The participants had special pages to do some journaling about what they had just listened to or dialogued about. There was also a special page dedicated to register insights. Figure 3 illustrates the scrapbook.

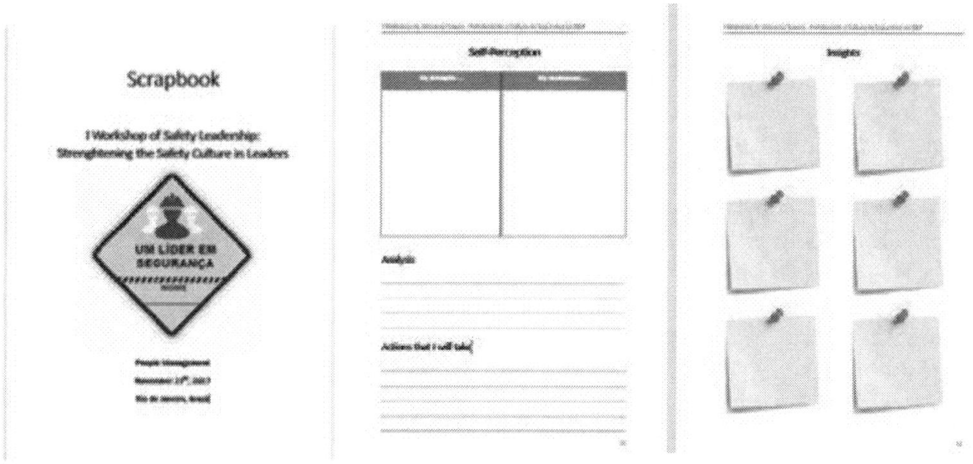

Figure 3. Pages from the Scrapbook used to stimulate self-consciousness and to plan actions

4 Exposing and Addressing Vulnerabilities

4.1 Control Room

We will now present the method used to address the issues managers pointed out, in order to transform the discussions into actions to strengthen the process safety leadership culture.

We prepared a special control room, where we designed a pathway decorated with images of accidents from the company and from the industry, and the number of fatalities. We also put an alarm sound playing during the route and videos showing accidents. These resources were intended to create an emotional ambiance to take the analysis to another level.

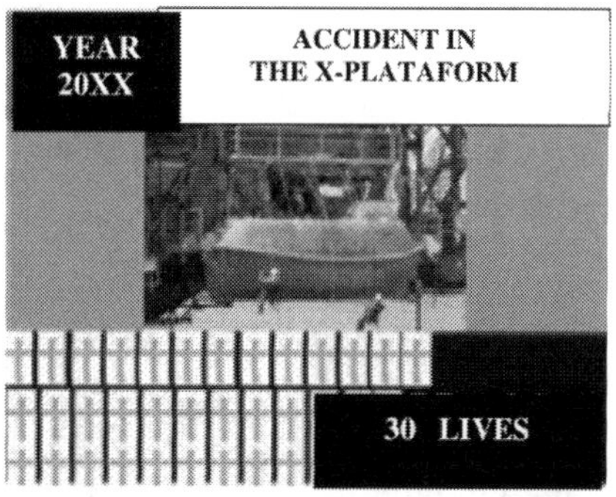

Figure 4. Picture of the Control Room and example of poster from accident used to decorate it.

Besides the planned journey passing several industry accidents, the Control Room was decorated with red material, to create an atmosphere of urgency. After going through the posters and watching the videos, the participants were invited to give their testimony of accidents they´ve been through.

The group tone got emotional as some of them shared unpleasant stories they had lived. The goal of inducing the reflection to an in-depth level was successfully achieved.

After that point, the participants went back to the regular room and dialogued about some questions that were in their scrapbook. The questions were related to the *Elements of a Strong Safety Culture* preconized by OGP (2013), mentioned in section 2 of this paper.

The methodology privileged social interaction among participants and the active listening of each other's considerations. Below is the list of the questions addressed to the participants.

Questions to Reflect and Dialogue:

1. Am I able to influence my operational team?
2. How many hierarchical levels are there between the operational team and me? To what extent does my speech actually makes sense to them?
3. How much of what is said by the operational team do I believe genuinely applies to me? Do I believe I receive good quality feedback?
4. Of the security actions implemented in my management, do I receive direct feedback on the effectiveness or otherwise of its implementation?
5. What degree of tolerance do I accept as a leader, regarding violations of safety standards?
6. Do I, as a leader, make it mandatory that the urgent inspection recommendations are executed immediately?
7. Do I, as a leader, take every opportunity to motivate, mentor and develop my team to enhance individual safety awareness and safety leadership?
8. Do I praise when I see a safe behavior?
9. Do I perform Behaviour Audits in a protocolled way, to meet the Scorecard, or do I do it because I believe in the effectiveness of what can arise from it?
10. Is Safety a value or is it a Priority for me as a leader?
11. Do I have the courage to step in and stop an activity when I see someone at risk? Do I always intervene?
12. Do my behaviours as a leader have safety as a foundation?
13. When I find that the environment is unsafe, do I privilege safety over other goals?
14. Safety comes first in my management?
15. How do I, as a leader, ensure that safety in my management is dynamic enough to ensure continuous improvement in operational procedures?
16. Thinking Safety as a chain, how can I, as a leader, be an effective link between operations and the company policy?

4.2 Results

In order to capture the precious discussions and share insights among the participants, in the final part of the workshop there was a collective presentation based on what each small group discussed about the questions above. Each group had to have at least one delegate, and others could participate if they wanted to.

Some of the findings reported were, for example: "*I have to go to the platform more often*"; *"Behaviour audits can be very useful, I should profit more from them*"; *"There are too many levels between my operational team and I. I have to get together with them on a regular basis."*

After sharing insights, all managers filled in the Personal Influence on Safety Matrix again. The group reported that, this time, the self-perception was lower, since many managers realized they need to invest more efforts and more quality time in Safety, in both dimensions – with oneself and towards others.

After that, each participant revised and complemented his individual action plan (figure 3). They also wrote a personal commitment as a Safety Process Culture Leader.

At the very end of the program, we asked everyone to do a check-out by sharing with the entire group one action of the action plan. This intended to create a deeper commitment, as well as possible insights to others. Also, it was a practice that could be used with the managers team.

According to the group evaluation, presented in the word cloud below, the program succeeded in strengthening the safety culture in leaders, who mentioned as the top benefits from the program: (*Safety as a*) Value; Communication; Life; Commitment; Example; Integration; Consciousness; and Synergy.

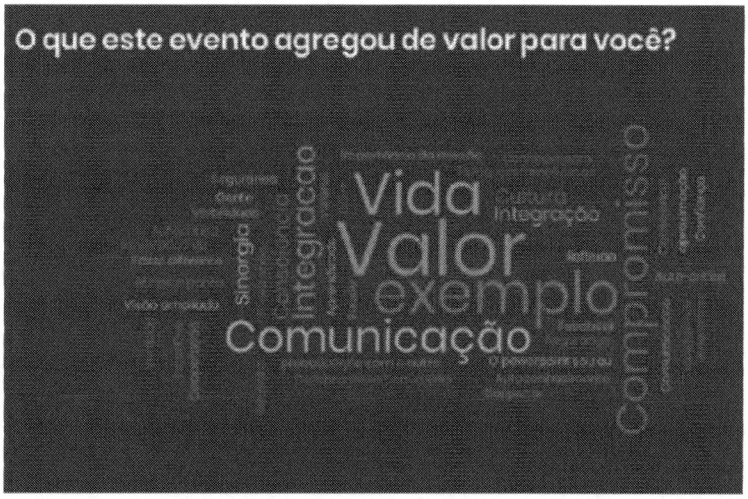

Figure 5. Evaluation: Word cloud containing the managers´ perception about the contribution brought by the program to participants.

5 Conclusions

The feedback given by the program´s participants was encouraging. It indicated that testimonies from partners, combined with relevant questions based on the daily problems managers face concerning safety, created a trust ambiance among managers.

The activities planned based on andragogic approaches favored dialogue and in-depth self-analysis. The use of the same instrument (Personal Influence on Safety Matrix) twice, as well as the individual scrapbook, facilitated the self-consciousness process and lead managers to review their initial assessment and to elaborate an action plan.

Leaders play a decisive role in strengthening the safety culture in the organizational world. Bringing about their vulnerabilities and having honest conversations with peers definitely helps managers to find ways to evolve from compliance to commitment.

Strengthening the safety leadership culture is a process that does not come to an end. It is an on-going process that must be continuously taken care of by leaders who value life and put safety first. Future projects that aim to implement effective actions in Safety Culture should take into consideration the company´s culture, as well as daily operational problems and safety as a core value to preserve life and business.

6 References

B. Brown, "*Daring Greatly: How the Courage to Be Vulnerable Transforms the Way We Live, Love, Parent, and Lead*", Penguin Random House, NY, 2015.

C. Migueles, J.R.B. Lafraia, G. C. Souza, "Criando o Hábito da Excelência: Compreendendo a Força da Cultura na Formação da Excelência em SMS", Quality Mark, Rio de Janeiro, 2014.

C. Taylor "*Walking the Talk: Building a Culture for Success*" Random House Business Books, London, 2015.

M. Muresan, "*Using Cybergogy and Andragogy Paradigms in Lifelong Learning*", 5th World Conference on Educational Sciences – WCES. Romania, 2013

OGP, *Shaping Safety Culture through Safety Leadership.* Report N° 452. International Association of Oil & gas Producers. October, 2013.

R. Charan, *Ram Charan on Leadership*,

Available at https://hbr.org/ideacast/2007/01/harvard-business-ideacast-27-r.html. Jan, 2007 (accessed Dec 4, 2018)

T. Kuppler, *Edgar Schein on Culture*,
Available at https://www.leadershipandchangemagazine.com/edgar-schein-on-culture/ Accessed on January 2, 2019.

W. Cohen, *Drucker's Top Tips to Successful Leadership*
Available at https://www.managementmattersnetwork.com/strategic-leadership/columns/druckers-top-tips-to-successful-leadership. Jun, 2018 (accessed Dec. 4, 2018)

7 Acknowledgments

I would like to thank Petrobras who fully supported this project, as well as Shell Brasil, for a very profitable partnership. A special thanks to my Human Resources, HSE and E&P work colleagues, who had a decisive role in making this project a reality.

548424 Development of Human Reliability Analysis Method for Oil and Gas Operations

The oil industry has grown in recent decades in terms of quantity of facilities and process complexity. However, human and material losses still occur due to major accidents at the facility. The analysis of these accidents reveals that many involve human failures that, if prevented, could avoid such accidents. These failures, in turn, can be identified, modeled and quantified through Human Reliability Analysis (HRA), which forms a basis for prioritization and development of safeguards for preventing or reducing the frequency of accidents. The most advanced and reliable HRA methods have been developed and applied in nuclear power plant operations, while the petroleum industry has usually applied Quantitative Risk Analysis (QRA) focusing on process safety in terms of technical aspects of the operation and equipment. A joint project from the B. John Garrick Institute for the Risk Sciences and a big company from the oil and gas industry is focusing on developing a HRA method focused on this industry operations, to bridge the existing gap between HRA and the Petroleum industry. The Phoenix HRA methodology was used as a basis, which has three layers i) a crew response tree (CRT), which models the interaction between the crew and the plant; ii) a human response model, modeled through fault trees, that identifies the possible crew failures modes (CFMs); and (iii) "contextual factors" known as performance influencing factors (PIFs), modeled through Bayesian networks. In addition to building on such a structure, the method development will rely on interviews with key personnel, visitations to installations, and analysis of past incidents data.

547327 Studying the Dependence of Operator Performance on Step Level Complexity

In the chemical, oil, and power industry, procedures are used by operators to perform their work correctly without compromising safety. Unfortunately, many procedures fail in adequately conveying information to their users. Heuristics tell us that technical jargon, vague language, and operator training all play a role in determining the effectiveness of procedures. To what extent these variables play a role is still not fully understood. Knowing this information will allow plants to more effectively write procedures. To counter this, researchers at Texas A&M's School of Public Health and Mary Kay O'Connor Process Safety Center are investigating the underlying factors that cause operators to deviate from procedures.

One possible factor may be that complex steps hinder an operator's ability to perform a task. This work aims to understand why certain steps are more complex than others. To do this, a set of procedures on how to operate a gas chromatography mass spectrometry (GCMS) machine is written. A study is performed in which participants use a set of procedures to analyze samples through gas chromatography. Using a scoring rubric, the complexity of each step is determined. The procedures have been designed to obtain an even distribution of complexity. Participants for this study are chemical engineering students with little to no prior experience in handling a GCMS machine. Thus, the experience, education, expertise, and training of the participants is controlled. The performance of each participant is reviewed and is grouped into one of four quadrants. For this study, a decrease in operator performance is expected as the complexity of each step increases.

548400 Prevent Them All or Stop Them Small: Tripod Best Practices in Reducing Chemical Accidents and Enhancing Response Preparedness

Prevent Them All or Stop Them Small: Tripod Best Practices in Reducing Chemical Accidents and Enhancing Response Preparedness

The connection between industry, government, and public safety, which we call the Tripod, is vital when implementing the 3-P's - Prevention, Protection, and Preparation. As identified by AICHE's Vision 2020 program, there are significant opportunities for organizations that address industrial process safety to work together. Responsible Collaboration between industry, regulatory and investigatory authorities, labor organizations, communities, and technical organizations promotes mutual understanding of risks, effects process safety and can reduce chemical accident risk in both small and large communities. Community based safety partnerships and leadership are fundamental to the success of such efforts.

For the past five years, a series of geographically focused Chemical and Ammonia Safety Days have cultivated and showcased chemical emergency prevention and planning efforts between Industry, NGOs such as the Ammonia Safety Training Institute (ASTI), and a myriad of public agencies including fire departments, health departments, LEPCs, SERCs, local regulators, and state and federal agencies such as EPA, CSB, DHS and OSHA. Within the context of Chemical Accidents and Releases, the focus of these efforts has been to Prevent Them All or Stop Them Small.

These Safety Day programs are designed for and with local community stakeholders from industry and first response agencies with a goal of seeding a permanent, engaged geographic collaboration for accident prevention and response readiness ownership and partnership. Agenda and curriculum elements include technical training focused on local risk, first responder field practicums and case studies, accident prevention program prioritization, practice and accountability, bowtie analyses of event propagation and management, and emergency response planning centered on the first 30 minutes and one-plan approach. The programs often include field demonstrations and exercises for fire departments and facility responders.

As a best practices demonstration, the presenters for this session will share the developed programs, recruitment methodologies and the demonstrated results of these efforts. In addition to Responsible Collaboration, other Vision 2020 principles embraced and promoted by the Safety Day Approach include Disciplined Adherence to Standards, Intentional Competency Development, Enhanced Application & Sharing of Lessons Learned, and Enhanced Stakeholder Knowledge. As a bonus, the societal theme of Harmonization of Standards is enhanced by the community of practice element of engaging federal, state and local regulators as well as industry stakeholders and process safety professionals.

548394 Risk Analysis Screening Tool – CCPS Project 280

A spreadsheet based Risk Analysis Screening Tool (RAST) is available through the Center for Chemical Process Safety (CCPS), courtesy of the Dow Chemical Company. The tool allows users to identify hazards and understand risks including causes, consequences and probabilities. RAST can be used for existing or proposed installations to facilitate MOC reviews, assess compliance with risk criteria, or evaluate proposed designs.

The users can input the type of equipment being used along with the equipment characteristics, chemical data, process conditions, and the plant layout. Using these inputs, the tool is used to help identify hazard scenarios and evaluate the consequences of representative loss of containment scenarios. RAST also allows the users to specify their own hazard scenarios based on their own knowledge of the process. Users can also input their own risk ranking matrix. With the hazard scenarios and the risk matrix, the user can select those scenarios of most concern and use the tool to create a Layer of Protection Analysis (LOPA) spreadsheet for the process.

As an example a comparison of RAST calculated scenario identification and consequence results to actual case studies from the CSB will be presented. Courses on how to use RAST, and how to adjust the pre-loaded default parameters are available.

550863 Impacts of Climate Change on Process Safety Management in Oil and Gas Operations

Impacts of Climate Change on Process Safety Management in Oil and Gas Operations

Chizaram Nwankwo

Coventry University

Project objectives and goals

The aim of this study is to analyse the impacts of climate change on process safety management in oil and gas operations. The objectives for this study include the following: -

- To evaluate past, present and future trends of climate change around various global regions
- To establish the relationship between climate change and process safety management in oil and gas operations
- To proffer adaptation and mitigation measures for climate change impacts on process safety management in oil and gas operations

In the last two decades, climate change has emerged as one of the most controversial topics that has been argued on several fronts (Hulme 2009). The idea of global warming and the negative impact of human-driven activity on the global weather has been fiercely debated (Cook *et al.* 2016). However, Field *et al.* (2012) argue that the facts and figures tend to reiterate the continuous occurrence and proliferation of climate change, as well as its increasing impacts in different industrial sectors. Indeed, the warming of the climate is unequivocal (Panteli and Mancarella 2015). According to IIGCC *et al.* (2017), 70% of global greenhouse gas (GHG) emissions are attributed to energy-related activities. The oil and gas industry contributes to 60% of these energy-related emissions due to production of fossil fuels as their main operational activity (Puig *et al.* 2015). Evidence from NASA (2018) suggests that there has been 1^0C increase in temperature which is close to the set mark in accordance to the Paris Agreement.

There are several case studies such as the flood in San Jacinto River, Houston Texas in 1984 and Coffeyville Refinery, Kansas in 2007. In the case of the San Jacinto flooding event in 1984, 8 pipelines were ruptured and 29 were undermined, 35,000 barrels of oil were spilled and ignited, while 547 people were injured and $23 million was incurred in damage costs (Matek 2017).

Hence, adaptive measures need to be applied to manage the severe consequences of climate change on oil and gas operations (Smith 2013). Such adaptive measures include more robust design specifications for oil and gas assets exposed to harsh climatic conditions (Cruz 2010). Ramasamy and Yusof (2015) suggest that this could be vital in offsetting the impacts of climate change on assets such as offshore rig platforms and equipment, transport pipelines, storage facilities etc. However, there is still little attention accorded to design and adaptation of oil and gas assets to offset the effects of climate change on them (Skjaerseth and Skodvin 2010). Therefore, this study seeks to analyse the impacts of climate change on deterioration of oil and gas assets.

Description of methods and results

Firstly, this study adopts a pragmatic paradigm since it explores the consequences of actions, is problem-centred, pluralistic, real-world practice oriented and allows for use of mixed models (Creswell and Creswell 2018). The area of investigation for this study is focused on impacts of climate change on oil and gas operations. The predominant approach taken for this study is a correlational research approach which tries to establish a relationship between climate change and oil and gas operations. The selected approach is due to the continuous occurrence of accidents in the oil and gas industry due to asset deterioration.

After thorough literature review of climate change trends and asset integrity issues in the oil and gas industry, this study identified a gap in knowledge that the oil and gas industry has not duly considered climate change during design of assets; hence posing a risk to oil and gas operations conducted under harsh climatic conditions. Therefore, the data collection for this study employed "concurrent triangulation" method of quantitative and qualitative data to ensure robust analysis of key variables such as climate change and oil and gas operations. Consequently, data collection tools for this study include questionnaires, interviews, archive data and documentary analysis. The data collection for this study was conducted over a period of 3 months.

Firstly, it involved piloting the questionnaire among academic and industry professionals to identify any errors, as well as testing and refining the data collection instrument. Next, the study was approved as a medium-risk project by Coventry University Ethics Committee, as primary and secondary data were collected from archive data, documents, questionnaires and interviews. The population sample for the study includes oil and gas industry professionals such as Mechanical Engineers, Facility Managers, Production and Well Engineers, Process Engineers, Health and Safety Officers, as well as Safety and Reliability Engineers. The questionnaires used for this study were designed using Bristol Online Survey (BOS) tool and covered questions around climate change and oil and gas operations. These data were predominantly collated and analysed using various software such as Zotero, Microsoft Excel and SPSS software. Statistical analysis conducted in this study include trend analysis, descriptive statistics, Pearson's correlation, Cronbach's alpha reliability analysis and Relative Importance Index (RII). Lastly, the findings of this study were benchmarked and justified with peer-reviewed academic literature.

Potential for application of results

This research could aid in curbing several accidents caused by asset deterioration due to increasing number of climate change events. It will help the oil and gas industry to avoid the enormous financial, legal and moral costs associated with accidents. Also, this study will provide a deeper insight on current climate change statistics and proffer adaptation and mitigation measures on how to deal with these problems. Similarly, it could broaden the scope of researchers to start incorporating impacts of climate change on various industrial sectors and subsectors asides the oil and gas industry.

References

Cook, J., Oreskes, N., Doran, P.T., Anderegg, W.R., Verheggen, B., Maibach, E.W., Carlton, J.S., Lewandowsky, S., Skuce, A.G., and Green, S.A. (2016) 'Consensus on Consensus: A Synthesis of Consensus Estimates on Human-Caused Global Warming'. *Environmental Research Letters* 11 (4), 048002

Creswell, J.W. and Creswell, D.J. (2018) *Research Design: Qualitative, Quantitative, and Mixed Methods Approaches.* 5th Edition (International Student Edition). Sage Publications

Cruz, A.M. (2010) *Vulnerability of Oil and Gas Infrastructure to Climate Change and Extreme Weather Events* [online] available from <http://indico.ictp.it/event/a09141/session/35/contribution/25/material/0/0.pdf> [5 March 2018]

Field, C.B., Barros, V., Stocker, T.F., and Dahe, Q. (2012) *Managing the Risks of Extreme Events and Disasters to Advance Climate Change Adaptation: Special Report of the Intergovernmental Panel on Climate Change*. Cambridge University Press

Hulme, M. (2009) *Why We Disagree about Climate Change: Understanding Controversy, Inaction and Opportunity*. Cambridge University Press

IIGCC, CERES, and IGCC (2017) *Global Climate Disclosure Framework for Oil & Gas Companies* [online] available from <http://www.iigcc.org/files/publication-files/Global_Climate_OG_DisclosureFramework.pdf> [14 November 2017]

Matek, B. (2017) *CHARACTERISTICS OF OIL AND GAS PIPELINE ACCIDENTS CAUSED BY CLIMATE CHANGE INTENSIFIED HURRICANES* [online] Baltimore, Maryland: Johns Hopkins University. available from <https://jscholarship.library.jhu.edu/bitstream/handle/1774.2/46088/Benjamin%20Matek.pdf?sequence=1&isAllowed=y> [26 June 2018]

NASA (2018) *Global Climate Change* [online] available from <https://climate.nasa.gov/> [9 May 2018]

Panteli, M. and Mancarella, P. (2015) 'Influence of Extreme Weather and Climate Change on the Resilience of Power Systems: Impacts and Possible Mitigation Strategies'. *Electric Power Systems Research* 127, 259–270

Puig, D., Haselip, J., and Naswa, P. (2015) *Adaptation to Climate Change in Colombia's Oil and Gas Industry: Recommendations to Promote Risk Management* [online] UDP-CRD-OLADE-01(EN). Denmark: UNEP DTU Partnership. available from <www.unepdtu.org> [19 September 2018]

Ramasamy, J. and Yusof, S.M. (2015) 'A Literature Review of Subsea Asset Integrity Framework for Project Execution Phase'. *Procedia Manufacturing* [online] 4, 79–88. available from <http://linkinghub.elsevier.com/retrieve/pii/S2351978915011336> [14 May 2017]

Skjaerseth, J. and Skodvin, T. (2010) *Climate Change and the Oil Industry: Common Problem, Different Strategies*. Manchester University Press

Smith, M.H. (2013) *Assessing Climate Change Risks and Opportunities for Investors: Oil and Gas Sector* [online] Analysis of Climate Change Risks and Opportunities. Technical Report. ANU and Investor Group on Climate Change. available here. [5 April 2018]

Further Information

Chizaram Nwankwo (E: nwankwo4@uni.coventry.ac.uk)

546246 Study of the Effects of Wind Speed and Surface Roughness Under Stable Atmospheric Condition Using Aloha Model and CFD Model

Study of the effects of wind speed and surface roughness under stable atmospheric condition using ALOHA model and CFD model

Many substances released in chemical processes are toxic, flammable, or even both. When these substances accidentally release to the atmosphere, human health and the environment could be threatened. As results, a reliable quantitative risk analysis tool is crucial to provide decision makers useful information on downwind concentration and hazards. ALOHA consequence model and Computational Fluid Dynamics (CFD) numerical method are widely used in chemical industries. Validation of these models under different scenarios is a necessary preliminary step before applying them to risk analysis. CFD models generally have acceptable concentration predictions on dense gas dispersion. ALOHA model is verified with DEGADIS model, but there is lack of study on effects of wind speed and surface roughness on dense gas dispersion. This paper compared field data from the Kit Fox experiments, which were CO_2 gas releases under stable atmospheric conditions with increased roughness, with CFD and ALOHA models' predictions. Also, statistical performance indicators were utilized to evaluate the accuracy of ALOHA model. Statistical performance indicators showed that ALOHA had systematic errors for dense gas dispersion with increased roughness under stable atmospheric conditions. Comparing effects of wind velocity profile and surface roughness values in CFD model and ALOHA model, this paper provided a better understanding of effects of key parameters in both ALOHA and CFD models. Furthermore, engineers would have better quantitative risk assessments under stable atmospheric conditions when they are using ALOHA consequence model and CFD model.

548222 Consequence Assessments and Collaboration to Streamline Project Siting

Collaboration is often sought after across disciplines, but all too often disregarded in favour of linear project development. For siting and consequence, as with other areas where design interacts with environment, health and safety constraints, this often leads to more costly design due to increased project iterations.

Stantec recently completed a consequence assessment for a compressor station to assess flammable extents during an emergency blowdown. At the same time, the client was experiencing siting issues due to sound emissions from the same process. The facility was grappling with competing factors: blowdown the facility in time to avoid catastrophic failure of their infrastructure, avoid excessive flammability extents from the blowdown stack location, and avoid extraneous noise exposure to nearby receptors. Using transient consequence modelling tools to model the blowdown sequence, Stantec was able to simultaneously and efficiently assess the flammable extents and sound emissions from the blowdown operations.

Consequence modelling for flammability was completed using a transient pipeline blowdown modelling tool to simulate the mass, energy, and momentum throughout the pipeline system. The model first simulated normal operating conditions, and then simulated the blowdown to estimate mass flow rate, liquid mass fraction, and release temperature from the blowdown stack. Several sizes of reducers upstream of the blowdown stack were introduced to estimate the pressure drops in the piping that would be required to appropriately reduce noise levels but also achieve full depressurization of the facility in time to meet safety criteria.

Consequence modelling of flammability was completed using SLAB, while the predicted change in sound pressure level from the blowdown stack was completed using Cadna acoustic modelling software. By combining these two analyses, the assessment costs streamlined for both required studies. In addition, the acoustics study benefited from additional insight provided by the consequence modelling tool.

548348 CFD Simulation of the Effect of Wind Speed and Release Elevation on Dispersion of Flammable Fluids at LNG Export Facilities

Facility siting studies for LNG export terminals are required by federal regulations to assess the potential hazard from accidental releases of flammable fluids. Federal regulations require under 49 CFR 193.2059 the evaluation of flammable vapor-gas dispersion distances from releases from flammable fluids in the facility for a range of representative wind conditions based on historical weather data expected to produce the maximum dispersion distances. A parametric CFD study will be conducted to evaluate not only the effect of wind speed but also the effect of the release elevation on the dispersion distance from flashing and jetting fluids. Hazard distances will be compared between the 2D software PHAST and the 3D software FLACS for a set of exemplar fluids found at LNG export facilities. The ability of the FLACS software to handle detailed 3D geometry provides for the evaluation of the effect of obstructions, congestion, confinement and mitigation measures on the dispersion distances. Release scenarios typical of LNG export facilities will be presented, with a focus on jetting and flashing releases of LNG, ethylene, propane, mixed refrigerants and heavy hydrocarbons.

546616 A Comparative Study of the Optimization Models for the Gas Detector Placement in Process Facilities

Gas detector network is an important layer of protection in process facilities for prevention gas release accidents of turning into fire and explosion disasters. But traditional standards just provide basic principles of the installation of gas detectors. Recently mathematical programming gradually became an important method to solve the problem. When using the mathematical programming method to optimize the detector layout, different optimization strategies will result in different optimization target models and detector layouts. In this study, three models have been developed and contrasted, namely minimal detection time P-Median model (MITP), minimal leakage concentration P-Median model (MICP), and minimal individual risk P-median model (MIRP), for the optimization of the gas detector placement based on stochastic programming. And as a basis, possible release scenarios were identified. In order to achieve the goal of using finite discrete scenarios to represent all scenarios effectively and quantitatively, clustering analysis was used to filter similar scenarios and select representative leak scenarios. The input data of these models was obtained from computational fluid dynamics. Furthermore, an adaptive particle swarm optimization (APSO) algorithm was applied to solve these models. The gas detector network optimization in a diesel hydrogenation unit was carried out as a case study. Results demonstrate that the MICP model is superior to the MITP model and MIRP model in describing the integrity of hazardous gas leakage and diffusion distribution. But the cumulative detection time of its optimized layout is slightly longer than that of MITP. Moreover, the control effect of MIRP model on personal risk caused by hazardous gas leakage was increased by 24.2%, although the minimum number of detectors was slightly higher than that of MICP model.

549411 Reducing Industrial Risk in a Converging It/OT World: 20 Experts Share Their Strategies

Rapid adoption of Industrial IoT requires tight integration between IT and OT. CISOs and corporate OT leaders are now scrambling to develop strategies, secure budget, and implement solutions to secure their mission-critical industrial control systems against burgeoning cyber threats.

Effective integration of IT and OT cybersecurity strategies is essential to address risks successfully. To secure industrial facilities and ensure safe, reliable production, IT and OT security teams – traditionally two separate disciplines with different priorities – must collaborate and share cybersecurity and risk management best practices.

In a series of interviews in late 2017 and early 2018, more than 20 different experts on the front lines of OT cybersecurity risk mitigation shared practical advice for making control systems more secure. These experts represented a diverse range of industries including oil and gas, chemicals and refining, and power generation.

In their interviews, they shared their insights into how to leverage IT/OT convergence and develop an effective OT cybersecurity program that meets challenges in real-world ICS environments. Technology plays an important role, but as these experts also point out, more technology isn't always the solution. Just as important is the need to establish effective communication and trust between cybersecurity professionals and OT engineers.

Security project teams must make a concerted effort to create alignment between business management, the cybersecurity community, and OT professionals. Without alignment, security initiatives will fail to attain the support required to ensure success.

In this presentation, attendees will learn:

- Practical, real-world recommendations from OT cybersecurity experts
- Best practices for breaking down barriers and building trust between cybersecurity professionals and OT engineers
- How to align business management and plant operations on cybersecurity strategies

549417 The Intersection of OT Security and Process Safety

The rise of cyberattacks on critical infrastructure creates troubling concerns for security, economic stability, and personnel and public safety. However, many companies are still relatively unprepared to successfully defend their facilities from cyberattacks launched by increasingly sophisticated threat actors. Safe production is at risk.

Safety focuses on incident prevention and measures risks to personnel. Cybersecurity focuses on protecting systems from threat actors and measures risk based on the threat landscape. The increased network connectivity that exists within industrial facilities today means that cybersecurity is now closely linked to the safe operation of industrial facilities and their processes. Successful defense of cyber-physical systems from nation-state attacks requires close coordination and collaboration between cybersecurity and process safety. Every safety initiative must now consider cyber risk. Every cybersecurity initiative must now consider safety.

Considerable overlap exists between risk analysis in safety and cybersecurity. Safety and cybersecurity disciplines can learn much from each other. However, safety and cybersecurity standards and corporate organizational structures remain siloed in far too many companies. This session focuses on the importance of executing a cohesive strategy for both safety and cybersecurity to ensure safe and reliable operations.

In this session, we will:

- Examine the existing cyber threat landscape
- Discuss how established process safety best practices can be applied to OT cybersecurity
- Provide practical guidance for improving cyber defenses and reducing safety and environmental cybersecurity risks

551385 Process Safety at Offshore Aerodromes (Helidecks)New Application for Process Safety

Offshore air transport is an importance activate of the oil & gas industry. In this way, the management of air operations is essential for maintaining the safety and reliability. The present work has the objectives of divulging the methods used by
Petrobras to audit maritime units with helideck and to demonstrate the contribution of such methods to the safety of process in the air operations. To achieve these objectives, the paper was based on the Center for Chemical Process Safety (CCPS) publication "Guidelines for the Safety of Risk-Based Processes", and aims to share the Bow tie methodology for the qualitative analyzes of risks inherent in Air Operations, focusing on the barriers typically faced by oil and gas industry, in order to contribute to the process by providing a graphical and structured approach to the required barriers, both to avoid catastrophic consequences and to prevent events recognized as precursors.

Thermodynamics and Computational Methods for Relief Valves

By Umesh Mathur, P.E.
umesh.mathur@aggreko.com , mathuru@gmail.com
Aggreko Process Services
3732 Magnolia Street, Pearland, Texas 77584, USA

15th Global Congress on Process Safety (AIChE, March, 2019, New Orleans)

1. INTRODUCTION

The computation of flow through a relief valve is one of the most important aspects of safety system design. This paper discusses the fundamental question of choosing the proper thermodynamic basis for such calculations. We also discuss the discrepancies we have observed in the literature on this crucial question and attempt to dispel some of the misconceptions we have found.

Our concern with this issue arose from a study of the relevant literature on this subject, which shows that the flow of a compressed gas (or gas-liquid mixture) through a rigidly held valve or hole in a pipeline must be treated as an irreversible, isenthalpic process. This reasoning flows simply from the First Law of Thermodynamics as applied to a flow process.

For example, as early as 1947, Hougen and Watson[1] discussed "Free Expansion" of gases (p. 551) and noted as follows:

"The unrestrained expansion of a gas is known as free expansion. Under conditions of no restraint no work is done, and under adiabatic conditions no heat is added. Free expansion under flow conditions is commonly known as throttling or as the Joule-Thompson effect referred to in Chapter XII.

The Joule-Thompson effect is measured experimentally by expanding the gas slowly and in steady flow through a well insulated porous plug; in this way, potential work is lost, and no heat is allowed to enter or leave the system through the walls. The fluid flows reversibly into and out of the process, but the expansion step is completely irreversible. Since no heat is added or lost, the process takes place under conditions of constant enthalpy, according to Equation (VII-11), page 207. However a change in internal energy results if any change in flow energy pV occurs."

Similarly, in a discussion of valves and throttles, Elliott and Lira[2] observe as follows:

"A throttling device is used to reduce the pressure of a flowing fluid without extracting any shaft work and with negligible fluid acceleration. Throttling is also known as ***Joule-Thompson expansion*** *in honor of the scientists who originally studied the thermodynamics. An example of a throttle is the kitchen faucet. Industrial valves are modeled as throttles....Changes in kinetic and potential energy are small relative to changes in enthalpy as we just discussed. When in doubt, the impacts of changes in velocity can be evaluated as described in Example 2.9. The amount of heat transfer is negligible in a throttle. The boundaries are not expanding, and there is also no mechanical device for transfer of work, so the work terms vanish. Therefore, a throttle is isenthalpic."*

Finally, in a discussion of nozzles, Klein and Nellis[3] observe that:

"...The potential energy terms of the inlet and outlet flows are not usually significant. Nozzles neither require nor produce power ($W_{in} = W_{out} = 0$) and are very nearly adiabatic. Note that the kinetic energy terms associated with the flows, particularly the outlet flow, cannot be neglected as the purpose of the nozzle is to increase the kinetic energy of the fluid. With these simplifications, an energy balance on the nozzle becomes:

$$H_{in} + \tfrac{1}{2}V_{in}^2 = H_{out} + \tfrac{1}{2}V_{out}^2 \qquad (4\text{-}54)"$$

In the light of these assertions by these noted authorities, and others[6][10], it is hard to understand how the flow through a relief valve, or a hole in a vessel that is open to the atmosphere, can be treated as isentropic (***reversible*** adiabatic) in nature, as opposed to the isenthalpic (***irreversible*** adiabatic) basis that is consistent with free expansion. This is because an isentropic expansion process would necessarily show an outlet enthalpy for the gas from the nozzle or valve (or an orifice in a punctured vessel) to have a lower enthalpy than the inlet, the difference in such enthalpies being equal to the work done, for example on a rocket or airplane. However, a relief valve or vessel remains stationary and absolutely no work is done during a free expansion by the gas flowing through.

Nevertheless, several standard references have used isentropic formulas for free expansion[4],[5] in stationary valves, or nozzles, from the inlet to the throat. We believe that this methodology is flawed as it violates the steady flow energy equation (i.e., the first law of thermodynamics written for an open system). This procedure would result in an erroneous prediction of outlet gas temperature when entropy is held constant. This, in turn, would predict an erroneous outlet gas density and thus, for a desired gas relief rate, the required cross-sectional area for the valve throat and also any conduit or pipe downstream of the valve.

The practical consequence of this anomaly in predicting outlet gas properties (especially the gas density at the throat) would be that the relief valve sizing could be incorrect. Therefore, it is important that the proper thermodynamic basis (isenthalpic expansion) be used in all such computations.

Conversely, for a given relief valve, errors in throat density based on an isentropic path would also affect the relief rate. Both types of errors could lead to potentially serious process safety concerns.

For these reasons, when calculating the release rate for gas at a high pressure through a relief valve, it is important to be able to use the proper basis for gas enthalpy (and therefore accurate computations of fluid thermodynamic properties). The same considerations apply to the gas release rate calculations resulting from a rupture in a pipeline or vessel containing a gaseous mixture under pressure.

To help clarify the proper methodology for a relief valve - or a hole in a pressurized vessel - the computations that should be performed are described below. We also show a few simple examples. Note that the possibility of sonic velocity at the throat, or choking, of the valve (or at the orifice in a ruptured vessel) is explicitly considered. In both such cases, the procedure described below is subject to the following assumptions:

- The opening can be represented by a convergent nozzle of exit area equal to the area of the throat of the relief valve, or hole in the case of pipeline rupture

- The compressed gas pipeline, or leaking container, is a large reservoir in which upstream pressure and temperature are maintained constant
- The compressed gas undergoes an irreversible, adiabatic process of expansion through the nozzle.

The flow through such a device from either (a) the inlet to the throat, or (b) from the throat to the outlet follows the steady flow energy balance equation (Equation 9.21, Himmelblau & Riggs[6]):

$$\Delta E = Q + W - \Delta(H + KE + PE) \qquad (1)$$

Where:

ΔE = energy accumulation (zero at steady state)

Q = net heat transfer into the system (very close to zero for a relief valve)

W = net work done by the system (zero for a relief valve)

ΔH = change in enthalpy of the system

ΔKE = change in kinetic energy of the system

ΔPE = change in potential energy of the system (zero for a relief valve)

Clearly, for steady flow through an orifice, valve (or relief valve), ΔE, Q, W, and ΔPE are all zero. Therefore, we are left with:

$\Delta(H + KE) = 0$, or

$\Delta H = -\Delta KE$, or

$$H_{out} = H_{in} - \Delta KE \qquad (2)$$

In other words, the outlet enthalpy is equal to the inlet enthalpy minus the change in kinetic energy in going through the throat or orifice. (It should be noted that the same logic would apply for discharge through an ordinary valve, a relief valve, or an orifice plate).

If there is no change in velocity, the outlet enthalpy will be the same as that of the inlet gas. This phenomenon is generally referred to as a Joule-Thompson (J-T) expansion. For an ideal gas, enthalpy is a function of temperature only; therefore, a J-T expansion would result in no change in temperature. However, for most real gases, a J-T expansion results in a decrease in temperature. There are a few notable exceptions, however, such as hydrogen or helium gas for which the J-T coefficient has the opposite sign[10]. On the other hand, if the gas speeds up, its kinetic energy will increase and, therefore, its enthalpy will decrease in accordance with equation (2) above. This kinetic energy increase would result in a decrease in outlet gas enthalpy, and therefore temperature, even for an ideal gas.

For ideal gases, the computation of enthalpy for ideal gases is straightforward as it is a function of temperature only. For real gases, however, enthalpy is also a function of pressure, temperature, and composition and use of an equation of state (EOS) becomes necessary[2] to calculate the enthalpy departure from the ideal, as discussed below.

COMPRESSIBILITY FACTOR AND ENTHALPY FOR REAL GASES

For real gases, enthalpy is found using a suitable equation of state (EOS), such as the Soave-Redlich-Kwong (SRK), Peng-Robinson (PR), or Lee-Kesler (LK). This procedure may be summarized as follows:

At the given upstream pressure (P), temperature (T), and composition, solve the EOS for the compressibility factor (z). This requires solving a cubic polynomial in z when using any cubic EOS, such as the SRK or PR, using either an explicit analytical or numerical trial and error method. For the LK, however, a trial and error procedure is always required as the LK EOS requires solving the 8-constant Benedict-Webb-Rubin[2] (BWR) equation, which is not a simple cubic in z, once for the simple fluid and again for the reference fluid.

Using the appropriate thermodynamic expressions for the EOS, use this value of z to calculate ΔH_d, the enthalpy departure from ideal.

Calculate the ideal gas enthalpy, H_0 (function of temperature T_0 only).

Find the real gas enthalpy using:

$$H_r = H_i - \Delta H_d \qquad (3)$$

Where:

H_r = real gas enthalpy
H_i = ideal gas enthalpy
ΔH_d = enthalpy departure from ideal

This procedure must be followed for calculating enthalpy at both upstream and downstream conditions.

The gas velocity at the throat may be sonic or sub-sonic. Also, as the gas speeds up in passing through the valve throat or hole (since the cross-sectional area is lower than that of the pipe), its kinetic energy increases and this results in a reduction of the enthalpy (Elliott & Lira[2], Eq. 2.53):

$$H_t = H_0 - \Delta KE \qquad (4)$$

Where:
H_0 = upstream gas enthalpy[J/kg]
H_t = exit gas enthalpy at the throat or hole[J/kg], and
ΔKE = change in kinetic energy[J/kg]

ΔKE is found as follows (Elliott & Lira's Eq. 2.53):

$$\Delta KE = (u_t^2 - u_0^2)/2 \qquad (5)$$

Where:

ut = throat (or hole) velocity[m/s], and

u_0 = upstream velocity[m/s].

Clearly, the outlet and inlet enthalpies are equal only if the change in velocity is negligible.

RELEASE RATE CALCULATION

The computation of the release rate from a pipe rupture should be done subject to the following equations:

Bernoulli's Equation

This equation relates upstream and downstream conditions for an arbitrary flow geometry, as follows:

$$P_0 + \rho_0.u_0^2/2 + \rho_0.g.z_0 = P_e + \rho_e.u_t^2/2 + \rho_t.g.z_t + \Delta P_{friction} \quad (6)$$

Where the subscripts o and e refer to the inlet and exit planes, respectively, and

P	= pressure[Pa]
ρ	= density[kg/m^3]
u	= velocity[m/s]
g	= gravitational constant[m/s^2]
z	= elevation above datum[m]
$\Delta P_{friction}$	= friction losses[Pa]

The friction losses have been ignored for simplicity in this presentation.

Obviously, for a hole, or throttle (or relief) valve, the elevations z_0 and z_t are the same.

Sonic Velocity

The sonic velocity in any medium[2] may be expressed as:

$$u_s = (\gamma\ .\ z\ .\ R\ .\ T\ /\ M)^{0.5} \quad (7)$$

Where:

u_s	= sonic velocity[m/s]
γ	= gas C_p/C_v ratio[]
z	= gas compressibility factor[]
R	= gas constant = 8314.46[J/(kgmol.K)]
T	= temperature[K]
M	= gas molecular weight[kg/kgmol]

In Equation (7), properties at the throat must be used for choked flow calculations. For ideal gases, the compressibility factor z is equal to 1. For real gases, however, the compressibility factor z is generally not equal to 1 and can be found only using an appropriate EOS.

C_p, C_v, and $\gamma=C_p/C_v$ ratio

Similarly, for ideal gases, the value of γ is found from the value of C_p, which is a function of temperature only, and the value of C_v is found from the following:

$$C_v = C_p - R \quad (8)$$

For real gases, however, the value of γ again requires use of an equation of state to calculate the ΔC_{pd} and ΔC_{vd} departures from ideality. The real gas C_p and C_v values are found using the ideal values and these calculated departures from ideality, by subtraction, in a manner similar to that described above for enthalpy:

$$C_{p1} = C_{p0} - \Delta C_{pd} \quad (9)$$

Where

C_{p1} = real gas C_p[J/(kg.K)]

C_{p0} = ideal gas C_p[J/(kg.K)]

ΔC_{pd} = departure in C_p from ideal gas value[J/(kg.K)]

Similarly, $C_{v1} = C_{v0} - \Delta C_{vd}$ (10)

In Equation (10),

C_{v1} = real gas C_v[J/(kg.K)]

C_{v0} = ideal gas C_v[J/(kg.K)]

ΔC_{vd} = departure in C_v from ideal gas value[J/(kg.K)]

In a conduit, orifice, or valve, the actual gas velocity, u_e, can never exceed the sonic limit, u_s. Note that u_s, in general, is different at the upstream conditions compared to the value at throat conditions. The upstream and throat temperatures and pressures of the gas may be different owing to the kinetic energy effect, Equation (4).

At a sufficiently low upstream pipeline or vessel pressure, the hole velocity would be lower than the sonic velocity limit. Above some critical limit for the upstream pressure, however, the maximum discharge flow is limited by the sonic velocity criterion at the hole.

Density

Gas density is computed using:

$\rho = P . M / (z . R . T)$ (11)

Where:

ρ = density[kg/m^3]

P = pressure[Pa]

M = gas molecular weight[kg/kgmol]

z = gas compressibility factor[]

R = gas constant = 8314.46[J/(kgmol.K]

T = temperature[K]

The gas compressibility factor is 1 for ideal gases; it is calculated from the EOS for real gases.

1.2 Velocity

The velocity in a pipe or hole (or valve throat) is found using:

$u = W / (\rho . A)$ (12)

Where:

u = velocity[m/s]

W = mass flow rate[kg/s]

ρ = gas density[kg/m^3]

A = cross sectional area[m^2]

For a circular pipe, the cross-sectional area is defined as:

$A = \pi.D^2/4,$ (13)

Where:
D = inner diameter[m]

For non-circular cross sections, the equivalent diameter D_e is defined using the concept of the hydraulic radius which is defined as:

D_e = 4 (Cross-sectional-area/Wetted-perimeter) (14)

It can readily be seen that this definition holds for a circular pipe also: $4 \cdot \pi \cdot D^2 / 4 / (\pi \cdot D) = D$

As noted earlier, for ideal gases, the compressibility factor z is equal to 1. For real gases, however, finding the gas density ρ requires first calculating z from the EOS. This is done using the procedure as described above.

It should be noted that conditions at the throat of a valve (or at the hole in a pipe) are not known. Therefore, the computation of the gas flow rate must be done using an iterative procedure. When using Microsoft Excel, this is best performed using the Solver Add-in. Solver is a non-linear optimization technique that can simultaneously adjust the unknown quantities and solve the problem by enforcing the constraints imposed by the Bernoulli Equation (6) for throat velocity, and the Steady-Flow Energy Equation (1) for hole enthalpy. Also, the calculated throat velocity from the Bernoulli equation can never exceed the sonic velocity limit, Equation (7), and this constraint must be added explicitly when the upstream pressure exceeds a threshold value.

When solving for an unknown gas flow rate, if the upstream pressure is sufficiently low, hole velocity could remain sub-sonic. However, if the upstream pressure is sufficiently high, the pressure at the hole and also the gas mass flow rate would vary in a manner that obeys the sonic velocity limit at the hole.

Example

CASE (A): Methane is flowing through a 300[mm] diameter, above-ground pipeline at an absolute pressure of 46[bar] and a temperature of 300[K]. The pipeline develops a hole with an equivalent diameter of 50[mm]. Calculate the mass rate of leakage to atmosphere and also the velocity of the gas jet emerging from the hole. The pressure inside the pipeline may be assumed to be constant owing to continuous supply from the source.

CASE (B): Rework Case (A) assuming that the pressure in the pipeline is 1.5[bar] instead of 46[bar].

Case (A) P_o = 46[bar]

PROBLEM: Solve for release rate from a hole in a pipeline - sonic velocity at throat.

WE SHOW THE FOLLOWING DETAILED SOLUTION FOR THE IDEAL GAS CASE ONLY. Note that the solution of the real gas case requires use of complex computer programs to solve the chosen equation of state (e.g., Lee-Kesler) iteratively for the required thermodynamic quantities. Since these routines are imbedded in an outer loop for the unknowns (T and P at the throat, and the mass flow rate), the computational burden is too great for hand calculations.

The POLYMATH software[7] was used for solving this problem. All variables are defined in Table 1. The problem statement for Case (A) for the ideal gas case is as follows:

 # Initial guesses

$T_t(0) = 300$ # Initial guess for T_t[K]
$W(0) = 20$ # Initial guess for W[Kg/s]
$P_t(0) = 46$ # Initial guess for P_t[bar]

 # Aly-Lee ideal gas enthalpy constants:
A = 33298 # Ideal gas enthalpy constant A
B = 79933 # Ideal gas enthalpy constant B
C = 2086.9 # Ideal gas enthalpy constant C
D = 41602 # Ideal gas enthalpy constant D
E = 991.96 # Ideal gas enthalpy constant E
F = -120280000 # Ideal gas enthalpy constant F

 # The following three functions must be made equal to zero by adjusting the three unknown variables W[kg/s], T_t[K], and P_t[bar]:
$f(T_t) = h_t - h_{t2}$
$f(W) = u_t - u_{t2}$
$f(P_t) = u_t - u_s$

 # Initial guess ranges for non-linear equations
W(min) = 1
W(max) = 50
$T_t(min) = 200$
$T_t(max) = 400$
$P_t(min) = 10$
$P_t(max) = 46$

This problem requires solving the three simultaneous, non-linear algebraic equations (defined above) by varying T_t (the hole temperature), W (the mass flow rate), and P_t (the hole pressure):
$f(T_t) = 0$
$f(W) = 0$
$f(P_t) = 0$

In this case, the upstream pressure is high enough to cause the velocity at the hole to reach the sonic limit. This problem requires solving three simultaneous non-linear algebraic equations. The POLYMATH solution shown below for Example A was obtained using a software package called POLYMATH[7]. Note that any other similar package can also be used. We verified this numerical solution using the MATHCAD[8] and EXCEL[9] software packages also.

NOTE: The numerical solutions for Cases (A) and (B) are shown in Table 1, following the problem statement for Case (B).

Case (B) $P_0 = 1.5$[bar]

PROBLEM: Solve for release rate from a hole in a pipeline. Assume that the throat velocity is sub-sonic. This means that the outlet pressure is fixed at 1[atm] = 1.01325[bar].

When the outlet pressure is fixed, we are left with two non-linear equations. Also, for consistency with the methodology in (A) above, we use a very large upstream diameter to make the inlet velocity very small.

NOTE: If the calculated outlet velocity exceeds the sonic limit, we would revert to the problem formulation under (A) above. (This would add an equation requiring outlet velocity to equal sonic velocity, and thus again require solving three simultaneous non-linear equations.)

As noted earlier, the solution of the real gas case requires use of complex computer programs to solve the chosen equation of state (Lee-Kesler, in these examples) for the required thermodynamic quantities.

In this case, the upstream pressure is low enough so that the velocity at the orifice does not reach the sonic limit. Accordingly, this problem requires solving only two simultaneous non-linear algebraic equations. The solution shown below was obtained using a software package called POLYMATH. Note that any other similar package can also be used. As with Case (A), we verified this numerical solution using the MATHCAD and EXCEL software packages also.

The POLYMATH problem statement for Case (B) is as follows:

$f(T_t) = h_t - h_{t2}$

$f(W) = u_t - u_{t2}$

 # Initial guesses

$T_t(0) = 300$ # Initial guess for T_t[K]

$W(0) = 20$ # Initial guess for W[Kg/s]

Using POLYMATH, the solution for Case (A) shows that the gas release rate, W, for an ideal gas is 17.16[kg/s]. As noted above, identical results were obtained using the Mathcad and Excel software packages.

TABLE 1: SONIC OR SUB-SONIC FLOW THROUGH A VALVE OR HOLE

VARIABLES AND EQUATIONS

Variable	Description & Units	Explicit equations
A	Ideal gas enthalpy constant A	A = 33298.
B	Ideal gas enthalpy constant B	B = 79933.
C	Ideal gas enthalpy constant C	C = 2086.9
D	Ideal gas enthalpy constant D	D = 41602.
E	Ideal gas enthalpy constant E	E = 991.96
F	Ideal gas enthalpy constant F	F = -120280000.
P_t	Throat pressure [bar]	P_t = 1.01325
kcal_j	Conversion: 1[kcal] = 4184[J]	kcal_j = 4184
M	Gas molecular weight[kg/kgmol]	M = 16.0
T_0	Inlet gas temperature[K]	T_0 = 300
P_0	Inlet gas pressure[bar]	P_0 = 1.5
z_0	Inlet gas compressibility factor[]	$z_0 = f(T_0,P_0,x)$
$(c_p/c_v)_0$	Inlet gas C_p/C_v ratio[]	$(c_p/c_v)_0$ = 1.32
R	Gas constant[J/(kgmol.K)]	R = 8314.46
D_0	Diameter of pipe[m]	D_0 = 0.3
D_t	Diameter of Throat[m]	D_t = 0.05
h_0	Inlet enthalpy[kcal/kg]	h_0 = (A*T_0 + B*C /tanh(C/T_0) - D*E / tanh(E/T_0) + F)/M/kcal_j
rho_0	Inlet density[kg/m^3]	rho_0 = P_0*M*1.e5/R/T_0
A_0	Pipeline cross-sectional area[m^2]	A_0 = 3.14159/4*D_0^2
A_t	Throat flow area[m^2]	A_t = 3.14159/4*D_t^2
h_t	Throat enthalpy[kcal/kg]	h_t = (A*T_t + B*C/tanh(C/T_t) - D*E/tanh(E/T_t) + F)/M/kcal_j
u_0	Upstream velocity[m/s]	u_0 = W/rho_0/A_0
rho_t	Throat density[kg/m^3]	rho_t = P_t*M*1.e5/R/T_t
u_t	Throat velocity[m/s]	u_t = W/rho_t/A_t
u_s	Sonic velocity at throat[m/s]	u_s = (z_0*$(c_p/c_v)_0$*R*T_t/M)$^{0.5}$
u_{t2}	Throat velocity from Bernoulli Eq.[m/s]	u_{t2} = (2/rho_t*(P_0*1.e5 + rho_0*u_0^2/2 - P_t*1.e5))$^{0.5}$
dh_{KE}	ΔKinetic Energy = Δ(u^2)/2[kcal/kg]	dh_{KE} = (u_{t2}^2 - u_0^2)/2/kcal_j
h_{t2}	Calc. enthalpy at throat = inlet enthalpy - Δ(u^2)/2[kcal/kg]	h_{t2} = h_0 - dh_{KE}
T_t	Temperature at throat[K]	$f(T_t) = 0$
W	Mass flow rate[kg/s]	f(W) = 0
P_t	Pressure at throat[bar]	$f(P_t) = 0$

TABLE 1: SONIC OR SUB-SONIC FLOW THROUGH A VALVE OR HOLE (CONTINUED)

CASE (A) AND CASE (B) SOLUTIONS (USING POLYMATH)

Variable	CASE (A) Value (SONIC)	CASE (B) Value (SUB-SONIC)
A	33300.0	33300.0
B	79930.0	79930.0
C	2086.9	2086.9
D	41600.0	41600.0
E	991.96	991.96
F	-120300000.0	-120300000.0
P_t	27.71686	1.01325
kcal_j	4184.0	4184.0
M	16.0	16.0
T_0	300.0	300.0
P_0	46.00	1.50
z_0	1.00	1.00
$(c_p/c_v)_0$	1.32	1.32
R	8314.46	8314.46
D_0	0.30	0.30
D_t	0.05	0.05
h_0	226.2098	226.2098
rho_0	29.50683	0.96218
A_0	0.07069	0.07069
A_t	0.00196	0.00196
h_t	205.2641	210.3263
u_0	8.22527	7.71172
rho_t	20.86568	0.73254
u_t	418.7383	364.6543
u_s	418.7383	427.2981
u_{t2}	418.7383	364.6543
dh_{KE}	20.94576	15.88352
h_{t2}	205.2641	210.3263
T_t	255.6215	266.1833
W	17.15555	0.52433
P_t	27.71686	1.01325

CASE (A) RESULTS

This problem was also solved using the Lee-Kesler equation of state to calculate non-ideal gas properties, using a suitable computer program. This solution is provided below, with the ideal gas results quoted alongside for comparison:

	CASE (A)	
	IDEAL GAS	REAL GAS
Inlet pressure[bar]	46.00	46.00
Inlet temperature[K]	300.00	300.00
Calculated mass flow[kg/s]	17.28	18.25
Inlet velocity[m/s]	8.27	8.09
Throat velocity[m/s]	424.65	421.48
Sonic velocity at throat[m/s]	424.65	421.48
Inlet Enthalpy[kcal/kg]	225.61	214.87
Delta Kin. Energy[kcal/kg]	21.54	21.22
Outlet Enthalpy[kcal/kg]	204.07	193.64
Throat pressure[bar]	27.32	25.73
Throat temperature[K]	254.27	250.83

It should be noted that these results are quite similar for this case because the behavior of the gas (methane) is close to ideal at the conditions specified. However, these results would very likely differ significantly if, for example, the inlet temperature was considerably lower and the pressure much higher in which case the gas behavior would be markedly non-ideal.

CASE (B) RESUTS

Case (B) was also solved using the Lee-Kesler equation of state to calculate non-ideal gas properties. This solution is provided below, with the ideal gas results quoted alongside for comparison:

	CASE (B)	
	IDEAL GAS	REAL GAS
Inlet pressure[bar]	1.5	1.5
Inlet temperature[K]	300.0	300.0
Calculated mass flow[kg/s]	0.524	0.525
Inlet velocity[m/s]	7.71	7.70
Throat velocity[m/s]	364.7	364.2
Sonic velocity at throat[m/s]	435.06	435.3
Inlet enthalpy[kcal/kg]	226.21	226.21
Delta Kin. Energy[kcal/kg]	15.88	15.85
Outlet Enthalpy[kcal/kg]	210.33	210.36
Throat pressure[bar]	1.01325	1.01325
Throat temperature[K]	266.18	266.25

As might be expected, the gas mass flow rate is far lower for Case (B) compared to Case (A). Also, for Case (B), it was found that the conditions at the throat are virtually identical for the ideal versus the non-ideal gas.

2. ADDITIONAL EXAMPLES

To illustrate the importance of the issues discussed in this paper, we show a few additional examples for the differences in our numerical results compared to others.

2.1 Gas Release From A Hole In A Vessel

A 0.1[in] hole forms in a tank containing nitrogen at 200[psig] and 80[°F]. Determine the mass flow rate through this leak (Crowl[4], Example 4-4, p. 143).

The solution for this problem using the Lee-Kesler EOS was 4.26e-2[lb/s], compared to 3.86e-2[lb/s] in the reference (gas velocity at the throat was sonic). This difference is over 10% and not negligible.

2.2 Flow Through A Nozzle

Air at p_0=7.0e5[Pa] and T_0=293[K] discharges to the atmosphere. Compute the discharge mass flux, G, the pressure, Mach number, and velocity at the exit[5].

Using the Lee-Kesler EOS, the results are as follows:

Quantity	Lee-Kesler (this article)	Ideal gas (isentropic)[5]	Difference
Mass flux[kg/(m^2.s)]	1589	1650	-3.68%
Exit temperature[K]	238.5	244	-5.5 [K]
Exit pressure[Pa]	212041	370000	-42.7%
Mach number	1.0	1.0	-
Sonic velocity[m/s]	338	313	+8%
Velocity[m/s]	338	313	+8%

Note that the nozzle is not predicted to be choked when the downstream pressure is above 2.12[bar] (kinetic energy effects must be included), as opposed to 3.7[bar] in the original reference.

If a relief valve is to be designed for this scenario, the size would be under-predicted somewhat using the isentropic formulas used in the original reference. However, this discrepancy could be significantly larger in cases where an isentropic expansion incorrectly predicts partial condensation of the gas at the throat. In that situation, the density of the two-phase mixture at the throat could be far lower than that of the uncondensed gas (for the isenthalpic case) and the valve could be under-sized significantly.

3. CONCLUDING REMARKS

It has been shown that calculation of the release rate from a relief valve (or a hole in a vessel) should be done using an isenthalpic (i.e., irreversible adiabatic) path for the flow process. Using an isentropic thermodynamic path for this problem violates the First Law of Thermodynamics, as expressed for steady flow in a stationary relief valve or a hole in a pipeline. This could lead to an incorrect design both when sizing relief valves and estimating release rate from an existing valve or hole in a vessel.

In addition, it is important to use the proper equation of state to estimate real gas properties reliably when compared to the ideal gas assumption, especially when gas conditions are far from ideal. The Bernoulli equation (for throat pressure), and the sonic velocity limit (when flow is choked at the throat) must also be included for a complete definition of the problem. If the upstream pressure is low enough and the throat is un-choked, the isenthalpic methodology - as described in this paper - should again be followed to ensure a theoretically sound and consistent basis. Iterative computations for non-ideal cases (real gases) generally are too intensive for hand calculations and require use of computerized chemical engineering software.

REFERENCES

1. Hougen, O.A., K. M. Watson: "Chemical Process Principles" (John Wiley & Sons, New York, and Chapman and Hall, London, 1st Edition, Part II, p. 551, 1947, and 2nd Edition with R.A. Ragatz, Vol. II, p. 653, 1959)
2. Elliott, J.R., and C. T. Lira: "Introductory Chemical Engineering Thermodynamics" (Pearson Education, Inc. publishing as Prentice Hall, pp. 68-69, 2nd Edition, 2012)
3. Klein, S. and G. Nellis: "Thermodynamics" (Cambridge Univ. Press, 1st Edition, p. 168, 2011)
4. Crowl, D. A., and Louvar, J.F.: "Chemical Process Safety" (Pearson Education, 3rd Edition, pp. 130-131, 2011)
5. Green, D. W., and R.H. Perry: "Perry's Chemical Engineers' Handbook" (McGraw-Hill, 8th Edition, Section 6, pp. 23-24, 2008)
6. Himmelblau, D.M., and J.B. Riggs: "Basic Principles and Calculations in Chemical Engineering" (Pearson Education, 8th Edition, Fig. 9.18, Item 3, p. 525, 2012)
7. Cutlip, M.B., and M. Shacham: "Problem Solving in Chemical and Biochemical Engineering with POLYMATH, Excel, and MATLAB" (Prentice Hall, 2nd Edition, 2007)
8. Mathcad 2001 Professional: "Mathcad User's Guide With Reference manual", (Mathsoft Engineering and Education, Inc., 2001)
9. Freund, S.M., M. Jones, and J.L. Starks: "Microsoft Excel 2013: Complete (Shelly Cashman Series)", (Course Technology, 1st Edition, 2013)
10. Lewis, G.N., and M. Randall: "Thermodynamics", (McGraw-Hill, 2nd Edition, Revised by K.E. Pitzer and L. Brewer, p. 48, 1961)

548434 Engineered Solutions for Effective Flare System Analysis

Relief and flare system documentation is an important part of Occupational Safety and Health Administration (OSHA) Process Safety Management (PSM) 29 CFR 1910.119. Flare system analysis (sizing or rating) can be performed using commercially available software. Effective flare system analysis requires reliable Process Safety Information (PSI) such as process and instrumentation diagrams (P&IDs), relief system evaluation, Heat and Material Balance (H&MB), relief valve data sheets, isometrics, and disposal system loads. Good engineering methodologies can be implemented to effectively design header systems and rate existing header for any change. Some methodologies include use of dynamic simulation to predict realistic relief loads, reboiler pinch analysis for applicable systems, and selection of global relief scenarios. This paper will describes some methodologies and a combination of instrumented systems for effective flare system analysis. Effective flare analysis helps save capital expenditure spent on flare headers and possible flare load reduction yields reduced radiation footprint on a facility.

548364 Flares Are the End of the Line; How to Permit Correctly

INTRODUCTION

-

Flare is one of the key equipment in any refinery or LNG facility. Many relief valves relieve to flare, and if the flare is not permitted corrected – and functional - the plant is down. You cannot operate the plant without an active flare or flare field. The permitting of a flare is one of the prime activities in the permitting process. But the permit is dependent on the type of flare. This presentation covers various Federal Regulations involved in the permitting process, whether general or specific to each type of flare.

FLARE TYPES AND THEIR DESIGN FUCTURES

-

There are mainly three types of flares. The most common and most economical one is an elevated flare. There are elevated flares as high as 660 ft. The can be classified many ways; self-supported, guy wire supported, derrick supported, demountable type etc. One of the most important is the type of burner, whether a pipe tip or "furnace" burners. The second type is a ground flare with rows of furnace burners arranged in arrays at grade and surrounded with a wind fence. The last type is an enclosed flare with furnace burner enclosed with a tall stack. This type can be large as 70 ft in diameter and 150 ft tall. The flare type on a given project is selected based on the plot availability, location of the plant, sparing philosophy etc.

PERMITTING REGULATIONS

This presentation will discuss cover various Federal regulations that are applicable to all types of flares and the various parameter that the regulators are needing when apply for the permit. The presentation will cover the required destruction efficiency of the flaring gases, exit velocity for an elevated flare, alternate means of emission limitation for ground flares. In particular, the tip velocity, burner type, backpressure, emission testing and smokeless capacity are specific to each type of flare.

WHAT YOU WILL LEARN

-

- This paper will provide sufficient information to ensure the correct type of flare is selected and permitted
- This paper will cover various flares types employed in refinery and in LNG plants
- The limitations of these flare
- Importance of flare in any facility
- Application Federal Regulations
- Parameter required in the permitting applications
- Ground flare testing to establish minimum destruction efficiency
- Alternate means of emission limitation requirements in lieu of testing

548232 A Case Study on Application of HAZOP for a Brewery

Abstract

Beer is the most consumed alcoholic beverage in the world, and most popular beverage after water and tea. It's an alcoholic beverage produced by the saccharification of starch, often derived from the malted grains such as barley and wheat, and the fermentation of the resulting sugar. The brewing process can be broken into four main constituents before the bottling line, Mashing, Lautering, Wort boiling and Fermenting. Breweries commonly rely on ammonia refrigeration to keep their supplies cold. However, some brewers fail to realize that ammonia is one of the dangerous chemicals covered under 29 CFR 1910.119, OSHA's regulation for preventing or minimizing the release of toxic, reactive, flammable or explosive chemicals. "Safety matters at all levels", even for companies that may not be PSM level facilities.

A HAZOP study identifies hazards and operability problems. The study involves investigating how the plant might deviate from the design intent. It is based on the principle that several experts with different backgrounds can interact and identify more problems and failure scenarios working together and combining their results. This paper presents a review of consequences identified in brewery HAZOP. It also outlines some of the recommendations and action items identified from the case study of a brewery.

Keywords: Process Safety Management (PSM), PHA, HAZOP, Brewery, Ammonia, OSHA.

548410 A Basic Review of Fundamental of Thermodynamics and Its Application in the Analysis of Consequences and Accident Investigation

Once the stage of hazard identification has been concluded in a risk analysis, the next stage corresponds to select and specify the risk scenarios that will be modeled their effects or damages associated. This stage correspond to the consequences analysis and usually takes place in three basic steps: a) source term definition, b) dispersion model and d) effect or damage model.

Depending on the risk scenario to be modeled, several thermodynamic fundamentals or principles are applied in each of these steps supporting or sustaining the behavior and expected results of the consequences analysis model. In this work we propose to review the main thermodynamic fundamentals that support each of the models applied in each of the three steps that define a consequence analysis. This review of basic thermodynamic principles defines a set of different thermodynamic basic principles to apply to representative and different cases of consequence analysis. These basic thermodynamic principles are applied in the consequence modeling of a pair of cases of accident inventions to corroborate their validity.

548331 Evaluating Safe Location Criteria for Vents and Relief Devices Discharging to Atmosphere

In some process technologies, it may not be feasible to dispose the effluents from safety valves and vents to a closed system and discharging directly to the atmosphere can be the only alternative. Some examples include High Pressure Polyethylene Processes, Olefins and Storage units. The discharge can be directed to the atmosphere if:

- Local regulations permits it;
- Does not harm the environment;
- Does not harm the neighborhood;
- And is directed to a Safe Location.

The criteria for defining what a safe location is can vary, but it must include at least a consequence assessment of the flammable, toxic and thermal radiation effects. Some guidance can be found on industry standards and engineering references such as API 521 and the CCPS Guidelines for Pressure Relief and Effluent Handling Systems

One aspect to discharge hydrocarbons to a Safe Location is to evaluate the dispersion of the discharged hydrocarbons and ensure that it is diluted bellow the lower flammable before reaching the ground level or elevated structures. These areas are typically with people circulation and ignition sources. Another aspect that must also be evaluated is the thermal radiation generated due to an ignition of the discharge affecting personnel or damaging equipment or other assets. Lastly, if there is a possibility of the flammable mixture reaching confined spaces, the explosion possibility must also be assessed.

API 521 gives some guidance on chapter 5.8 for estimating the dispersion and concentration profile from a typical relief device discharge. It also describes that if the vapor relief velocity is high (> 30 m/s), the air entrance rate is also high and the vapors typically would be diluted below lower flammable limit in a very short distance.

This phenomenon explains the standard recommendation of locating the discharge of the relief device at least 3 m above any adjacent equipment, building or structure located within 15 m horizontally from the discharge point.

Experience has shown, however, that for some special cases these typical recommendations does not apply, such as:

- Heavy gas releases;
- Cryogenic releases;
- Control valve releases;
- Discharge piping with unconventional design (rain cap, horizontal, etc.);
- Discharge location within a building wake.

Adding to the complexity of the analysis, the relief devices can also be fitted with steam or other purge gas injection that could positively affect the dispersion.

After some case studies, which included the analysis of High Pressure Polyethylene plants Vent Stacks and Olefins unit relief and venting devices, Braskem developed an internal engineering standard with guidance and criteria for evaluating existing projects and new designs in order to ensure they comply with safe location. The guidance includes:

- Relief cases selection;
- Simplified and rigorous dispersion analysis;
- Consequence assessment endpoints;
- Thermal radiation effects to assets;
- Liquid, vapor and two-phase releases;
- Process Hazard Analysis;
- Diluting/snuffing steam;
- New systems and retrofitting criteria;

Applying the criteria in the developed standard allows the evaluation if the relief is in a safe location or not. If the discharge is not in a safe location, a risk ranking and recommendations to change the system must be made. One kind of a typical recommendation is to install the relief in a higher location. However, for some cases, achieving Safe Location is not possible and changes in the system are needed, such as relocating to a closed system or design changes to eliminate the relief scenarios associated with that relief device.

The development of an internal engineering guideline ensures a standard approach throughout the company for evaluating the Safe Location of hydrocarbon discharges and improves the quality of the risk evaluation during the PHA cycles.

548407 Experimental Study of Electrostatic Hazard inside Scrubber Column Using Response Surface Methodology

Experimental Study of Electrostatic Hazard inside Scrubber Column using Response Surface Methodology

Jingyao Wang[a,b], Yue Sun[a,b], Xiaodan Gao[a,b], Benjamin Wilhite[a,b], M. Sam Mannan [a,b]

[a]Mary Kay O'Connor Process Safety Center, Artie McFerrin Department of Chemical Engineering, Texas A&M University, College Station, Texas 77843-3122, USA

[b] Artie McFerrin Department of Chemical Engineering, Texas A&M University, 3122 TAMU, College Station, Texas 77843, United States

Abstract

Electrostatic ignition has triggered significant fire/explosion incidents in liquid processing industries including chemical, oil/gas, food, pharmaceutical and transportation. In this work, the impact of operating flowrate and temperature on the electrostatic energy accumulation in scrubber columns was studied. An *in-situ* charge measurement apparatus combined with a particle scattering laser function was installed inside a pilot-scale scrubber column, which can measure the aerosol particle size distribution, concentration and charge simultaneously. Four electrostatic ignition scenarios were considered including discharge from a single aerosol droplet, aerosol clouds, ungrounded conductor charged by the aerosol and ungrounded conductor charged by the liquid. Response surface methodology was applied to construct a quadratic empirical model correlating the electrostatic energy response with the flowrate and temperature. The obtained electrostatic response shows that corona from a charged single droplet was too small for safety concern. However, the ungrounded conductor polarized by charged aerosol or liquid can accumulate significant energy that is higher than the minimum ignition energy of some flammable chemicals. Temperature has two-sided effect where high temperature generates more concentrated charged aerosol but also facilitates the charge relaxation in the liquid phase. In all scenarios, increasing gas/liquid flowrate can cause significant electrostatic energy accumulation and can increase the hazards of many incidents. To reduce the electrostatic hazards, it is recommended to reduce the gas/liquid interaction flowrate and eliminate the presence of ungrounded conductors in liquid process operations.

Keywords: Scrubber column design, Electrostatic hazards, Response surface methodology, Process safety, Hazard identification, Aerosol hazards.

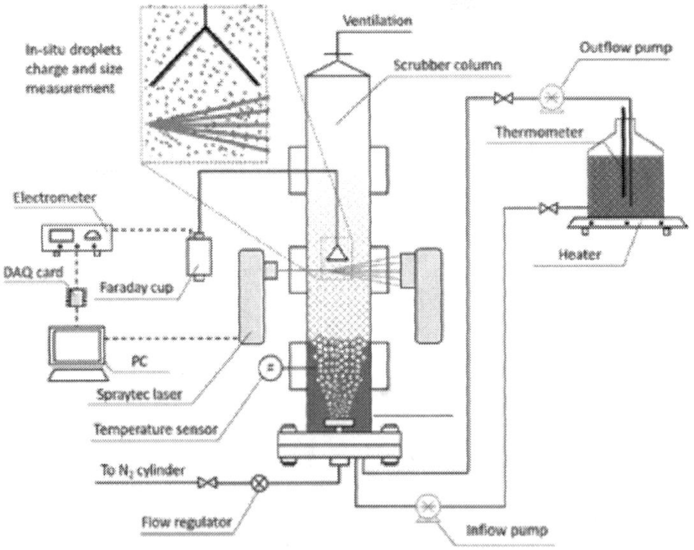

550852 Process Safety in Oil Refineries: Vacuum Distillation Column Hazop Study

Errors that can occur in the identification studies for hazards and risks which is one of the elements of a process safety management cause resulting major accident having undesirable long - term effects on human and environment. For preventing major accidents, identification and analysis of process hazards due to equipment failures, design or montage defects, improper maintenance, misuse, material defects, metal fatigue, corrosion, welding defects, human failure etc. are the main principles of process safety management. The aim of present study is to investigate possible accidental risks occurring in Vacuum Distillation Column of oil refining within contributing to the process safety frame.

In this cross - sectional study, semi quantitative HAZOP analysis was used for identifying accidental risks. By means of HAZOP studies 66 deviations, 238 deviation causes, 287 consequences, 651 barriers and 113 recommendation were identified. In the determination of accidental risk in the vacuum distillation column by applying semi quantitative HAZOP analysis method, effective results were obtained. The most common causes of deviations in the column have been found as control loop failure, leakage and mechanical failure. It is stated that the most common accidental risks are as a portion of 66.9 % operational deterioration, fire, explosion, release known as major accident risk. It is identified that the accidental risks mostly cause existence damage and the highest accidental risk occurs in the "vacuum column overhead system" and "bottom of the vacuum column system" 87.7 % of 113 recommendations are modification in the design, 5.3% is about modifications in the process safety management system.

551179 API Process Safety Site Assessment Program - Trends and Good Practices in Process Safety

The Process Safety Site Assessment Program is a collaborative industry effort managed by API and is focused on improving process safety in petroleum refining and petrochemical facilities. The program has conducted over 65 process safety assessments and 38 HF RP-751 assessments since its inception in 2012. The assessments are conducted by independent and credible third-party teams of industry-qualified process safety expert assessors using industry-developed protocols.

The presentation will review trends, both positive and low scoring, seen in the extensive benchmarking data collected. Good practices observed in the assessments will be identified and discussed. The emphasis will be to have the attendees leave with an understanding that typically some site(s) has developed a way to handle challenging areas of process safety.

545445 Emerging Trends in Process Safety Management Relating to Contractor Workforce Compliance

Following a Six Sigma Black Belt process improvement session conducted 2016 to 2018 for petrochemical complexes along the Gulf Coast the presentation identifies several emerging trends for Best Practices in industrial workforce compliance.

Safety Councils work for petrochemical plants and their contractors to facilitate innovative ways to reduce incident rates resulting in a 95% reduction in recordables. 2017 marked an evolution in Safety Council deliverables, slowly migrating away from site specific training to develop new workforce compliance and risk management processes. Working within the industry we developed several internationally recognized Best Practices for contractor compliance, risk management, labor cost management. ISTC collaborated with our partners bringing years of quantitative data to identify problems that are developing, have occurred in the workplace and formulating solutions, cost structures and most importantly compliance management to combat workforce issues. The presentation, best practices in contractor compliance highlights the journey from identifying the problems, and the associated issues with hiring, compliance, auditing, accident response and even Department of Homeland Security, CFATS inspections. Beyond the logistics of building solutions for a particular problem, the presentation reflects on how companies were able to share their issues among peers without violating confidentiality or company procedures to solve what was to be discovered as an industry trend. Using Six Sigma tools, the units were able to identify root problems, collaborate on a solution, model cost recovery and implement the solutions to the contractor workforce with a minimum of disruption to the business at hand.

Análisis Cuantitativo De Riesgo (QRA) Como Herramienta Para La Toma De Decisión Sobre La Infraestructura En Sistemas De Transporte De Fluidos Peligrosos

Daniel Mauricio Moncada Saavedra
CENIT -Transporte y Logística de Hidrocarburos
Cll 113 #7-80 Piso 13 Bogotá – Colombia
daniel.moncada@cenit-transporte.com

Maria Isabel Montoya Rodriguez
CENIT -Transporte y Logística de Hidrocarburos
Cll 113 #7-80 Piso 13 Bogotá – Colombia
maria.montoya@cenit-transporte.com

Prepared for Presentation at
American Institute of Chemical Engineers
2019 Spring Meeting and 15th Global Congress on Process Safety
New Orleans, LA
March 31 – April 3, 2019

Análisis Cuantitativo De Riesgo (QRA) Como Herramienta Para La Toma De Decisión Sobre La Infraestructura En Sistemas De Transporte De Fluidos Peligrosos"

Daniel Mauricio Moncada Saavedra
CENIT -Transporte y Logística de Hidrocarburos
Cll 113 #7-80 Piso 13 Bogotá – Colombia
daniel.moncada@cenit-transporte.com

Maria Isabel Montoya Rodriguez
CENIT -Transporte y Logística de Hidrocarburos
Cll 113 #7-80 Piso 13 Bogotá – Colombia
maria.montoya@cenit-transporte.com

Palabras clave:

Abstract

La pérdida de contención en activos utilizados para el transporte de hidrocarburos puede ocasionar impactos considerables a personas, medio ambiente, económicos y de reputación. Con el fin de mitigar la ocurrencia este tipo de eventos, se implementan estrategias sobre los activos, siendo la identificación y análisis de riesgo un elemento clave durante el proceso.

El presente análisis utilizó un enfoque cuantitativo, tanto para probabilidad como para consecuencia, con el fin de poder estimar el riesgo que genera la operación de una línea de transporte en un área de alta consecuencia clasificado como sitio identificado, planteando posibles alternativas de control y mitigación del riesgo.

Como resultado del estudio realizado se pudo establecer el nivel de riesgo actual asociado a la operación de la línea en el sitio de interés. Como actividades de mitigación para reducir el riesgo asociado se identificó: la reparación de los defectos, así como el realineamiento y reubicación de la tubería.

1 Introducción

Para el presente análisis se implementó un cálculo de la probabilidad de falla a partir del análisis de integridad estructural del sector de interés. Adicionalmente, se consideró el análisis de la susceptibilidad que se presenta a partir de las amenazas presentes, la cuales son identificadas a partir de las actividades de monitoreo e inspección desarrolladas sobre la infraestructura. El cálculo de la interacción entre las amenazas a la integridad; así como el resultado final de probabilidad de falla, se puede realizar mediante diferentes enfoques, el presente estudio utilizo cálculo de probabilidad de excedencia y aplicación de redes bayesianas para la estimación.

Para el cálculo de la consecuencia se buscó verificar la posible afectación a personas, producto de la ignición del producto transportado. Dentro de los sucesos finales se consideró: incendio de piscina, incendio por chorro de fuego, llamarada y la sobrepresión debida a la explosión. El cálculo de estos efectos se realizó en el sitio donde se presenta la perdida de contención, así como también para el caso en el que el hidrocarburo es transportado por un cuerpo de agua o se desplaza en tierra.

Una vez determinada tanto la probabilidad de que se desarrollen los sucesos finales identificados como la consecuencia que cada uno de ellos puede generar, se procedió a realizar el cálculo del riesgo expresando el resultado en términos de riesgo social e individual, comparando el resultado contra criterios de tolerabilidad referentes en la industria. Para la condición actual establecida se identificaron alternativas las cuales buscaban disminuir el nivel de riesgo, implementando actividades de mitigación del nivel de riesgo.

2 Análisis de Riesgo Cuantitativo

3 Frecuencia Final de Ocurrencia de los Sucesos Finales

En el sector bajo interés se tienen que la línea es susceptible a las siguientes amenazas:

3.1 Análisis de Condición.

En este aparte se relacionan los resultados de análisis de probabilidad de falla, obtenidos en el sector de interés, PK 43+000 al PK 45+000 del poliducto, para las amenazas de corrosión externa, corrosión interna, geotecnia y daños por terceros involuntarios:

3.1.1 Corrosión Externa (CE).

A continuación, se describe el estado de salud de las barreras de CE del poliducto en el sector solicitado (ver Ilustración 1), así como el análisis de probabilidad de excedencia de las anomalías ILI identificadas con la corrida del 2014. En la Ilustración 2 se observa la disponibilidad de la URPC y se reporta que la tubería cuenta con recubrimiento en buen estado. Bajo las condiciones y premisas consideradas para el cálculo y proyección de la POE (Probabilidad de excedencia; entre ellas una rata de corrosión de 2-3 mpy aproximadamente; no hay riesgo de excedencia de los criterios de profundidad y ERF (Factor Estimado de Reparación) de las anomalías ILI hasta el año 2021 en el sector 43+590 y 42+746 (Anomalías ILI con DR(Distancia de registro): 2646.76, 3490.08, 3492.38, 3492.71) (Ilustración 2).

Ilustración 1. Perfil de riesgo y salud de barreras CE.

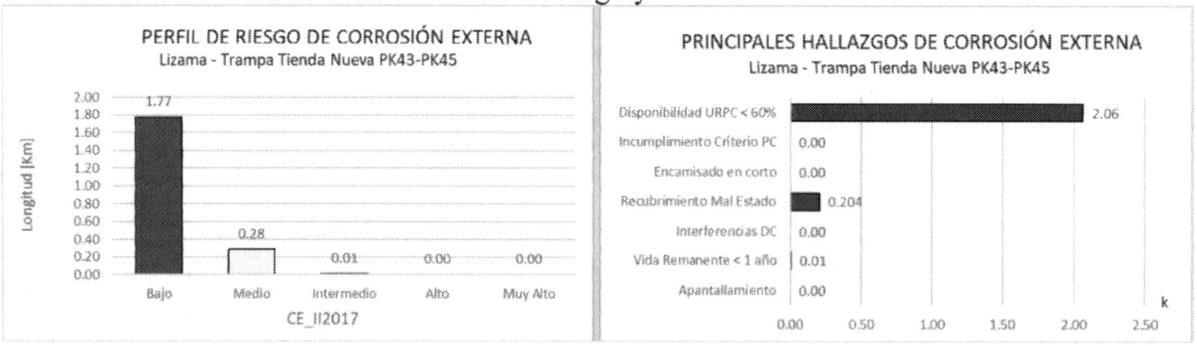

Ilustración 2. Análisis de POE de anomalías de CE en el tiempo.

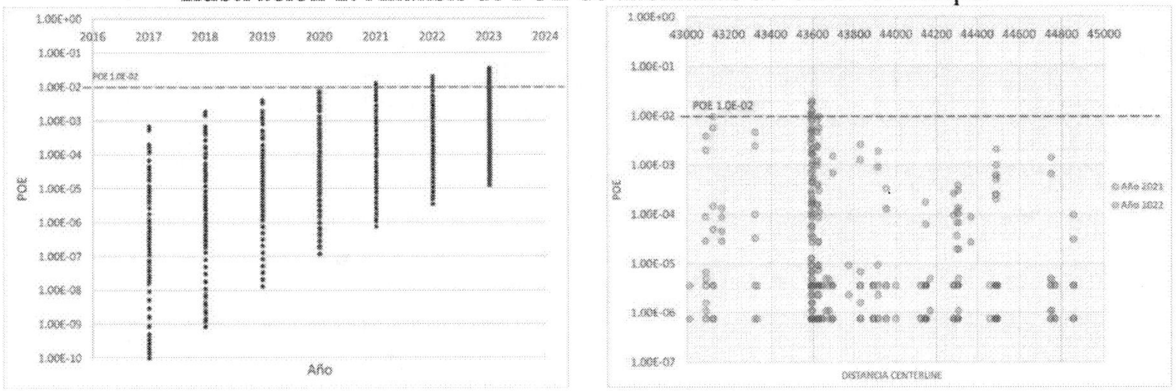

3.1.2 *Corrosión interna (CI).*

A continuación, se describe el estado de salud de las barreras de CI de la tubería en el sector solicitado (ver Ilustración 3), así como el análisis de probabilidad de excedencia de las anomalías ILI identificadas en la corrida del 2014. De acuerdo con los resultados de valoración de riesgo, se presenta un correcto estado de salud de las barreras de corrosión interna. Adicionalmente, de acuerdo con las condiciones y premisas consideradas para el cálculo y proyección de la POE, no hay riesgo de excedencia de los criterios de profundidad y ERF de las anomalías ILI en el mediano plazo (Ver Ilustración 4).

Ilustración 3. Perfil de riesgo y salud de barreras CI.

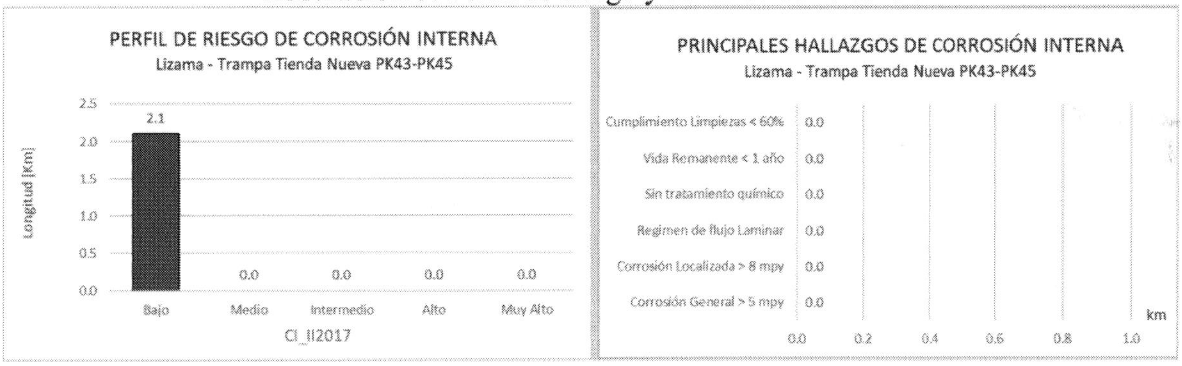

Ilustración 4. Análisis de POE de anomalías de CI en el tiempo.

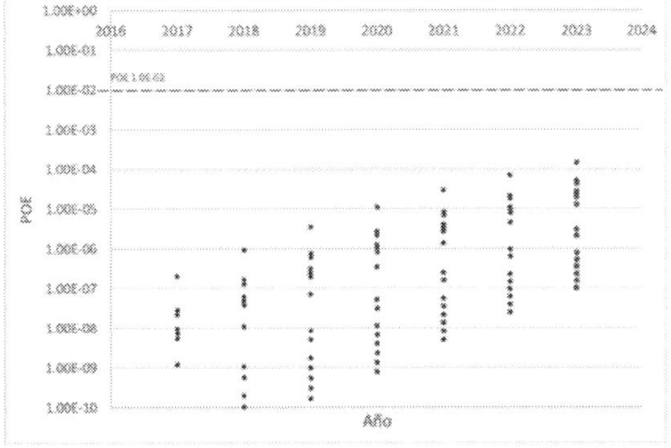

3.1.3 Geotecnia.

En el sector de interés se reporta hallazgo de geotecnia en la distancia centerline 44643 – 44703 (PK 44+400), donde se presenta proceso erosivo incipiente. Actualmente, no se identifica avance de la condición, razón por la cual, se continuará seguimiento a través de la inspección visual del derecho de vía (IVDDV).

Ilustración 5. Perfil de riesgo y salud de barreras de Geotecnia.

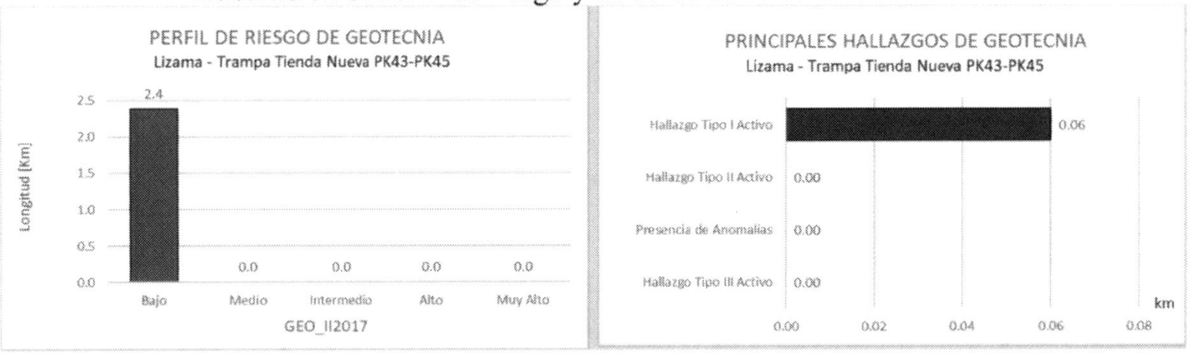

3.1.4 *Daños por Terceros Involuntarios (DPT).*

Debido a la condición de la tubería y a la gestión realizada por el operador para el control de la amenaza, ésta presenta una susceptibilidad baja de ser golpeada por un tercero sobre el derecho de vía. A continuación, se relaciona el perfil de riesgo de la amenaza de DPT en el sector analizado:

Ilustración 6. Perfil de riesgo y salud de barreras de Daños por Terceros Involuntarios.

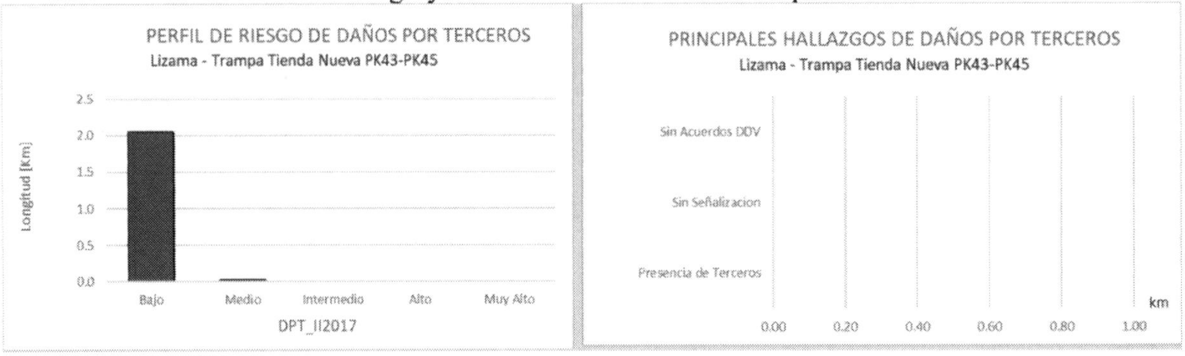

De acuerdo con el análisis ejecutado, el evento más probable obedece a una pérdida de contención generada por la amenaza de corrosión externa; donde el análisis estructural identifica que el modo de falla se daría por fuga. Sin embargo, las anomalías encontradas no presentan riesgo de excedencia de los criterios de profundidad y ERF hasta el año 2021.

Con el fin de poder estimar la probabilidad de falla de la línea en el sector de interés, se realizó el análisis estructural considerando principalmente defectos por perdida de espesor, bajo estas premisas se obtuvo un valor de probabilidad de excedencia de los criterios de profundidad y ERF de 7.09E-4. Del análisis de condición se concluye que el iniciador más probable en este sector es la falla de la tubería por fuga asociado a corrosión externa, para el cual a continuación se presenta el análisis de riesgo.

4 Análisis de consecuencia

A continuación, se presenta el análisis de consecuencia para los sitios en los cuales se ha identificado que se puede llegar a presentar una pérdida de contención.

4.1 Identificación de escenarios

4.1.1 Selección de sustancias.

En el poliducto se transportan las siguientes sustancias: Gasolina, GLP, Jet A1 y B2

4.1.2 Identificación de iniciadores.

Los iniciadores representan las situaciones iniciales de pérdidas de contención o exposición que se pueden presentar en los das diferentes equipos. Los iniciadores pueden ser de dos tipos, genéricos o específicos. Los iniciadores genéricos, involucran una pérdida de contención debida a causas no específicas del proceso u operación, como pueden ser corrosión, errores de construcción, fallas de la soldadura, etc. Los iniciadores específicos se caracterizan por tener causas concretas debido a las condiciones de proceso, el diseño, los materiales utilizados, etc.

A continuación, se describen los iniciadores tenidos en cuenta para cada uno de los sitios en los cuales se identificó la inestabilidad de taludes.

- **Sector PK 43+590.** Este sitio tiene la menor distancia desde la infraestructura a la escuela bajo análisis, 30 metros aproximadamente, sumado a esto en las cercanías de este se ha identificado la presencia de anomalías las cuales, bajo las condiciones actuales de operación y deterioro hacia el año 2021, pueden exceder los límites de profundidad critica a la falla.

- **Sector PK 43+805 – PK 44+110.** En este sitio se tiene que ante la posible pérdida de contención del producto se puede llegar a presentar la formación de una ruta de derrame en tierra, por la cual se desplazaría el producto hacia el interior de la Escuela.

4.1.3 Identificación de sucesos finales.

La identificación de los sucesos finales que pueden tener lugar después de que se da el iniciador, se debe realizar utilizando los árboles de eventos o sucesos. Dichos árboles permiten ver el comportamiento de la sustancia tras la descarga y estimar la probabilidad de ocurrencia de cada suceso, teniendo en cuenta los sucesos intermedios que condicionan la evolución del suceso final.

Esta identificación de escenarios se lista en Tabla 11 para los escenarios que involucran sustancias inflamables tales como Gasolina o Nafta. Los posibles sucesos finales identificados en el poliducto son: derrame continental, incendio de piscina, incendio de chorro y llamarada.

Tabla 1. Identificación de escenarios.

Escenarios	Descripción	Iniciadores	Sustancia	Tipo de pérdida
Afectación directa				
E1	Poliducto	L3	Gasolina	Incendio de piscina
				Incendio de chorro
				Llamarada
E2	Poliducto	L3	GLP	Incendio de piscina
				Incendio de chorro
				Llamarada

Escenarios	Descripción	Iniciadores	Sustancia	Tipo de pérdida
Afectación directa				
Afectación indirecta				
E3	Poliducto	L3	Gasolina	Incendio de piscina
				Llamarada

4.2 Estimación del área de afectación por los efectos que involucran producto

4.2.1 Condiciones para la simulación.

Para determinar las consecuencias, es necesario definir las condiciones propias del entorno de la infraestructura que servirán como base para la simulación de los efectos. Estos parámetros son presentados en la Tabla 2. Las estabilidades atmosféricas fueron determinadas a partir de la rosa de vientos basada en datos estadísticos y modelos meteorológicos, la cual es presentada en la **¡Error! No se encuentra el origen de la referencia.**.

Tabla 2. Condiciones atmosféricas

Parámetro	Valor	Estación de Referencia
Temperatura promedio (°C)	31	Colosal
Humedad relativa promedio (%)	83	
Rugosidad el terreno (m)	0,08	
Estabilidad atmosférica más probable y velocidad del viento (m/s)	2F	
Estabilidad atmosférica F y velocidad del viento (m/s)	2F, 5D	

4.2.2 Distancias de afectación directa

Las distancias de afectación representan las longitudes dentro de las cuales se pueden manifestar los efectos correspondientes a los niveles de afectación, para cada suceso final (incendios de piscina o llamaradas). Se denominan distancias de afectación directas, ya que los sucesos finales de que dependen se originan y desarrollan su efecto directamente en el sitio en el que se produce la pérdida de contención del producto

4.2.3 Estimación de Descargas.

Para la simulación de los posibles efectos producto de una pérdida de contención, es necesario conocer cuánto producto podría derramarse en cada punto de estudio sobre la infraestructura. El volumen derramado corresponde al estimado teniendo en cuenta el llenado de la tubería, las condiciones de operación y el tiempo de respuesta operativo, es decir, el tiempo de detección más el tiempo de actuación (generalmente el seccionamiento del sistema o parada de bombeo).

Tabla 3. Tiempos de actuación

Equipo	Tiempo de Detección (min)	Tiempo de Actuación (min)
	Fuga	Fuga
Poliducto	30	180

Tabla 4. Volumen y Tasa de descarga

Volumen (US bls)	Tasa de descarga (bls/hr)
59	22

4.2.4 *Simulación de Efectos de los Sucesos Finales.*

En este numeral se presentan los resultados de las distancias de afectación directa, obtenidas para cada uno de los niveles de afectación asociados a los sucesos finales. En la **¡Error! No se encuentra el origen de la referencia.** se presentan las distancias de afectación obtenidas para el evento incendio de piscina, es importante resaltar que este depende en gran medida del área de empozamiento que se forma una vez se produce la descarga del producto.

En la **¡Error! No se encuentra el origen de la referencia.** se presentan las distancias de afectación para incendio de chorro obtenidas durante las simulaciones, es importante resaltar que la distancia obtenida depende de las propiedades químicas de la sustancia y de las condiciones en las cuales se produce la descarga.

Tabla 5. Distancia de afectación incendio de piscina.

Equipo / Abscisado	Iniciador	Sustancia	Volumen Derrame	Niveles de Afectación por Incendio – Incendio de Piscina						
				1.6 Kw/m²	5.0 Kw/m²	7.3 Kw/m²	9.5 Kw/m²	14.5 Kw/m²	20.9 Kw/m²	37.5 Kw/m²
			(Bbl)	Distancia (m)						
K43+590	Rotura mínima	GLP	59	46,94	28,63	24,58	21,98	18,01	14,55	8,38
K43+590	Rotura mínima	Gasolina	59	94,63	49,69	35,17	27,20	20,98	20,98	-

Tabla 6. Distancia de afectación incendio JET FIRE.

Equipo / Abscisado	Iniciador	Sustancia	Volumen Derrame	Niveles de Afectación por Incendio – Chorro de fuego						
				1.6 Kw/m²	5.0 Kw/m²	7.3 Kw/m²	9.5 Kw/m²	14.5 Kw/m²	20.9 Kw/m²	37.5 Kw/m²
			(Bbl)	Distancia (m)						
K43+590	Rotura mínima	GLP	59	45,9	33,61	30,93	29,33	27,13	25,23	23,42
K43+590	Rotura mínima	Gasolina	59	44,43	31,20	28,46	26,82	24,58	22,96	20,84

En la **¡Error! No se encuentra el origen de la referencia.** se presentan las distancias de afectación por llamarada; las cuales están relacionadas con la volatilidad del producto junto con las condiciones climáticas de las zonas del evento.

Tabla 7. Distancia de afectación Llamarada.

Equipo / Abscisado	Iniciador	Sustancia	Volumen de Derrame	Niveles de Afectación por Llamarada más probable		Niveles de Afectación por Llamarada más estable	
				LII	LII/2	LII	LII/2
			(Bbl)	Distancia (m)		Distancia (m)	
K43+590	Rotura mínima	GLP	59	24,12	58,53	26,48	66,7
K43+590	Rotura mínima	Gasolina	59	-	-	-	-

4.2.5 *Distancias de Afectación Indirecta*

Las distancias de afectación indirecta representan las longitudes de potencial afectación por sucesos finales que se originan en una trayectoria o ruta de derrame (en tierra o en agua) o, a partir de estancamientos o acumulaciones de producto. En este numeral se presentan los resultados de las distancias de afectación obtenidas para cada uno de los niveles de afectación asociados a los sucesos finales que se pueden presentar en la ruta de derrame. Estos resultados corresponden a la radiación generada por piscina de fuego, evento que en los cuerpos de agua depende de la sección del cuerpo de agua, así como por el frente de llama generado por el volumen derramado.

En la Tabla **88 y ¡Error! No se encuentra el origen de la referencia.**9 se referencian los corredores de afectación generados por incendio de piscina sobre las rutas de derrame en tierra identificadas en el poliducto, en el sector comprendido entre los PK 43+000 - PK 45+000.

Tabla 8. Distancias de afectación indirectas POOL FIRE.

Abscisado	Medio de Transporte	Sustancia	Incendio						
			Incendio en piscina						
			1.6 Kw/m^2	5.0 Kw/m^2	7.3 Kw/m^2	9 Kw/m^2	14.5 Kw/m^2	20.9 Kw/m^2	37.5 Kw/m^2
			Distancia (m)						
Km 43 - 45	Ruta en tierra	Gasolina	82,02	54,68	27,34	24,61	19,14	13,67	8,20

Tabla 9. Distancias de afectación indirecta FLASH FIRE

Abscisado	Medio de Transporte	Sustancia	Niveles de afectación por llamarada más probable		Niveles de afectación por llamarada más estable	
			LII	LII/2	LII	LII/2
			Distancia (m)		Distancia (m)	
Km 43- 45	Ruta en tierra	Gasolina	2,14	4,74	8,64	32,24

4.2.6 Áreas de Afectación

En sistemas de ductos, el área de afectación directa se define por un corredor medido desde el eje central del ducto. Este corredor está dado por la mayor distancia probable de afectación obtenida en la simulación de efectos para todas las sustancias analizadas. Por otro lado, el área de afectación indirecta está dada por varios corredores a lo largo de las posibles rutas de derrame tanto en suelo como en agua. En la Ilustración 77 se presenta el área de afectación identificada.

Ilustración 7. Área de afectación identificada.

4.2.7 Identificación de Elementos Vulnerables

Para calcular el riesgo de la población respecto de la posible afectación producida por la pérdida de contención descrita, se requiere el número y localización de las personas, en la Tabla 100 se presenta la población identificada al interior de la escuela.

Tabla 10. Presencia de población Escuela

Presencia de Población							
Escuela - Dia				Escuela - Noche			
Niños	63	Profesores	6	Niños	0	Profesores	2

5 Evaluación del Riesgo.

A continuación, se presentan los resultados del análisis de riesgo el cual tuvo en cuenta la información presentada en los numerales anteriores asociada a las posibles causas de falla, los modos en que puede darse cada falla identificada, la frecuencia de ocurrencia de la falla, la probabilidad de ocurrencia de los sucesos finales, y finalmente, la estimación de la magnitud de las consecuencias de tales sucesos finales en el entorno en términos cuantitativos.

5.1 Riesgo Individual.

Los resultados de riesgo individual expresados en número de fatalidades por año (o año-1), sí permiten establecer zonas de aceptabilidad del riesgo según los diferentes criterios que se utilizan en la industria. En la Ilustración 8 se presentan los resultados de riesgo individual en el sector comprendido entre los PK 43+000 y 45+000.

Ilustración 8. Riesgo Individual.

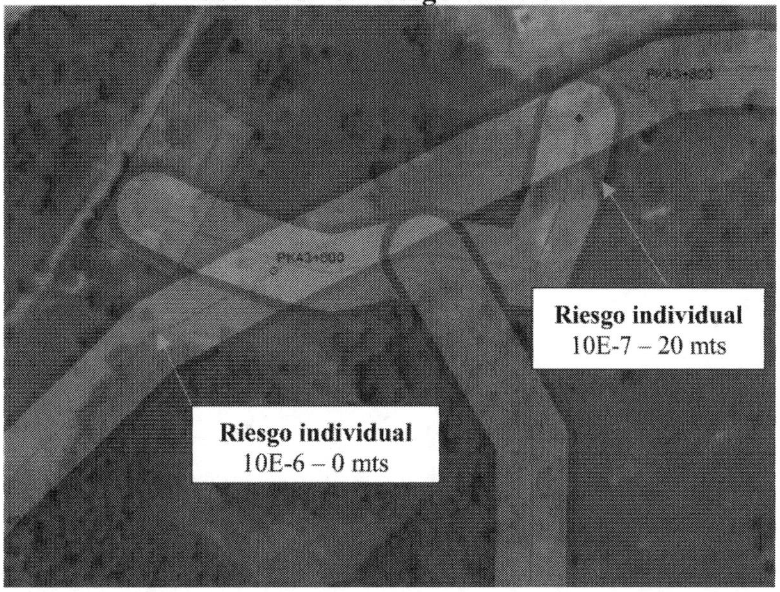

5.2 Riesgo Social

El riesgo social representa el número de fatalidades que pueden llegar a ocurrir en los sitios evaluados en el poliducto en el PK 43+000 – PK 45+000, al revisar la información contenida en la Ilustración 9 se puede apreciar como el riesgo se mantiene en la región tolerable, muy cercana al límite inferior ALARP. El evento dimensionante corresponde a una llamarada en la ruta de derrame, con el ducto transportando gasolina, la cual puede llegar a afectar a ocho (8) personas.

Ilustración 9. Riesgo social Escuela PK 43+000 – PK 45+000.

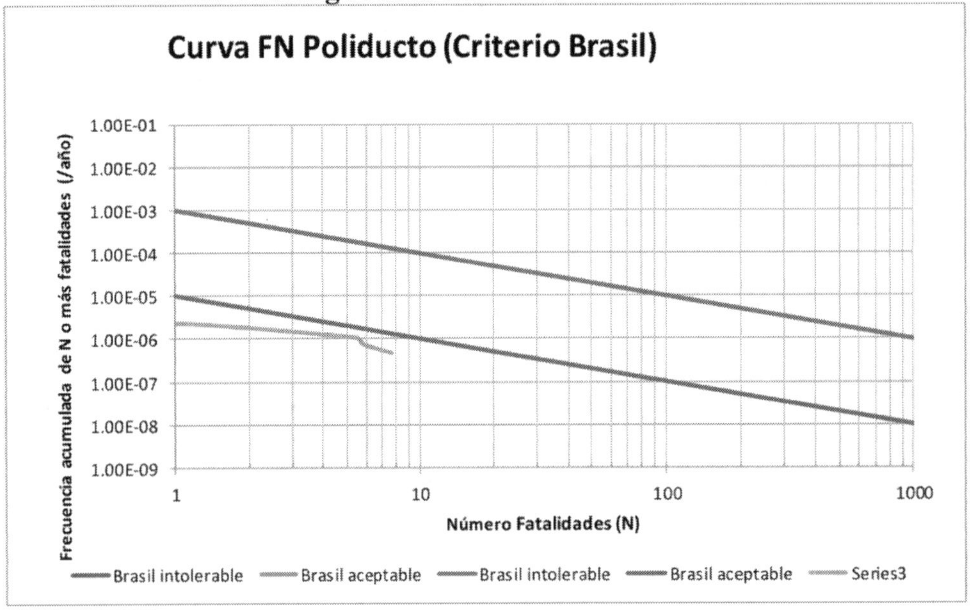

6 Alternativas de mitigación de Riesgo

Considerando los resultados del análisis de riesgo cuantitativo sobre el Poliducto para el sector PK 43+000 - PK 45+000, se tiene que ante una rotura total el evento dimensionaste corresponde a una llamarada con una distancia de afectación de 135 metros, para el caso de una fuga por corrosión el evento dimensionaste corresponde a una llamarada con una distancia de afectación de 32,24 metros, finalmente como resultado de una perdida de contención se puede llegar presentar la formación de una ruta de derrame en tierra la cual ingresaría a la escuela, con una distancia de afectación de 8.6 metros generada por una llamarada, tal como se presenta en la Ilustración 10.

Ilustración 70. Distancias de afectación

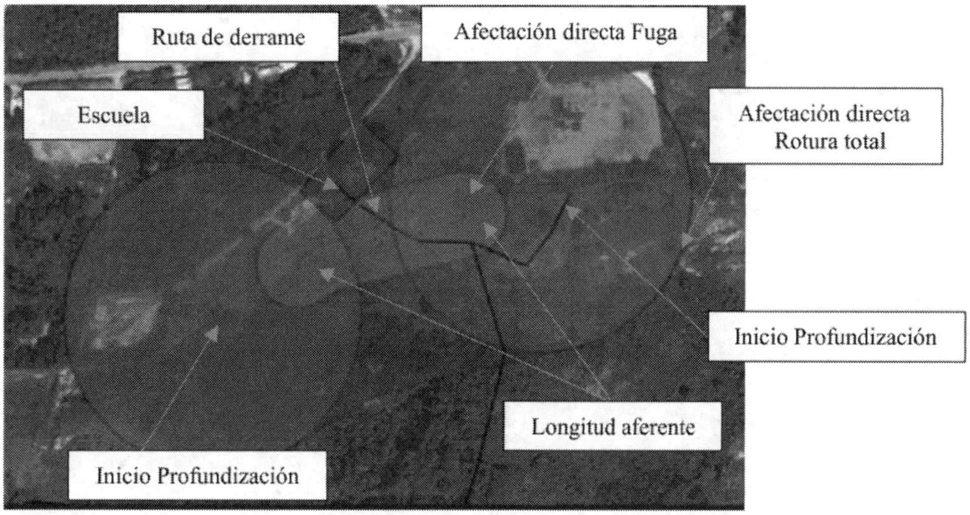

Teniendo en cuenta que el evento corresponde al presentado en la ruta de derrame, se hace necesario considerar la evaluación de alternativas de mitigación de riesgo. Con base en lo anterior, se consideró que la actividad de realineación y profundización cubra la longitud aferente de esta ruta, la cual es de 155 metros e inicia en la coordenada 74°33'35.466"O - 6°9'8.213"N y finaliza la coordenada 74°38'15.025"O - 6°9'8.737"N. Lo anterior, tiene como objeto de eliminar la ruta de derrame actual que conduce a la escuela en caso de una falla y perdida de contención, dado que el vertimiento se genere en sentido contrario al actual.

Sumado a lo anterior es necesario considerar una distancia de protección adicional para el dado caso en el que se llegue a presentar la rotura total de la línea, esta distancia se establece a partir del radio de impacto del evento llamarada, localizando los puntos sobre la línea que no presentan afectación a la escuela. A partir del análisis se estableció que esta distancia es de 316 metros e inicia en la coordenada 74°31'33,57146"O - 6°9'8,22442"N y finaliza en la coordenada 74°32'28,98241"O - 6°9'9,70756"N, esta distancia cubre la longitud aferente descrita anteriormente.

Dado que la alternativa de mitigación tiene como fin eliminar la afectación indirecta y el impacto directo sobre la escuela a continuación se presentan dos alternativas de trazados propuestos, que aseguran que las rutas de derrame en caso de una perdida de contención tendrán dirección opuesta a la ubicación de la escuela.

Ilustración 80. Opción 1 realineamiento entre PK 43 + 695 a PK 44 + 205.

Ilustración 91. Opción 2 realineamiento entre PK 43 + 625 a PK 44 + 460 de 752 metros.

En la ilustración 12 se presenta las dos alternativas propuestas y la distancia de separación del trazado actual de tubería, así como de la escuela.

Ilustración 102. Alternativas propuestas para mitigación de riesgo

7 Conclusiones

Del análisis realizado se tienen las siguientes conclusiones.

- El máximo número de fatalidades a generar es de ocho (8) y es generado por una llamarada en la ruta de derrame en tierra la cual, en el dado caso de que se presente el evento, ingresaría a las instalaciones de la escuela.
- La mayor susceptibilidad de la línea es por la amenaza de corrosión externa, especialmente en algunos sectores en los cuales el recubrimiento se encuentra en mal estado, esta condición junto con las anomalías encontradas en la línea, impactaron la frecuencia de falla establecida para el sector.
- En términos generales el nivel de riesgo se encuentra en la zona tolerable, muy cercana al límite inferior ALARP.

8 Recomendaciones.

- Es recomendable realizar actividades tendientes a disminuir el volumen disponible a desplazar en la ruta de derrame, para ello se puede realizar la profundización de la línea lo cual además de disminuir el efecto de los eventos por JET FIRE y FLASH FIRE disminuye la cantidad de producto disponible a desplazar al favorecer la evaporación de este.

- A partir del análisis realizado se tiene que la distancia a realinear y profundizar es:

  Opción 1 realineamiento entre PK 43 + 695 a PK 44 + 205 con una longitud de 516 metros
  Opción 2 realineamiento entre PK 43 + 625 a PK 44 + 460 con una longitud de 752 metros.

- Con ambas alternativas se elimina la afectación indirecta y la afectación directa, sobre la escuela, estas distancias cubren todos los posibles eventos que se pueden presentar y tienen afectación sobre la escuela en mención.

- Se pueden construir cortacorrientes los cuales eviten que la ruta de derrame ingrese o haga su tránsito en las cercanías de la escuela, junto con la construcción de estructuras de retención las cuales detengan el avance del producto mientras se realizan actividades de recuperación de este.

- Es recomendable rehabilitar el recubrimiento en aquellos sectores en los cuales se encuentra en mal estado, junto con la atención de aquellas anomalías que sean críticas.

9 Referencias

[1] Fingas, M. (2011). Oil Spill Science and Technology. Kidlington, UK: Gulf Professional Publishing Elsevier.

[2] D.O.T. Cornell Law School. Obtenido de 49 CFR 192.905 - How does an operator identify a high consequence area?: https://www.law.cornell.edu/cfr/text/49/192.905 Accesado 15 Enero, 2019

[3] D.O.T. Cornell Law School. Obtenido de 49 CFR 192.5 - Class locations: https://www.law.cornell.edu/cfr/text/49/192.5 Accesado 15 Enero, 2019

[4] Yellow Book. TNO. Methods for the calculation of Physical Effects Due to releases of hazardous materials (liquids and gases) – Third edition Second revised print 2005

[5] Purple Book. TNO. Guidelines for quantitative risk assessment – First edition 1999/2005

Referencias no citadas:

[6] ASME.ASME B31.8S. Managing System Integrity of Gas.New York, USA.2005.

[7] ASME.PIPELINE TRANSPORTATION SYSTEMS FOR LIQUID HYDROCARBONS AND OTHER LIQUIDS ASME B31.4-2002. New York, USA.2002.

[8] API.API Std 1160. Managing System Integrity for Hazardous Liquid Pipelines. Washington, USA.2008.132 Pags.

548036 Best Practice in MOC and Pssr Automation for Opex

Management of Change (MOC) and Pre Start-up Safety Review (PSSR) are still the most challenging elements of OSHA's Process Safety Management (PSM) standard and the EPA's Risk Management Program (RMP) rule. Now MOC and PSSR are requirements of the US Department of Interior's Bureau of Safety and Environmental Enforcement's Safety and Environmental Management Systems (SEMS). Even though the PSM standard has been around since 1992 and the industry has been managing change for several decades, we still can get it wrong, sometimes with disastrous results. The mindful or diligent efforts of many working in concert are necessary to ensure that change is identified, analyzed and executed in a quality way. But just when we get the process on paper and get the workforce up-to-speed, we have employee turnover, complacency sets in and sometimes there is regulatory change.

Many high-profile companies have addressed this challenge by creating and automating standard processes across their manufacturing or production facilities. MOC and PSSR are two of these major processes that have been successfully automated or **digitized**. Technology has been used to **automate** each step of the MOC workflow, enabling the collection of change ideas, formalizing the analysis of the ideas, and documenting the approval and execution of the change. Reports and dashboards can provide visibility of the health or current state of your MOC and PSSR processes. The use of **software** can enable the standardization of best practices.

On a well-meaning quest to standardize on a MOC/PSSR software solution and process, many companies are stifling creativity and innovation. Operational Excellence cannot be achieved in such ways. This paper discusses how the chemical process industry has defined the MOC and PSSR best management practices and how they should be automated for operational excellence.

Chemical Process Safety Requirements in the Fire Code

Ronald Melucci
Engineering Planning and Management, Inc.
959 Concord Street
Framingham, MA 01701
rbm@epm-inc.com

Prepared for Presentation at
American Institute of Chemical Engineers
2019 Spring Meeting and 15th Global Congress on Process Safety
New Orleans, LA
March 31-April 3, 2019

Ronald Melucci
Engineering Planning and Management, Inc.
959 Concord Street
Framingham, MA 01701
rbm@epm-inc.com
Keywords: NFPA Fire Code, International Fire Code, Process Safety

Chemical Process Safety Requirements of the Fire Code

Abstract

When most Chemical Process Safety professionals think about regulation and enforcement, they think about the applicable Occupational Safety and Health Administration (OSHA) and Environmental Protection Agency (EPA) regulations. However, the majority of US states adopt NFPA 1 or the International Fire Code (IFC) as their state fire code which contain requirements for chemical process safety. These requirements are applicable for processes of all sizes and levels of complexity below PSM and RMP thresholds and they address both fire and non-fire related chemical hazards. These requirements are enforced by municipal or state fire officials/marshals, who are becoming more aware of these requirements and chemical process safety in general. This paper will review some of the code requirements with a focus on requirements that may be commonly overlooked in the industry.

Introduction

The OSHA Process Safety Management (PSM) and EPA Risk Management Program (RMP) regulations, enforced by OSHA and EPA, respectively, are familiar to most safety professionals who work in industries that include processing of large quantities of chemicals. However, the applicable fire code requirements related to chemical process safety are often more restrictive than the federal regulations and are generally lesser known to such safety professionals.

The two model fire codes adopted by states in the US, are: *NFPA 1, Fire Code,* published by The National Fire Protection Association (NFPA), and the *International Fire Code* (IFC) published by the International Code Council (ICC). The states that do not adopt the IFC will typically use NFPA 1 and each state may also institute their own NFPA 1 amendments. Local amendments are almost always more restrictive than the model code verbiage, and many states have legislation that prohibits fire code amendments that reduce the level of safety/standard of care intended by the model fire code.

Both of these model fire codes contain many requirements that are similar to PSM and RMP regulations and contain references to them for additional guidance. Generally speaking, the model fire codes do not regulate all hazardous materials, but focus on hazards that are acutely dangerous to people or property. The PSM and RMP regulations address these hazards as well but additionally include materials that present a risk of chronic health or environmental hazards.

However, many states and municipalities have local fire code amendments that often include lower chemical quantity thresholds, such as those of the Commonwealth of Massachusetts as described in the following sections. The enforcement of these requirements is therefore not limited to OSHA and the EPA, as is commonly believed. The requirements are also enforceable by the local (municipal) fire authorities through the fire code references to these regulations.

Fire Code Requirements

The Commonwealth of Massachusetts currently adopts NFPA 1 – 2015 edition, with amendments, as 527 CMR 1.00: *Massachusetts Comprehensive Fire Safety Code*. Since 2006, 527 CMR has included chemical process safety requirements that are more restrictive than the federal regulations. The impetus for this was several accidents that occurred involving chemical process quantities below the OSHA and EPA thresholds. These Massachusetts (MA) amendments have been added to Chapter 60 of NFPA 1 – 2015 edition and include specific reference to OSHA PSM and EPA RMP regulations. Chemical processes with quantities above the PSM/RMP thresholds are defined in MA as "Category 5" processes. However, the 527 CMR amendments have also defined processes with lower quantities, i.e. Category 1, 2, 3 and 4, each with specific and unique requirements as described herein.

The NFPA 1 committee is currently reviewing proposals to adopt requirements with a "tiered" quantity threshold approach similar to MA. This is in recognition of the fact that severe hazards often exist, and severe accidents have occurred, in chemical processes with quantities below the PSM/RMP thresholds. It is likely that such a "tiered" approach will be incorporated into NFPA 1 and subsequently adopted by many US states. A preview of the applicable NFPA 1 and 527 CMR requirements are summarized herein.

Massachusetts Amendments to NFPA 1, Section 1.12.8.34.2 introduces the requirements as: *Process or Processing of Any Hazardous Material at any Facility* and it states the following:

(1) "A permit shall be required for the process or processing of any hazardous material at any facility identified in this *Code* as Category 2 through Category 5, as described in Chapter 60.
(2) The Authority Having Jurisdiction (AHJ) may require technical assistance in accordance with *1.15* to evaluate the adequacy of a Category 3 or Category 4 facility process safety conditions, programs, procedures, and practices undertaken at the facility, but only after a notice of denial has been properly served upon the person making application."

As part of the MA amendments to NFPA 1 – 2015 edition, Section 60.8, ***Hazardous Material Process or Processing,*** was added to 527 CMR. Section 60.8 applies to both new and existing facilities that process hazardous materials, with a list of exceptions such as:

- Motor vehicle service stations
- Refrigeration systems using a refrigerant other than ammonia or LPG,
- Certain wastewater treatment operations
- Storage of hazardous materials in atmospheric vessels.

Definitions:

NFPA 1 defines "**Hazardous Materials**" as "A chemical or substance that is classified as a physical hazard material or a health hazard material, whether the chemical or substance is in usable or waste condition."

OSHA and EPA define "**Process**" as "Any use, storage, manufacturing, handling, or the on-site movement of hazardous chemicals, or combination of these activities." The MA state fire code defines "Process" as "A sequence of operations in which the sequence can be inclusive of physical operations such as heating, cooling, mixing, distilling, compressing, and pressurizing, and chemical operations, such as polymerization, oxidation, reduction, and other chemical reaction processes. The sequence can involve but is not limited to: preparation, separation, combination, purification, or any actions that cause a change in state, energy content, or chemical composition."

MA Amendments to NFPA 1 Section 60.8 categorizes hazardous material processes into five (5) categories, as follows:

Table 1: Hazardous Material Threshold Quantities in the MA Amendments to NFPA 1

Category	**Quantity (gallons)**
Category 1	< 2.5
Category 2	2.5 – 60
Category 3	60 – 300 (or Group H)
Category 4	300 and not Category 5
Category 5	≥ the threshold quantities stated in 29 CFR 1910.119

The MA Amendments also add Section 60.8.5.1 to NFPA 1, which requires that facilities operating hazardous material processes must maintain the required documents and procedures for periodic inspection and review by the head of the Fire Department to remain in compliance with the state fire code. Requirements for each process category are as follows:

Table 2: Outline of Regulatory Requirements in the MA Amendments to NFPA 1

Category	Requirements
Category 1 (Vessel Capacity ≤ 2.5 gal)	• OSHA 1910.1450 (Occupational exposure to hazardous chemicals) • OSHA 1910.1200 (Hazard communication to employees) • Policy per NFPA 1, Ch. 66 (Flammable liquids) and Ch. 67 (Flammable solids) • Requirements of NFPA 1, Section 60.1 – 60.5
Category 2 (Vessel Capacity ≤ 60 gal)	• All requirements of Category 1 • Permit required per NFPA 1, Sections 1.12.8.34.2 and 60.8.1.1
Category 3 (Vessel Capacity ≤ 300 gal OR Group H Occupancy)	• All requirements of Category 1 & 2 • Category 3 Hazard Evaluation • Requirements of NFPA 1, Section 60.6
Category 4 (Vessel Capacity ≥ 300 gals AND not Category 5)	• All requirements of Category 1-3 • Limited Safety Program (PSM "lite")
Category 5 (Vessel Capacity ≥ the threshold quantities stated in 29 CFR 1910.119)	• All requirements of Category 1-4 • Process Safety Management (PSM) Program 29 CFR 1910.119 (OSHA) • Risk Management Program (RMP) 40 CFR Part 68 (EPA)

Category 1 – Requirement Details

Category 1 processes, as stated in Table 2 above, must possess documentation of compliance with OSHA 1910.1450, *Occupational Exposure to Hazardous Chemicals in Laboratories*, OSHA 1910.1200, *Hazard Communication,* maintain and implement a policy in compliance with NFPA 1, Ch. 66, *Flammable and Combustible Liquids*, and Ch. 67, *Flammable Solids*, demonstrate compliance with NFPA 1, Section 60.1.5.1, *Emergency Action Plan*, and Sections 60.5.1.4.3.2(5) through 60.5.1.4.3.2(9) and comply with permitting requirements in Sections 1.12 and 60.8.4 of the MA Amendments to NFPA 1.

OSHA 1910.1450 and 1910.1200

Compliance with OSHA 1910.1450 requires facilities to set and adhere to exposure limits, monitoring plans, hygiene plans, precautions, training, medical plans, and other plans, as found to be acceptable by the AHJ, to react to and minimize potential exposure to hazardous chemicals. OSHA 1910.1200 requires hazard communication methods for employees such as inventory documentation, hazard programs, labels, safety data sheets, and training.

NFPA Chapter 66 and 67

In addition to demonstrating compliance with OSHA standards, facilities meeting Category 1 requirements must implement and maintain a written policy in compliance NFPA 1, Ch. 66 and Ch. 67.Chapter 66 is divided into 29 sections, most of which contain multiple requirements. Section 66.6 contains the prescriptive requirements for fire and explosion prevention and risk control, and Section 66.6.4 goes on to list requirements related to a facility's hazards analysis, which is required to be reviewed to ensure that fire and explosion hazards are addressed by fire prevention, fire control, and emergency action plans. The AHJ should be consulted with for the development of the hazards analysis to ensure that the extent of the fire prevention and control is deemed acceptable or an engineering evaluation of the operation and application of fire protection and process engineering principles is soundly applied. The hazards analysis must include, but not be limited to, analysis of the following:

(1) Fire and explosion hazards
(2) Emergency relief from process vessels
(3) Applicable facility design requirements in Chapters 17, 18, 19, 28, and 29 of NFPA 30
(4) Applicable requirements for liquid handlings, transfer, and use, as covered in Chapters 17, 18, 19, 28, and 29 of NFPA 30
(5) Local conditions/exposures
(6) Emergency response capabilities of the local emergency services

Additional guidance for conducting a fire risk assessment can be found in NFPA 551, *Guide for the Evaluation of Fire Risk Assessments.* Facility management must repeat the hazards analysis when any of the following occur:

(1) Changes occur in the materials in process
(2) Changes occur in process equipment
(3) Changes occur in process control
(4) Changes occur in operating procedures or assignments

This is consistent with the requirements of OSHA 1910.119 Sections (e) and (l). Not all sections of Chapter 66 may be related to processes; however, Section 66.17 is for processing facilities. Section 66.17 includes prescriptive requirements for processing facilities such as location of process vessels and equipment (Section 66.17.4), containment, drainage, and spill control (Section 66.17.10), and management of operations hazards (Section 66.17.15). Later sections of Chapter 66 include additional requirements for dispensing, handling, transfer, and use of liquids (processes) (Section 66.18) and specific operations such as recirculating heat transfer system, vapor recovery and vapor processing system, solvent distillation units, and cooking oil storage tank systems in commercial kitchens (Section 66.19).

NFPA 1 Chapter 60

In addition to Chapters 66 and 67 of NFPA 1, Category 1 processes must also adhere to the requirements of 60.1 – 60.5 of NFPA 1 and the MA Amendments to NFPA 1. Depending on whether the quantities of materials at the facility exceed the MAQ or not, determines which requirements of Section 60.1 are applicable (Section 60.1.3).

Section 60.1.5 of NFPA 1 requires that an emergency action plan, consistent with the available equipment and personnel, shall be established to respond to fire and other emergencies in accordance with requirements set forth in NFPA 1. Section 10.8 of NFPA 1 includes additional requirements and guidance on generating emergency action plans, such as what must be included in the plans, purpose, responsibilities, etc. Section 60.4 of NFPA 1 contains requirements related to permissible storage and use locations and sets prescriptive requirements on the maximum allowable quantities for each individual control areas based on the area's occupancy type.

Section 60.5 contains the fundamental requirements for hazardous materials, including requirements for personnel training, ignition sources, systems, equipment, and processes (Section 60.5.1.4, 60.5.1.5, and 60.5.1.6 respectively). In Massachusetts, Section 60.5 includes multiple requirements that come from both NFPA 1 and the MA Amendments to NFPA 1. Despite Category 1 processes having a vessel capacity much less than Category 2-5 processes (Category 1 processes have a vessel capacity ≤ 2.5 gallons), there are numerous NFPA and state fire code requirements that must be followed. Therefore, the importance of fire protection engineers and their understanding of the NFPA and state fire codes cannot be understated for processing facilities. Each subsequent category process contains additional fire code fire protection/chemical process safety requirements of increasing complexity.

Category 2: Requirement Details

Category 2 processes must comply with all Category 1 requirements and additionally submit for a permit to the AHJ, as required per 1.12.8.34.2 and 60.8.1.1 of MA Amendments to NFPA 1. Permits submitted to the AHJ must comply with the requirements of Section 1.12 of NFPA 1 – 2015 edition. The nature of the operations and materials at the facility determine the permit requirements. Depending on the nature of the process, the AHJ may require technical assistance to evaluate the adequacy of facility process safety conditions, programs, procedures, and practices undertaken at the facility, but only after a notice of denial has been properly served upon the person making application.

Category 3: Requirement Details

While meeting all the requirements of Category 1 and 2 processes, Category 3 processes must also provide the following documentation and evaluations:

- Compliance with the requirements of Category 1 and 2 processes;
- Category 3 hazard evaluation; - per Section 60.8.2
- Implementation of process safety controls to mitigate hazards of normal and abnormal conditions identified in the hazard evaluation;
- Maintain documentation for review for a minimum of two (2) years;
- Third-party review (if required by the fire department/AHJ);
- Post-Incident Analysis per 60.8.6 (to be initiated within 48 hours of incident);

MA Amendment to NFPA 1, Section 60.8.2.2 provides requirements and details for the hazard evaluation that is required for Category 3 processes. **Hazard evaluations** are meant to identify hazards, including adjacent vessels that contain hazardous materials, and determine the required preventative, protective, and safety control measures in conformance with **r**ecognized **a**nd **g**enerally **a**ccepted **g**ood **e**ngineering (and safe work) **p**ractices (RAGAGEP) associated with a particular process or condition. Section 60.8.2.5 requires that a person with *specialized knowledge* beyond that of an average person, about risk assessment, process hazard analysis, and/or process safety management principles, for the process or processes being evaluated *conduct the necessary evaluations*. Section 60.8.2.5 further emphasizes the importance of both chemical process safety specialists and fire protection engineers for analysis of hazardous processing facilities.

In addition to generating a hazard evaluation, Category 3 processing facilities (vessel capacity ≤ 300 gal OR Group H Occupancy) must also meet the requirements of NFPA 1, Section 60.6, *Emergency Action Planning, Fire Risk Control and Chemical Hazard Requirements for Industrial Processes.* Where quantities of materials in use require compliance with Protection Level 1, Protection Level 2, Protection Level 3, or Protection Level 4 based on materials exceeding the Maximum Allowable Quantity (MAQ) in the following categories shall comply with the requirements of Chapter 7 of NFPA 400, *Hazardous Materials Code*:

(1) Unpackaged organic peroxide formulations that are capable of explosive decomposition in their unpackaged state
(2) Oxidizer Class 3 and Class 4: solids and liquids
(3) Pyrophoric solids, liquids, and gases
(4) Unstable reactive Class 3 and Class 4: solids, liquids, and gases
(5) Highly toxic solids, liquids, and gases
(6) Water-reactive liquids, Class 3

NFPA 400, Chapter 7, *Emergency Planning, Fire Risk Control, and Chemical Hazard Requirements for Industrial Processes,* contains requirements and guidance on topics such as:

- Process review and plan preparation
- Operating and maintenance procedures
- Safety reviews
- Incident investigation plans

Section 7.2.3 contains specific requirements for Emergency Action Planning that facilities must comply with. Some requirements of Section 7.2.3 include:

- Procedures for notifying and coordinating with the fire department and other applicable agencies
- Procedures and schedules for conducting drills
- Alternative measures for occupant safety
- Maintenance of fire protection equipment

Lastly, in lieu of complying with Chapter 60 in its entirety, occupancies containing high hazard Level 1 to high hazard Level 5 contents shall be permitted to comply with Chapter 10 of NFPA 400, as stated in NFPA 1, Section 60.7. MA Amendments to NFPA 1 further states that complying with NFPA 1, Section 60.7, is subject to an independent review in accordance with Section 1.15 and a copy, including its recommendations, shall be submitted to the Building Official.

Category 4: Requirement Details

Category 4 processing facilities (vessel capacity ≥ 300 gal) must meet:

All the requirements of Category 1, 2, and 3 processes

And also create and maintain a **Limited Safety Program (PSM "lite")**. The Limited Safety Program documents evaluation, policy and/or required procedures to ensure compliance with the following requirements of the MA Amendment to NFPA 1:

(1) Process information including, but not limited to, MSDS for the chemicals and products being processed, process chemistry, piping and instrumentation diagram, safety relief design, process control safety alarms and interlocks;

(2) Facility suitability including, but not limited to, the *Building Code* compliance, electrical hazard classification, ventilation design, fire alarm and fire protection, spill containment and control;

(3) A process hazard safety analysis including, but not limited to, effects in the event of failure, suitable administrative and engineering controls to minimize failure and to control unanticipated releases, and emergency responses to safeguard life and property;

(4) Written procedures, including routine operating and maintenance, as well as precautionary, shut-down and emergency response measures;

(5) A written training program for operating and maintenance personnel and outside contractors whose work or activity may affect process safety;

(6) A written records management protocol, which tracks any changes, including but not limited to changes to chemicals, equipment, operating procedures training program. Such records shall include the date of such change and the name of the manager responsible for such change; and an internal review at a maximum every three years.

These requirements are similar to those found in OSHA 29 CFR 1910.119, *Process Safety Management of Highly Hazardous Chemicals*.

Category 5: Requirement Details

Category 5 processing facilities must provide evaluation and documentation per 29 CFR 1910.119 OSHA, *Process Safety Management of Highly Hazardous Chemicals*, and 40 CFR Part 68 EPA, *Risk Management Program Rule*, all while meeting the requirements for Categories 1-4. OSHA's *Process Safety Management* (PSM) *of Highly Hazardous Chemicals* was issued in February 1992 and contains 14 elements that focus mainly on worker safety and catastrophic loss of containment. The Process Hazard Analysis (PHA) is the "heart" of the standard and is the most time consuming. The PHA must be completed every five (5) years and *is required to be completed by someone knowledgeable with the PHA methodology*. Additionally, compliance audits for the standard must be completed every three (3) years and may be completed in-house or by a third party (preferred). EPA's *Risk Management Program* (RMP) *Rule* consists of 12 elements and is focused on public safety and the environment. The program contains Chemical Accident Prevention Provisions (CAPP) and Accidental Release Prevention/Risk Management Plan Rule. The RMP must be submitted to the EPA every fire (5) years or if there are any changes.

Conclusion

Some fire code requirements for chemical process safety apply to processes with quantities much lower than the PSM/RPM thresholds. Failure to comply with these requirements can result in fire department citations, fines, remedial work and in extreme cases, shutdown of operations. Between 2013 and 2018, 214 OSHA PSM violations were recorded in Massachusetts alone, which resulted in over $485,000 in penalties. There are far more non-PSM facilities than PSM facilities in MA, so the magnitude of the problem can be seen by simple extrapolation. Failing to comply with the fire code and the federal requirements can also result in significant property loss, injuries and fatalities with much higher costs than the violation penalties. Compliance with the local fire code and federal regulations should be determined by experienced chemical process safety specialists and fire protection engineers to minimize the risk of significant loss.

Fire protection engineering and science is its own discipline and chemical process safety is its own discipline usually practiced by the chemical engineering field. Finding ways to create awareness and bridge the gap between the two disciplines, and sharing chemical process safety management concepts with those in the fire protection arena, will likely improve the oversight, understanding and enforcement of process safety resulting in better overall safety. There may be a way of joining these disciplines into one overarching safety category and therefore giving a better sense of how interrelated they really are.

546769 Poly-Butadiene Formation in Olefins Debutanizer Condenser

Abstract Title : Poly-butadiene formation in Olefins Debutanizer condenser

Discipline : Process Safety

Topic : Poly-butadiene formation in Debutanizer condenser

Author Name : Ali Bazroun

Participating team: Fahad AlShammari, and Eldridge Monette, Bertrand, Da, Cruz, Fahad AlSaad, Abdullah Al-Ismail

ABSTRACT

In one of Sabic olefins plant Plant, the condenser of the debutanizer tower was fouling. This was evident through the pressure increase in the tower and the cooling water control valve reaching the maximum opening (69%) limited by the mechanical stopper. The condenser process service is butadiene, which has the tendency of polymerizing and formation of popcorn, which could plug relief piping or accumulate in the condenser until the vessel shell is over stressed causing loss of containment and leading to fire or explosion.

The risk assessment team evaluated that the risk as significant, meaning that it could result into fatality and multiple lost time injury with a likelihood of 10E-3. With the propagating shell deformation, the likelihood could increase to L3 (10E-2) over a longer period. The risk could become C2-L3 and remain Significant. To ensure effective risk mitigation, a major plant modification was executed by constructing a new process line that bypasses the debutanizer tower and sends the material to the flare. The project execution was remarkable as it was completed in about a week. 15 tons of C4 is being flared per. This, however, allowed the olefins unit to mitigate the risk effectively without any interruption of the production. The change was not done before in the local industry, and it reflected how effective measures can be taken to mitigate risk through strong leadership and excellent engineering. Interim mitigations, moreover, were in place until the modification was completed: for example, inhibitor injection in the exchanger to inhibit the popcorn formation as soon as possible, and an emergency procedure in case loss of containment. Without the effective mitigation, Sabic operation team would have had to shutdown the olefins unit for over 2 months. This olefins unit feeds two poly ethylene that would shutdown consequently, causing significant losses over the mentioned period.

After isolating the exchanger, borescope test proved the formation of polymer. A root cause analysis was initiated, along with cleaning the tower and fabricating a new exchanger. The details of the RCA will be shared in the full paper.

Industrial process plant safety by collaboration of a Safety Instrumented System and a Dynamic Simulator

Tadaaki Ando
Yokogawa Corporation of America
12530 West Airport Blvd, Sugar Land, TX 77478, USA
Tadaaki.Ando@us.yokogawa.com
(Tadaaki.Andou@jp.yokogawa.com)

Yasuhiko Yamashiro
Yokogawa Electric Corporation
2-9-32 Nakacho, Musashino-shi, Tokyo, 180-8750 Japan
Yasuhiko.Yamashiro@jp.yokogawa.com

Prepared for Presentation at
American Institute of Chemical Engineers
2019 Spring Meeting and 15th Global Congress on Process Safety
New Orleans, LA
March 31 – April 3, 2019

Industrial process plant safety by collaboration of a Safety Instrumented System and a Dynamic Simulator

Tadaaki Ando
Yokogawa Corporation of America
12530 West Airport Blvd, Sugar Land, TX 77478, USA
Tadaaki.Ando@us.yokogawa.com
(Tadaaki.Andou@jp.yokogawa.com)

Yasuhiko Yamashiro
Yokogawa Electric Corporation

Keywords: Industrial process plant, SIS (Safety Instrumented System), Dynamic Simulator

Abstract

Safer HMI design for operators' quick and appropriate response is very important. Allowed maximum delay for operator's action from the alarm is not same all the cases because it is depends on many factors, i.e. ambient temperature, density of material, O2, and etc. So design of SIS includes HMI should be based "worst case". However, the 'worst case' is difficult to be found.

A "Dynamic Plant Simulator" can precisely simulate behaviors of the real process by visualizing how a hazard is diffused at various stages of a chemical reaction. And combination of DCS/SIS/HMI simulator with "Dynamic Plant Simulator", safety logic design is optimized about trip point and alarm threshold, and HMI design is modified safer manner.

The effective result of application of "cyber model by Dynamic Process Simulator" to a vinyl acetate monomer (VAM) plant and its value for operator training will be discussed.

1 Introduction

Industrial process plants in Japan have been seeking for the highest productivity and optimization of their processes, same with any process plant all over the world. These plants were renovated and expanded as time passed for the purpose, the plants themselves became complicated and troublesome as a result. Few years ago, severe process plant incidents occurred in Japan. The most probable causes of these severe incidents are out of "incorrect judgement and operation" and "shortage of prediction for technical aspects" during exceptional manual work that has been done by incidents. Understanding the hazardous situation by risk analysis and designing safety measures utilizing a SIS (Safety Instrumented System) are important and effective for resolving "incorrect judgement and operation" and "shortage of prediction of technical aspect". Safety measure includes human intervention via SIS man-machine interface which showing alarm, event, and process data.

The safety measure and its function are developed on the desk by brainstorming or so-called "hazard and operability study (HAZOP)." The verification of the safety function hardware could be established by reliability based analysis, SIL calculation in other word. However, the verification of safety logic is the issue because it has only been verified "on the desk". Especially the determination of the criteria for the emergency shutdown is the most important, but it is not possible to be confirmed with "try and error" on a running plant.

2 Safe HMI Design

Safer HMI design with considering different conditions and situations in the real process for operators' quick and appropriate response is very important. To design it, knowing the behavior of the process itself and related factors which affect the process is necessary.

For example, a reaction speed varies on the ambient temperature, density of material, O_2, catalyst, and etc., so allowed maximum delay for operator's action from the alarm is not same all the cases. So design of SIS includes HMI should consider "worst case". However, finding the worst-case scenario with how much is the allowed operator reaction is difficult because of many factors.

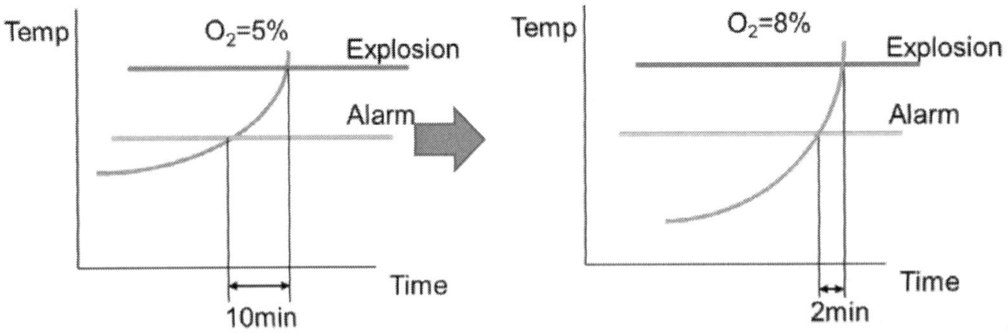

Fig.1 Example: Difference of O_2 density vs. allowed reaction time

Another example is the 'thermal runaway'. Even if feeding of material, heat, and catalyst were stopped by ESD because of a high temperature of a reactor, reaction continues due to a feed-forward loop of heating, means reaction generates heat to continue the reaction itself.

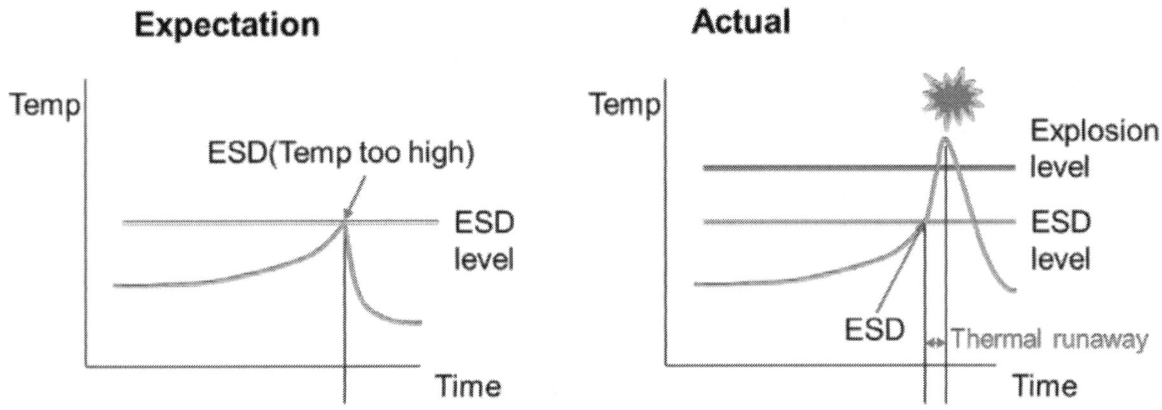

Fig.2 Even after ESD, thermal runaway leads the process to explosion

Thus, considered 'worst case' could not be a 'worst', or could be 'too much pessimistic'.

A "Dynamic Plant Simulator" can precisely simulate behaviors of the real process by visualizing how a hazard is diffused at various stages of a chemical reaction. And combination of DCS/SIS/HMI simulator with "Dynamic Plant Simulator", safety logic design is optimized about trip point and alarm threshold, and HMI design is modified safer manner.

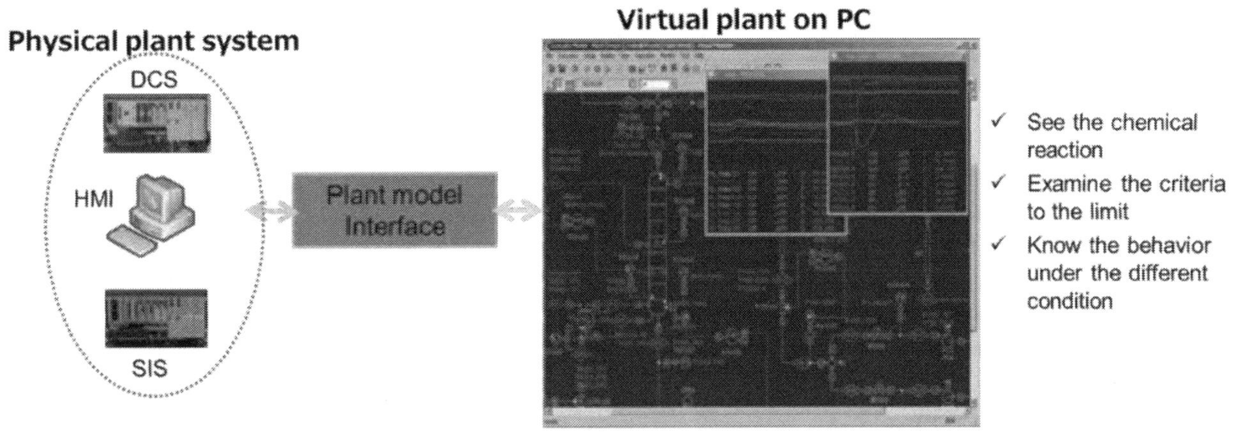

Fig.3 Dynamic Plant Simulator

3 "Dynamic Plant Simulator" at VAM (Vinyl Acetate Monomer) process

"Dynamic Plant Simulator" with "DCS/SIS/HMI simulator" was applied for several shutdown conditions of VAM plant. "Dynamic Plant Simulator" can simulate a variation of factors like ambient temperature and thermal runaway, because is based on a theory of chemical reaction. And "DCS/SIS/HMI simulator" is copying a real function and operation of DCS/SIS/HMI.

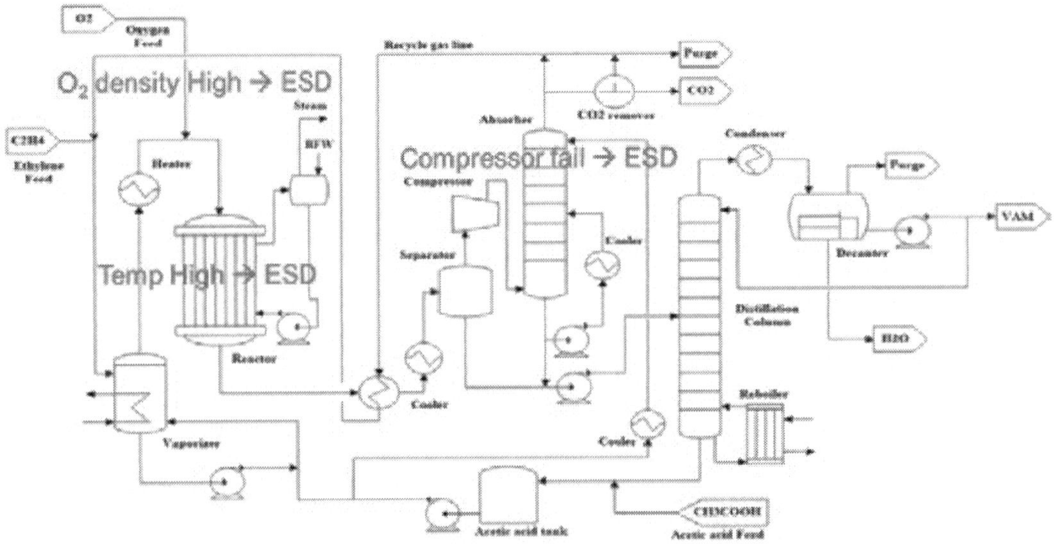

(Main Reaction) $C_2H_4 + CH_3COOH + 1/2O_2 \rightarrow CH_2{=}CHOCOCH_3 + H_2O$

(Side Reaction) $C_2H_4 + 3O_2 \rightarrow 2CO_2 + 2H_2O$

Fig.4 VAM process model

3.1 Shutdown by high temperature to avoid a damage to reactor

Reactor temperature must be monitored and controlled lower than 200 Degrees C (392 Degrees Fahrenheit) to avoid a damage of the reactor. If the temperature reached to a set point for emergency shutdown, emergency shutdown action stops feeding of Ethylene and O_2.

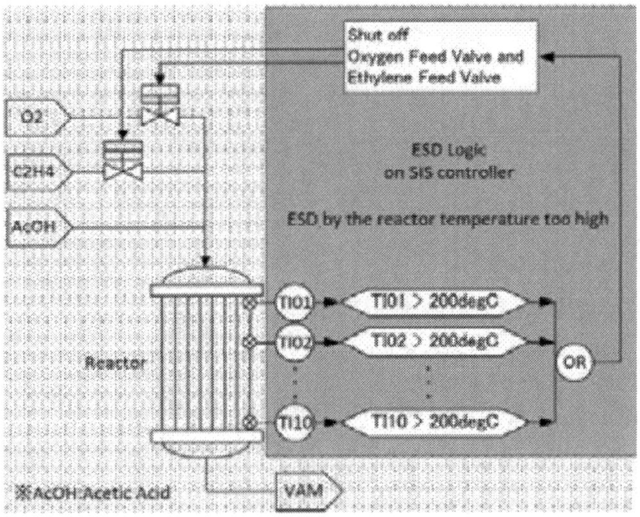

Fig.5 ESD logic for high temperature shutdown

When 190 Degrees C (374 Degrees Fahrenheit) is selected as the set point to shutdown, the result of simulation is, the reactor temperature must become over 200 Degrees C due to a runaway reaction in the reactor. It means reactor must be damaged. When 180 Degrees C (356 Degrees Fahrenheit) is selected, the simulation shows the reactor temperature can be kept lower than 200 Degrees C, so no damage to the reactor.

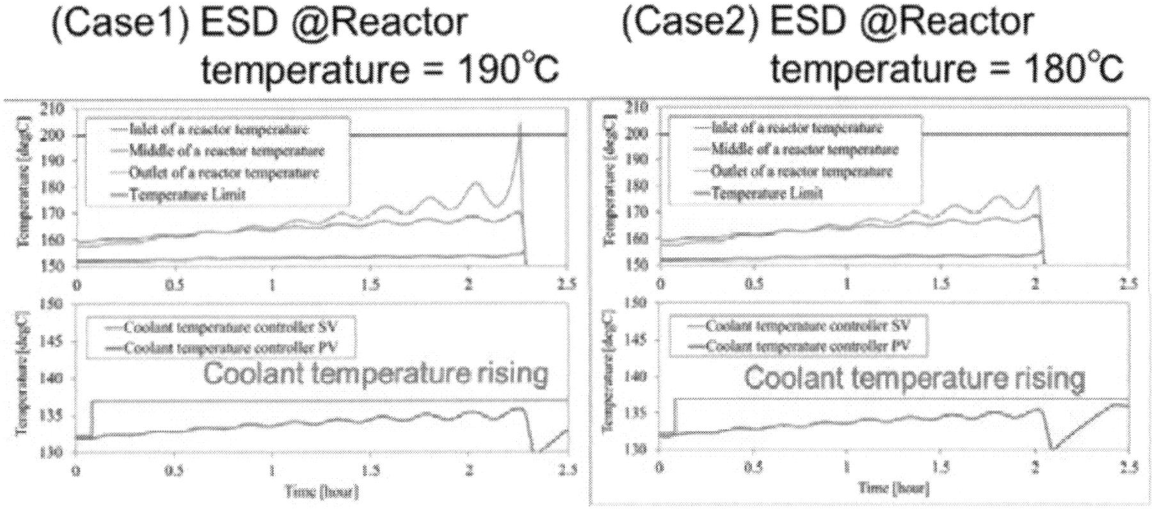

Fig.6 Simulation results

And simulation suggested the reactor temperature must be monitored in multipoint, middle of the reactor and top of the reactor. When the reaction temperature is lower than 160 Degrees C (320 Degrees Fahrenheit), temperature of a middle of the reactor is higher than other points, but when the reaction temperature is higher than 160 Degrees C, the temperature of top of the reactor is higher than other points.

3.2 Shutdown by high O_2 density to avoid explosion

If the density of O_2 in the reactor exceeds 8%, there is a risk of explosion. However, the O_2 density varies depends on other factors, i.e. feed pressure of ethylene to the reactor. Simulation has been done about O_2 density when the feed pressure of ethylene is suddenly changed. Simulation of some cases of dumping the feed pressure to lower side which increase the density of O_2 in the reactor were done.

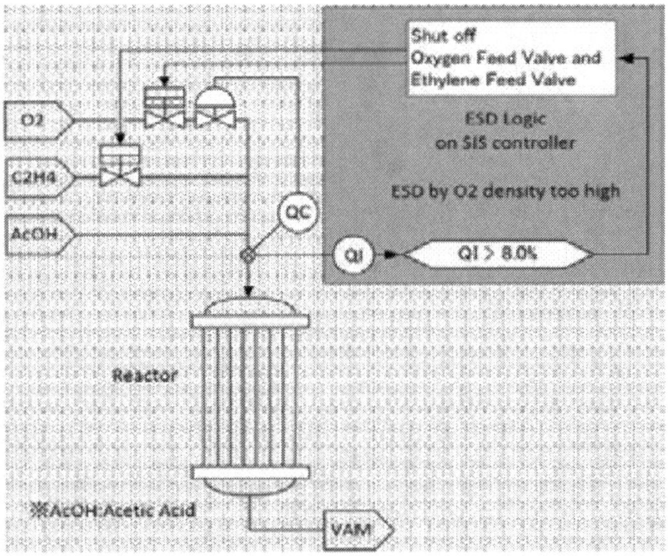

Fig.7 ESD logic for high O_2 density shutdown

When the feed pressure of ethylene dumped 10%, density of O_2 in the reactor doesn't reach to 8%. So there is no risk of explosion. However, the dumping of ethylene feed pressure is 20%, O_2 density will be higher than 8%, then an emergency shutdown must be activated. Shutdown does affect to business.

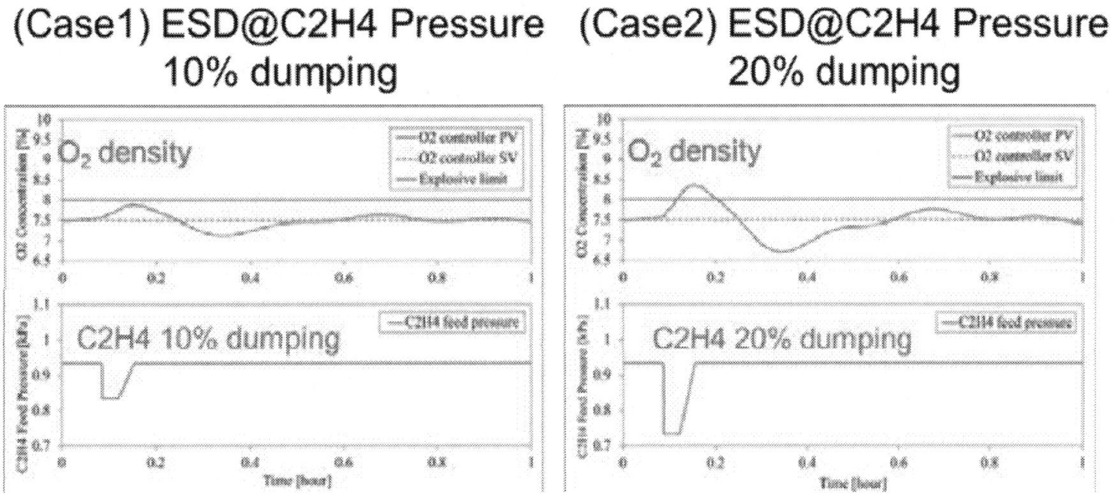

Fig.8 Simulation results

If there is a possibility of dumping ethylene feeding pressure, feeding O_2 in normal operation should be lower to avoid the risk of reactor explosion. On the other hand, lower O_2 density leads lower reaction speed, less production. Finding a best balance with safety and productivity is the key.

Using dynamic process simulator, many factors of process can be manipulated not only within normal operation range, but also close to abnormal, emergency situation or more un-realistic range, and results can be confirmed. This type of "try-and-error" approach is not possible on the real process, but possible in the "cyber" world to find a best balance of setting for maximize production with keeping or improving safety.

4 "Dynamic Plant Simulator" for Operator training

The training of operators is important in many reasons. Especially training for response to critical / emergency situation is indispensable, because operator rarely or never face such a situation and consequent secondary action to DCS/SIS or response to alarm flooding in a real plant. Therefore operator are impossible to familiarize with their required response in right procedure to critical / emergency situation. If operators are not capable or not knowing how to response such a situation, risk must propagate real hazard or accident. Furthermore, operators need to re-start the process after confirming the cause of emergency doesn't exist anymore and checking all the conditions of materials are steady, status of equipment are ready to re-start. This procedures are also not normal for operators.

“Dynamic Plant Simulator” realizes operator training with “equal to real” environment. Operators will be trained for correct and timely manner response until proficient. And operators will have more understanding for process itself by playing on the simulator.

5 Conclusion

The effectiveness of “Dynamic Process Simulator” for safer design of DCS/SIS/HMI is confirmed on VAM plant. The best balance of safety and productivity can be found and optimization of safety action is created by simulation. The created new design of DCS/SIS/HMI is possible to use an actual system directly. And operator training with “Dynamic Process Simulator” will increase a total safety of the plant.

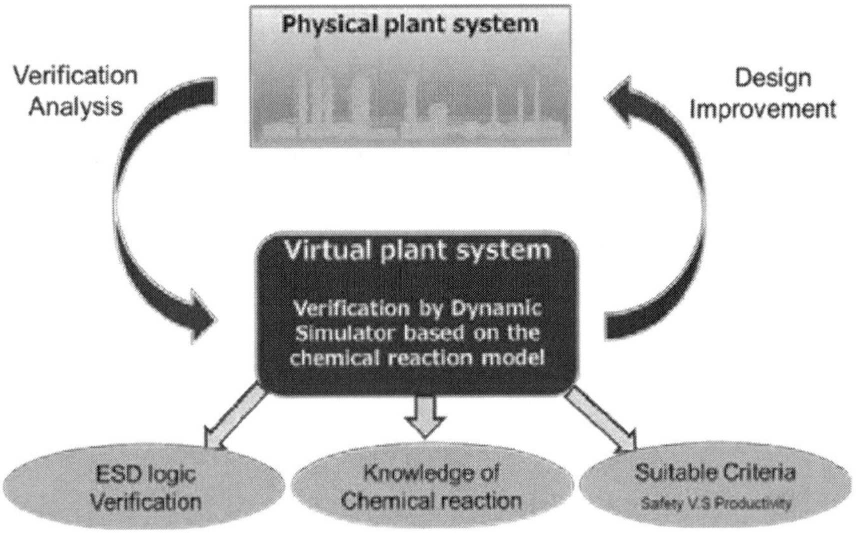

Fig.9:Virtual plant system realizes well balanced plant operation

Analysis of Chemical Effluent Discharge through the Emergency Relief System using Air-Water Properties

Basar Ozar
Jensen Hughes
600 W Fulton St. Suite 500
Chicago, IL 60661
bozar@jensenhughes.com

Jens Conzen
Jensen Hughes
600 W Fulton St. Suite 500
Chicago, IL 60661
jconzen@jensenhughes.com

Damian D. Stefanczyk
Jensen Hughes
3610 Commerce Dr. Suite 817
Baltimore, MD 21227
dstefanczyk@jensenhughes.com

Prepared for Presentation at
American Institute of Chemical Engineers
2019 Spring Meeting and 15th Global Congress on Process Safety
New Orleans, LA
March 31 – April 3, 2019

Analysis of Chemical Effluent Discharge through the Emergency Relief System using Air-Water Properties

Basar Ozar
Jensen Hughes
600 W Fulton St. Suite 500
Chicago, IL 60661
bozar@jensenhughes.com

Jens Conzen
Jensen Hughes
600 W Fulton St. Suite 500
Chicago, IL 60661
jconzen@jensenhughes.com

Damian D. Stefanczyk
Jensen Hughes
3610 Commerce Dr. Suite 817
Baltimore, MD 21227
dstefanczyk@jensenhughes.com

Keywords: two-phase flow, hydrodynamic load, emergency relief system, chemical effluent

Abstract

A sudden opening of a pressure relief device (typically a pressure relief valve or a rupture disk) during an overpressure event will cause the chemical effluent to accelerate through the emergency relief system (ERS) piping. The rapid change of momentum of the effluent creates transient hydrodynamic loads on the piping. ASME B31 in the United States and European Directive 2014/68/EU requires these transient loads to be considered along with other applicable loads (i.e., dead weight, thermal) in the system design. Additionally, the pressure distribution along the piping may be of importance, particularly when other vessels are connected to the same ERS. The back pressures generated on these vessels should be evaluated and compared with the vessels' MAWP (maximum allowable working pressure) as well as the set pressure of any isolation devices. As such, a detailed time history of the effluent pressure and hydrodynamic loads are required to perform a proper evaluation. The effluent may be composed of proprietary chemical mixtures containing various materials. Therefore, the exact material properties of the effluent during the transient might not always be known. In this study, a previously published methodology is used for calculating the dynamics of the effluent and the resultant hydrodynamic loads for a typical ERS by using the RELAP5 (Reactor Excursion and Leak Analysis Program) software. Air-water properties are assumed for the effluent. Key output parameters, such as pressure drop and hydrodynamic loads, are then compared with established hand calculation methods using the actual effluent properties. Given the mechanical nature of the hydrodynamic load event, it was found that the use of air-water properties provides a reasonable representation of the important characteristics of the actual effluent discharge in mechanical design space.

1 Introduction

Pressure build up in a system may occur as a result of a process upset, instrument failure or fire. These upset scenarios can result in either, or both, vaporization due to exothermic runaway reactions or generation of gases due to the decomposition process, which are of particular interest in terms of system and public safety. The extent of the potential hazards caused by such incidents are generally determined by the rate and amount of the energy being generated, the chemical substance makeup, and the quantity of the effluent.

The vessels or systems, which contain such chemicals (e.g. chemical reactors or storage tanks), are typically protected by an ERS that prevents from over-pressurization of the equipment as a consequence of the vaporization of chemicals and/or generation of gases. The process for designing of an ERS was previously discussed by Conzen and Wood (Ref. [1]) and Ozar (Ref. [2]). The design process consists of a process hazard analysis (PHA), defining credible and limiting upset scenarios, adiabatic calorimetry testing, vent sizing calculations, fluid transient analysis, pipe stress analysis, and a dispersion analysis. The central task in the design process is the fluid transient analysis (Refs. [1] and [2]), which is the main focus of this study.

The ERS is typically equipped with either a rupture disk or a pressure relief valve (PRV). Some systems may have both pieces of equipment. The sudden opening of the relief device during an overpressure event will typically result in a two-phase discharge and cause the chemical effluent to accelerate through the ERS piping. The rapid change of momentum of the effluent creates transient hydrodynamic loads on the piping. ASME B31 (Ref. [3]) in the United States and European Directive 2014/68/EU (Ref. [4]) requires these transient loads to be considered along with other applicable loads (i.e., dead weight, thermal) in the system design. Additionally, the pressure distribution along the piping may be of importance, particularly when other vessels are connected to the same ERS. The back pressures generated on these vessels should be evaluated and compared with the vessels' MAWP as well as the set pressure of any isolation devices. As such, a detailed time history of the effluent pressure and hydrodynamic loads are required to perform a proper evaluation.

The two-phase behavior of the chemical effluent has been studied extensively. Following earlier developments by Epstein et al. (Ref. [5]) and Grolmes and Leung (Ref. [6]), a simplified correlation was suggested by Leung (Ref. [7]) to predict flashing choked flow in a frictionless duct or nozzle. The correlating parameter, ω, was based on an isenthalpic (constant enthalpy) assumption.

Leung and Grolmes (Ref. [8]) integrated the momentum equation for homogenous two-phase flow in a horizontal duct using "ω" as a closure relationship for two-phase mixture specific density. Leung and Epstein (Ref. [9]) extended the derivation to an inclined duct configuration. Leung (Ref. [10]) provided the derivation of the reaction forces during a two-phase discharge from ERS piping.

Computational methods have also been used by various industries to investigate two-phase flow behavior in pipes and the hydrodynamic loads due to a sudden acceleration or deceleration of the fluid and the propagation of pressure waves. RELAP5 is a tool for analyzing small-break loss of coolant accidents and system transients in nuclear power plants. The program has the capability to model one dimensional single and two-phase fluid flow and heat transfer problems. The software has undergone extensive review and validation by the US Nuclear Regulatory Commission (Ref. [11]). It has also been benchmarked (Refs. [12], [13] and [14]) for predicting pressure wave propagation and hydrodynamic loads in piping systems. Refs. [15] and [16] provide a discussion about some of the nuclear power plant applications. However, RELAP5 is limited to non-condensable gas-water and steam-water two-phase flow applications.

This study provides a discussion about using the analytical methods developed by Leung and Grolmes (Ref. [8]) and Leung (Ref. [10]) to demonstrate the similarity between the two-phase mixture of common industrial chemical effluents and air/steam-water two-phase flow. Computational and analytical calculations for three different discharge scenarios are also provided. The comparisons between the computational and analytical results are used to demonstrate the applicability of using air/steam-water properties to simulate chemical effluent discharge to reasonably predict the hydrodynamic loads and pressure distribution in the ERS following the actuation of a pressure relief device.

2 Theoretical Background

The governing steady-state conservation equations for one-dimensional homogenous two-phase flow in a horizontal duct with a constant flow area are (Ref. [8]):

Mass

$$G = \text{constant} \tag{1}$$

Momentum

$$vdP + G^2\left(vdv + \frac{4fv^2dL}{2D}\right) = 0 \tag{2}$$

Energy

$$h_0 = h + \frac{G^2v^2}{2} \tag{3}$$

G, v, P, f, L, D and h are the mass flux, specific volume, pressure, two-phase friction factor, length of the duct, hydraulic diameter of the duct, and specific enthalpy, respectively. The specific enthalpy and specific volume are expressed in terms of quality (x) and phasic properties, i.e., $h_0 = h_{f0} + x_o h_{fg0}$, $h = h_f + xh_{fg}$, and $v = v_f + xv_{fg}$. Subscript 0 denotes stagnation conditions and fg the difference between vapor and liquid properties.

Epstein et al. (Ref. [5]) and Grolmes and Leung (Ref. [6]) suggested a simple relation between the two-phase specific volume and pressure as given by:

$$\frac{v}{v_0} - 1 = \omega\left(\frac{P_0}{P} - 1\right), \tag{4}$$

where ω is a non-dimensional parameter. Leung (Ref. [7]) correlated ω for flashing choked flow in a frictionless duct or nozzle. The correlating parameter was shown to be in the following simplified form:

$$\omega = \frac{x_0 v_{fg0}}{v_0} + \frac{C_{f0} T_0 P_0}{v_0} \left(\frac{v_{fg0}}{h_{fg0}} \right)^2 \tag{5}$$

C_{f0} is the heat capacity of the liquid. The last term on the right-hand side is zero for non-flashing gas-liquid two-phase flow and the non-dimensional parameter, ω, is provided by:

$$\omega = \frac{x_0 v_{fg0}}{v_0} \tag{6}$$

Leung and Grolmes (Ref. [8]) considered a two-phase horizontal discharge from a large reservoir (Figure 1). They integrated Equation 2 over the length, L, by substituting Equations 4 and 5 and using the isenthalpic (adiabatic and reversible) flow approximation. The momentum equations after integration yields to:

$$N \equiv 4f\frac{L}{D} = \frac{2}{G^{*2}}\left[\frac{\eta_1 - \eta_2}{1-\omega} + \frac{\omega}{(1-\omega)^2} ln\left[\frac{(1-\omega)\eta_2 + \omega}{(1-\omega)\eta_1 + \omega}\right]\right] - 2ln\left[\frac{(1-\omega)\eta_2 + \omega}{(1-\omega)\eta_1 + \omega}\left(\frac{\eta_1}{\eta_2}\right)\right] \tag{7}$$

The non-dimensional parameters are defined by:

$$\eta_1 \equiv \frac{P_1}{P_0}, \eta_2 \equiv \frac{P_2}{P_0}, \eta_3 \equiv \frac{P_3}{P_0}, \ G^* \equiv G/\sqrt{\frac{P_0}{v_0}}, N \equiv 4f\frac{L}{D} \tag{8}$$

η, G^* and f are the dimensionless pressure, dimensionless mass flux and two-phase friction factor, respectively. Leung and Grolmes (Ref. [8]) recommended a range between 0.003 and 0.005 for the friction factor.

G^* and η_1 are related via Bernoulli's equation for compressible fluid (Ref. [8]):

$$G^* = \frac{\sqrt{-2[\omega ln\eta_1 + (1-\eta_1)\omega - (1-\eta_1)]}}{\omega\left(\frac{1-\eta_1}{\eta_1}\right) + 1} \tag{9}$$

The exit condition can be either subcritical, in which case $\eta_2 = \eta_3$, or critical where the local choking statement would require (Leung and Grolmes, Ref. [8]):

$$G^* = \frac{\eta_2}{\sqrt{\omega}} \tag{10}$$

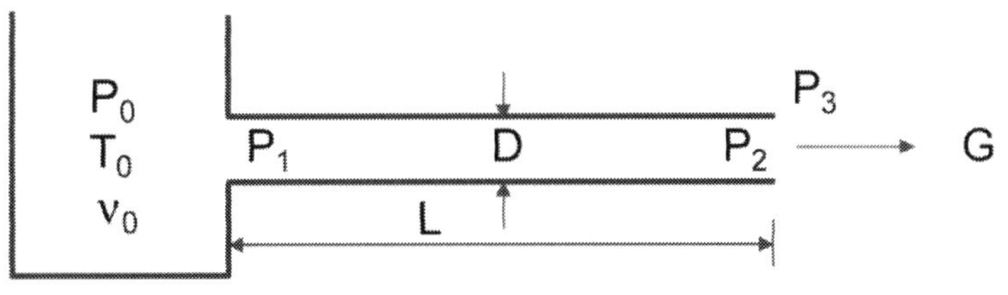

Figure 1. Two-phase horizontal discharge from a large reservoir.

With $\eta_1 = \eta_2 = \eta$, the above two equations satisfy the following transcendental equation (Leung and Grolmes, Ref. [8]):

$$\eta^2 + (\omega^2 - 2\omega)(1-\eta)^2 + 2\omega^2 ln\eta + 2\omega^2(1-\eta) = 0 \tag{11}$$

Similarly, the generalized thrust expression (T_R) is provided by Leung (Ref. [10]):

$$\frac{T_R}{P_0 A_e} = \frac{G^2}{P_0 \rho_0}\left(\frac{v_2}{v_0}\right) + \left(\frac{P_2}{P_0} - \frac{P_3}{P_0}\right) \tag{12}$$

The first term represents the fluid momentum expulsion rate and the second term represents the exit plane pressure (imbalance) force. For the case of unchoked exit, the exit pressure imbalance term vanishes, and the trust is only dependent on the exit momentum rate (Ref. [10]):

$$\frac{T_R}{P_0 A_e} = \frac{2[-\omega ln\eta_2 - (\omega-1)(1-\eta_2)]}{\omega\left(\frac{1}{\eta_2}-1\right)+1} \tag{13}$$

This approximate analysis shows that for two-phase flow in ducts (e.g. piping systems) ω is the sole independent variable and hence is the scaling parameter.

3 Transient Fluid Analysis using RELAP5

3.1 Description of the benchmark scenarios

A chemical reactor with a volume of 2.7 m^3 is considered for the benchmark scenarios. The reactor vessel is rated for 3 bars-g. The reactor is equipped with an emergency relief nozzle and a rupture disk of 41.1 mm inner diameter (DN40, 3.6 mm thickness). The nozzle is located near the top of the vessel. The piping downstream of the rupture disk has an inner diameter of 80.9 mm (4 mm thickness) and is 20 m long. It is a straight and horizontal section of piping.

Three different relief scenarios are used for the benchmarks. A summary of the postulated scenarios is provided below:

- Scenario 1: Polyurethane (PUR) and toluene diisocyanate mixture at 60 °C.
- Scenario 2: N-methyl pyrrolidone, formic acid, trimellitic anhydride and methylene diphenyl diisocyanate (MDI) at 90 °C.
- Scenario 3: Acetone under-fire scenario with vessel initially filled to 20% of its volume.

The chemical mixtures for Scenarios 1 and 2 are gas (CO_2, carbon dioxide) generating reactions. Alternatively, the scenarios are also called gassy reactions. A foamy mixture is expected inside the reactor vessel for these scenarios. Therefore, the chemical effluent discharging through the ERS is a two-phase gas-liquid mixture.

Scenario 3 is the boiling of a solvent due to a fire load applied on the vessel. The level of the boiling mixture is low, and liquid acetone is not entrained into the ERS. Only acetone vapor discharges through the ERS for this scenario.

3.2 Description of the computer model

The calculation was performed using the RELAP5/MOD3.3 (Patch 05) computer code, which is a light water reactor transient analysis code developed for the United States Nuclear Regulatory Commission (US NRC). It is suitable for the simulation of a wide variety of hydraulic and thermal transients in both nuclear and nonnuclear systems involving mixtures of steam, water, non-condensable gases and solutes under single phase and two-phase conditions (Ref. [17]).

A RELAP5 model was developed for the simulation of the venting scenarios from the reactor vessel into the ambient. The model was used to determine the hydrodynamic loads resulting from the sudden opening of the rupture disk. The mass flow rate of the two-phase effluent and the pressure distribution along the ERS piping were also calculated. Figure 2 provides an overview of the nodalization diagram of the model.

Time dependent volumes (TDVs) were used for the pressure boundary conditions and the thermo-physical condition of the fluid at those boundaries. Pipe components (P) were used to model the piping. Single junctions (SJs) connected the components such as pipes with pipes or pipes with TDVs. The valve junctions (V) was similar to the SJs but with the additional ability of the junction to open and close.

The length-to-diameter ratios (L/D) were set close to 1.0 for each pipe node and the time step was selected to be below the acoustic Courant limit. This is recommended for simulations involving propagation of acoustic waves and more generally for models where rapid fluid momentum changes are anticipated. The rupture disc was assumed to open linearly over a period of 10 msec, which is consistent with typical data (Ref. [18]). A steady-state calculation was performed for the first 5 seconds of each simulation to initialize the model. The transient portion of the simulation was initiated at 5 seconds, when each rupture disc started to open and was fully open at 5.01 sec. No heat transfer through the pipe walls was considered for the hydrodynamic fluid analysis. This is often a reasonable assumption for industrial processes due to the short duration of the relief event.

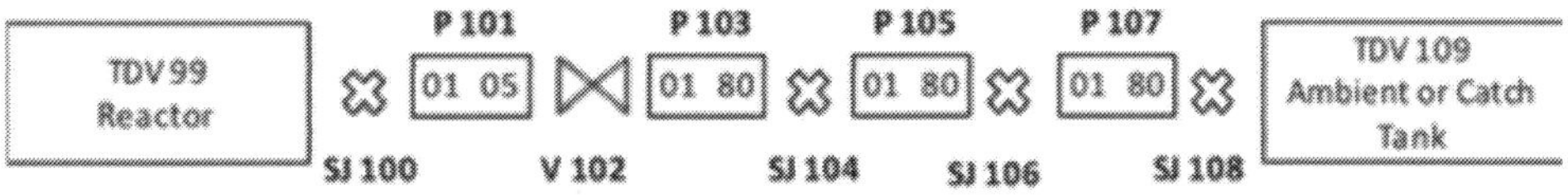

Figure 2. Nodalization diagram for the RELAP5 model.

The hydrodynamic loads (also commonly referred to as fluid force) on the modeled piping were computed through command files by using the RELAP5 output data. The generalized force equation in one-dimensional form can be resolved by a pipe segment bounded by two elbows as shown in Equation (14) (Ref. [19]). Note that the sign convention for the force is considered positive in the nominal direction of the flow as defined within the physical system.

$$F_W = -\frac{\mathrm{d}}{\mathrm{dt}} \int \rho A V dx \tag{14}$$

F_W is the hydrodynamic load, ρ is the fluid density, A if the inner flow area of the pipe, V is the fluid velocity, and x is the distance along the pipe. For pipe segments with an open end an additional force component needs to be considered, which is regarded as the thrust load. Examples of open ended pipe segments are pipes that connect to pressure vessels, large manifolds, or a catch tank. The total thrust load is calculated as shown in Equation 15, which is made up of the momentum flux (due to fluid momentum expulsion) and pressure terms:

$$F_T = -(P_E - P_A)A - \dot{m}_E V_E, \tag{15}$$

where P_E, P_A, $\dot{m}_E$ and V_E are the pressure at the pipe exit, pressure of the reservoir the open piping is connected to, and the mass flow rate and fluid velocity through the pipe exit, respectively.

3.3 Boundary conditions for the RELAP5 model

A summary of the upstream boundary (reactor vessel) conditions for the RELAP5 models of the three scenarios are provided in Table 1. 10% over-pressure (10% greater than the set pressure) was used for Scenarios 1 and 2, which is typical for relief system sizing of reactive chemicals. 21% over-pressure was used for Scenario 3 per API 520's (Ref. [20]) recommendation for under-fire scenarios. Two-phase gas liquid discharge was assumed for Scenario 1 and 2, where foamy mixtures inside the reactor vessel were established. Single-phase gas discharge was assumed for Scenario 3. The gas-liquid mixture of chemical effluent was simulated by using the air-water properties. The quality values of the air-water mixtures were adjusted so that the ω value of the gas-liquid mixture of the chemical and that of the air-water mixture would match. The quality and the ω value of chemical mixtures were determined based on the initial fill ratio and the homogenous two-phase mixture assumptions. A single-phase acetone vapor release scenario (Scenario 3) was simulated by using air. The determination of the choking condition at the rupture disk was made based on the analytical model, which was described in Section 2. The downstream boundary conditions were set to be at typical atmospheric conditions (air at 20 °C and 1 bar-a).

Table 1. Upstream Boundary Conditions for the RELAP5 Model

Upstream Boundary Conditions	Scenario 1	Scenario 2	Scenario 3
Pressure	4.4 bar-a	4.4 bar-a	4.8 bar-a
Flow regime	Two-phase gas-liquid	Two-phase gas-liquid	Single phase vapor
Simulant	Air-water	Air-water	Air
ω	0.085	0.336	Not calculated
Choking at the burst disk	No	Yes	Yes

4 Results and discussion

The time history of the mass flow rate at the rupture disk calculated using RELAP5 for Scenario 1 is provided in Figure 3. The flow quickly accelerates following the opening of the rupture disk. The effluent experiences the frictional and form losses only for the sections of the piping, which it has already occupied at a given time. Therefore, the transient mass flow rate is slightly greater than the steady-state mass flow rate while the effluent is filling up the ERS piping. Once the entire ERS piping is filled, the mass flow rate achieves a steady-state value.

Figure 4 shows the pressure-time histories downstream of the rupture disk and at the end of the ERS piping for Scenario 1. The pressure downstream of the rupture disk gradually increases to its steady-state value as the effluent fills the piping. The pressure at the exit of the piping remains near atmospheric pressure (downstream boundary condition).

Figure 5 shows the dynamic load calculated on the ERS piping. The total load and its constituents are provided. There is a spike for the wave load as the rupture disk opens. This spike is caused by the depressurization wave traveling from the rupture disk to the reactor vessel. Typically, this section of the piping is filled with gas or vapor for which the speed of sound is well known. Therefore, the initial spike is captured accurately and the speed of sound of the two-phase mixture (which is typically unknown) does not have a significant role on the results. The wave load continues to be the dominant mechanism of the load as the effluent continues to fill up the ERS piping. Once the piping is filled up, the flow momentum term replaces the wave load. The flow momentum is due to the discharge of the effluent from the ERS piping into the ambient or a catch tank and it is the first term on the right-hand side of Equation 15. The pressure force is zero since the exit of the ERS piping is at the same pressure as the ambient; there is not choking at the exit of the piping.

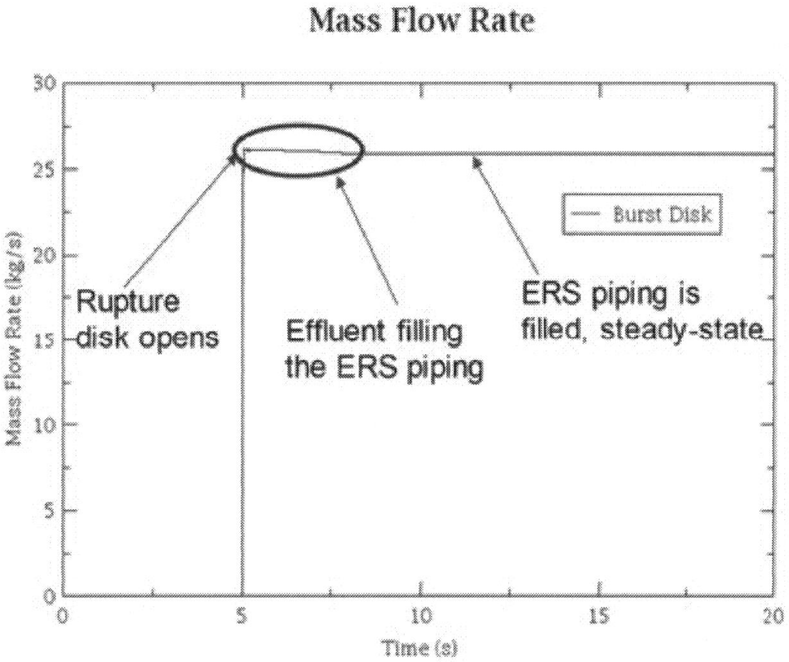

Figure 3. Mass flow rate for Scenario 1.

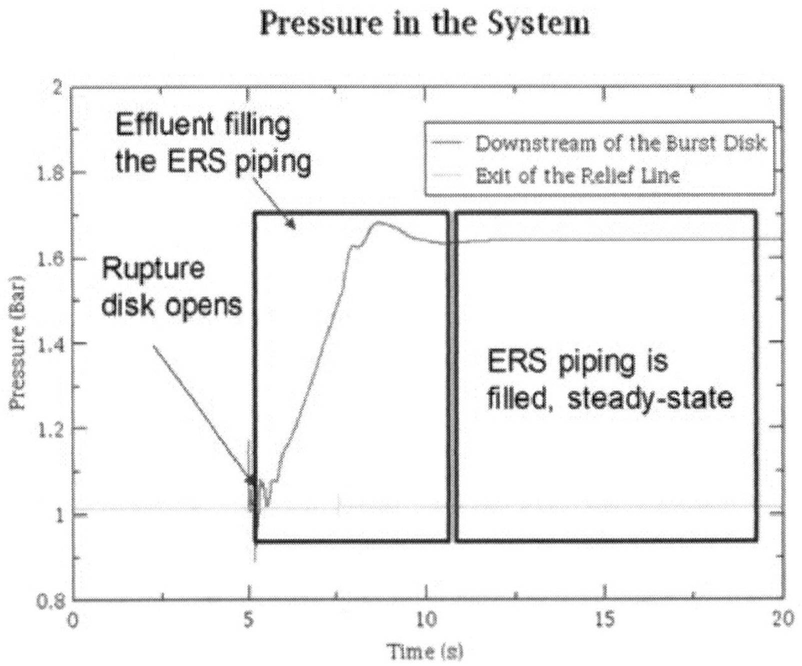

Figure 4. Pressure in the ERS piping for Scenario 1.

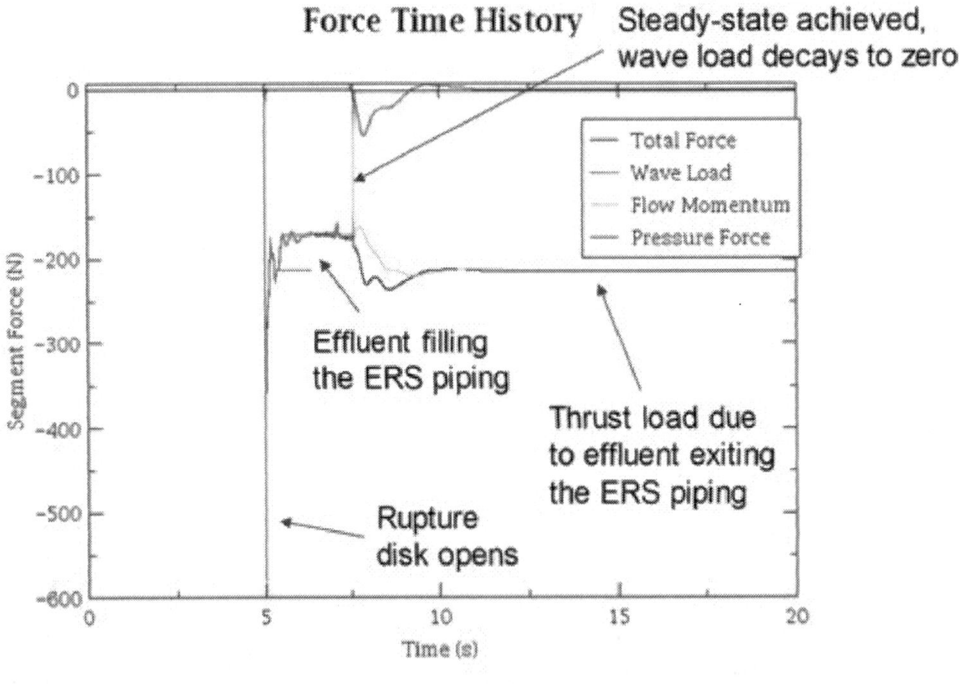

Figure 5. Dynamic loads for Scenario 1.

It is interesting to note that the wave load (while the ERS piping is filling up) and the load due to flow momentum (at steady-state) are almost similar. This is explained using the mathematical relations for the wave and thrust loads. The wave load for an effluent while filling up an empty piping (using Equation 14) is mathematically same as the flow momentum term in Equation 15.

The steady-state pressure, mass flow and thrust load values were compared between RELAP5 calculations and the analytical model, which was provided in Section 2. Actual chemical properties were used for the theoretical model. The comparisons between the two methods are provided in Table 2. The results demonstrate that the two-phase chemical effluent can be simulated with air/steam-water properties as long as the ω values are matched for these mixtures.

Table 2. Comparison between calculations using air-water (RELAP5) and actual two-phase chemical properties

Parameter	Scenario 1	Scenario 2	Scenario 3
Pressure downstream of the burst disk (bar-a)	1.6\1.7	1.6\1.7	1.9\1.6
Mass flow rate (kg/s)	25.9\26.1	16.3\16.4	1.2\1.2
Steady-state thrust load (N)	212.5\191.4	176.3\199.9	219.3\235.0

Black: RELAP5 results   Red: Theoretical model

In the chemical manufacturing industries, the effluent can be composed of proprietary chemical mixtures containing various materials. Therefore, it can be challenging to know the exact material properties at any given time during an emergency relief event. The method presented in this study requires only the stagnation properties of the chemicals at the venting conditions for predicting key parameters, such as the pressure distribution and wave loads during the transient. These properties are typically established by adiabatic calorimetric testing or similar characterization of the chemicals. The method presented in this study may also be used with other commercially available software, which can predict transient fluid dynamics and wave loads.

5 Conclusions

In this study, a previously published methodology is used for calculating the dynamics of the effluent and the resultant hydrodynamic loads for a typical ERS by using the RELAP5 software and assuming air-water properties for the effluent. Key output parameters, such as pressure drop and hydrodynamic loads, are then compared with established analytical methods using the actual effluent properties. The following conclusions were observed:

- The fluid transient analysis is a key aspect of the design of the emergency relief systems. It provides the wave loads required for the dynamic pipe stress analysis. It also provides key parameters such as the pressure distribution, mass and volumetric flow rates through the system, and wave loads. These key parameters are often used to determine back pressure effects on other vessels, for chemical dispersion analysis and dynamic pipe stress analysis.

- The analytical approach shows that there is a similarity between the air/steam-water and the gas/vapor-liquid chemical mixtures if the "ω" values of the two mixtures are matched. This allows for an emergency relief event to be simulated with air/steam-water properties.

- Similar pressure distribution in the piping system and loads can be achieved through the similitude of "ω". Therefore, "ω" of the air/steam-water mixture should be matched with the "ω" of the chemical effluent at the stagnation conditions. The quality of the air/steam-water mixture can be adjusted until a similar "ω" is achieved.

- The sonic velocity of the effluent is typically not known. Also, it is practically difficult to match the sonic velocity of the air/steam-water mixture with the sonic velocity of the chemical effluent, which is a major requirement for tracking pressure waves and wave loads. This study demonstrates that the emergency relief scenarios are pipe filling events, which only require the momentum flux to be calculated for a reasonable prediction of the initial wave load. Therefore, the sonic velocity of the effluent is not required to capture the initial wave load.

- The effluent can be composed of proprietary chemical mixtures containing various materials. Therefore, it can be challenging to know the exact material properties at any given time during the transient. The method presented in this study may be used with most commercially available software and air/steam-water properties for a reasonable representation of the important characteristics of the actual effluent discharge in mechanical design space.

- The study needs to be extended further to identify applicability and limiting conditions for this method.

6 References

[1] Conzen, J., Wood. R.G., (2015), "Modeling and Computation of Reaction Forces on Relief Piping During Depressurization", AIChE 11th Global Congress of Process Safety.

[2] Ozar, B., (2018), "Analysis of Chemical Effluent Discharge through the Emergency Relief System using Stream-Water Properties," DIERS Fall Meeting, 15-17 October, Houston, USA.

[3] ASME. *ASME Code for Pressure Piping, B31.* The American Society of Mechanical Engineers, New York, NY, 2016.

[4] "Directive 2014/68/EU of the European Parliament and of the Council," Official Journal of the European Communities, May 2014.

[5] Epstein, M., Henry, R.E., Midvidy, W., Pauls, R., (1983), "One-dimensional Modeling of Two-phase Jet Expansion and Impingement," Thermal-hydraulics of Nuclear Reactors, 2nd International Topical Meeting Nuclear Reactor Thermal-hydraulics, Santa Barbara, CA.

[6] Grolmes, M.A., Leung, J.C., (1984), "Scaling Considerations for Two-phase Critical Flow," Multi-phase Flow Heat Transfer III, Part A: Fundamentals, T.N. Veziroglu and A.E. Bergles, eds., 549, Elsevier, Amsterdam.

[7] Leung, J.C., (1986), "A Generalized Correlation for One-Component Homogenous Equilibrium Flashing Choked Flow," American Institute of Chemical Engineers, Vol. 23, pp. 1743-1746, October.

[8] Leung, J.C., Grolmes, M.A., (1987), "The Discharge of Two-phase Flashing Flow in a Horizontal Duct," American Institute of Chemical Engineers Journal, Vol. (33), p. 524.

[9] Leung J.C., Epstein, M., (1990) "The Discharge of Two-phase Flashing Flow from an Inclined Duct," Transactions of the ASME, Vol. 112, pp. 524-527, May.

[10] Leung, J.C., (1992), "Reaction Forces During Two-phase Discharges," American Society of Mechanical Engineers Journal of Fluids Engineering, Vol. (114), p. 689-692.

[11] Information Systems Laboratories, (2016), "RELAP5/MOD3.3 (Patch03) code manual volume I to VIII, NUREG/CR-5535/Rev P5-Vol I to VIII", prepared for USNRC.

[12] International Agreement Report, (2007), "Simulation of the Propagation of Pressure Waves in Piping Systems with RELAP5/MOD 3.2.2: Comparison of Computed and Measured Results", TÜV SÜD - NUREG/IA-0206, June.

[13] Sokolowski, L., Koszela, Z., (2012), "RELAP5 Capability to Predict Pressure Wave Propagation Phenomena in Single- and Two-Phase Flow Conditions", Journal of Power Technologies 92 (3), page 150 – 165.

[14] Miller, J.S., Ramsden, K.B., (2005), "Validation of a Thermal Hydraulic Computer Code to Perform Two-Phase Multi-Component Force Calculations for Structural Evaluations", Proceedings of 13th International Conference on Nuclear Engineering (ICONE13), 16-20 May 2005, Beijing, China.

[15] Ozar, B., Henry, C.E., Harvill, R., Norton, D.A., (2012), "Analysis for Low Pressure Cooling Injection System Suction Hydrodynamics for a Boiling Water Reactor," Proceedings of 20th International Conference on Nuclear Engineering (ICONE20), July 30-August 3, 2012, Anaheim, CA.

[16] Henry, C.E., Lim, J., Ozar, B., (2012), "Investigation of Hydrodynamic Loads Associated with Pyrotechnic Valve Actuation," Proceedings of 20th International Conference on Nuclear Engineering (ICONE20), July 30-August 3, 2012, Anaheim, CA.

[17] Riemke, R.A., Davis, C.B., Schultz, R.R., (2009), "Recent Hydrodynamics Improvements to the RELAP5-3D Code", Proceedings of the 17th International Conference on Nuclear Engineering, ICONE17-75380, July 12-16.

[18] Health and Safety Executive, "Offshore Technology Report 2002/023: Testing and Analysis of Relief Device Opening Times," Pipeline Simulation and Integrity Ltd., Yorkshire/UK.

[19] Moody, F.J., (1990), "Introduction to Unsteady Thermofluid Mechanics", John Wiley and Sons, New York, Section 7.4, p. 365-366.

[20] API Standard 520, (2008), "Sizing, Selection, and Installation of Pressure-relieving Devices in Refineries," 8th edition.

547437 Evaluation of the Thermal Stability of Styrene in Contact with Impurities

Safe handling and storage of reactive monomers have long been an issue for polymer manufactures. The unsaturated double bond may self-react to initiate unintended polymerization reactions under many scenarios, for example, completely depletion of inhibitors, exposure to fire or external heat, and in contact with impurities. Usually, a trace amount of incompatible species is able to trigger unintended reactions and lead to catastrophic consequences. Thus, understanding how the impurities promote self-reaction is expected to provide valuable guidance to safe monomer-handling practices. Our goal in this study is to quantify the contamination effects of some most common contamination chemicals on styrene self-polymerization. Thermal analysis tools including DSC and ARSST are applied to examine the exothermic characteristics of styrene mixed with contaminating substances at different concentration levels and mixing ratios. From the preliminary results, we observed the onset of self-polymerization shifted to a lower temperature when in contacted with water and also alkaline. A trace amount of concentrated strong acid initiated unwanted exothermic activity even at room temperature. Self-heating rate was also elevated when diluted acid was present. Additional research is needed to understand what is the impact in of other types and concentrations of contaminants such as rust, metal ions. Our study demonstrates significant implications in the prevention of runaway incidents during transportation and storage of styrene.

549038 A Mixed-Integer Programming Framework for Placement of Fire Detectors in Chemical Processing Facilities

In chemical processing plants, gas and fire detectors provide an important layer of protection for personnel. However, given the large variety of chemical and physical sensors and number of factors influencing their capabilities to successfully detect hazards, the decision of where to place sensors is challenging. Gas detectors (e.g. ultrasonic, catalytic bead) must be in contact with gas leaks to be effective, while fire detectors (e.g. point infrared, open path infrared, optical) require a visual path to the fire, free of obstructions, for successful detection. Numerical optimization techniques provide an opportunity to overcome these challenges and determine effective sensor placement.

The optimal gas detector placement problem was formulated by [1] as a stochastic program based on the p-median facility location problem and relied on scenarios generated from CFD simulations within a petrochemical facility. Their results showed that coupling coverage with expected detection time using a scenario-based method outperformed the existing coverage-only approaches and has inspired formulations based on MINLP approaches for non-uniform failure probabilities [2].

But while gas detectors must be placed in the path of gas leaks for effective detection, fire detectors must be placed within unobstructed visual distance of flames. The optimal placement of visual fire detectors in processing plants can be particularly difficult due to the large number of physical obstructions (e.g. piping, valves, tanks).

This work focuses on an optimization formulation for placement of fire (visual) detectors. The scenario-based, p-median approach demonstrated by [1] for gas detectors is translated to a maximum coverage model for fire detectors in a 3D space with consideration to failure probabilities. We show that this approach allows for the placement of fire detectors for a fixed sensor budget while under uncertainty. We demonstrate the effectiveness of this formulation on several case studies. We further discuss how the framework is readily scalable for larger problem sizes and amenable to various objectives.

We would like to thank the Purdue Process Safety and Assurance Center for their support in this work.

References:

[1] Legg, S. W., Benavides-Serrano, A. J., Siirola, J. D., Watson, J. P., Davis, S. G., Bratteteig, A., and Laird, C. D. (2012). A stochastic programming approach for gas detector placement using CFD-based dispersion simulations. Computers and Chemical Engineering, 47:194–201.

[2] Liu, J., and Laird, C.D., A Global Stochastic Programming Approach for the Optimal Placement of Gas Detectors with Nonuniform Unavailabilities, to appear in Journal of Loss Prevention in the Process Industries, 2017.

546855 Advance Alarm Management Applied to DCS Improvement in a Petrochemical Plant: A Case of Study

Alarms use in process industry is fundamental for avoiding process accidents in which people, the environment or process equipment could be damage. Currently, Carbon Black production plants are equipped with distributed control systems (DCS) for setting alarms values and to provide indication of abnormal events to the control rooms operators. However, the basic control system could not meet the needs against the different production conditions and the presence of improper alarms could send a bad signal to control rooms and it could be the start of a process safety event.

In the present study, an alarm system assessment was performed in a petrochemical plant, located at Mamonal industrial park, in Cartagena – Colombia; based on data analysis, it was observed that when process conditions changes occured, the DCS was not be able to adjust alarms settings automatically, according to the load fluctuation and/or process conditions. In order to solve this problem, in first place the DCS performance was studied and it was reported an alarms list, which provided information regarding to the alarm flooding phenomena, alarm chattering and others alarm system issues.

On the other hand, the information earned and alarm rationalization methodology were used for establishing an action plan in order to improve DCS performance. Results showed the utility of these tools and it was identified the need to perform betters in the alarm system logic to avoid flooding alarm phenomena and to make operations safer.

548361 Overcoming Motivational Biases in Hazops

Hazard and Operability Studies (HAZOPs) are one method of assessing the risk in a facility. Based on the risk ranking of a scenario, recommendations may be required to reduce the risk. Often, these recommendations require additional capital expenditure and/or operational resources. In aging plants where resources are not readily available or in greenfield projects where there is a tight timeline to start up, there is sometimes a bias to risk rank scenarios such that there is no need to add additional safeguards. Motivational biases can outweigh common sense such that scenarios are not ranked according to their real risk. We have developed a method to overcome some of these motivational biases with the use of a polling system to engage all participants in the HAZOP. Participants are asked to vote anonymously on the severity and likelihood for a given scenario, then the results are discussed. During this discussion time, arguments for all the different risk rankings are discussed, including those voted on by the minority in the room. This facilitates an open discourse about the real risks for a given scenario rather than shifting the attention to the most vocal person in the room, who may be strongly influenced by a desire to reduce spending or minimize the effort required for maintenance. Following this discussion a second poll is performed to reach a consensus. We have found this to be an effective method for reducing motivational biases and ensuring that the right decisions are made based on the most logical argument instead of based on budget or time constraints. Tracking polling results increases the accountability of participants and makes them less likely to skew the risk ranking. The requirement for consensus ensures that all opinions are heard and considered, even if they are not popular. Together these make the polling system an effective method for overcoming motivational biases in HAZOPs and ensuring that safe decisions are made.

546515 Integrating Culture into Process Safety Assurance

Purpose: PETRONAS Chemicals Group Berhad (PCG) has embarked on a group-wide culture transformation, introducing the concept of the Results Pyramid® and a Culture of Accountability® i.e. a workplace culture where people take accountability to think and act in the manner necessary to achieve results.

Methodology: This paper will discuss two such activities anchoring on the Results Pyramid® that have been deployed across several plants and units within PCG.

The first activity is an enhancement of existing assurance programmes, where culture subject matter experts are assigned to observe plant leadership teams and front-liners during interviews and site verifications conducted by the technical assessors. The observations are synthesised into belief statements, which are presented together with the technical findings.

The second activity is Power Up, a session designed to incorporate powerful and specific feedback exchange between various teams, culminating in a specific list of actions to improve delivery of each team's activities in order to achieve the shared key results. This workshop can be done either as a follow-up to an assurance activity, in preparation for a project/turnaround activity or as a part of a periodic performance review.

Results: Examples of observations, feedback exchanges and specific actions from these activities will be shared, together with the impact on process safety leading and lagging indicators. Recommended practices and lessons learnt for continuous improvement will also be included.

Conclusion: Applying the Results Pyramid® concept to the design and implementation of Process Safety-related programmes has helped to identify the underlying leadership, mindset and behaviour issues impacting their success. By doing so, more specific interventions can be put in place to ensure full compliance to safety requirements and maximise effectiveness of these programmes.

548442 Emergency Plans Integration By Diferent Companies

In Brazil, different companies with pollution potential must have an emergency plan. However, each company analyzes its risks and contingency measures separately, based on the Conama 398 law. With the establishment of the Area Plan, companies from the same territory with pollution potential caused by oil in waters began to analyze how to combat a major accident quickly and effective together. This work shows how the integration of the individual contingency plans of each company and the cooperative result of the area plan in territory of significant scope and considerable environmental sensitivity.

No Brasil, diferentes empresas com potencial de poluição devem ter um plano de emergência. No entanto cada empresa analisa seus riscos e medidas de contingência separadamente, com base na lei Conama 398. Com a instituição do Plano de Área, empresas do mesmo território com potencial de poluição causadas por óleo em águas passaram a analisar como atender conjuntamente de forma rápida e eficaz a um grande acidente em seu território de abrangência. Este trabalho mostra como foi estruturada e organizada a integração dos planos de emergência individuais de cada empresa e o resultado cooperativo do plano de área em território de abrangência significativa e considerável sensibilidade ambiental.

548111 Moving Back to Flammable Refrigerants - Facility Siting and Personnel Safety Considerations

"Everything old is new again – Paul Allen"

Not just a catchy tune from the 1970's, this jingle summarizes a current technological shift in the world of refrigeration. Historically, refrigeration cycles used toxic and/or flammable materials as the working fluid. As the technology matured and made its way into transportation and consumer applications, the working fluid also evolved. Halogenated hydrocarbons, such as chlorofluorocarbons (CFCs), were developed as replacements to conventional refrigerants such as ammonia and propane. CFCs, which may have toxic effects, are desired for their tunable working range and for being non-flammable. However, the environmental drawbacks to the use of CFCs has been well researched and documented. As we embark into the future, where climate change and environmental quality are at the forefront of concern to consumers and industry alike, we are seeing a move back to flammable refrigerants. Everything old is, indeed, new again.

New construction and expansion projects can certainly consider and plan for the use of flammable refrigerants in commercial and industry facility design and siting, but what about retrofit applications? What are the risks and hazards associated with converting a CFC refrigeration system, designed and sited for the use of non-flammable refrigerants, to a system now utilizing flammable refrigerants? Beyond the need for additional fire protection systems, what risks are posed to the facility and personnel? In this paper, we seek to answer these questions by comparing the hazards and risks of mid-scale refrigeration systems using either non-flammable or flammable refrigerants. We explore how facility siting constraints may change with respect to the use of flammable refrigerants, and perform a quantitative risk assessment (QRA) for an example retrofit project in order to assess the potential for increased risk that should be considered in such projects.

548210 Making Sense of Risk Results

Risk based studies are very useful, but at the same time, the results are extremely complicated. While we have a basic set of results that can be used to illustrate the risks to people, these are relatively simple for risk practitioners and sometimes too complicated for site managers.

FN curves show the risk to all people from all sources, and individual risk contours illustrate the risk from all scenarios to people across specific locations, but how do we translate these results into actions. Generally, the action is to relocate people to lower risk areas, or upgrade buildings to mitigate the impacts identified. However, this is not always feasible. What if the site footprint is limited; there are public receptors who cannot be relocated; or building upgrades are cost-prohibitive or infeasible?

We must better quantify and evaluate the **sources** of risk rather than focusing on mitigating the impacts from the events. We need to evaluate the type of risks, the potential initiating events, and then we can assess the potential methods to effectively mitigate the risks. Existing risk result presentations are often very poor at illustrating the sources of risk. To effectively identify the sources of risk requires that we carefully dissect the risk results, while maintaining their validity and intent.

This paper will focus on how we find these risk drivers and make enough sense of them to create actionable recommendations that protect people.

551205 Ammonia Dispersion Modeling for Ammonia Storage with Multi-Scenario Failure System

PT Pupuk Kalimantan Timur is one of the largest fertilizer companies in Southeast Asia where located in Bontang City, East Kalimantan Province, Indonesia. The total production capacity of PT Pupuk Kalimantan Timur are 2.7 million Tons / year Ammonia, 3.43 million Tons / year Urea and 350 thousand tons / year NPK.

To accommodate ammonia production, PT Pupuk Kalimantan Timur currently operates 3 ammonia tanks with a capacity of 2 x 25,000 tons and 1 x 50,000 tons. In order to manage the hazard risk contained in the ammonia tank, it is necessary to do a modeling to map the potential and ammonia leakage potential areas and the impact of their exposure to the workforce and the community around the factory area.

The results of this modeling are expected to be used by Pupuk Kalimantan Timur's management in carrying out corrective actions for occupational health and safety performance, especially in controlling the potential hazard of ammonia, especially in ammonia tanks.

In this paper modeling is done using the AERMOD® Modeling System software. AERMOD Modeling System was chosen because it can be used on:

1. U.S. EPA regulatory Gaussian plume air dispersion model.
2. For near-field impacts (50 km max).
3. Prefered model for complex terrain.
4. Improved treatment of urban effects.

For modeling, we choose the AS-T-101 ammonia tank which has the largest capacity, which is 50,000 tons. Ammonia distribution models will be simulated based on weather conditions and ammonia leak source scenarios that can occur in the AS-T-101 ammonia storage tank system. Ammonia leakage sources will be displayed in 5 scenarios, i.e:

1. 1 Vent opens
2. 1 PSV poping
3. 2 PSV poping
4. 3 PSV poping
5. 3 PSV and 1 vent poping

The AERMOD Modeling System program is used to make ammonia distribution models for each event scenario assuming an exposure radius of up to 10 km. Based on the results of the calculation and modeling of the potential distribution of ammonia from AS-T-101 using the AERMOD Modeling System, the highest exposure of ammonia is in the 5th scenario, which is up to 36611 μg / M3.

546560 De-Mystifying the Concepts of Hazardous Area Classification

Purpose

The purpose of this topic is to address and clarify major items of confusion associated with hazardous area classification assessments and compliance auditing in classified locations. The ultimate objective is developing documentation that provides a framework for the control of electrical energy ignition sources in environments made hazardous by the potential presence of flammable vapors.

Background

In February of 1992 Occupational Safety and Health Administration (OSHA) 29 CFR 1910.119 Process Safety Management of Highly Hazardous Chemicals became law in the United States and confusion reigned supreme. Over the next several years industry professionals began the process of sorting through the responsibilities, regulations and impending deadlines that had been placed squarely in their laps. Acronyms such as PSM (Process Safety Management), PSI (Process Safety Information), RAGAGEP (Recognized and Generally Accepted Good Engineering Practices), MOC (Management of Change), and PHA (Process Hazards Analysis) became commonplace in the vocabulary of process safety professionals nationwide.

Employers now were required to meet stringent deadlines regarding the creation and management of a plethora of process information associated with equipment in the process covered by this law, i.e. any covered process. This information would be used to support the exhaustive efforts needed to create a baseline PHA for each covered process by May of 1997. In addition OSHA required all PHA's be placed on a five (5) year life cycle requiring their review and revalidation. Along with the revalidation of PHA's came the task of also revalidating all process safety information on the same five (5) year cycle.

As a result, every willing and able bodied employee was thrust into the quagmire of supporting, participating in, facilitating, or conducting PHA's. The required information used to support PHA's included the materials of construction, Piping and Instrument Diagrams (P&ID's), Relief System design and design basis, ventilation system design, material and energy balances, safety shutdown systems, and last but not least **electrical classification**. Yes, "electrical classification" was and still is legally considered an integral part of required process safety information.

This abstract addresses the following and provides an interactive discussion regarding the three (3) major myths surrounding electrical classification:

- assigning a classification to a location
- proper use of the classification information
- keeping the documentation evergreen

  The following topics will be addressed and clarified:

- What Is Hazardous / Electrical Classification?
- The Process of a Hazardous Area Classification Assessment
- The Concept of Electrical vs Hazardous Area Classification
- The Compliance Assessment (where does it fit in the process of controlling electrical energy ignition sources)
- Common Compliance Assessment Issues

546108 Maintaining Safety Excellence By the Implementation of Process Safety Inspection in Badak LNG

Badak LNG as one of the most experienced LNG Producers in the world, has undergone safety culture change throughout its operation. The goal of this change is to achieve excellent safety performances, indicated by the lower rate of Lost Time Injury Frequency (LTIF), lower number of Loss of Primary Containment (LOPC) Tier-1, High Potential Incident and multiple Train trip. This change was triggered by the occurrence of a fatal incident in 2005 as well as lost time injury incident in 2006. Since then, Badak LNG has integrated Process Safety Management (PSM) into its Company Management System, strengthened the roles and responsibilities of Managements, and pursued for interdependent safety behaviour of each personnel. One of PSM implementations is through process safety inspection.

Process safety inspection is an activity to assess the current equipment integrity based on visual observations compared with equipment's standard condition. The standard condition checklist for process safety inspection in Badak LNG is based on Physical Barrier Assessment tools, owned by Det Norske Veritas (DNV). The assessed items are divided into 5 categories: Inherent Safety, Hazard Prevention Measure, Hazard Control, Mitigations, and Support System. The overall score is determined by total number of standard items in all categories compared to the total number of verified items in all categories. The minimum acceptable score is at 80%.

During the year of 2012-2017, the average score of process safety inspections in Badak LNG is 90%. As a result of these inspections, Badak LNG had also identified the need of improvement in some items such as blinds, pipework and flanges, passive fire protection, fire & gas detection system and lock-open/lock-close block valves. In overall, by the implementation of PSM, Badak LNG has aligned on the right path to maintain its safety excellence achievement in LNG business, such as zero LTIF since 2007, zero multiple Train trip since 2012, zero LOPC Tier-1 since 2016, and zero High Potential Incident in 2017.

548411 Using HAZOP/LOPA to Create an Effective Mechanical Integrity Program

Many people view the conduct of a HAZOP/LOPA to address regulatory requirements as a chore, and stop there. However, the implementation of a quality HAZOP/LOPA has the potential to provide a framework for addressing numerous safety and operational optimization issues at plants, including the formulation/refinement of the Mechanical Integrity Program. The purpose of this paper is to focus on the mechanical integrity program, illustrate how a quality HAZOP/LOPA can support the effective implementation of some of the new requirements for California Refineries, and optimize key elements of an effective Mechanical Integrity Program, e.g.:

- Inspection/testing methods

- Testing intervals

- Maintenance outage periods

- Repair prioritization and allowable outage

- Identification of low priority equipment

544678 *"Identifying Critical Process Parameter to Ensure Asset Integrity" or "Pulse of Mechanical Integrity Management System"*

The management of equipment begins with an awareness that ageing is not about how old the equipment is, but is about what is known about its condition and the factors that influence the onset, evolution and mitigation of its degradation. The presentation provides an overview of Integrity Operating Window (IOW) parameters of any process that are key to understand the asset degradation and prevent Process Safety Incidents.

IOW are the situations in the process industries where a process parameters (temp., pressure, dosing rate, conc. etc...) when gets out of operating windows could result in significant loss of primary containment leading to potential severe consequences.

Arrazi - SABIC developed an IOW system that covers IOW identification, associated risk, mitigation (interim – long term), stewardship/management of IOWs and elimination.

The presentation shall cover Insight concept of the Topic, major Process Safety accidents related to breeching IOWs, actual example, IOW Management Process, its influence on PM plans etc.

three key bullet points that we can include in the agenda:

- Knowing operating limits with associated mechanical integrity or process safety risk to ensuring proper actions to be in operating envelope
- Developing the engineering leadership to systematically guide the operators, superintendents and engineers to be in safe zones
- Dynamic inspection frequency based on plant conditions and sharing some industrial major incidents for this critical item

The Practice of Process Safety Analysis and Management System Based on Safety Reliability
(THE NEW TITLE:
The PHAMS system in SINOPEC)

Jiang Xue
SINOPEC Research Institute of Safety Engineering,
QingDao,China
Jiangxue.qday@sinopec.com

The PHAMS system in SINOPEC

Jiang Xue
SINOPEC Research Institute of Safety Engineering,
QingDao,China
Jiangxue.qday@sinopec.com

Keywords: Safety Barriers Reliability, PHAMS, HALOPA, SIL, Safety Management Online.

Abstract

The big data technique has been in trend for the process safety assessment. In China, it's the very need to transfer the qualitative safety management to quantitative safety management, which should rely on the professional risk analysis technique, as well as the standard safety reliability database. Based on the above, it's very important to integrate several professional risk analysis technique, as HAZOP, LOPA, SIL etc. to realize the integral risk identification and evaluation, grading and management plan making, supporting by the proprietary SINOPEC risk metrics and grading standards, as well as safety equipment reliability database,which is developed as the Process Safety Analysis and Management System(PHAMS) for SINOPEC, to support all SINOPEC branches to make dynamic risk analysis and management online

1 Introduction

With the smart factory's development, the static, periodic risk assessment and management can not satisfy the identification and real-time safety control of plant's risk events. The PHAMS which establishes the plant-wide dynamic risk analysis, can effectively solve the risk rating and monitoring of the unit and realize the plant-wide or corporate-wide management optimization.

The key data for HALOPA, SIL and Management module in PHAMS is the dependent protection barriers, which we rely on RBI correction, Personnel exposure probability and ignition probability to optimize the six mainly safety barriers' failure probability, simultaneously, do use of the Bayesian Correction Modelling to verify the instant failure probability on site, as well as the SIL identification, which can validly save the grading cost on time and personnel, and all of these are towards to the deeper research on dynamic risk analysis for PHAMS.

2 HALOPA Analysis System

2.1 SINOPEC RISK MARTIX

According to international common matrix criteria, as well as statistics from the SINOPEC accident data and the maintenance data, the frequency of risk occurrence and severity classes are coupled to the 8×7 matrix by a linear exponential model, and be assigned risk index to each risk level.

SINOPEC RISK MARTIX		Occurrence Frequency							
		1	2	3	4	5	6	7	8
		$\leq 10^{-6}$/y	10^{-6}-10^{-5}/y	10^{-5}-10^{-4}/y	10^{-4}-10^{-3}/y	10^{-3}-10^{-2}/y	10^{-2}-10^{-1}/y	10^{-1}-1/y	>1/y
Severity Classes	A	1	1	2	3	5	7	10	15
	B	2	2	3	5	7	10	15	23
	C	2	3	5	7	11	16	23	35
	D	5	8	12	17	25	37	55	81
	E	7	10	15	22	32	46	68	100
	F	10	15	20	30	43	64	94	138
	G	15	20	29	43	63	93	136	200

Figure 1 SINOPEC Risk Matrixp[1]

Based on this matrix, SINOPEC has established four-level risk list for establishing targeted management of branch network. For example, the executive board of the group is in charge of the red-level risk list, the branch directors is in charge of the orange-level risk list, etc.

2.2 HALOPA Analysis Modelling

2.2.1 System risk analysis based on equipment operation cycle

The HALOPA analysis model can be used to identify the hazard of the device or process unit, clarifying the initial event, process deviation, independent protective layer and consequence events, combined with the event tree analysis method to analysis the reduction influence of the occurrence frequency of initial events. The protection layer performs a failure frequency evaluation role to determine the final frequency of the on-site device risk events during each operating cycle under the coupling reduction of the existing protective layers at each level [2].

An important assumption of the model is that each protection layer is independent, that is, the failure of a single protection layer will not affect the risk reduction impact of other existing protection layers [3]. Each time a protective layer takes effect, the frequency of accidents is reduced by the risk reduction of the protective layer.

When the frequency of the initial event does not exceed 1 time/year or is less than twice the test frequency of the first protective layer, the evaluation of the consequence frequency can be referred to the following formula [4]:

$$f_I^C = f_I \times P_{enable} \times \prod_{j=1}^{J} PFD_{Ij} \quad (1)$$

where

f_I=Occurrence frequency of initial event I annually;

Penable=Moderation factors by enabling conditions and other conditional probabilities.

PFDIj=Failure probability of dependent protection layer j of initial event I.

When the frequency of the initial event exceeds 1 time/year or exceeds the test frequency of the first protective layer twice, the first protection layer failure shall be used as the initial event, and the calculation formula of the consequence frequency is as follows:

$$f_I^C = f_{\text{IPL\_first}} \times P_{enable} \times \prod_{j=2}^{J} PFD_{Ij} \qquad (2)$$

where

$\mathbf{f_{IPL\_first}}$= The failure probability of 1st dependent protection layer

The HALOPA analysis model in PHAMS designed a modified calculation model for equipment failure frequency and personnel exposure probability, and realized BPCS, SIS, physical protection, leakage storage and collection, fire and explosion suppression for equipment operation cycle and device in service. The system's six major types of protection layer failure frequency have the correction factors of risk reduction, and realize system risk assessment based on the device production cycle. For example, according to the system risk assessment of a diesel hydrogenation unit based on the PHAMS system, on-site data collection of various protective layers and key correction conditions under one cycle of the driving operation to date, online preliminary analysis of the risk overall value and grading of the feeding unit The results are as follows:

Figure 2 HALOPA Analytical Function Online

2.2.2 Dynamic risk assessment based on real-time data of safety control systems
According to the protection layer frequency analysis model, the failure probability of the protection layer lowering the frequency of a certain initial event can be regarded as θ , and the a priori risk probability distribution function is:

$$f_0(\theta) \propto \theta^a(1-\theta)^b \quad (3)$$

Its likelihood function $L(E|\theta)$ is

$$L(E|\theta) \propto \theta^m(1-\theta)^{n-m} \quad (4)$$

where

a, b=the prior distribution parameter of the failure probability of the protective layer;
m=The number of the posterior cumulative event occurred when the protective layer is valid;
n=The number of the posterior cumulative event occurred when the protective layer is failed.

With the posterior probability distribution function corrected by dynamic risk analysis as$f_1(\theta|E)$, the posterior probability distribution can be obtained by:

$$f_1(\theta|E) = \frac{f_0(\theta)L(E|\theta)}{\int f_0(\theta)L(E|\theta)d\theta} \propto \theta^a(1-\theta)^b\theta^m(1-\theta)^{n-m} \propto \theta^{a+m}(1-\theta)^{b+n-m} \quad (5)$$

Since m and n are parameters of the age change and the posterior probability distribution function seems to be continuous, the dynamic failure frequency of the protective layer at a certain moment can be obtained by the following formula[4-6].

$$P_i(\theta) = \int_{i-1}^{i} f(\theta)d\theta = (1-F_i(\theta|E)) - (1-F_{i-1}(\theta|E)) = F_{i-1}(\theta|E) - F_i(\theta|E) \quad (6)$$

In where $F_i(\theta|E)$ is $f_i(\theta|E)$ the cumulative failure frequency function as

$$F_i(\theta|E) = \int_0^i f(\theta|E)d\theta \quad (7)$$

And then bring in the following formula to get the dynamic risk frequency of the device or process unit:

$$f_I^C = f_I \times P_{enable} \times \prod_{j=1}^{J} P(\theta)_{Ij} \quad (8)$$

and

$$f_I^C = f_{IPL\_first} \times P_{enable} \times \prod_{j=2}^{J} P(\theta)_{Ij} \quad (9)$$

Where

$P(\theta)_{Ij}$=Posterior failure probability value of the jth independent protection layer

Based on this, the PHAMS system realizes the real-time update of the protection layer risk reduction frequency, and intelligently analysis the overall risk value and grading of the unit to plant.

3 SIL Intelligence Identification Online System

3.1 SIL Identification Modelling

The analysis data of the SIS protection layer of the HALOPA evaluation system in PHAM is updated and interconnected by means of the safety reliability database. After the HALOPA dangerous event using the same safety instrument loop is automatically brought into the SIL identification, the PHAMS system takes the same kind of consequences. C, to calculate without the cumulative frequency F^C before the SIF

$$F^C = \sum_{i=1}^{n} f_i^C \qquad (10)$$

where

$$f_I^C = f_I \times P_{enable} \times \beta \times \prod_{j=1}^{J} PFD_{IJ} \times \prod P_{condition} \qquad (11)$$

$$\beta = \frac{2^N - 1}{N} \qquad (12)$$

where

f_I^C= The occurrence frequency of the dangerous event I with consequence C

β= the common factors caused by several SIFs

N=the number of SIFs for the dangerous event I

Based on the modelling, the PHAMS system can intelligently group common risk events and assess the risk reduction factor RRF of SIF in each group [7] :

$$\text{RRF} \geq \frac{F^C}{TMEL} \qquad (13)$$

and

$$1 \times R_{\text{Factor}} < RRF_{SIL1} \leq 10 \times R_{\text{Factor}} \qquad (14)$$

$$10 \times R_{\text{Factor}} < RRF_{SIL2} \leq 100 \times R_{\text{Factor}} \qquad (15)$$

$$100 \times R_{\text{Factor}} < RRF_{SIL3} \leq 1000 \times R_{\text{Factor}} \qquad (16)$$

where

R_{Factor}= risk reduction factor between1 to 10.

Based on the SIL identification results of each group, the optimal solution is intelligently selected as the RRF of this safety instrument loop.

Figure 3 SIL Intelligence Identification Online

3.2 The SIL Identification Based on SINOPEC Risk Matrix in PHAMS

With the aid of data warehouse technology, PHAMS system can compare HALOPA assessment data which involved with SIS protection layers of the initial situation, the consequences classes and non SIS protective layer of data, and automatically take the residual risk of the events into the risk matrix. Then based on SINOPEC risk matrix, the SIL identification module can intelligently couple and assess the residual risk and RRF.

according to Sinopec's safety management focusing on high severity impact, PHAMS sets an independent TMEL requirements (as shown in the following table), and can automatically selects the most serious impact event as the final SIS identification data of this event. .

Table 1 TMEL of SIL Intelligence Identification[8-10]

Severity Classes	*Health & Safety (r/y)*	*Property Damage (r/y)*	*Social Impact (r/y)*
A	$\leq 1\times 10^{-1}$	≤ 1	≤ 1
B	$\leq 1\times 10^{-2}$	$\leq 1\times 10^{-1}$	$\leq 1\times 10^{-1}$
C	$\leq 1\times 10^{-3}$	$\leq 1\times 10^{-2}$	$\leq 1\times 10^{-2}$
D	$\leq 1\times 10^{-5}$	$\leq 1\times 10^{-4}$	$\leq 1\times 10^{-4}$
E	$\leq 1\times 10^{-6}$	$\leq 1\times 10^{-5}$	$\leq 1\times 10^{-5}$
F	$\leq 1\times 10^{-7}$	$\leq 1\times 10^{-6}$	$\leq 1\times 10^{-6}$
G	$\leq 1\times 10^{-7}$	$\leq 1\times 10^{-6}$	$\leq 1\times 10^{-6}$

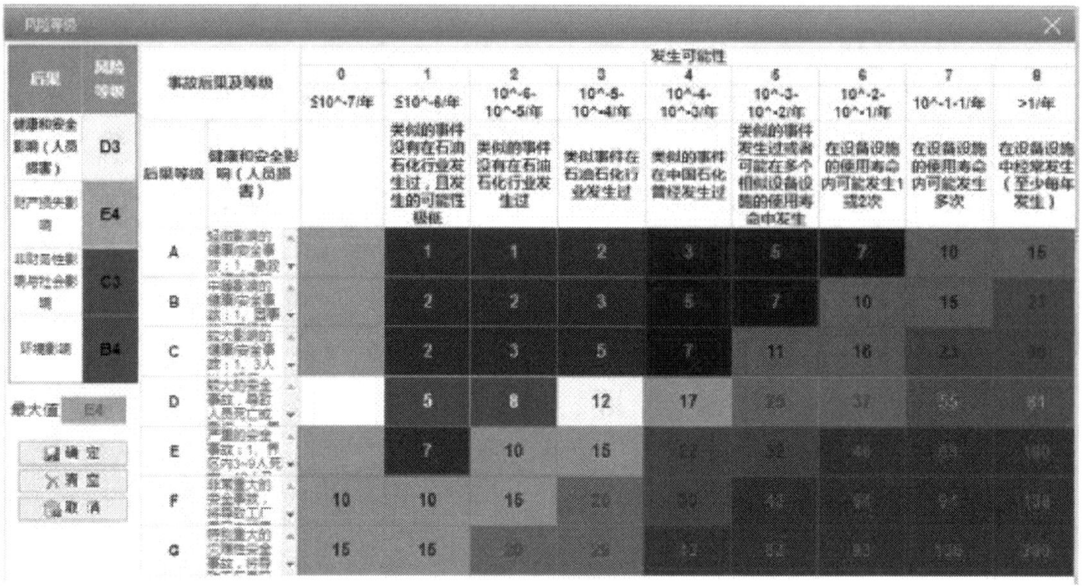

Figure 4 SIL Intelligence Identification on SINOPEC Risk Matrix

On the basis of SIL identification, the PHAMS system can also implement SIL verification based on multi-factor model, and evaluate the structural constraints, PFDavg and MTTF respectively. The verification result data is consistent with the accuracy requirements of IEC61508 and IEC61511.

4 Risk Management and Decision Support System

With the powerful safety and reliability database and expert experience database of PHAMS system, as well as the data mining of the result data of HALOPA and SIL system, the Risk Management and Decision Support System can implement the cumulative risk frequency calculation of independent devices and plants in construction or in service, to assess the overall risk based on different impact severities through the linear comparison [11-12].

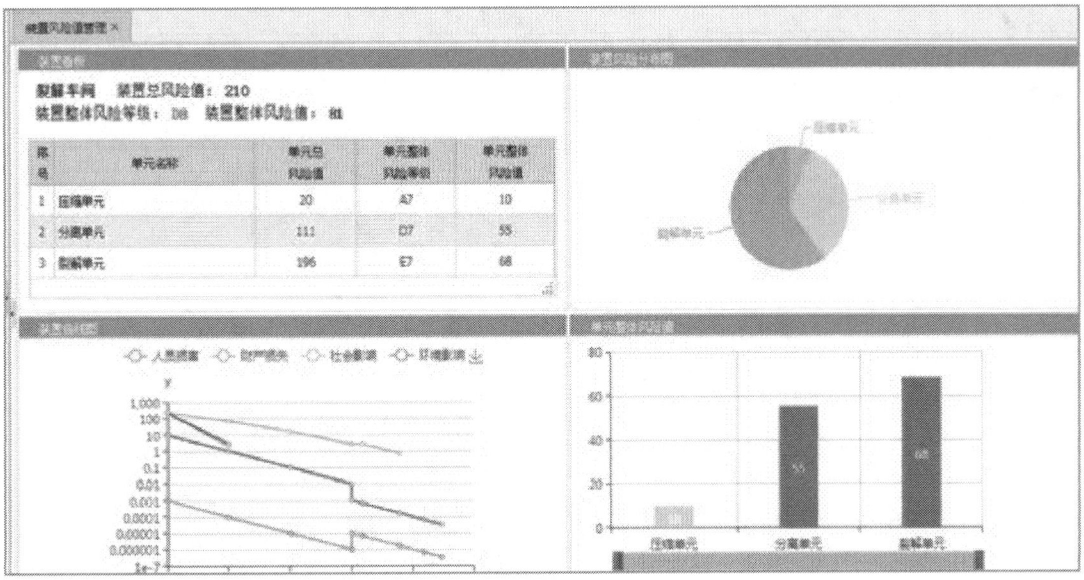

Figure 5 Overall Risk Rating and Risk Event Distribution Analysis of Ethylene Cracking Facility

5 Conclusion

1. PHAMS system sets up 9 categories of safety reliability database, expert experience data and Sinopec risk matrix, etc, which can support unified SINOPEC risk assessment and management standards, risk analysis template, process data storage and data mining for PHAMS system , as well as the interconnection and real-time update of data of each system.

2. HALOPA system implements plant overall risk analysis based on protection layer data, and performs data coupling evaluation from static analysis based on equipment cycle and dynamic risk correction to realize real-time evaluation of risk assessment technology and device operating conditions. The dynamic risk analysis research will be further strengthened to realize the system risk prediction function.

3. The SIL identificaiton and verification system can realize the evaluation of safety instrument function of in service and in construction. In the long run, it helps to improve the safety management level of the plant's overall safety instrument function

4. The risk management and decision support system can realize the overall risk assessment and risk distribution based on various evaluation results, through in-depth mining and statistical analysis, the PHAMS can realize the big data analysis management of Sinopec branches' risk management.

6 References

[1] The Bureau of Safe Production Supervision and Administration of SINOPEC. Guide for Safety Risk Identification and Assessment of SINOPEC[S].2018

[2] AQ/T 3054-2015. Guide for Application of Protective Layer Analysis Methods. Chapter 8： Scenery Frequency Calculation [S].

[3] Khan,Seyed Javad Hashemi,Nicola Paltrinieri,Paul Amyotte,Valerio Cozzani,Genserik Reniers.

Dynamic risk management: a contemporary approach to process safety management[J]. Faisal Current Opinion in Chemical Engineering.

[4] CCPS. Risk-based Process Safety [M]. 2013.

[5] Maryam Kalantarnia,Faisal Khan,Kelly Hawboldt.Dynamic risk assessment using failure assessment and Bayesian theory[J].Journal of Loss Prevention in the Process Industries . 2009 (5)

[6] Rakesh Ranjan,Sonam Singh,Satyanshu K. Upadhyay. A Bayes analysis of a competing risk model based on gamma and exponential failures[J] . Reliability Engineering and System Safety . 2015

[7]Zugravu-Soilita N. How does Foreign Direct Investment Affect Pollution? Toward a Better Understanding of the Direct and Conditional Effects[J]. Environmental & Resource Economics, 2017:1-46.

[8] Khakzad N, Yu H, Paltrinieri N, et al. Chapter 5 – Reactive Approaches of Probability Update Based on Bayesian Methods[J]. Dynamic Risk Analysis in the Chemical & Petroleum Industry, 2016:51-61.

[9] Villa V, Paltrinieri N, Khan F, et al. Towards dynamic risk analysis: A review of the risk assessment approach and its limitations in the chemical process industry[J]. Safety Science, 2016, 89:77-93.

[10] Mark Adrian van Staalduinen,Faisal Khan,Veeresh Gadag. SVAPP methodology: A predictive security vulnerability assessment modeling method[J]. Journal of Loss Prevention in the Process Industries.

[11] Podofillini L, Dang V N. Conventional and dynamic safety analysis: Comparison on a chemical batch reactor[J]. Reliability Engineering & System Safety, 2012, 106(4):146-159.

[12] R A F. Problems with risk analysis in the chemical industry. A detailed examination of the theoretical and practical problems faced by the risk analyst in the study of a chemical plant[J]. Plant/operations Progress, 1983, 2(3):185–190.

549550 Process Hazard Analysis

Process Safety Management System starts with managing the process related risks and from the first step of establishing a new process unit that is needed to evaluate the risks and define the necessary measurements. Process Hazard Analysis methodologies are the systematic tools which help to evaluate the risks in a process. For different needs there are several methodologies like DOW FETI, HAZOP, What if, HAZID, Fault Tree and Event Tree Analysis, Scenario Modelling, Facility Siting etc. Some of these methodologies are semi quantitative some are quantitative methodologies. In order to define and prevent major industrial accidents in our refinery we use DOW FETI, HAZOP, FTA and ETA and Scenario modelling. Moreover for some kind of projects we may also apply What if and HAZID analysis instead of HAZOP depending on the complexity. Facility Siting is another tool for risk related studies which is used to evaluate the risks for buildings in order to protect people inside the buildings. The workflow for a full unit PHA study starts with DOW FETI calculation which is used for risk ranking of al static equipment inside the unit. DOW FETI calculation is done in accordance with the equipment temperature, pressure, hold-up amounts, components and also reaction types if there is any. Then risk assessment process continues with HAZOP studies which is for scenario identification. In HAZOP, first step is chopping the unit into small parts to be evaluated which is called nodes. After that all nodes are evaluated using P&IDs and PFDs by keywords like temperature, pressure, flow, level etc. HAZOP helps to understand what kind of scenarios are possible inside the unit and these critical scenarios are taken through FTA and ETA studies. HAZOP is a semi quantitative risk assessment method where FTA and ETA are quantitative methods. FTA is applied for critical scenarios in order to find out the probability of top event then ETA is then applied in order to find out the probability of scenarios like fire, explosion, toxic dispersion etc. caused by this top event. FTA and ETA analysis all use the statistical PFD and failure rate values of all initiating events and barriers defined in HAZOP for the related scenario. Scenario Modelling is the following step which is used for understanding the effect of the probable scenario. The most probable scenarios coming from ETA is modelled to see the effect zones. These effect zones are used to plan Emergency Response Studies for all these potential major accidents.

548517 Site-Wide Asset Integrity Analytics & Benchmarking

All too often, reliability engineers focus on the equipment that is deemed as most critical or covered by some regulatory reference, while sometimes forgetting about those areas where something as simple as water can cause major issues. Non-destructive testing technicians and API inspectors alike can help facilities gather and compile data to perform predictive analytics and even benchmark sites with similar assets. This data can be used to identify assets where others may be seeing issues. This also allows facilities to compare reliability engineering effectiveness. This presentation will share some sample results of non-destructive testing and inspections of both regulatory required and non-required assets in the fuel ethanol production sector as well as an example of how that data can be assembled and shared across any sector to educate and learn from each other. Afterall, there are no secrets in safety!

548330 Reliability, Availability, and Maintainability (RAM). What Does It Provide?What Should You be Aware of before You Conduct a RAM Study?

Reliability, Availability and Maintainability (RAM) studies can significantly add value to your Process Safety Management (PSM) activities, especially with regards to Mechanical Integrity (MI). RAM studies are an excellent way to determine which equipment limits you from achieving optimum availability and throughput. Using industry failure rate data or even site-specific data can help you identify areas for improvement and determine how best to maintain your equipment. But RAM studies can go beyond purely theoretical studies by using the experience and knowledge gained from operating and maintaining similar facilities to develop a pragmatic approach to solving mechanical integrity-related issues. This paper will cover the pros and cons of site versus industry data and will present a case example of utilizing experience to better define the proposed problems and solutions. If done correctly, RAM studies can be more than statistics; they can be roadmaps to process reliability improvements.

546707 Driving Asset Integrity and Process Safety Consistency across a Multinational Organization

Mexichem is in the process of implementing world-class process safety management (PSM) and Asset Integrity Management (AIM) programs globally for their facilities. This is a major undertaking as Mexichem, to a large extent, has grown through acquisitions. Therefore, each plant has a safety culture and management system unique to that asset. As we all know, it's difficult to drive global consistency when dealing with different types of operations in various regions throughout the world under differing regulatory regimes. This is especially difficult when there is no common understanding of process safety. One of the first steps Mexichem has taken is to develop a "fit-for-purpose" management system with an emphasis on developing a common understanding of process safety and ensuring the culture is robust enough to support the system when it is implemented.

To aid Mexichem in this overwhelming task, Risktec Solutions were engaged to develop and deliver a bespoke set of interactive training workshops on a variety of PSM / AIM topics. The first of which was related to the identification and management of safety critical equipment (SCE). This presentation will outline the plan and methodology Mexichem and Risktec have used to transfer such knowledge and discuss some of the successes and challenges along with way.

548216 Alarm Floods and Their Connection to Plant Incidents and Near Misses

Investigations of many catastrophic plant events have shown alarm flooding to have been significant distractions for the operators just prior to the incident. Alarm floods have also been determined as a cause or contributing factor to smaller incidents and near misses.

In fact, many of the more recent findings by U.S. Chemical Safety and Hazard Investigative Board (CSB) include alarm management as one of the key issues in the investigative reports.

Even after several years of trying, many plants still struggle with controlling alarm floods. Many also have problems generating consistent and accurate alarm metric reports for the entire plant. Now that the ISA 18.2 Alarm Performance Metrics are considered "generally accepted good engineering practices," we should make haste to demonstrate where each of our plants fall (particularly the PSM plants) versus these metrics before the regulating authority moves in this area. If we are not able to prove we meet the metrics (particularly for alarm flooding) under all operating conditions, we should remediate.

Static rationalization can reduce your average number of alarms, but without controlling the alarm floods, there is no help for the operator when he needs it the most. For most, we must begin the process by recognizing that alarm floods occur most of all during a change of process state.

The means of controlling alarm floods has been available to us for many years. Unfortunately, only a select few have implemented the solution and are achieving the desired results today. These plants have achieved equal to or better than ISA 18.2 since before it was published.

This session will cover the justification for true alarm management from the safety and economic perspective with several practical steps that everyone should take if they are not currently meeting the ISA 18.2 specifications during all operating conditions.

548318 Do Incident Data & Lessons Learned Really Facilitate Process Improvement?

Regulations are a set of guidelines for the industry for protecting the people and environment, when it comes to process safety systems. While some companies tend to find that complying with regulatory requirements are always a challenge, on the other hand some tend to go overboard in their bid to stay compliant. In spite of the route taken, there are incidents – major, minor and near misses.

Other than the accumulation of this data, the investigations also result in lessons learned. Over the past four decades, lot of incident data and lessons learned has accumulated, been assimilated, and also analyzed. Complementing that are the audits, system reviews, hazard identification exercises, changes to regulations, and knowledge sharing forums.

It is a very efficient and effective mechanism to leverage these vast sources of current information developed, and implement its learnings, in the operations of existing processes. While the core concepts of safety are universal across the various industrial sectors, its application may differ. But once again, the lessons learned databases are effective sources of making this leveraging process more efficient.

This paper will discuss the some case studies of how the concept of lessons learned and the knowledge of the incident data has been leveraged by the authors in various manners ensuring success from one industry to another – example (1) Learnings from an incident due to gaps in the Emergency Response Plan, bringing about success at another chemical manufacturing site; (2) Incident data from a steel plant, when analyzed and applied to a chemical plant resulted in improved of effectiveness.

546673 Structured & Scrutinized Approach Towards HSE Excellence; A Success Story

Author: Usman Ali, Process Safety Engineer

Co-Author: M Ahsan Sarfraz, Unit Manager HSE

This paper studies the practices & programs that support reporting of Near misses, minor unsafe conditions, incidents and quantitative risk assessment strategies in our organization for analyzing and learning from past events of operational failures, injuries, safety and environmental incidents.

A near miss is a leading indicator to an incident that, if scrutinized appropriately, can prevent injuries and accidents. Some persons may be tempted to write off near misses as "no harm, no damage" situations, but organizations who track near misses, determine how and why they occurred, and take corrective action can prevent similar or more serious incidents from occurring in the future. Fatima Fertilizers Limited believes that, "We don't look at what happened; we look at what could have happened".

Keeping in mind the importance of near miss reporting, FFL initiates drive and implement systems for

1. Capturing and increasing near miss reporting at whole site.
2. Logging data for analysis and using this valuable data for identifying gaps in existing process safety system.
3. Implementing sustainable improvement plans and safety campaigns (Safe hands, Safe Routines, Job Cycle Checks, Safe Drive etc) in our process safety systems for making our site a safer workplace.

These actions has improved the site performance which is reflected as Site has now over 45 Safe Million Man Hours,

Total Record-able Injury Rate (TRIR) is below 0.2 since last 03 years and No Loss time injury since last 06 years.

548162 Events Report Database Analysis for SHE/PSM Programs Improvements

Incident databases contain records of events that have occurred in a petrochemical facility during recent years. The most of chemical industries report databases are filled with accident valuable data (frequency and severity). Harnessing data mining to explore incident databases, it was used data mining to analyze the incidents database of the studied plant. Inferent approaches are ineffective for low frequency, high consequence events because of their rarity. Given this information limitation, this work uses database analyses to forecast incident frequencies, their relevant causes, equipment involved, and their possible consequences, in specific petrochemical plants.

Databases information analysis, also help to avoid future accidents, thereby reducing the risk. More specifically, this study presents dynamic analyses of incidents in a petrochemical plant ubicated in Mamonal industrial park, in Cartagena- Colombia. Differences in the chemical-specific predictions with the overall predictions are demonstrated. Furthermore, value theory is used for consequence of rare events by formulating distributions for events over a threshold value. Finally, the accidents number is used to estimate the capital at risk within an industry utilizing the frequency and loss-severity distributions.

In conclusion, it was recommended to update de maintenance program (predictive, preventive and corrective) in cases of abnormal changes in production rates, comparing the current production levels, with the analyzed plant production forecast. Also, it was suggested to improve the report culture, due to gaps founded regarding to events classification.

549483 The Process Safety Analysis and Management System in Sinopec

The big data technique has been in trend for the process safety assessment. In china, it's the very need to transfer the qualitative safety management to quantitative safety management, which should rely on the professional risk analysis technique, as well as the standard safety reliability database. Based on the above, it's very important to integrate several professional risk analysis technique, as HAZOP, LOPA, SIL,BOWTIE etc. to realize the integral risk identification and evaluation, grading and management plan making, supporting by the proprietary SINOPEC risk metrics and grading standards, as well as safety equipment reliability database,which is developed as the Process Safety Analysis and Management System for SINOPEC, to support all SINOPEC branches to make dynamic risk analysis and management online

The key data for HAZOP, SIL and Management module in PHAMS is the dependent protection barriers, which we do focusing on the failure probability, at the same time, we do use of the Bayesian Correction Modelling to verify the failure probability of safety equipent and protection barriers on site, which can validly save the grading cost on time and personnel, all of this is towards to the deeper research on dynamic risk analysis for PHAMS.

Thanks to the PHAMS, we has set up and implemented several standards on HAZOP, LOPA and SIL for SINOPEC.

552170 Analysis and Improvement of Safety Integrity Level Considering Human Factors

Safety Instrumented System (SIS) is an irreplaceable active protection layer that prevents the propagation of process accidents and plays an important role in reducing the risk of accidents. In addition to the sensor response triggering the SIS, a human emergency button is often provided in practice to ensure that the SIS is triggered by personnel in the event of the sensor part failure. Due to human factors added, the emergency button will improve the reliability of the SIS. However, there is no research to study the human reliability of the SIS upon sensor part failure in a continuous operation chemical plant, and how human factors can improve and affect the overall reliability of the SIS still remain undiscovered. In this paper, a verification methodology was proposed which integrated with Cognitive Reliability and Error Analysis Method (CREAM) and Reliability Block Diagram (RBD) to emphasize human reliability analysis of the SIS verification. The operator behavior is subdivided into several steps and the Human Error Possibility (HEP) is calculated by a modified CREAM. Then the RBD method is used to verify the SIS reliability considering human factors. An SIS system of one reacting furnace was taken as a case study using this method to show how human factors can improve the reliability of the SIS. The proposed methodology allows SIS verification considering human factors to improve safety integrity level in real practice thus help stakeholders saving costs and making more decisions on SIS implementation.

551560 Ethics for Registered Engineers, a 1PDH Professional Ethics Presentation

LAPELS staff will discuss the Board, enforcement activities and investigative procedures, recent cases, the duties of a firm's supervising professional, ethical issues and other common questions.

552164 How Safety Culture Drives the Improvement of Process Safety Competencies in Wanhua

In the past two decades, Wanhua continuously focuses on safety improvement and technology innovations. Particularly from 2017, Wanhua has been promoting safety as the company's core management theme, with the aim of developing a comprehensive process safety competence system. This article addresses the history of company safety culture with several unique cases/practices, and how safety culture drives the improvement of process safety competences in Wanhua Chemical Group.

552166 How can we use Artificial Intelligence (AI) Technology to Assist Process Safety? Opportunities and Prospects

arely any technology like AI (artificial intelligence) technology has gained such a tremendous scientific and economic interest presently. It is often stated that AI will be the trend of the future, which is supposed to influence our daily life and lead to an industrial revolution in the near future. But for its specific application in the petrochemical industry, most people have no clue. This paper introduces the essence of AI and its main mature application scenarios. The possibility and necessity of applying AI to petrochemical industry are analyzed. Several application scenarios of AI assisted process security are described in detail£¬such as intelligent production process optimization, robot inspection, intelligent pipeline monitoring and analysis and evaluation etc. Combined with the problems in process safety management, the role of artificial intelligence in process safety management is demonstrated by examples. To make further progress towards a safer industry, AI technology can strengthen information management, improve process control-ability, prepare to obtain production data, reduce manual wrong operation, and ensure the intrinsic safety of the whole production process.

552167 Analysis of Shell Moerdijk Accident by Systems-Theoretic Accident Model and Processes and Bowtie Analysis

As a Systems-Theoretic Accident Model and Processes (STAMP) was used to analyze the Shell Moerdijk Accident by MIT¡¯s Nancy Leveson Professor as a bench-mark case. This research intended to use the Systems-Theoretic Process Analysis (STPA) to identify the hazards, safety requirements and constrains, unsafe control actions in the operational control structure in order to understand the flaws and context in decision making process. The results were also compared with the causes-incident-consequences-barriers by the Bowtie Analysis (BTA) to compare both methods and to demonstrate their usefulness. It could be summarized that STPA yielded many possible cases with components of different types and degrees of hazards while BTA limited by their possible cause-consequence links. While comparing to Casual Analysis based on STAMP (CAST), BTA displayed a simple but limited analysis without the possible extensions into the complex social-management-technological relationships.

551629 Ethics As an Underpinning to Safety

Ms. Grubbe will demonstrate and define ethics in support of good safety, and will show how ethics is eroding in academia, government and in industry. She will also suggest a way for academics and university industrial board members to consider inserting ethics into every engineering curricula.

551630 Case Study Reviews in Ethical Failures Contributing to Process Safety Incidents

Ms. Grubbe will use her knowledge and experience of several serious process incidents, and will tie the messages of the ethical basis of safe operations to key instances where ethical failures occurred. She will discuss the hard dynamic that occurs when an engineer chooses to remain silent vs. speak out about a noted issue.

551893 Ethical Issues in Safety- Optimizing Safety Culture and Leadership to Rise Above the Challenge!

In today's fast paced and demanding business environment engineers and operators are faced with competing priorities and pressures. During these complex times, everyday decisions must be made without compromising the Ethics of Safety and allowing safety priorities to rise to the top . During this session, the importance of a strong safety culture and leadership behaviors will be discussed as key to successfully navigating the Ethics of Safety. Practical short and long term solutions will be shared that can be applied across a broad spectrum of industries.

552168 Deficiencies and Prospects of Safety Management in China's Chemical Enterprises

There are unilateral management modes such as "personnel-based management", "accident-based management" and "hidden danger-based management", but lack risk-based system safety management concepts and methods in the safety management and monitoring of China's chemical industry, which is also the main reason for the low safety performance and frequent accidents in China's chemical industry. The defect of unilateral management modes and the advantages of risk-based system safety management methods are analyzed in this paper to forecast the transformation of safety management in China's chemical industry and the way of systematic safety management, which would enhance the safety performance level of China's chemical industry as a whole.

Lessons Learned from a Dust Hazard Analysis at a JV Site in China

Julie D. Kister, PE
Air Products
EH&S Process Safety
7201 Hamilton Boulevard
Allentown, PA 18195-1501
kisterjd@airproducts.com

Phil Parsons
Senior Consultant
Blast Effects Division
Baker Engineering & Risk Consultants Inc.
3330 Oakwell Ct
San Antonio, TX 78218
www.BakerRisk.com
pparsons@bakerrisk.com

Prepared for Presentation at
American Institute of Chemical Engineers
2019 Spring Meeting and 15th Global Congress on Process Safety
New Orleans, LA
March 31 – April 3, 2019

Lessons Learned from a Dust Hazard Analysis at a JV Site in China

Julie D. Kister, PE
Air Products
EH&S Process Safety
7201 Hamilton Boulevard
Allentown, PA 18195-1501
kisterjd@airproducts.com

Phil Parsons
Senior Consultant
Blast Effects Division
Baker Engineering & Risk Consultants Inc.

Keywords: DHA, Dust Hazard Analysis, China, Joint Venture, Coal, Gasification

Abstract

In September 2017 Air Products acquired a Coal Gasification Joint Venture Plant in China. Coal dust is a combustible dust. NFPA 652 "Standard on the Fundamentals of Combustible Dust"[4] requires that a Dust Hazard Analysis (DHA) be completed for a facility where materials have been determined to be combustible or explosive. This paper discusses the challenges and the lessons learned from conducting a DHA at a joint venture facility in China. Some of the challenges discussed include cultural differences encountered by a US based team conducting a dust hazard analysis at a site in China, using NFPA standards as a guideline at the site in China, dealing with language barriers in review meetings and technical documents, and the challenges of conducting a DHA at a joint venture site where solids handling involves multiple companies.

1 Introduction

Air Products signed an agreement in September 2017 to form a $1.3 billion joint venture (JV) in China.[1] The joint venture, called Air Products Lu'An (Changzhi) Co., Ltd., owns and operates an Air Separation Unit (ASU), gasification unit and syngas clean-up system in Changzhi City, Shanxi Province, China. The joint venture receives coal, steam and power from the coal company and supplies syngas to the coal company.[1] All gasification units were onstream starting in August 2018.[2]

Figure 1: Gasification Units at Air Products Lu'An (Changzhi)

There are four individual gasification units at Air Products Lu'An (Changzhi). Pulverized coal is used as the fuel for the gasification units.[3] As Air Products engineering teams working on the Lu'An project began taking a closer look at the design, the engineers realized that the presence of coal dust could create dust fire and explosion hazards. Air Products historical mix of products focuses mostly on gases and liquids. Internal information and expertise on handling combustible dusts was lacking. Therefore, the issue was referred to the Air Products Global Process Safety team.

After reading NFPA 652 "Standard on the Fundamentals of Combustible Dust"[4] and NFPA 654 "Standard for the Prevention of Fire and Dust Explosions from the Manufacturing, Processing, and Handling of Combustible Particulate Solids"[5], the team's course of action became clear. Air Products would need to perform a Dust Hazard Analysis (DHA) at the Lu'An facility to review the safety of the site equipment, solids handling procedures, and housekeeping for pulverized coal and coal dust. Air Products contacted consultants at BakerRisk to prepare and conduct the DHA. Because of the recent updates to NFPA 652[4], Air Products felt that the dust hazard expertise would be in the US. Therefore, a US based consultant was selected instead of a local Chinese consultant. Phil Parsons, a Senior Consultant at Baker Engineering & Risk Consultants Inc. with experience in handling combustible dusts, was brought on board to plan and lead the DHA. Phil resides on the relevant NFPA standards committees (i.e., NFPA 652[4] and 654[5]) so he was able to bring the necessary context to the DHA regarding interpretation and application of these standards. Air Products and BakerRisk signed a Nondisclosure Agreement (NDA) prior to the start of the preparation work for the DHA.

The DHA was performed at the Lu'An site in Changzhi City, Shanxi Province, China on December 4, 5, & 6, 2018. The International team consisted of engineers from the Lu'An joint venture site in China (Gasification Process Engineer, Air Products Site Safety Engineer, Gasification Plant Manager and the Technical manager), the Air Products Shanghai Office (Jerry Lu, Amy Shen and Yocky Li Global Operations Plant Process Engineering Group), the Air Products Global Headquarters in Allentown, Pennsylvania (Julie D. Kister EH&S Process Safety), and the Baker Engineering & Risk Consultants Inc. Office in San Antonio, Texas (Phil Parsons Senior Consultant).

1.1 Dust Fire and Explosion Hazards

Dust explosions may be classified as either *primary* or *secondary* in nature. Primary dust explosions typically occur within individual pieces of equipment inside process plants or similar enclosures (baghouse, cyclone, grinder, etc.), and are generally controlled by pressure relief venting through purpose-built ducting to the atmosphere. Secondary dust explosions are the result of dust accumulations (outside of equipment) within the facility, which are disturbed, suspended, and ignited by the primary explosion, resulting in a more dangerous uncontrolled explosion inside the workplace. Fatalities from dust explosions are normally the result of secondary dust explosions, however vulnerabilities from primary explosions can occur if equipment is not designed with the correct explosion protection. Nevertheless, control of fugitive dust emissions and housekeeping outside of equipment are a major focus of all combustible dust-related standards and are the key to reducing the potential for a catastrophic event.

2 Developing a Plan for the Dust Hazard Analysis

One of the first issues the team needed address was to determine what standards or codes needed to be followed. Air Products is a Global company with its headquarters in US. The Lu'An site is located in the Shanxi Province, China. The Chinese Code is GB15577 "Safety Regulations for Dust Explosion Prevention and Protection"[6]. The Basis of Design for the Lu'An site states that the hazards of dust explosions were considered when designing the site.[3] Three main steps were taken to prevent dust fires and explosions: 1) Handling of pulverized coal and hot/dry flyash (residue of coal combustion) under a nitrogen inert atmosphere. 2) Good housekeeping in areas where coal and flyash are handled. 3) Appropriate design that includes relief devices on storage silos.[3] Air Products policy is to follow the standard that is more stringent. The team felt that the requirements in the NFPA standards were more detailed and stringent.

2.1 Relevant Standards

NFPA 652[4] provides the general requirements for the management of combustible dust fire and explosion hazards and directs the user to the NFPA industry or commodity-specific standards as appropriate (i.e., NFPA 61[7], 484[8], 654[5], 655[9], and/or 664[10]). Where a commodity-specific NFPA standard exists (i.e., in this case, NFPA 654[5]), its requirements are to be applied in addition to those in NFPA 652.[4] Where a requirement in a commodity-specific NFPA standard differs from the requirement specified in NFPA 652[4], the requirement in the commodity-specific NFPA standard may be used.

2.1.1 NFPA 652[4] General Requirements

The owner/operator of a facility with potentially combustible dust is responsible for the following activities, each of which is discussed in the subsections below:

1) Determining the combustibility and explosibility hazards of materials in accordance with Chapter 5 (Hazard Identification);
2) Identifying and assessing any fire, flash fire, and explosion hazards in accordance with Chapter 7 (Dust Hazards Analysis [DHA]);
3) Managing the identified fire, flash fire, and explosion hazards in accordance with Section 4.2.4 (Compliance Options);
4) Communicating the hazards to affected personnel in accordance with Section 9.5 (Training and Hazard Awareness).

2.1.1.1 Hazard Identification

The owner/operator of a facility handling particulate solids is required to establish whether handling such materials poses fire and explosion hazards. A number of American Society for Testing and Materials (ASTM) methods are presented in NFPA 652[4] that the owner/operator may employ to determine if the dust handled is combustible.

2.1.1.2 Dust Hazard Analysis (DHA)

As defined by NFPA 652[4], a DHA is a systematic review to identify and evaluate the potential fire, flash fire, and/or explosion hazards associated with the presence of combustible particulate solids in a process or facility and shall be performed / updated every 5 years. The DHA shall include the following:

1) Identification and evaluation of the process or facility areas where fire, flash fire, and explosion hazards exist;
2) Where such a hazard exists, identification and evaluation of specific fire and deflagration scenarios to include the following:
    i. Identification of safe operating ranges;
    ii. Identification of the safeguards that are in place to manage fire and deflagration hazards;
    iii. Recommendation of additional safeguards where warranted, including a plan for implementation.

Each part of the process where combustible dust is present shall be evaluated, and the evaluation shall address the following:

1) Intended and unintended potential combustible particulate solid transport between

parts of the process system;

2) Potential fugitive combustible dust emissions into a building or building compartments;
3) Potential for deflagration.

Each part of the process that contains a combustible particulate solid and that can potentially include both of the following conditions should be considered a *fire hazard* and should be documented as such:

1) Oxidizing atmosphere;
2) Credible ignition source.

A *deflagration* hazard should additionally be considered if a credible suspension mechanism exists in an enclosed environment. Threshold housekeeping dust accumulation levels and non-routine dust accumulation levels (e.g., from a process upset) should be in accordance with the relevant commodity-specific NFPA standard (i.e., $^{1}/_{32}$-inch for the Luan facility).

2.1.1.3 Compliance Options

Managing the identified fire, flash fire, and explosion hazards shall be performed by one of the following methods:

1) A prescriptive approach in accordance with the following sections of NFPA 652[4]:
    a. Chapter 5: Hazard Identification (see Section 2.1.1.1),
    b. Chapter 7: DHA (see Section 2.1.1.2),
    c. Chapter 8: Hazard Management: Mitigation and Prevention. This section of NFPA 652[4] provides general prescriptive requirements to be satisfied for safe operation,
    d. Chapter 9: Management Systems, which includes operating practices and procedures, inspection and maintenance, training and hazard awareness, and management of change, among others.

2) A performance-based approach in accordance with NFPA 652[4] Chapter 6.

The above assessments should be applied in conjunction with any prescriptive provisions of applicable commodity specific NFPA standards. The ***prescriptive approach*** was undertaken for this study.

2.1.1.4 Training and Hazard Awareness

NFPA 652[4] places significant emphasis on the need to provide employee training on the hazards associated with handling combustible dust. Employees, contractors, temporary workers, and visitors shall be included in a training program, according to the potential exposure and potential hazards to which they might be exposed or could cause. Job-specific training shall ensure that employees are knowledgeable about fire and explosion hazards of combustible dusts and particulate solids in their work environment. Where explosion protection systems are installed, the training of personnel shall include the operations and potential hazards presented by such systems. Refresher training shall be documented and provided periodically (annually, at a minimum). New-hire workers shall be trained before commencing work in the facility.

3 Preparing for a DHA in China

Once the team decided to focus on following NFPA 652[4] and 654[5], the next phase involved planning for the DHA. The joint venture was established following the design phase and design reviews of the site. Since the design phase was past, the team wanted the site to have some operational experience before the DHA started. This delay would allow time to see if coal dust was accumulating in the production facility and to allow time to see if the equipment had any reoccurring issues. The team began actively planning for the DHA in April 2018. In addition to the DHA, a full plant hazard review was being conducted. The DHA was planned as a separate meeting that focused only on the dust hazards. The documentation of the DHA will eventually be included with the full plant hazard review. The DHA will be reviewed every five years as part of the hazard review. The hazard review was conducted through fall 2018 and winter 2019. The DHA was scheduled in December 2018 in-between scheduled hazard review sessions. This timing enabled the engineering team to be onsite for both the DHA and the hazard reviews.

NFPA 652[4] requires that the site determine the combustibility and explosibility hazards of materials of concern. General combustibility and explosibility data are available for coal dust. However, the team requested combustibility and explosibility data for the specific type of coal used at Lu'An. The coal at Lu'An is mixed with limestone to help it flow better. The addition of limestone can increase the ignition energy, which makes the coal dust harder to ignite. Therefore, the addition of limestone can reduce the risks from a release. The Lu'An site sent a pulverized coal sample and a flyash sample a laboratory to be tested in July 2018. A Chinese laboratory was selected to avoid having to ship the samples internationally. Specific American Society for Testing and Materials (ASTM) methods are listed in NFPA 652[4] to determine the combustibility and explosibility hazards of materials of concern. This type of testing was available at the laboratory in China.

The results from the pulverized coal/limestone sample show it to be a combustible dust. The DHA conducted at Lu'An was based on the combustible properties of this pulverized coal sample. If the coal is changed (source of coal, amount of limestone mixed with coal etc.), the coal will need to be retested. The potential to retest was noted in the DHA documentation. The flyash sample testing showed that the flyash was not combustible. As a result, the DHA focused on the coal handling sections of the site and did not review sections that handle flyash. If the flyash composition changes as a result of a process change, a flyash sample should be retested as well.

The next step was to obtain the necessary documents for the DHA. When dealing with a 14 or15-hour time difference among all parties, email was a helpful tool. The DHA team was able to email the technical documents back and forth between the Lu'An site in China, the Air Products Shanghai Office, the Air Products office in Allentown, Pennsylvania, and the BakerRisk office in San Antonio, Texas using the working hours at those sites. Most of the documents were already prepared in both Chinese/ Mandarin and English. Longer technical documents, like the Basis of Design[3], were created as separate Chinese and English documents. The P&ID and other technical documents were a mix of both Chinese and English on the same document. The documents had the information written in Chinese and an English translation included. The US based team found that the app Google Translate[11] worked well for cases where the English translation was not included.

The team held several teleconferences in preparation for the DHA. These meetings were held in the morning in the US and in the evening in China. Engineers from teams on both continents were willing to call into meetings that were outside of normal working hours. For additional teleconferences, the engineers in US meet and relayed information to China by email, and the engineers in China meet and relayed information to the US.

Since members of the team needed to travel internationally to attend the DHA meeting, a lot of advance work and preparation was required. Although advance work was emphasized, in retrospect this is an area that had room for improvement. The gasification process is complex with many units. The team's time on site was very limited. Understanding the equipment, process, and having a working knowledge of the P&IDs was an area of advance preparation that could have been improved. This would have enabled the team to better use the time available for the Gasification Unit tour and the time available to ask the plant engineering staff questions, providing more overall efficiency

4 Language Barrier and Cultural Differences

The language barrier provided challenges for both the technical documents and for the onsite meetings. Three engineers on the team, Jerry Lu, Amy Shen, and Yocky Li from the Air Products Shanghai Office, speak both Chinese/ Mandarin and English fluently. The fact that they were technical contributors to the meeting as well as being able to help with translations was key to the success of the DHA. The starting and ending meetings for the DHA included all the engineering staff at the Lu'An site. These meetings were done with a structured translation. The information was first given in English by the consultant Phil Parsons. Jerry Lu then translated the information into Chinese. For the final meeting, it was important to translate the information into Chinese because an overview of the recommendations that the site will receive as part of the final DHA report was included. This format was very effective, although it took additional time. The meetings took about two and one-half times longer with the translation than if it was done in one language.

The site tour and the detailed meetings of the analysis were done with a smaller group. There were three engineers from the Lu'An site in addition to the engineers from the Shanghai Office who participated in the full review. All three spoke some English. Generally, each discussion would start in English. As the conversation became more involved and more technical issues were discussed, it would switch to Chinese. When the conversation was complete, it would switch back to English. The information was captured in a PHAWorks[12] file which was written in English. The fact that three members of the team were fluent in both English and Chinese ensured that information was not lost in the translation. This less structured translation process took extra time but was not as time consuming as doing the full structured translation. The team worked well together and was able to avoid becoming frustrated with the language barriers. In all, it was estimated that the DHA took approximately 50% longer to complete than it would if all team members spoke the same language.

Figure 2: Gasification Unit Toured on December 4, 2018

A comment that was made during the one of our discussions at lunch was that "the Chinese culture is focused on service". The US team found this to be true throughout the trip and in the DHA meetings. The engineers at the Lu'An site and from the Shanghai office wanted the DHA to be a success and worked hard to help achieve it. The engineering team from the US was worried about being seen as the 'outsiders who were there to tell everyone what to do'. However, the engineering staff at the Lu'An site was very interested in making their site safer and taking advantage of having help with the DHA. The engineering staff was dedicated to attending the meetings and provided much helpful information and insight to the process. The US team was given an extensive tour of the Gasification Units and the Lu'An staff ensured that all questions were answered. They understood the need to finish the meetings in the time scheduled and worked hard adhere to that schedule. They were also very hospitable to the US team throughout the trip. They made sure lunch was provided every day, even though two of the three days on site the team worked though the scheduled lunch break. They made dinner plans every night, helped with travel arrangements, and in general made sure the trip was a success.

The three words that describe the team's approach to dealing with language and cultural barriers and performing an effective analysis are: Ownership, Understanding, and Patience. The entire team had ownership for the project. For the US side, the team focused on holding a successful DHA and increasing the safety of the site. For the Chinese side, the team had ownership for the Lu'An site and wanted to work to improve the site and make it safer. Several non-English speaking operators and other key personnel were present during the DHA meetings. They provided invaluable inputs but never once became deterred or irritated at the amount of back and forth translation required to convey a particular message. Working together for a common goal was key to dealing with the language barrier. There were many potential barriers to success: a language barrier, pre-trip planning was done with a 14 to15- hour time zone difference, the two engineers on the US Team had never traveled to China, three companies were involved (Air Products, BakerRisk, and the joint venture Air Products Lu'An), but all team members demonstrated an understanding of these challenges and were patient in working through the challenges. In particular, patience was key in navigating complex combustible dust hazard phenomenon discussions in two languages with a lot of repetition through translation to ensure not only that the hazards were accurately defined, but also that practical mitigation solution recommendations were developed.

5 DHA Findings and Recommendations

The three major findings of the DHA were: 1) The majority of the coal handling system was determined to be safe due to the use of nitrogen purge and the outside location of the Gasification Units. Having the units located outdoors significantly reduces the possibility of a secondary explosion by minimizing areas of confinement. 2) The high ignition energy of the coal powder reduces the relative potential for ignition to occur, although a recommendation was made to have a pure coal sample tested to ensure hazards associated with the coal handling process prior to the limestone addition were being adequately mitigated. 3) Relatively modest recommendations were developed to improve safety associated with combustible dust fire/explosion hazards.

Figure 3: Good housekeeping and outdoor location of Gasification Units reduce dust fire and explosion hazards

An area of concern found by the DHA is the solids handling of the customer where they handle the coal prior to the addition of the limestone (i.e., the ignition sensitivity of the coal dust is likely much lower than that acquired from the sample test data). The customer's scope extends into the Gasification Unit with a process area that is located within the process confines of each of the four Gasification Units. This is a concern for two main reasons. First, since the customer operates a crushed coal conveyor within a rigid enclosure inside the gasification process area, a secondary explosion in one of these enclosures would result in significant consequences to not only personnel, but also to process infrastructure resulting in potential significant loss of production. Second, if an ignition source such as a smoldering ember is generated in the coal supply area (outside of Air Products control), the ember could end up in the Air Products process presenting a potential combustible dust fire or explosion hazard.

Figure 4: Dust layer on piping in enclosed supplier shed

The two recommendations developed to address this concern are: 1) Open communications to encourage the customer to establish a robust housekeeping program to meet the intent of NFPA 654[5]. 2) Open communications to encourage the customer to establish a robust preventative maintenance (PM) program to inspect and follow manufacturers recommended PM practices on all equipment, particularly moving components. These two recommendations are challenging because Air Products Lu'An cannot directly implement them and so must rely on maintaining open communications with the customer. Air Products Lu'An will need to maintain continuous communications with the customer about housekeeping and maintenance.

In addition to the recommendations that involved opening communications with the customer, there are also a number of recommendations focusing on the Air Products Lu'An scope. These recommendations include the following items: (this is not a complete list of recommendations but it provides an overview of general areas that were a focus of the DHA report) 1) Develop a housekeeping program at Lu'An that meets the intent of NFPA 654[5] with a focus on ensuring that dust is cleaned up as a priority (e.g. prior to the end of shift) and a focus on proper cleaning technique such as using an electrically classified vacuum for cleaning up coal dust. 2) Make improvements to the coal receiving vessels and area for each gasification unit. 3) Develop best practices for maintenance and hot work. Hot work permits should include the following requirements: Ensure that all combustibles (e.g., coal dust, filter socks) are removed from the baghouse prior to hot work being performed. All entry/exit openings to the baghouse (e.g., inlet, outlets) are to be blinded with non-combustible material. Fire watch must be maintained for two hours post any hot work performed.

6 Path Forward

The DHA was completed during the scheduled time in December 2018. The path forward involves documenting the results and closing out the recommendations. A final report was issued in February 2019 and the DHA was documented in PHAWorks.[12] The Lu'An site is currently undergoing a full plant hazard review. The DHA will be incorporated with the hazard review, which will ensure that it will be regularity revalidated. The DHA recommendations will be maintained and tracked along with the recommendations for other parts of the hazard review to ensure that they are completed in a timely manner. One challenge will be maintaining continuous communications with the customer about housekeeping and maintenance.

7 Conclusion

As companies become more global, the need to hold safety meetings with international teams will increase. Several key findings from our experience with a conducting an international Dust Hazard Analysis at Lu'An are:

- Standards and codes to follow must be agreed upon before the start of the meetings.
- When members of the team travel internationally to attend the meeting, the time onsite is limited and precious. Preparation and advance work are key to completing the meetings in the scheduled time.
- Duration to complete the DHA with international team members took approximately 1.5 times longer than the duration if all team members spoke the same language. This time must be allotted for in planning the meetings.
- Team members who are both bi-lingual and technical contributors to the meetings help ensure that information and technical details are not lost due to translation.
- All team members should have ownership for the project. Having a common goal motivated everyone to work through issues.
- Language and cultural differences are not difficult to overcome if the right personnel are selected to participate. Patience is a virtue.

8 References

[1] Air Products CorpNews, Air Products and Lu'An Clean Energy Company to form $1.3 Billion Joint Venture, Published September 10, 2017, Accessed on January 14, 2018.

[2] Air Products CorpNews, Air Products' New Milestone – Lu'an Train 1 Syngas Facility Onstream, Published September 9, 2018, Accessed on January 14, 2018.

[3] Basis of Design, Lu'An CTL Project, SHELL PROJECTS & TECHNOLOGY, GLOBAL SOLUTIONS UPSTREAM, August 14, 2013.

[4] NFPA 652 – Standard on the Fundamentals of Combustible Dust, 2019 Edition, National Fire Protection Agency.

[5] NFPA 654 – Standard for the Prevention of Fire and Dust Explosions from the Manufacturing, Processing, and Handling of Combustible Particulate Solids, 2017 Edition, National Fire Protection Agency.

[6] GB15577 - Safety Regulations for Dust Explosion Prevention and Protection, 2018.

[7] NFPA 61 - Standard for the Prevention of Fires and Dust Explosions in Agricultural and Food Processing Facilities, 2017 Edition, National Fire Protection Agency.

[8] NFPA 484 - Standard for Combustible Metals, 2019 Edition, National Fire Protection Agency.

[9] NFPA 655 - Standard for Prevention of Sulfur Fires and Explosions, 2017 Edition, National Fire Protection Agency.

[10] NFPA 664 - Standard for the Prevention of Fires and Explosions in Wood Processing and Woodworking Facilities, 2017 Edition, National Fire Protection Agency.

[11] Google Translate, Commercially Available Translation App.

[12] PHAWorks, Commercially Available Software Program for Safety Reviews, PrimaTech.

552169 Research on Leak Detection of Ethylene Spherical Tank Based on Convolution Neural Network

The leakage of ethylene spherical tank will pose a serious threat to safety production. However, it is difficult to detect the leakage in the initial stage. This paper studied on a method of automatic leakage detection using convolution neural network. It built a recognition model based on convolution neural network with the scene images of normal and leakage conditions from different angles. The model used maximum pooling method and ReLU activation function for optimize the network performance. The test results show that the proposed method and model can realize the automatic leakage detection effectively of ethylene spherical tank.

550776 Application of Safety Management Information System in Hsse Management System of Sinopec

During the construction and application of HSSE management system, there are some problems, such as low efficiency of operation, ununified standard implementation and low degree of control. The information technology based on HSSE management system is able to solve the typical problems in current implementation and operation of the system. This article investigated the application of information system in various energy and chemical enterprises, introduced the design structure, service functions of the Safety Management Information System and its role in HSSE management. The design method of operation process is illustrated through the risk control and hazard indentification, with the establishment of this system, the operation effectiveness of HSSE management system can be improved.

Comparación de los resultados de los análisis de riesgo y consecuencias en sistemas de transporte de refinados y GLP

Carlos Humberto Nieves
CENIT Transporte y Logística de Hidrocarburos
Cll 113 #7-80 Piso 13 Bogotá – Colombia
carlos.nieves@cenit-transporte.com

Maria Isabel Montoya
CENIT Transporte y Logística de Hidrocarburos
Cll 113 #7-80 Piso 13 Bogotá – Colombia
maria.montoya@cenit-transporte.com

Daniel Mauricio Moncada
CENIT Transporte y Logística de Hidrocarburos
Cll 113 #7-80 Piso 13 Bogotá – Colombia
daniel.moncada@cenit-transporte.com

Julian Mauricio Florez Rodriguez
CENIT Transporte Y Logística De Hidrocarburos
Calle113 # 7 - 80 Torre, AR Calle 113 #7-80, Colombia, Bogotá
jmflorezr@unal.edu.co

Prepared for Presentation at
American Institute of Chemical Engineers
2019 Spring Meeting and 15th Global Congress on Process Safety
New Orleans, LA
March 31 – April 3, 2019

Comparación de los resultados de los análisis de riesgo y consecuencias en sistemas de transporte de refinados y GLP

Carlos Humberto Nieves
CENIT Transporte y Logística de Hidrocarburos
Cll 113 #7-80 Piso 13 Bogotá – Colombia
carlos.nieves@cenit-transporte.com

Maria Isabel Montoya
CENIT Transporte y Logística de Hidrocarburos

Daniel Mauricio Moncada
CENIT Transporte y Logística de Hidrocarburos

Julian Mauricio Florez Rodriguez
CENIT Transporte Y Logística De Hidrocarburos

Keywords: Transporte GLP y Derivados, HCA, Consecuencias, Rutas de Derrame

Abstract

El transporte de refinados por ductos es uno de los mayores desafíos de la industria de hidrocarburos en regiones con topografía compleja, como Colombia, donde los ductos deben atravesar diversas regiones montañosas y accidentadas, lo que además de incrementar la complejidad asociada a los peligros geotécnicos, se suma a las numerosas posibilidades de rutas de derrame aumentando así la posibilidad de alcanzar centros poblados y generar graves consecuencias.

En Colombia una parte importante de la infraestructura de transporte de hidrocarburos atraviesa la región Andina, y dado que el consumo de productos refinados ocurre como una prioridad en los centros urbanos, la infraestructura recorre la Cordillera de los Andes, que forma la región más poblada de Colombia, incluidas las ciudades más importantes. Debido a las altas consecuencias que una falla en la tubería en esta región tendría directa o indirectamente a través de la ruta del derrame; es necesario valorar el riesgo incluyendo tanto las consecuencias directas como indirectas.

Es así como, la estimación de consecuencias indirectas en el caso de transporte de refinados muy inflamables como la gasolina o la nafta, se ha convertido en un reto; ya que no solo se debe incluir la estimación de la ruta de derrame si no los posibles sucesos finales que pueden tener lugar en la misma como un incendio o una llamarada. La inclusión de este tipo de consecuencias en la estimación final del nivel de riesgo puede llegar a generar resultados iguales e incluso mayores a los del transporte de GLP. El propósito de este documento es realizar una comparación de los resultados de riesgo para un sistema que transporta diferentes productos y analizar las diferentes situaciones encontradas con respecto a los niveles de consecuencias directas, indirectas y riesgo.

1 Introducción

La industria es uno de los sectores más importantes para el desarrollo económico y social de cualquier país. El sector industrial facilita la modernización de los países al recibir transferencia de tecnología y buenas prácticas en procesos productivos, y es, en general, uno de los sectores con mayor capacidad de inversión, de innovación y de generación de empleo, a la vez que es también uno de los principales actores en la mejora de la competitividad de los países. No obstante, así se obtengan unas ventajas también se derivan una serie de riesgos "como los denominados accidentes mayores", los cuales, a raíz de una pérdida de contención, ya sea a causa de un mal diseño, mantenimiento, eventos naturales o de terceros, pueden generar unas consecuencias no deseadas sobre las personas, el medio ambiente y/o los bienes. Ejemplo de este tipo de riesgos son los incendios, explosiones o dispersiones de gases tóxicos, que como en el caso de Bhopal cobraron la vida de alrededor de 2500 personas. Es por ello por lo que es necesario trabajar en pro de la seguridad de dichas instalaciones y del territorio en donde las mismas se encuentran.

En los últimos años, el gobierno colombiano ha reforzado su legislación con la expedición de la Ley 1523 de 2012 "Por la cual se adopta la política nacional de gestión del riesgo de desastres y se establece el Sistema Nacional de Gestión del Riesgo de Desastres y se dictan otras disposiciones" para avanzar hacia un sistema de gestión de riesgo de con un mayor enfoque en la prevención; en este sentido también ha establecido las directrices para la elaboración del plan de gestión de riesgo de desastres de las entidades Públicas y Privadas (decreto 2157 de 2017), donde el 45% de los requerimientos se enfocan en actividades encaminadas al conocimiento del riesgo.

Es así como para las diferentes entidades públicas y privadas, entre ellas CENIT, se vuelve necesario conocer las consecuencias potenciales asociadas a su operación como trasportador de hidrocarburos, entre ellas las asociadas al transporte de productos como GLP, gasolina y nafta. Comúnmente se tiene la concepción de que el transporte de productos como el GLP, dado su alto nivel de inflamabilidad y mayores radios de impacto, genera mayores consecuencias de las que se derivan con productos como la gasolina o la nafta. Sin embargo, esta no es una conjetura muy aceptada en el caso de transporte por ductos, donde la posible existencia de sucesos finales como incendio de piscina tardías y/o llamaradas sobre las rutas de derrame generan consecuencias indirectas que deben ser incluidas.

2 Areas de alto impacto

El transporte de hidrocarburos y sus derivados generan diferentes niveles de riesgo de acuerdo al producto transportado, esto debido a los sucesos finales que se generan, estos sucesos finales tienen diferentes radios de impacto en función del volumen liberado y de las condiciones constructivas de la tubería lo cual son determinantes para la detección de Áreas de Alto Impacto (D.O.T, 2019).

Las áreas de alto impacto determinan las zonas de posible afectación existentes para dicho sistema de transporte, sin embargo, estas afectaciones no tienen la misma proporcionalidad de impacto negativo, en el caso de la población esta tiene una clasificación de acuerdo al Class Location (D.O.T, 2019), generando una ponderación de los efectos negativos que se pueden tener a lo largo del poliducto.

Las simulaciones para el cálculo de radios de impacto potenciales generan mayores radios para productos como el GLP que para derivados como la Nafta y la Gasolina, lo cual conlleva a mayores afectaciones directas a personas por la presencia de sucesos finales como la llamaradas y sobrepresión. Sin embargo, en el caso de la Nafta y la gasolina, la existencia de sucesos finales como incendios de piscina tardías y llamaradas sobre las rutas de derrame que pueden alcanzar centros poblados generan consecuencias indirectas que deben ser incluidas.

Las consecuencias entonces, pueden ser directas e indirectas, en el caso del GLP solo se consideran consecuencias directas producto de los sucesos finales, en el caso de la Gasolina y Nafta se tienen directas e indirectas las cuales en un dato consolidado pueden generar un nivel de consecuencias y riesgo similar al del GLP, esto indicaría que priorizar el transporte de alguno de los productos puede ser indiferente frente al riesgo total que genera la Tubería.

2.1 Caso de estudio

Para el análisis comparativo del riesgo generado por el transporte de GLP versus gasolina o nafta, se utilizó un Poliducto de 370 Kilómetros.

La relación de los radios de afectación de GLP son un 84.3% superiores a los de Gasolina en el poliducto lo cual implicaría un daño potencial mayor en el transporte del GLP.

La operación con GLP implica un aumento del radio de impacto potencial de la Tubería, pero en algunos casos este radio de impacto potencial puede ser similar al generado por la gasolina y la población puede ser afectada de manera similar.

En otras ocasiones el radio de impacto potencial del GLP genera mayores afectaciones, pero las rutas de derrame de la gasolina pueden generar efectos similares los cuales son equipotenciales en las consecuencias a la población o al medio ambiente, como se observa en la figura 1.

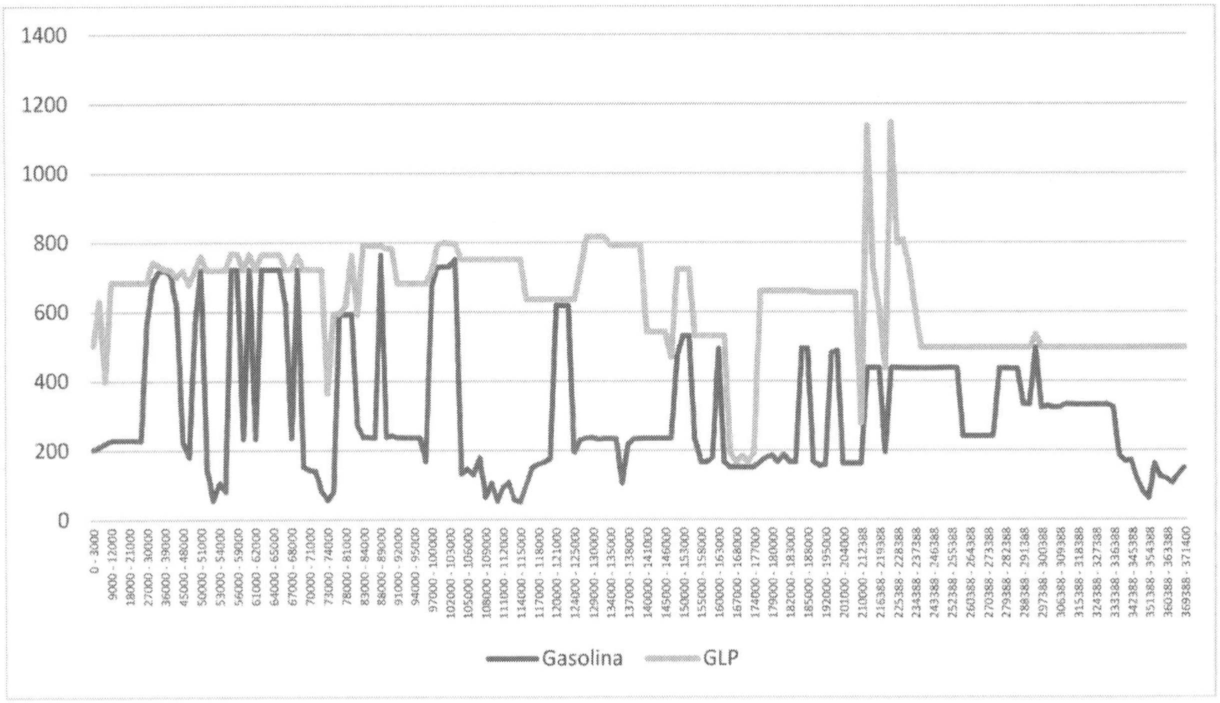

Figura 1. Radios de Impacto Potencial GLP y Gasolina

Se tiene afectaciones a población en 33 centros poblados, de los cuales 2 corresponden a asentamientos dispersos tipo class location II, 27 a Tipo III y Cuatro Tipo IV desde un análisis de transporte de GLP.

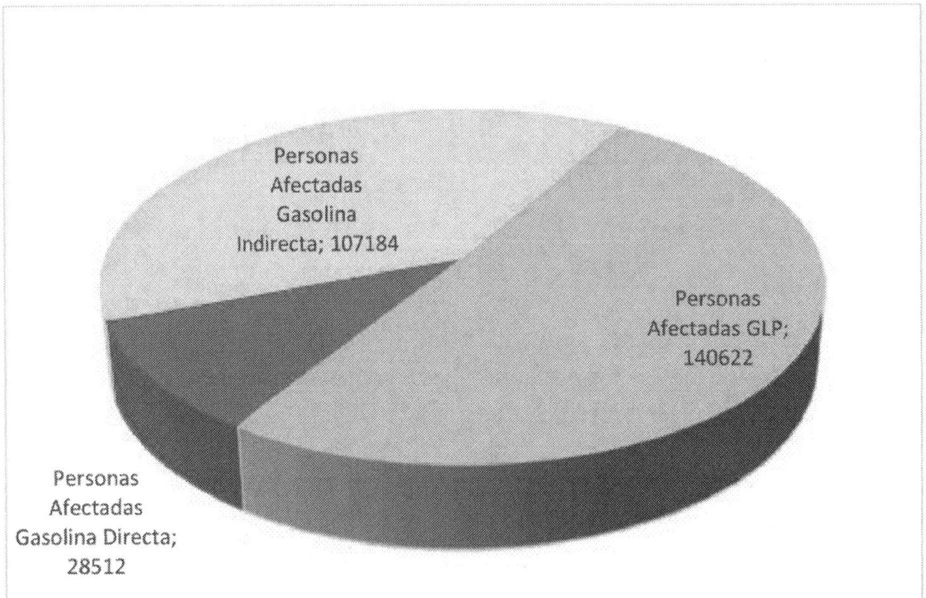

Figura 2. Total, Personas Potencialmente afectadas en el sistema por GLP y Gasolina

De acuerdo a la tabla 1 y al consolidado de la figura 2, los efectos producidos por el GLP y la Gasolina son potencialmente similares en la afectación a personas, lo cual implica que el transporte de GLP no hace variar significativamente la afectación en la población respecto al transporte de gasolina.

Tabla 1. Afectación sobre la población dentro del Radio de Impacto Potencial

Población	RIP Gasolina (mts)	RIP GLP (mts)	GLP	Personas Afectadas GLP	Afectación Gasolina	Personas Afectadas Gasolina Directa	Personas Afectadas Gasolina Indirecta	Tipo Class Location
1	229.5	684.19	Directa	268	Directa	100	223	II
2	561.75	680.12	Directa	384	Directa	384		III
3	617.62	680.12	Directa	286	Directa	286		III
4	238.11	792.95	Directa	1125	Directa	60		III
5	237.04	682.35	Directa	420	Directa	80		III
6	235.67	724.69	Directa	18144	Directa	32	5525	IV
7	233.65	792.6	Directa	1310	Indirecta	32	2189	III
8	723.41	792.6	Directa	400	Directa	400		III
9	723.41	792.6	Directa	1248	Directa			III
10	723.2	792.6	Directa	0	Directa	24		III
11	165.57	660.8	Directa	295	Directa	295		II
12	493.17	660.8	Directa	224	Directa	224		
13	165.5	660.8	Directa	1233	Directa	1233		III
14	192.89	660.8	Directa	9977	Directa	1354	27009	III
15	165.57	660.8	Directa	21204	Directa	4896	6210	IV
16	154.3	656.22	Directa	2300	Directa	2300	600	III
17	161.58	656.22	Directa	21261	Indirecta	28	41466	III
18	437.55	1137.54	Directa	836	Directa	380		III
19	436.74	1137.54	Directa	2496		0		III
20	437.6	726.93	Directa	988	Directa	128		III
21	241.14	602.15	Directa	910	Directa	637		III
22	240.94	602.15	Directa	52	Directa	30		III
23	436.42	1147.77	Directa	676	Directa	252		III
24	435.83	1147.77	Directa	24516	Directa	2104		IV
25	434.96	1147.77	Directa	964		496		III
26	434.96	1147.77	Directa	60	Directa	16		III
27	534	798.64	Directa	9111	Directa	8241	23517	III
28	332.05	726.75	Directa	7232	Directa	4192		IV
29	167.78	602.12	Directa	1032		60		III
30	170.57	495.73	Directa	1160		0		III
31	160.85	495.73	Directa	201	Directa	16		III
32	117.65	495.73	Directa	465		32	445	III
33	148.05	495.73	Directa	9844	Directa	200		III

Podemos focalizar el análisis en cuatro grupos para las 33 poblaciones involucradas (ver figura 3), el primero desde la Población 2 hasta la 4, donde los efectos son similares debido a la distancia a la que se encuentran las poblaciones de la tubería y el radio de impacto potencial RIP toca parcialmente los asentamiento nucleados que tienen baja densidad de población, el segundo grupo corresponde a las poblaciones 8 y 10 donde la distancia del radio de impacto potencial de la gasolina es muy cercana a la del GLP y tener transporte de GLP como en el caso de la población 8 no le afecta el nivel de consecuencias, el tercer grupo corresponde a la población 14 hasta la 17, donde se tiene un radio mayor de GLP pero los efectos sobre la población son similares debido a las afectaciones indirecta generadas sobre las rutas de derrame y la posibilidad de ignición tardía, el cuarto grupo desde la población 24 hasta la 33 donde radios mayores de impacto potencial del GLP generan aumento en el riesgo de estas poblaciones.

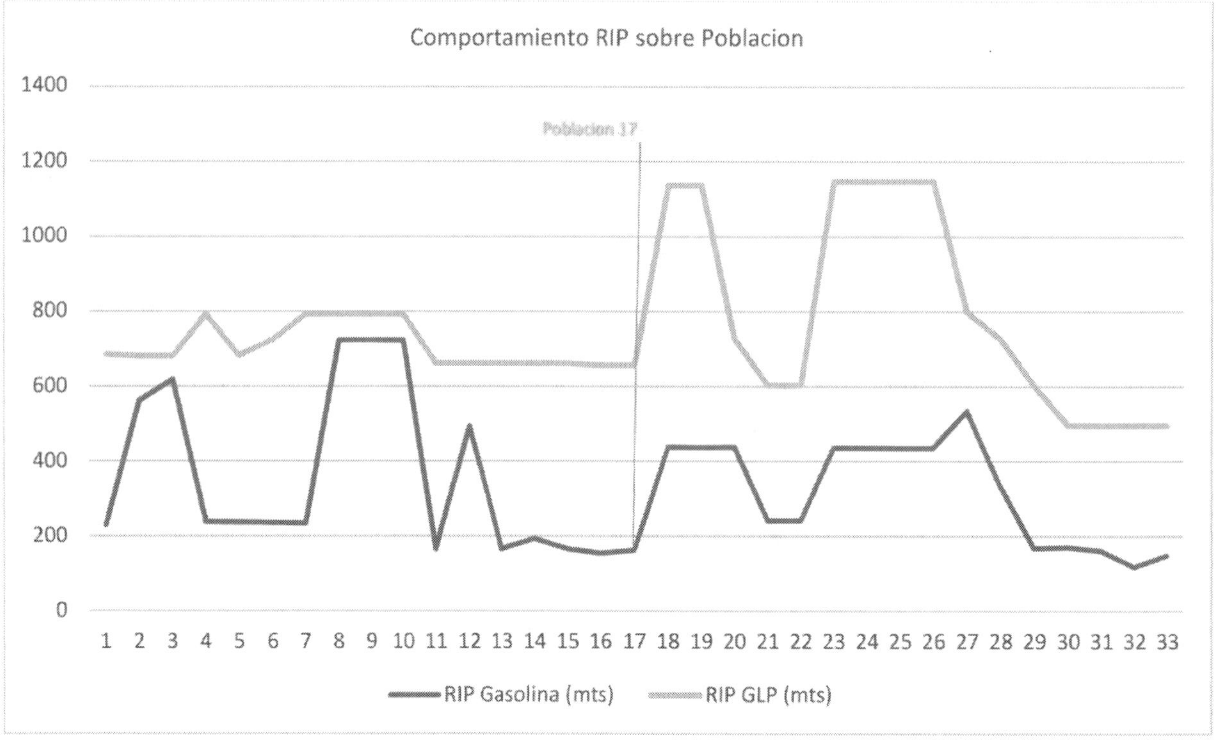

Figura 3. Radios de Impacto Potencial en las Poblaciones

Estos grupos generan patrones de comportamiento que definen la incidencia en el nivel de riesgo del ducto:

1. Baja densidad de población dentro de los Radios de Impacto Potencial del GLP, no reducen su riesgo por la suspensión del transporte de este fluido.
2. Poblaciones en donde los radios de impacto potencial para el GLP y la Gasolina son similares, en este caso la operación con o sin GLP no genera disminución en el Riesgo final del poliducto, esta situación se presenta en zonas de pendientes altas y crestas en el relieve topográfico.

3. Poblaciones en donde los radios de impacto potencial generados por el GLP son mayores que los de la Gasolina, pero la topografía genera rutas de drenaje que pueden llevar el fluido liberado a dichos centros poblados, tienen efectos negativos similares sobre la población. Esto sucede en zonas de pendiente moderada con tendencia a alta (ver figura 4).
4. Poblaciones en donde los radios de impacto potencial generados por el GLP son mayores que los de la Gasolina y se encuentran en zonas de topografía baja, tienen un mayor riesgo debido al transporte de GLP.

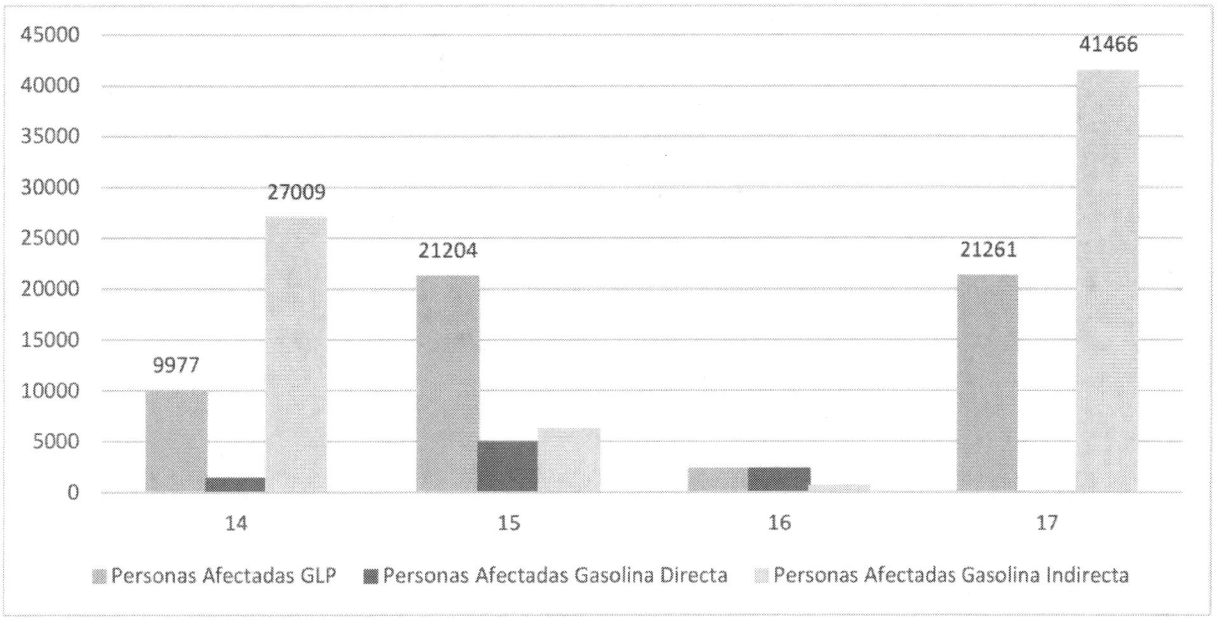

Figura 4. Área de Alto Impacto Tercer grupo de Análisis

Garantizar la integridad de los activos para el transporte de GLP y Gasolina, en este caso la tubería, para cualquiera de los dos productos transportados mediante el mantenimiento preventivo (ITPM), (Center for Chemical Process Safery, 2019) evitara afectaciones de carácter catastrófico sobre la población y el medio ambiente.

2.2 Análisis detallado población 17

El sector de la Población 17 (ver figura 5) debido a su ubicación espacial y la topografía de pendiente moderada de la zona presenta una afectación directa por GLP de 21261 personas, con gasolina tiene 28 personas debido a que sus radios de impacto potenciales no llegan hasta donde se ubica la población, sin embargo, esta cifra se aumenta y sobrepasa la afectación por GLP debido a las consecuencias indirectas generadas por las rutas de derrame que cruzan la población y pueden tener una ignición tardía.

Figura 5. Área de Alto Impacto Población 17

La ignición tardía se puede generar como producto de la perdida de contención en la tubería, y su llegada a una de las fuentes hídricas que pasan por la población sirviendo de vehículo para el transporte. El fluido transportado en un suceso de rotura total del poliducto puede mantener las condiciones de inflamabilidad para que en algún momento durante su recorrido por la fuente superficial de agua, pueda encontrar una fuente de ignición que inicie una llamarada o un incendio de piscina y desencadenar un incendio en todo el recorrido de la fuente de agua dentro del casco urbano con afectaciones significativas a la población.

3 Conclusion

Las afectaciones indirectas a personas generadas por rutas de derrame sumadas a las directas en sistemas de transporte de refinados como la gasolina pueden llagar a tener una magnitud igual o muy cercana a las que genera el transporte de productos como GLP, estos tienden a tener una serie de patrones aplicables como que en zonas de pendientes medias a altas debido a los accidentes topográficos y la generación de rutas de derrame se pueden tener afectaciones de magnitud similar y que las barreras naturales pueden disminuir los efectos del radio de impacto potencial del GLP. Esta situación no es igual en zonas de pendientes bajas en donde no se presentan barreras naturales, los radios de impacto potencial del GLP generan mayores afectaciones a la población.

En zonas de pendiente baja también se generan rutas de derrame las cuales debido a las depresiones tienden a formar piscinas e inundar lagunas, las que alcanzan fuentes superficiales de agua y tienden a llegar a cuerpos de agua de mayor caudal, velocidad y ancho de rio, lo cual facilita a que el fenómeno se minimice.

4 References

[1] M.I. Montoya, C.H. Nieves, J.E. Camargo. Perspectiva de la regulación de accidente mayor en Colombia y el rol de la industria en la gestión de riesgos. Chemical Engineers 8th Latin American Conference on Process Safety Buenos Aires, Argentina September 11-13, 2018.

[2] A. Chaparro, M.I. Montoya, H. J. Palacios, M. Ramirez. Definición de rutas de derrame sobre corrientes hídricas superficiales en derrames de hidrocarburos. 5th Latin American Conference on Process Safety Cartagena de Indias, Colombia. August 12-14, 2013

[3] L.E. Zabala, M.I. Montoya, E. Sarmiento, C. Cuaical. K.L De la Cruz, H. Fajardo. Modelamiento de consecuencias en ductos de transporte de hidrocarburos, bajo la perspectiva de accidente mayor. 5th Latin American Conference on Process Safety Cartagena de Indias, Colombia. August 12-14, 2013

Web resources:

[4] Center for Chemical Process Safery. An AIChE Technologycal Community. Obtenido de https://www.aiche.org/ccps/topics/elements-process-safety/manage-risks/asset-integrity-and-reliability/introduction. Accesado 15 enero, 2019.

[5] D.O.T. Cornell Law School. Obtenido de 49 CFR 192.905 - How does an operator identify a high consequence area?: https://www.law.cornell.edu/cfr/text/49/192.905 Accesado 15 Enero, 2019

[6] D.O.T. Cornell Law School. Obtenido de 49 CFR 192.5 - Class locations: https://www.law.cornell.edu/cfr/text/49/192.5 Accesado 15 Enero, 2019

Additional references not cited:

[7] ASME.ASME B31.8S. Managing System Integrity of Gas.New York, USA.2005.

[8] ASME.PIPELINE TRANSPORTATION SYSTEMS FOR LIQUID HYDROCARBONS AND OTHER LIQUIDS ASME B31.4-2002. New York, USA.2002.

[9] API.API Std 1160. Managing System Integrity for Hazardous Liquid Pipelines. Washington, USA.2008.132 Pags.

Desarrollo de Planes de Emergencia para Proyectos Costa Afuera de Apertura a la Inversión Externa, GOM

Presenter/Principal Author: Yaneira E. Saud, CSP
ERM
840 West Sam Houston Parkway North, Suite 600, Houston, Texas
yaneira.saud@erm.com

Ricardo Lopez, Jose L. Monroy
ERM
ricardo.lopez@erm.com; jose.monroy@erm.com;

Prepared for Presentation at
American Institute of Chemical Engineers
2019 Spring Meeting and 15th Global Congress on Process Safety
New Orleans, LA
March 31 – April 3, 2019

Desarrollo de Planes de Emergencia para Proyectos Costa Afuera de Apertura a la Inversión Externa, GOM

Presenter/Principal Author: Yaneira E. Saud, CSP
ERM
840 West Sam Houston Parkway North, Suite 600, Houston, Texas
yaneira.saud@erm.com

Ricardo Lopez, Jose L. Monroy
ERM

Palabras Clave: plan de respuesta a emergencias, peligros, consecuencias, escenarios, sistema de comando de incidentes, costa-afuera

Resumen

La globalización y demandas cada vez más exigentes de aplicación de mejores prácticas para la gerencia efectiva de riesgos ha requerido una actividad mayor en el área de planes de respuesta a emergencias. Las instituciones financieras promueven estándares según la Corporación International de Financiamiento (IFC por sus siglas en Inglés) para la aprobación de recursos para proyectos de inversión capital. El resultado es un conjunto de requerimientos basados en riesgo, armonizados con las mejores prácticas de planificación para emergencias.

Por ejemplo, México en su programa de Licitación de Bloques para inversión externa ha desarrollado un marco regulatorio que contiene los elementos de regímenes existentes como el Safety Case en Europa y el Sistema de Gerencia de Seguridad y Ambiente para USA costa afuera (SEMS, por sus siglas en Inglés). Por ejemplo, la respuesta a emergencias en el Golfo de México (GOM) debe seguir guías específicas como el Sistema de Comando de Respuesta (ICS, por sus siglas en Inglés) establecidos por el Sistema Nacional de Gestión de Incidentes en USA (NIMS por sus siglas en Inglés). Una situación similar se ha observado en proyectos en Colombia, Venezuela, Surinam, Trinidad y Tobago y otros proyectos en países con salida hacia el Mar Caribe, el Pacifico o el Atlántico.

Por otra parte, la regulación local en México para estudios de impacto ambiental (EIA, por sus siglas en inglés) establecen los requerimientos de identificación y análisis de impactos, incluyendo incidentes mayores como reventón de pozo y derrame de tasa máxima estimada al suelo o cuerpos de agua. Conocer que información extraer simplifica la captación de data requerida y la configuración y planeamiento de respuesta a emergencias cumpliendo requerimientos internacionales de comando, recursos, y comunicaciones puente con las entidades involucradas.

La EIA en su apartado específico de Análisis de Riesgos, incluye tres elementos principales: Identificación de peligros, análisis de consecuencias y la definición de planes de contingencia, los cuales son desarrollados y evaluados como parte del Sistema de Administración de Riesgos (SASISOPA,

Sistema de Administración de Seguridad Industrial, Seguridad Operativa y Protección al Medio Ambiente).

Esta presentación discute estrategias para la extracción de información de los estudios de impacto ambiental, y análisis de escenarios de pérdida de contención para desarrollar planes de emergencia basados en riesgo máximo estimado. También desarrolla todos los elementos de un plan de emergencias completo y efectivo.

1 Introducción

Los planes de respuesta y control de emergencias (RCE) se mantienen como un elemento crítico en el ciclo de vida de las instalaciones costa afuera, por ejemplo: parte integrante de los estudios de impacto ambiental y las medidas de mitigación; la aprobación de fondos de inversión externa, la aprobación de permisos de operación por parte del ente regulador local, la fase de pre-comisionado, arranque y operación de las instalaciones extendiéndose hasta su vida útil y final abandono o desmantelamiento.

La legislación local frecuentemente se ancla en mejores prácticas internacionales y exige los planes RCE para otorgar permisos de exploración y producción "el que"; sin embargo, no proporciona información sobre el contenido y expectativas de los planes RCE, es decir "el cómo" no cuenta con guías de implementación, las cuales quedan a discreción del patrocinador del proyecto o el Operador.

2 Requerimientos de Planes RCE

2.1 Instituciones Financieras y Mejores Prácticas

En el caso de proyectos de apertura a la inversión externa, en general son las instituciones financieras las que promueven un grupo base de estándares para la redacción de planes RCE orientados a la protección de la salud, el medio ambiente, y la seguridad. Como ejemplo de este tipo de requerimientos, podemos citar las guías publicadas por la IFC para la aprobación de recursos para proyectos de inversión capital. El resultado es un conjunto de requerimientos basados en riesgo, armonizados con las mejores prácticas de planificación para emergencias en aguas y tierras compartidas. Según estas guías, cada proyecto debe "...identificar, tan pronto como sea posible, los peligros que un proyecto conlleva para el medio ambiente, la salud y la seguridad, así como otros riesgos asociados al mismo, en el funcionamiento de la instalación o en el ciclo del producto". Las guías también establecen, entre otros temas, la necesidad de comprender la probabilidad de ocurrencia y la gravedad de los riesgos de medio ambiente, salud y seguridad, tomando como base la naturaleza de las actividades a desarrollar, y las posibles consecuencias de un manejo inapropiado de los riesgos.

El paso a seguir en este contexto, es la comparación y armonización de estos requerimientos con los establecidos por la legislación local, y si es del caso, por las políticas corporativas de cada empresa patrocinadora del proyecto. Como complemento final de este ejercicio, se recomienda realizar un análisis de brechas en relación con mejores prácticas globales, tales como las establecidas por el Instituto Americano de Petróleos (API 1174) 0, la Organización International para la Estandarización (ISO 22320) [2] y la Asociación Regional de Empresas del Sector Petróleo, Gas, Biocombustibles en Latinoamérica y el Caribe (ARPEL) [3] entre otras entidades.

2.2. Empresa patrocinadora del proyecto

La empresa patrocinadora, interesada, o parte de una asociación estratégica es frecuentemente una compañía establecida de Petróleo y Gas con guías técnicas bien establecidas para todos los elementos de un sistema de manejo de riesgos de los procesos, incluyendo planificación y respuesta a emergencias. Algunas compañías cuentan con especialistas en el área de planificación y control de incendios, fugas toxicas y otras emergencias tecnológicas y han desarrollado herramientas prácticas de estimación de cantidad y tipo de producto derramado, cálculo de recursos para enfriamiento o extinción y zonificación de la emergencia por zonas de peligrosidad. En el pasado muchos eventos accidentales han cobrado la vida del personal de primera intervención cercano al área de suceso. Es por eso que identifica la "zona caliente" donde requerimientos especiales de estrategias de acercamiento, protección personal y descontaminación deben ser provistas. Otras empresas o proyectos menos experimentados y que participan en una nueva inversión, deben hacerse de esos recursos y mejores prácticas en la medida que estos estén disponibles en la literatura o en actas de Conferencias donde se aborde el tema.

2.4 Legislación Local: Caso Legislación Mexicana

En el año 2013 se genera en México una reforma constitucional en la que se apertura a la inversión nacional y extranjera la asignación de contratos para las siguientes actividades específicas [4]:

1. Reconocimiento y exploración superficial, y la exploración y extracción de hidrocarburos;

2. Tratamiento, refinación, enajenación, comercialización, transporte y almacenamiento del petróleo;

3. Procesamiento, compresión, licuefacción, descompresión y regasificación, así como el transporte, almacenamiento, distribución, comercialización y expendio al público de gas natural;

4. Transporte, almacenamiento, distribución, comercialización y expendio al público de petrolíferos; y

5. Transporte por ducto y el almacenamiento que se encuentre vinculado a ductos de petroquímicos.

A partir de esta reforma se creó la Agencia Nacional de Seguridad Industrial y de Protección al Medio Ambiente del Sector Hidrocarburos (ASEA) [4]; dicha agencia es un órgano desconcentrado de la Secretaría de Medio Ambiente y Recursos Naturales (SEMARNAT), la cual era hasta ese momento la institución nacional encargada de la gestión y regulación de aspectos ambientales para todos los sectores industriales.

Entre las atribuciones de la ASEA se incluyen: regular, supervisar y sancionar en materia de Seguridad Industrial, Seguridad Operativa y Protección al Medio Ambiente en relación con las actividades del Sector Hidrocarburos (el Sector).

En el tema específico de análisis de riesgo, toda actividad del Sector deberá presentar previo a la instalación, una Evaluación de Impacto Ambiental (EIA) [5] y un Análisis de Riesgo [6]; dichos documentos deben presentarse a evaluación y autorización por parte de la ASEA.
La realización de los análisis de riesgo del Sector de acuerdo a la guía publicada por ASEA para tal fin, incluyen de manera general:

- Identificación de peligros;
- Jerarquización de riesgos;
- Análisis de consecuencias;
- Análisis de vulnerabilidad; y
- Sistemas de seguridad y medidas para administrar los escenarios de riesgo.

También se encuentra en proceso por parte de ASEA un Protocolo de Respuesta a Emergencias en las actividades del sector hidrocarburos [7]. Las disposiciones propuestas definen y establecen medidas técnicas en la formulación de los protocolos para hacer frente a las emergencias o situaciones de riesgo crítico con motivo del desarrollo de las actividades del Sector Hidrocarburos; considerando los escenarios determinados en su análisis de riesgo, así como aquellos que se presenten por motivo de factores externos. Los lineamientos para la elaboración de estos protocolos incluyen:

- Escenarios de riesgo (insumo principal)
- Fenómenos de tipo geológico, hidrometeorológico, sanitarios, y socio-organizativos
- Siniestros externos de incidencia directa o indirecta a la Instalación
- Plan de Atención de Emergencias Interno (PAEI)
- Plan de Atención de Emergencias Externo (PAEE)
- Programas de capacitación, entrenamiento, simulacros y mantenimiento de equipos de Emergencia;
- Plan de acción de atención a recomendaciones derivadas del Análisis de Riesgo para el Sector Hidrocarburos (ARSH)
- Listas de verificación de acciones para la atención de la Emergencia

Los resultados de dicho análisis deberán ser tomados en cuenta para la implementación del Sistema de Administración (SASISOPA), también evaluado y autorizado por ASEA. EL SASISOPA deberá estar conformado por dieciocho elementos, incluyendo de manera específica el elemento XIII. Preparación y Respuesta a Emergencias, en dicho elemento se debe documentar de manera general [8][9]:

1. Aquellos mecanismos para identificar situaciones potenciales de emergencia como resultado de su proceso de evaluación de impactos ambientales y riesgos, tales como fugas, derrames, incendio y/o explosión, de las sustancias peligrosas y actividades de alto riesgo; y
2. Cuantificar escenarios de riesgo aplicando metodologías probadas de simulación para estimar las consecuencias e identificar zonas de riesgo para las instalaciones, el personal, la población y el medio ambiente.

Y de manera particular:

1. Incluir elementos de respuesta para cada situación potencial de emergencia identificada, disponiendo de los recursos necesarios para controlar o hacer frente al evento, tales como recursos financieros y humanos preparados, capacitados y, en su caso, certificados, servicios médicos, equipamiento, sistema contra incendio, sistemas de contención, rutas de evacuación, equipo de protección personal y medios de comunicación, entre otros.

2. Disponer en las instalaciones de una organización con responsabilidad y autoridad definida para responder y controlar una emergencia integrada por quienes tienen la responsabilidad de operar y

mantener las instalaciones y acorde a su autoridad y responsabilidad operativa, la cual debe estar capacitada y entrenada para actuar de manera segura y controlada ante un evento de emergencia.

3. Establecer un centro de operación a emergencias el cual debe estar en una zona segura y dotada de los medios y recursos necesarios, tales como medios de comunicación, equipos de cómputo, servicios de internet, simuladores, planos y diagramas de la instalación, hojas de datos de seguridad; en general, información de la seguridad del proceso, así como el propio plan de respuesta a emergencias, planes de contingencia por explosiones, fugas y derrames de sustancias peligrosas e Hidrocarburos, donde la organización realice sus actividades de coordinación, comunicación y toma de decisiones para la respuesta y control de una emergencia.

4. Facilitar el resguardo de instalaciones después de una emergencia para que las posibles evidencias no sean alteradas, de tal manera que permita disponer de elementos para llevar a cabo la investigación y análisis del accidente y evaluar las condiciones de integridad de las instalaciones.

5. Formular programas de simulacros de respuesta a emergencias y evacuaciones para ejecutarse periódicamente con la participación de todo el personal involucrado o afectado potencialmente por la emergencia. Cada simulacro debe ser evaluado por personal competente para identificar y corregir debilidades. Se deben mantener procedimientos y medios para establecer la coordinación con las autoridades competentes en la atención a emergencias.

6. Garantizar la adquisición y disponibilidad de equipos, materiales y sistemas para la atención de emergencias certificados y su inclusión dentro de los programas de mantenimiento.

7. Considerar los requerimientos de estándares nacionales e internacionales para el diseño, construcción, selección, pruebas y operación de equipos de respuesta a emergencia y contingencias ambientales.

2.5 Legislación Regional: Caso Golfo de México

Mientras que un evento local cuya respuesta está dentro de los límites de la instalación se conoce como un TIER 1, el mecanismo general del Plan Nacional de Contingencia (PNC) en México, es el equivalente a un TIER 2; mediante el cual se da comunicación a la Secretaría de Marina (SEMAR) de la presencia de un evento con potencial afectación más allá de los límites de las instalaciones de perforación y/o producción de pozos. Así mismo como a las unidades de Protección Civil en las ciudades costeras del Golfo de México (principalmente de los estados de Veracruz, Tabasco y Campeche).

Dichas instituciones, son las encargadas en primera instancia de proveer de los servicios y transporte necesario para minimizar los efectos de una emergencia. Sin embargo, los recursos tanto humanos como de equipamiento pueden ser limitando dependiendo del caso de emergencia (p.e. descontrol de pozo), por lo que en ese caso se activaría el TIER 3, mediante el cual empresas particulares encargadas de la atención de emergencias, brindarían soporte tanto a las autoridades mexicanas como al operador, movilizando equipo en la mayoría de las ocasiones desde las costas del Estado de Texas en los Estados Unidos.

En el caso concreto de notificaciones de emergencias dentro de la zona de permiso/exclusión de operaciones de perforación o producción de hidrocarburos, la empresa debe utilizar el "Sistema de Información de Incidentes y Accidentes (SIIA)" del gobierno de México, para informarle de manera detallada a la "Agencia Nacional de Seguridad Industrial y de Protección del Medio Ambiente del Sector de Hidrocarburos (ASEA)" de la ocurrencia de incidentes/accidentes, y las acciones que se tomaron.

3 Fuentes de Información para un Plan RCE

Los planes RCE cuentan con diferentes fuentes de información, las cuales es necesario identificar y extraer para proporcionar los elementos que serán los puntos de partida en la definición de los escenarios potenciales de riesgo. A continuación, se presentan las fuentes de información y una breve descripción de los elementos.

3.1 RCE en el Ciclo de Vida de una Instalación

Los proyectos de inversión de capital y apertura a la inversión externa típicamente se rigen por mejores prácticas de gerencia de proyectos con fases de desarrollo e hitos de decisión. Desarrollar plan y preparación RCE requiere estar presente, interactuar y extraer información relevante de ingeniería en cada fase de proyecto. Esto se ilustra en la Figura 1.

Figura 1. ERP en Fases de Proyecto

3.2 Bases de Diseño

Las bases de diseño y las filosofías de aislamiento, contención, detección y respuesta ante un escape de producto peligroso o contaminante son muy importantes para entender el proceso y estimar/describir escenarios probables de emergencia en caso de que los sistemas preventivos lleguen a fallar y se produzca un evento tope de pérdida de contención primaria o secundaria.

- Sistema instrumentado de Seguridad (SIS, por sus siglas en inglés) / Sistema de paro de emergencia (ESD, por sus siglas en inglés)
    - Sistema automático de protección a las instalaciones por condiciones anormales de proceso como sobrepresiones y manual mediante acción de un operador al detectarse una condición anormal de proceso, llevando a las unidades de proceso a una condición segura sea por la interrupción o el aislamiento.
- Sistema de detección de gas y fuego (F&G, por sus siglas en inglés)
    - Sensores de Detección de fuego

  - Sensores de Detección de mezcla de gas combustible.
  - Alarmas audibles y visuales
  - Estaciones manuales

3.3 Estudios de Impacto Ambiental (EIA por sus siglas en Inglés)

Los EIA son contratados típicamente durante la etapa de ingeniería básica, cuando se tiene suficiente información sobre el proyecto y sus impactos potenciales al ambiente y las personas.

- Identificación de peligros y jerarquización de riesgos – escenarios de fugas, derrames, incendio y/o explosión de sustancias peligrosas y actividades de alto riesgo;
- Análisis de consecuencias - Determinación de radios potenciales de afectación mediante la aplicación de modelos matemáticos de simulación considerando los casos más probables y los peores casos.
  - Para el caso de pozos costa afuera, simulación de radiación térmica y sobrepresión con el propósito de definir ubicación de un pozo de alivio por descontrol de pozo. Y consideración de dos tiempos estimados de respuesta, uno para la movilización de un "capping stack" o tecnología similar y el necesario para la perforación de un pozo de alivio; con la finalidad de estimar la magnitud de la fuga al mar
- Representación de los resultados de radios potenciales de afectación en planos a escala; y
- Análisis de vulnerabilidad con la identificación de áreas de interés (personas, población, medio ambiente e infraestructura).
- Siguiendo la regulación local en México y Latinoamérica, el EIA debe contar con una sección dedicada a planificación y respuesta a emergencias, incluyendo alianzas con terceros acreditados para la respuesta táctica de derrames en cuerpos de agua. Las labores de recolección, tratamiento, transporte de un derrame de hidrocarburo son complejas y onerosas necesitado empresas y personal calificado, así como acuerdos de ayuda mutua.

3.4 Identificación de Peligros (HAZID por sus siglas en Inglés)

HAZID es la herramienta inicial para identificar y organizar en un Registro de Riesgos los peligros de una instalación costa afuera, el formato es tabular y permite documentar los peligros que aplican a las fases de diseño, construcción, comisión o puesta en marcha, operaciones/mantenimiento y abandono/desmantelamiento. Cada peligro tiene asociado una causa, consecuencia, medidas de prevención y salvaguardas intrínsecas al diseño, medidas de mitigación previstas; y la contribución de estudios formales de riesgo, y sistemas administrativos de manejo de riesgos tomando en cuenta el factor humano. Con frecuencia un HAZID se complementa con diagramas causa efecto tipo "Bowtie" para ilustrar el proceso de evaluación de riesgos desde causas o eventos iniciadores que desencadenan un evento tope (p.e. pérdida de contención primaria o secundaria) y que últimamente resultan en consecuencias con detrimento a personas, el ambiente, la reputación y loas activos de las empresas e interesados involucrados en el proyecto. Los "bowties" permiten distinguir con claridad aquellos elementos críticos de seguridad para prevenir o mitigar el evento tope, y los responsables de su funcionamiento de acuerdo a sus especificaciones de diseño. Es común aplicar matrices cualitativas de riesgo para analizar con un grupo de expertos y conocedores del proyecto el riesgo resultante de combinar la consecuencia potencial y la posibilidad o frecuencia de ocurrencia de eventos tope o escenarios.

3.5 Análisis de Riesgos del Proceso (PHA por sus siglas en Inglés)

Los PHA y en especial aquellos sistemáticos y estructurados como el HAZOP (Hazards and Operability por su siglas en Inglés) son un importante fuente de información para los escenarios de RCE. Es imposible planificar asertivamente para un evento de peligrosidad, posibilidad de ocurrencia y magnitud desconocida. Para el efecto, la guía CCPS "Guidelines for Technical Planning for on-site Emergencies" [10] presenta un listado de técnicas y referencias para adelantar la evaluación de riesgos, usando análisis de peligros en procesos orientados a la planeación de respuesta a emergencias. Esta guía también describe en detalle la priorización de incidentes de planificación de emergencia para evaluación de consecuencias.

Los escenarios de emergencia se conforman en un PHA/HAZOP, desarrollándose progresivamente a partir de eventos iniciadores aunados a la falla de sistemas preventivos y de mitigación en virtud de la máxima consecuencia estimada.
Los sistemas preventivos y de mitigación pueden estar conformados por hardware (pasivo, activo o continuo) respuesta humana o una combinación de estas. Estos sistemas también se conocen como elementos críticos de seguridad (SCE, por sus siglas en inglés) o barreras legítimas y se definen como aquellos elementos del equipo o las estructuras cuya falla podría provocar un accidente grave o cuyo propósito es evitar que ocurra o limitar las consecuencias de un incidente importante.

En el caso exploración y producción los SCE adaptados a los pozos son componentes capaces de prevenir o mitigar las consecuencias de un evento de control de pozos con una posible pérdida de contención. Esta definición adaptada se vincula con la funcionalidad del pozo de contener o aislarlo y, por lo tanto, equivale a las barreras del pozo.

Es precisamente este enfoque de barreras, el que se toma en cuenta dentro de la evaluación HAZOP dentro de la lista de palabras guía de la metodología. Así la desviación a alguna de las barreras, generalmente destinadas a contener y aislar, en caso de un potencial accidente. Podría generar consecuencias indeseables a ser documentadas.

Y posteriormente haciendo un complemento del análisis con alguna herramienta cualitativa de determinación de riesgo, como una matriz de jerarquización; será posible determinar el nivel potencial de riesgo (alto, medio o bajo) de los peligrosos mayores.

3.6 Análisis de Consecuencias y Análisis Cuantitativo de Riesgos (QRA por sus siglas en Inglés)

Con el soporte de una estimación numérica es posible determinar la probabilidad y magnitud de que un evento se presente con los denominados contornos de riesgo, de aquellos escenarios previamente identificados.

La estimación de consecuencias es fundamental, pues después de desastres como el Deepwater Horizon (2011), las empresa patrocinadora, interesada, o parte de una asociación estratégica; tienden a evaluar los escenarios de RCE dando más énfasis a la preparación ante una máxima consecuencia o pérdida estimada y menos énfasis en los sistemas de prevención y control que resultan en una frecuencia muy baja de ocurrencia.

En muchos casos, si se cuenta con los recursos es deseable completar un QRA con determinación más precisa de frecuencias de escenarios, esto habilita afinar un enfoque es basado en riesgo y permite a la gerencia de proyecto tener elementos de decisión para aprobar mayores recursos para respuesta, o estudiar la incorporación de medidas de ingeniería para la reducción del riesgo.

4 Tabla de Contenido de un ERP Basado en Mejor Prácticas

Un plan de RCE efectivo debe incluir, como mínimo la siguiente información que representa en general una excelente practica para este tipo de documentos:

1 Introducción: Objetivo, alcance, descripción de la instalación
2 Sistema de Comando de Incidentes (ICS por sus siglas en inglés): Estructura, definición y asignación de funciones y responsabilidades, localización del puesto de mando con el equipo de respuesta a disposición.
3 Centro Unificado de Comando (UCS por sus siglas en inglés), en caso de necesitarse coordinación multi-organizacional: Estructura, definición y asignación de funciones y responsabilidades, localización del puesto de mando.
4 Escenarios de accidentes: Escenarios de accidentes probables que puedan afectar la salud, el medio ambiente y la instalación, y que incluyan factores de riesgo para cada uno, en términos de probabilidad de ocurrencia y nivel de consecuencias esperadas
5 Niveles de emergencia: Clasificación y requisitos específicos por nivel
6 Plan de acción para cada tipo de accidente: Incluye como mínimo listas detalladas de personal e instituciones a contactar y notificar, lista de recursos disponibles para atender la emergencia, listas de chequeo de acciones de respuesta, formularios suministrados por el ICS, listas con tácticas detalladas de respuesta, fichas de datos de seguridad de los materiales (MSDS por sus siglas en inglés)
7 Disponibilidad de recursos: Recursos disponibles para atender la emergencia, cómo accederlos y descripción de actividades en curso para mantenerlos en óptimo estado (por ejemplo, equipos y herramientas)
8 Entrenamiento y competencias: Entrenamiento requerido para cada función de los integrantes del sistema de respuesta, frecuencia del entrenamiento
9 Simulacros y ejercicios de práctica: Incluye descripción de los ejercicios, frecuencias de los mismos, quién debe asistir, y proceso de evaluación y crítica al terminar los simulacros de respuesta de emergencia
10 Reuniones y actividades de planeación: Incluye relaciones con la comunidad y compromisos de soporte externos
11 Revisión de gerencia
12 Documentación de soporte: Incluye descripción de alarmas, rutas de evacuación y escape, diagramas de localización de equipo de emergencia, certificación de operadores de limpieza al terminar la emergencia, otros requerimientos regulatorios

5 Etapas de un Plan de Respuesta a Emergencia

Las diferentes etapas asociadas con la respuesta a incidentes y emergencias están ilustradas en la Figura 3.

Figura 3. Etapas de un Plan RCE

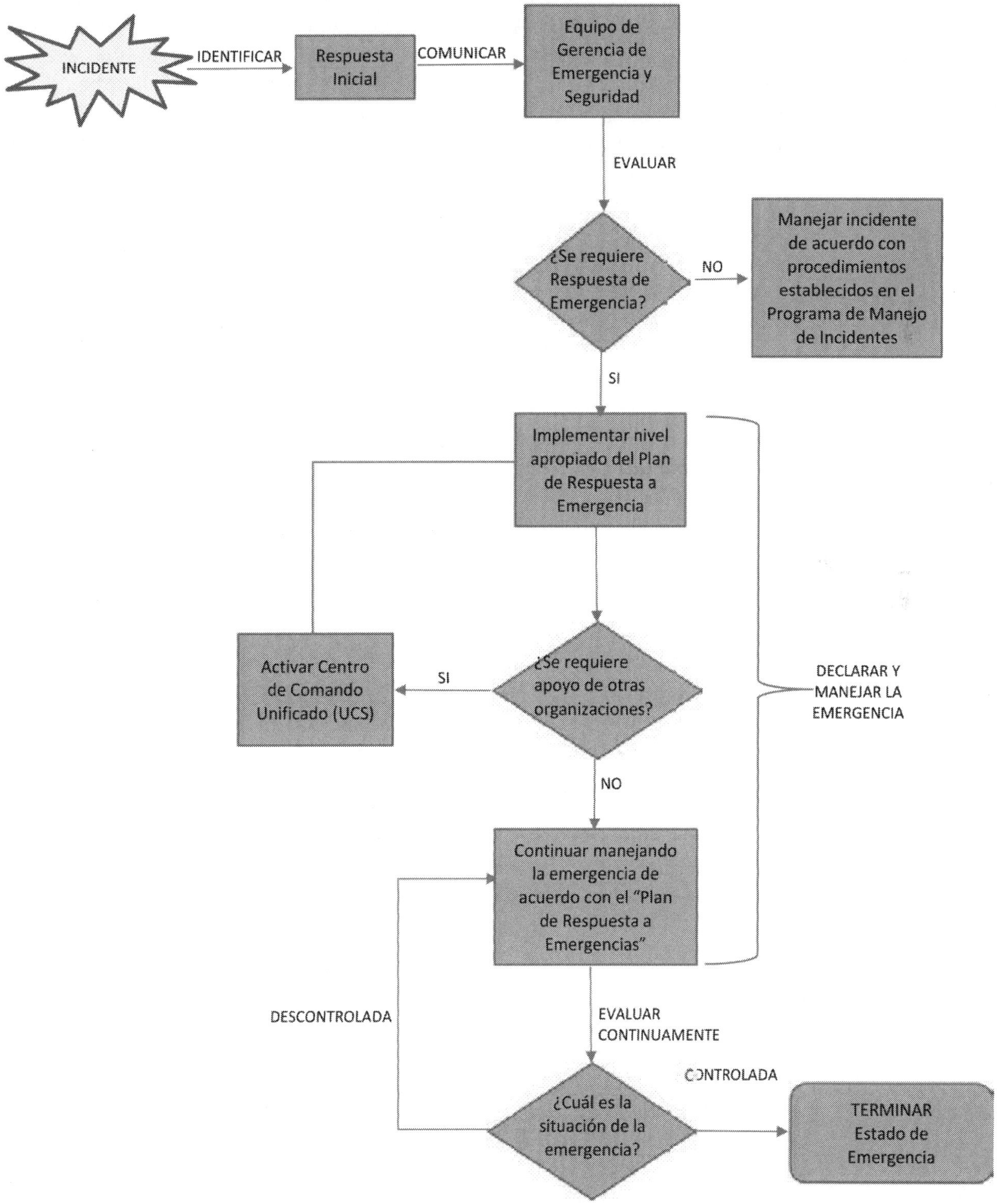

6 Implementación y Accesibilidad del plan RCE

Una parte muy importante de la implementación de un plan RCE se relaciona con su publicación y accesibilidad por parte de todo el personal involucrado en el manejo de una emergencia. Esto se logra programando entrenamientos periódicos en el contenido y uso del documento y los recursos disponibles para atender una emergencia, lo cual hace parte del plan en como tal. Para el efecto, se deben tener copias del plan disponibles en cualquier momento que se necesiten y sin depender de servicios eléctricos lo cuales pueden no estar disponibles durante la emergencia y etapas de control. El enfoque tradicional es imprimir varias copias en papel y tenerlas almacenadas en sitios estratégicos de la instalación (p.e. Sala para Centro Unificado de Comando), así como tener versiones digitales en formato PDF en la red corporativa de la empresa. Sin embargo, esta solución no es la ideal por razones obvias, ya que es bastante impráctico imprimir muchas copias del mismo documento que pueden quedar desactualizadas en cualquier momento, sin contar con la necesidad de acceder físicamente a sus sitios de archivo durante una emergencia. Algunos problemas similares se pueden observar con la opción de tener copias del documento en formato PDF, ya sea en computadores aislados o en la red interna de la empresa.
Una solución alterna que se propone es convertir el documento principal a formato HTML fuera de línea, con enlaces (hipervínculos) a otras secciones, diagramas y otros documentos soporte o de referencia citados por el RCE. El documento principal (en formato HTML) y los archivos de soporte (en formato HTML, PDF, etc.) se pueden almacenar en una sola carpeta en tabletas digitales tipo Android o iOS (Apple), las cuales se entregarían a cada uno de los potencialmente involucrados en un plan de respuesta de emergencia. Estos archivos son de muy fácil actualización cuando se requiera, y las tabletas se pueden tener muy cerca del lugar de trabajo en caso de necesidad.

7 Conclusión

Los planes de RCE son una parte fundamental para el análisis y mitigación de impactos de los proyectos de inversión de capital, en especial aquellos de apertura a la inversión externa. La regulación no ofrece guías de desarrollo, pero claramente ancla el desempeño de RCE a mejores prácticas en la materia a nivel global. La ocurrencia de eventos de accidentes en GOM han incorporado muchas lecciones aprendidas en esta área y la literatura y sesiones tecnológicas compartidas en la industria consolidan los aspectos de RCE que deben tomarse en cuenta de las más tempranas etapas del diseño. Un plan RCE bien desarrollado y mantenido en condición "siempre verde" de actualización continua es muy efectivo para limitar las consecuencias de un evento de pérdida de contención. Sin embargo, un plan RCE deficiente se torna en un pasivo que aflora indefectiblemente en las inspecciones de autoridades competentes y las auditorias de sistemas de manejo de riesgos de los procesos para el Golfo de México; p.e. SASISOPA en México y SEMS (Sistema de Manejo de Seguridad y Ambiente por sus siglas in Inglés) en Estados Unidos.

8 Referencias

[1] Recommended Practice for Onshore Hazardous Liquid Pipeline Emergency Preparedness and Response, API RP 1174, American Petroleum Institute, First Edition, December 2015

[2] Security and resilience - Emergency management- Guidelines for incident management, ISO 22320, International Organization for Standardization, Second Edition, 2018-11

[3] Developing Emergency Response Programs: Natural, Man-made and Operational Risks, Regional Association of Oil and Natural Gas Companies in Latin America and the Caribbean (ARPEL), April 2004

[4] Ley de la Agencia Nacional de Seguridad Industrial y de Protección al Medio Ambiente del Sector Hidrocarburos. Diario Oficial de la Federación (DOF) de México. 11 de agosto de 2014 http://www.diputados.gob.mx/LeyesBiblio/pdf/LANSI_110814.pdf [Accedido 6 Feb 2019]

[5] Ley General del Equilibrio Ecológico y la Protección al Ambiente de México, 31 de octubre de 2014 http://biblioteca.semarnat.gob.mx/janium/Documentos/Ciga/agenda/DOFsr/148.pdf [Accedido 6 Feb 2019]

[6] Guía para la Elaboración del Análisis de Riesgo para el Sector Hidrocarburos, SEMARNAT - ASEA, 2018 https://www.gob.mx/cms/uploads/attachment/file/343905/GUIA_ANALISIS_DE_RIESGO.pdf [Accedido 6 Feb 2019]

[7] Lineamientos para la Elaboración de los Protocolos de Respuesta a Emergencias en las Actividades del Sector Hidrocarburos, Anteproyecto de ASEA, 2019 http://www.cofemersimir.gob.mx/portales/resumen/46031 [Accedido 15 Feb 2019]

[8] Disposiciones administrativas de carácter general que establecen los Lineamientos para la conformación, implementación y autorización de los Sistemas de Administración de Seguridad Industrial, Seguridad Operativa y Protección al Medio Ambiente aplicables a las actividades del Sector Hidrocarburos que se indican. Diario Oficial de la Federación (DOF) de México. 13 de mayo de 2016.

[9] Disposiciones administrativas de carácter general que establecen los Lineamientos en materia de seguridad industrial, seguridad operativa y protección al medio ambiente para realizar las actividades de Exploración y Extracción de Hidrocarburos en Yacimientos No Convencionales en tierra. Diario Oficial de la Federación (DOF) de México. 16 de marzo de 2017

[10] Guidelines for Technical Planning for on-site Emergencies, Center for Chemical Process Safety (CCPS) – AIChE, First Edition, 1995

Enfoque multi-amenaza para abordar peligros Natech

Adriana María Mesa Gómez
Departamento de Ingeniería Química, Universidad de los Andes
Carrera 1E #19A-40, Bogotá (CO)
am.mesa@uniandes.edu.co

Joaquim Casal
Departamento de Ingeniería Química, Universitat Politècnica de Catalunya
Campus Diagonal-Besòs. Avinguda Eduard Maristany, 10-14, Edifici I, Planta 5a. 08019 Barcelona, España.
joaquim.casal@upc.edu

Felipe Muñoz Giraldo
Departamento de Ingeniería Química, Universidad de los Andes
Carrera 1E #19A-40, Bogotá (CO)
fmunoz@uniandes.edu.co

Prepared for Presentation at
American Institute of Chemical Engineers
2019 Spring Meeting and 15th Global Congress on Process Safety
New Orleans, LA
March 31 – April 3, 2019

Enfoque multi-amenaza para abordar peligros Natech

Adriana María Mesa Gómez
Departamento de Ingeniería Química, Universidad de los Andes
Carrera 1E #19A-40, Bogotá (CO)
am.mesa@uniandes.edu.co

Joaquim Casal
Departamento de Ingeniería Química, Universitat Politècnica de Catalunya

Felipe Muñoz Giraldo
Departamento de Ingeniería Química, Universidad de los Andes

Keywords: natech, riesgo, desastre, evento natural, multi-amenaza

Abstract

Los eventos naturales como huracanes, relámpagos, inundaciones, terremotos o deslizamientos de tierra, entre otros, pueden afectar instalaciones de proceso originando la pérdida de contención de materiales peligrosos. A la combinación de este tipo de eventos naturales y sus posibles efectos se les conoce como eventos Natech -Natural Hazard Triggering Technological Disasters, por sus siglas en inglés-. Las pérdidas de contención generadas por eventos Natech pueden provocar incendios, explosiones o la emisión de nubes tóxicas a la atmósfera, con un impacto potencial significativo en el medio ambiente y la población localizada en las áreas vecinas. En las últimas décadas, la frecuencia de ocurrencia de estos eventos ha aumentado, activando el interés de los investigadores por el estudio de nuevas metodologías de análisis de riesgos para prevenir y mitigar posibles daños en instalaciones de proceso, poblaciones y medio ambiente. Es esencial por tanto conocer el progreso realizado en las metodologías de análisis de riesgos de los eventos Natech, identificando las lagunas existentes para poder desarrollar una metodología con una mayor fiabilidad. Por lo tanto, en este artículo se presenta una comparación de las metodologías existentes de análisis de riesgos Natech bajo una perspectiva multi-amenaza, así como de las metodologías a priori propuestas por organizaciones mundiales y locales. El análisis ha permitido concluir que los métodos multi-amenaza se centran en la mitigación del riesgo en áreas urbanas sin tener en cuenta el riesgo Natech. En el caso de los documentos de gestión técnica, se encontraron algunas metodologías que incorporan el riesgo Natech en los procesos de gestión de riesgos; esto pone de manifiesto que, en la actividad industrial, las prácticas de prevención y mitigación se han aplicado a los eventos Natech.

1 Introducción

Los eventos naturales pueden representar amenazas importantes para las unidades de proceso de diferentes industrias, ya sea costa afuera o costa adentro. Asimismo, algunos peligros derivados de eventos como tornados, huracanes, relámpagos, terremotos, entre otros; pueden afectar instalaciones industriales produciendo pérdida de contención de equipos de proceso que almacenan materiales peligrosos. Las emisiones consecuencia de las pérdidas de contención ocasionadas pueden provocar incendios, explosiones o emisión de nubes tóxicas [1, 2], lo que conlleva a un impacto significativo en el medio ambiente y la población cercana. Estos eventos naturales que tienen implícito un riesgo o peligro tecnológico son conocidos como Natech -Natural Hazard Triggering Technological Disasters- [3].

De hecho, los eventos Natech han producido grandes impactos en diferentes partes del mundo y se han convertido en una cuestión de gran relevancia para la sociedad moderna. Aunque hasta el momento se considera que la probabilidad de que ocurran dichos eventos es baja, tienen un impacto importante y un alto nivel de complejidad en términos de gestión de riesgos, porque son el resultado de eventos en cascada [4] o efecto dominó [2]; cabe resaltar que el efecto dominó ha sido responsable de varios accidentes catastróficos que han afectado a la industria química y de procesos, así como infraestructura petrolera [2, 5]. Adicionalmente, los escenarios de accidentes de alto impacto y baja probabilidad -HILP-, como los eventos tecnológicos desencadenados por peligros Natech en sitios industriales, están generando una gran preocupación; sin embargo, el análisis de estos eventos es un desafío relativamente nuevo que enfrentan las personas involucradas en la gestión del riesgo de desastres y puede ser particularmente complicado. Su complejidad resalta la importancia de estudiar y analizar eventos pasados, enfatizando que las lecciones aprendidas de investigaciones de accidentes históricos deberían adoptarse mientras se preparan para posibles incidencias futuras. Así, los métodos de investigación de accidentes ayudan a comprender mejor el curso de los eventos y las causas que los conducen [6].

Por otra parte, el concepto de multi-amenaza se resume en una combinación de dos o más factores de amenaza manifestados de manera aislada, simultánea o por reacción en cadena, para producir un suceso disparador de un desastre -definición tomada de la Comisión Nacional de Prevención de Riesgos y Atención de Emergencias, Costa Rica- [7]. En contexto con lo anterior, actualmente el número de personas afectadas por peligros derivados de eventos naturales está incrementando ya que la industria está creciendo en muchas regiones del mundo sujetas a múltiples peligros de esta naturaleza. Así, los responsables de la toma de decisiones se enfrentan al desafío no solo de mitigar los peligros y riesgos individuales, sino también los peligros y riesgos múltiples, que deben incluir la consideración de sus interrelaciones.

Teniendo en cuenta lo anterior, en este artículo se ha desarrollado una comparación de las metodologías existentes de análisis de riesgos Natech bajo una perspectiva multi-amenaza y las metodologías a priori propuestas por organizaciones mundiales y locales. Como resultado final, se obtuvo que los métodos multi-amenaza se centran en la mitigación del riesgo en áreas urbanas sin tener en cuenta el riesgo Natech. Por esta razón, la Oficina de las Naciones Unidas para la Reducción del Riesgo de Desastres -UNISDR- estableció, para hacer frente a la tendencia de aumento en intensidad y frecuencia de los desastres en los últimos años, el Marco de Sendai para la Reducción del Riesgo de Desastres 2015-2030 [8] llamado a fortalecer los modelos de reducción del riesgo de desastres, las evaluaciones, los mapas, el monitoreo y los sistemas de alerta temprana multi-peligro, y promover la realización de estudios integrales multi-peligro.

2 Comparación de metodologías bajo un enfoque multi-amenaza

A continuación se presenta una comparación de las metodologías existentes relacionadas con peligros Natech bajo una aproximación multi-amenaza. El alcance de la búsqueda se limita a metodologías encontradas en la literatura académica y desarrolladas por organizaciones gubernamentales y no gubernamentales para la gestión del riesgo a nivel industrial.

3.1 Peligros Natech con una perspectiva multi-amenaza -literatura académica-

En general, los peligros que representan los eventos naturales han tomado relevancia en los últimos años, teniendo en cuenta las afectaciones que han sufrido las poblaciones a lo largo del tiempo por los impactos de este tipo de eventos. Se han generado y divulgado algunos trabajos bajo enfoques multi-amenaza que permiten identificar, prevenir y/o mitigar los posibles perjuicios que se puedan tener. En términos de análisis cualitativos, [9] presentó una amplia visión general, caracterización y visualización del papel de 18 tipos de procesos antropogénicos en la activación, influencia e interacción con 21 peligros naturales. Aunque este trabajo no está enfocado al riesgo Natech, sus resultados resaltan las interacciones que pueden ocurrir entre los procesos desarrollados por el hombre en un lugar determinado y los eventos naturales a los cuales la población en ese sitio es vulnerable.

En términos de análisis semicuantitativo, algunos trabajos analizan riesgos naturales bajo un enfoque multi-amenaza. Dentro de estos análisis, [10] realizaron un estudio de caso utilizando una metodología de riesgos múltiples en la que evaluaron el riesgo individual asociado a inundaciones, terremotos y deslizamientos de tierra. Adicionalmente, [11] desarrollaron una metodología semicuantitativa con la cual cuantificaron, clasificaron y representaron en un mapa la distribución de peligros en el área costera de Campania (Italia). Los autores evaluaron el efecto de los peligros individuales en el área. El análisis de peligros múltiples se realizó mediante una matriz de interacción que permite saber si los eventos naturales objeto de estudio presentan alguna interacción o no. Para este estudio de caso, se realizó un análisis de peligros múltiples para la interacción de eventos naturales como inundaciones, terremotos, deslizamientos de tierra, erupción volcánica y erosión (Figura 1).

Además, para abordar las amenazas planteadas por los factores climáticos en combinación con las presiones antropogénicas locales y regionales que afectan a los ecosistemas y actividades marinas, [12] desarrollaron una metodología de evaluación de peligros múltiples, aplicándola al mar Adriático para el escenario de referencia 2000-2015. Los autores realizaron un análisis centrado en el peligro antropogénico teniendo en cuenta algunos peligros naturales. Aunque el tratamiento es similar al análisis cualitativo presentado por [9], la metodología semicuantitativa presenta menos subjetividades. Finalmente, [9] presentó una metodología semicuantitativa a escala local, desarrollada para aumentar la efectividad de la planificación del uso de la tierra relacionada con la gestión de riesgos, particularmente en relación con los riesgos múltiples que afectan al mismo territorio.

Los cuatro análisis semicuantitativos que se han presentado están dirigidos a la minimización de los riesgos en las áreas urbanas, es decir, no tienen ciertamente un enfoque de riesgo Natech. Sin embargo, metodologías como la propuesta por [9] pueden tener una contribución significativa en este campo, ya que tienen implícito el tema de la planificación del uso del suelo, que va de la mano con la ubicación de las instalaciones y los parques industriales.

Figura 1. Matriz de interacción entre amenazas [11] .

En términos de análisis cuantitativo se destaca el trabajo desarrollado por [13], quienes desarrollaron una metodología para analizar los riesgos naturales de múltiples procesos. Se toman en cuenta los eventos naturales de derrumbes, nevadas y flujo de escombros. En esta metodología, el riesgo se calcula como una función basada en parámetros de peligro, vulnerabilidad, probabilidad de impacto espacial, probabilidad de impacto temporal, probabilidad de ocurrencia estacional y daño potencial. Se calculan así los riesgos de los procesos individuales y, al final, cada uno de ellos se combina dentro de un mapa de riesgos de múltiples peligros. Se llegó a la conclusión de que los enfoques de peligros múltiples no solo se usan para obtener un resumen del riesgo general, sino que también tienen una gran importancia en la planificación efectiva de las contramedidas. Esta metodología se aplica a áreas urbanas y no tiene tampoco un enfoque de riesgo Natech.

Finalmente, se encontró un marco que incluye tres etapas. El propósito de los autores [14] fue desarrollar un marco teórico para una evaluación coherente de riesgos múltiples. En el mismo, el primer nivel es un diagrama de flujo cualitativo que guía al usuario para determinar si se requiere un enfoque de riesgo múltiple para el problema en cuestión. El segundo nivel es un enfoque semicuantitativo simplificado para explorar más a fondo la necesidad de una evaluación detallada. El tercer nivel es un análisis cuantitativo de riesgos múltiples basado en redes bayesianas. Se abordan los componentes clave de este marco, como la evaluación de los peligros en cascada, la estimación de la vulnerabilidad dependiente del tiempo y la elección del nivel requerido de sofisticación (Figura 2).

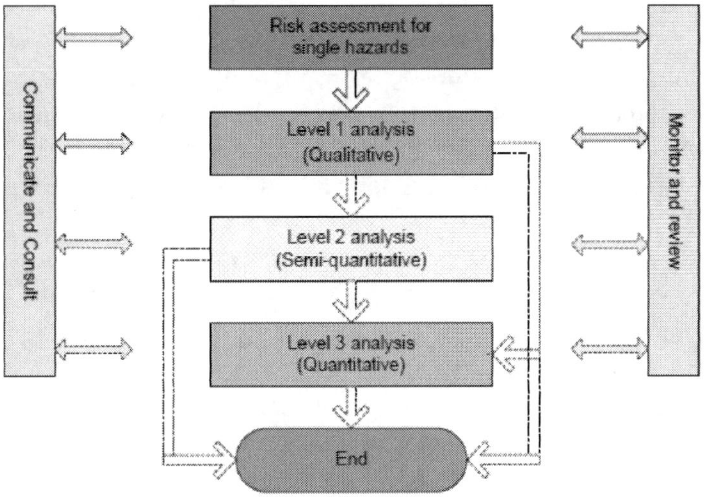

Figura 2. Diagrama de flujo multietapa [12].

En general, el número de trabajos realizados para analizar riesgo Natech bajo una perspectiva multi-amenaza es limitado. Si bien se ha incrementado el interés por obtener nuevas herramientas que permitan abordar este tipo de peligros, las metodologías disponibles se enfocan a la gestión del riesgo social sin tomar en consideración la interacción que puedan tener los eventos naturales múltiples en las industrias y sus posteriores efectos sobre la población y el medio ambiente. Sumado a esto, la relación que pueda presentarse entre eventos naturales se analiza teniendo en cuenta vulnerabilidades estáticas, es decir, se analizan los eventos de forma individual y no como un conjunto, dejando de lado la dinámica que se podría presentar en esta variable si se tuviera en cuenta la interacción entre los eventos.

3.2 Peligros Natech con una perspectiva multi-amenaza -gestión industrial-

Algunas metodologías industriales toman en cuenta factores externos como eventos naturales dentro de su análisis de riesgo. En el caso del American Petroleum Institute -API-, la metodología propuesta en API RP 581 -Inspección basada en riesgos RBI- [15] para la inspección basada en riesgos en instalaciones de proceso, solo considera la frecuencia de terremotos dentro del análisis de amenaza del clima y fuerzas externas. Con lo anterior, se identifica una brecha en esta metodología ya que, dentro de la evaluación de riesgos de los activos de una instalación en el sector de petróleo y gas, los eventos naturales no se consideran riesgos tangibles. Por otra parte, en API RP 752 -Gestión de riesgos asociados a la ubicación de los edificios permanentes de la planta de proceso- [16] se proporciona orientación adicional a las empresas que operan en instalaciones cubiertas por la norma OSHA PSM para la gestión adecuada de riesgos, teniendo en cuenta explosiones, incendios y emisiones de materiales tóxicos para edificios nuevos y existentes destinados a ser ocupados; sin embargo, esta práctica recomendada excluye las instalaciones de almacenamiento ya que dichas instalaciones solo están ocupadas de forma intermitente.

El Instituto Americano de Ingenieros Químicos, Centro para la Seguridad de Procesos Químicos -CCPS-, una organización que identifica y aborda las necesidades de seguridad de procesos en las industrias química y farmacéutica, ha publicado una serie de documentos de orientación que se utilizan ampliamente en estas industrias. En el libro Guidelines for Safe Storage and Handling of Reactive Materials [17], esta organización afirma que el almacenamiento de sustancias químicas crea potenciales peligros que incluyen lesiones, enfermedades, daños al medio ambiente, daños a la propiedad e interrupción de la actividad, y señala que los desastres naturales, incluidas las inundaciones, pueden desencadenar estos peligros. Además, CCPS enfatiza que las inundaciones tienen el potencial de liberar sustancias químicas al medio ambiente. Para un mejor conocimiento del riesgo de inundación, las compañías acuden a los recursos de la Agencia Federal para el Manejo de Emergencias -FEMA-, incluidos Flood Insurance Studies [18] y Flood Insurance Rate Maps [19].

Actualmente, en Estados Unidos las consecuencias derivadas de eventos Natech se evalúan siguiendo las pautas del Plan de Gestión de Riesgos -RMP- [20], al que se le ha asignado la metodología del Programa de Gestión de Riesgos de la EPA. Los resultados de esta metodología se presentan como informes resumidos y mapas de riesgo interactivos. La herramienta se puede utilizar para la planificación del uso del suelo y de emergencias.

Los anteriores son documentos de gestión de riesgo Natech desarrollados por instituciones de los Estados Unidos. En contraste con este enfoque, en la Unión Europea los riesgos asociados a accidentes químicos graves están regulados por la Directiva Seveso, que se aplica a las actividades industriales que usan, manejan o almacenan sustancias peligrosas específicas. La Directiva Seveso requiere que algunas instalaciones que fabrican o almacenan determinadas cantidades de productos químicos peligrosos generen un informe de seguridad que demuestre que la instalación tiene una política de prevención de accidentes graves, que se han identificado escenarios de incidentes y se han mitigado los riesgos. Además, esta directiva requiere que las instalaciones aborden el aumento del riesgo derivado de la ubicación geográfica de una planta. Con el tiempo y después de una serie de incidentes, la Unión Europea actualizó la Directiva Seveso en 2012 para abordar los riesgos climáticos extremos de manera más explícita. Actualmente, la Directiva Seveso requiere que las empresas identifiquen y evalúen periódicamente los peligros ambientales, como inundaciones y terremotos, en la seguridad de la instalación.

Finalmente, bajo una perspectiva transversal, dentro de los Principios Rectores para la Prevención, Preparación y Respuesta de Accidentes de Productos Químicos [21], la Organización para la Cooperación y el Desarrollo Económicos –OCDE- incluyó en 2015 un apéndice en el que aborda los peligros naturales que activan los accidentes tecnológicos, y consiste en una serie de enmiendas a los Principios Rectores y en la adición de un nuevo capítulo que proporciona una guía más detallada sobre la prevención, preparación y respuesta de Natech. Toma en cuenta los resultados del Taller sobre Riesgos Naturales que Desencadenan los Accidentes Tecnológicos -Natech Risk Management- realizado en 2012, en Dresde, Alemania, que se llevó a cabo bajo los auspicios del Grupo de Trabajo de la OCDE sobre Accidentes de Productos Químicos -WGCA-.

En resumen, es evidente que los esfuerzos para tener una gestión de los riesgos derivados de eventos naturales en la industria han aumentado. Muchas organizaciones gubernamentales y no gubernamentales han desarrollado planes para incluir este tipo de riesgo en los planes de la industria para mitigar las consecuencias que pueden ser provocadas por el impacto de un evento natural. Sin embargo, estos esfuerzos deben incrementarse para abordar todos los eventos naturales que pueden afectar a las industrias y, por lo tanto, tener planes para la prevención y mitigación de las consecuencias para las poblaciones y entornos circundantes. Adicionalmente es importante mencionar que, dentro de los documentos de gestión industrial, los enfoques desarrollados no tienen en cuenta la interacción de dos o más eventos naturales en un mismo lugar, es decir que estas herramientas se limitan a un análisis de riesgo Natech bajo una perspectiva de amenaza individual.

3 Conclusiones

Analizando la literatura académica, se ha observado que existen limitaciones en las metodologías de análisis de riesgos con un enfoque de peligros múltiples para eventos Natech, ya que las metodologías encontradas se centran en la prevención y mitigación de consecuencias en áreas urbanas y no en instalaciones industriales. Además, aquellas que tienen en cuenta peligros múltiples manejan vulnerabilidades estáticas, lo que significa que la interacción de eventos naturales no se tiene en cuenta.

Bajo una perspectiva de gestión industrial, se han encontrado algunas metodologías que toman en cuenta los eventos Natech dentro de los procesos. Organizaciones como API, CCPS, EPA, FEMA y OCDE han hecho un esfuerzo para incorporar en sus metodologías el análisis de eventos naturales como peligros reales. Sin embargo, las herramientas disponibles a nivel industrial permiten la gestión de riesgos bajo una perspectiva de amenaza individual y no bajo una perspectiva de multi-amenaza.

Es evidente, por tanto, especialmente teniendo en cuenta el desarrollo de la industria en países afectados por severos eventos naturales, que es necesario un esfuerzo adicional para poder disponer de metodologías más potentes, que incluyan los escenarios Natech desde un punto de vista multi-amenaza.

4 Agradecimientos

Este trabajo se ha realizado como una primera fase dentro de la tesis doctoral *propuesta metodológica para la realización de análisis multipeligro en unidades de proceso*, llevada a cabo en el Grupo de Investigación de Diseño de Productos y Procesos (GDPP) de la Universidad de los Andes y en el Centro de Estudios del Riesgo Tecnológico (CERTEC) de la Universidad Politécnica de Cataluña, bajo el patrocinio de Colciencias mediante la convocatoria 785 para formación de investigadores nacionales en estudios de doctorado. Por esta razón queremos agradecer a estas tres instituciones por su asesoría y patrocinio en el desarrollo de esta investigación.

5 Referencias

[1] J. Casal, "Chapter 1 - Introduction,"in *Evaluation of the Effects and Consequences of Major Accidents in Industrial Plants (Second Edition)*, J. Casal, Ed.: Elsevier, 2018, pp. 1-24.

[2] J. Casal, "Chapter 10 - Domino Effect,"in *Evaluation of the Effects and Consequences of Major Accidents in Industrial Plants (Second Edition)*, J. Casal, Ed.: Elsevier, 2018, pp. 405-437.

[3] A. M. Cruz and E. Krausmann, "Vulnerability of the oil and gas sector to climate change and extreme weather events," (in English), *Climatic Change,* Article vol. 121, no. 1, pp. 41-53, Nov 2013.

[4] K. R. D. S. Nascimento and M. H. Alencar, "Management of risks in natural disasters: A systematic review of the literature on NATECH events," *Journal of Loss Prevention in the Process Industries,* vol. 44, pp. 347-359, 2016/11/01/ 2016.

[5] A. Necci, V. Cozzani, G. Spadoni, and F. Khan, "Assessment of domino effect: State of the art and research Needs," *Reliability Engineering & System Safety,* vol. 143, pp. 3-18, 2015/11/01/ 2015.

[6] A. Chakraborty, A. Ibrahim, and A. M. Cruz, "A study of accident investigation methodologies applied to the Natech events during the 2011 great east Japan earthquake," *Journal of Loss Prevention in the Process Industries,* 2018/01/11/ 2018.
[7] CNE, "Glosario oficial de conceptos sobre desastres," ed. Costa Rica: Litografía e Imprenta Universal, S.A., 2010, p. 24.
[8] UNISDR, "Marco de Sendai para la reducción del riesgo de desastres 2015-2030," 1 ed. Ginebra, Suiza: United Nations 2015, p. 40.
[9] J. C. Gill and B. D. Malamud, "Anthropogenic processes, natural hazards, and interactions in a multi-hazard framework," *Earth-Science Reviews,* vol. 166, pp. 246-269, 2017/03/01/ 2017.
[10] C. J. van Westen, L. Montoya, L. Boerbom, and E. badilla Coto, "Multi-hazard risk assessment using GIS in urban areas: a case study for the city of Turrialba, Costa Rica," in *Proceedings of the Regional Workshop on Best Practices in Disaster Mitigation: Lessons Learned from the Asian Urban Disaster Mitigation Program and Other Initiatives* Bali, Indonesia, 2002, pp. 120-136.
[11] T. De Pippo, C. Donadio, M. Pennetta, C. Petrosino, F. Terlizzi, and A. Valente, "Coastal hazard assessment and mapping in Northern Campania, Italy," *Geomorphology,* vol. 97, no. 3, pp. 451-466, 2008/05/15/ 2008.
[12] E. Furlan, S. Torresan, A. Critto, and A. Marcomini, "Spatially explicit risk approach for multi-hazard assessment and management in marine environment: The case study of the Adriatic Sea," *Science of The Total Environment,* vol. 618, pp. 1008-1023, 2018/03/15/ 2018.
[13] R. Bell and T. Glade, "Multi-Hazard Analysis in Natural Risk Assessments," in *International Conference on Computer Simulation in Risk Analysis and Hazard Mitigation* Rhodes, Greece, 2004, vol. 9, pp. 197-206.
[14] F. Nadim *et al.*, "D5.2 Framework for multi-risk assessment . MATRIX project," 2013.
[15] *API RP 581 - Risk-Based Inspection (RBI).* American Petroleum Institute, 2016.
[16] *API RP 752 - Management of Hazards Associated with Location of Process Plant Permanent Buildings.* American Petroleum Institute, 2009.
[17] CCPS. *Guidelines for Safe Storage and Handling of Reactive Materials.* Center for Chemical Process Safety, American Institute of Chemical Engineers. New York, NY. Wiley, 2010.
[18] FEMA. *Flood Insurance Studies.* Federal Emergency Management Agency, 2018.
[19] FEMA. *Flood Insurance Rate Maps.* Federal Emergency Management Agency, 2015.
[20] EPA, "Risk Management Plan (RMP) Rule," ed. United States: US Environmental Protection Agency, 2015.
[21] OECD. *Addendum number 2 to the OECD guiding principles for chemical accident prevention preparedness and response to address natural hazards triggering technological accidents (NATECHs).* Organización para la Cooperación y el Desarrollo Económicos, 2th ed., Paris, Francia, 2015.

Control Avanzado de Procesos (APC): Una herramienta de entrenamiento y aseguramiento del desempeño en la implementación de seguridad de procesos con base en riesgo (RBPS)

Andres F. Valencia
WEST Advanced Solutions LLC
433 North Loop W. Houston TX, 77008
andres.valencia@west.net.co

Prepared for Presentation at
American Institute of Chemical Engineers
2019 Spring Meeting and 15th Global Congress on Process Safety
New Orleans, LA
March 31 – April 3, 2019

Control Avanzado de Procesos (APC): Una herramienta de entrenamiento y aseguramiento del desempeño en la implementación de seguridad de procesos con base en riesgo (RBPS)

Andres F. Valencia
WEST Advanced Solutions LLC
433 North Loop W. Houston TX, 77008
andres.valencia@west.net.co

Keywords: Advanced Process Control (APC), Risk Based Process Safety (RBPS), Training and Performance Assurance.

Abstract

Dentro de la gestión de la seguridad de procesos con base en riesgo (RBPS), el entrenamiento y el aseguramiento del desempeño es el elemento clave que sirve como herramienta para dar a empleados y contratistas, de forma general o específica, los fundamentos y conocimientos que requieren para desarrollar su trabajo de forma segura.

El Control Avanzado de Procesos (APC) reúne un amplio rango de estrategias de control y tipos de tecnología. Sin importar si el APC estará basado en Control Multivariable (MVC), un Modelo de Control Predictivo (MPC), o Lógica Difusa (FL), entre otras estrategias, su desarrollo e implementación requiere de un conocimiento claro y robusto del proceso donde será aplicado, el cual deberá ser reflejado en un modelo dinámico del mismo.

El objetivo de este trabajo es mostrar los beneficios y las herramientas que se pueden derivar a partir de la implementación de APC en el Sistema Básico de Control de Proceso (BPCS) de una planta, para fortalecer el entrenamiento y aseguramiento del desempeño de operadores, haciendo uso de los modelos dinámicos de proceso con sus restricciones, límites y objetivos de control.

1 Introducción

El entrenamiento y aseguramiento del desempeño es el elemento dentro del modelo de la gestión de la seguridad de procesos con base en riesgo (RBPS) que da a empleados y contratistas de una organización, de forma general o específica, los fundamentos y conocimientos que requieren para desarrollar su trabajo de forma segura [1].

Este elemento siempre irá necesariamente ligado con factores comportamentales del personal, y deberá ser evaluado periódicamente para identificar necesidades de reentrenamiento. Cuando se plantea hacer algún cambio en el proceso de producción, será también necesario impartir el entrenamiento apropiado a los grupos de interés sobre el manejo de las nuevas condiciones y estados que pudiesen originarse, siendo unos de los pasos claves del manejo del cambio (MOC) [2].

El Control Avanzado de Procesos (APC) reúne un amplio rango de estrategias de control que requieren del modelo dinámico del proceso. De manera convencional, su desarrollo normalmente está ligado a la integración de componentes adicionales de hardware y aplicativos de software dentro de la pirámide de automatización de una planta, y al conocimiento claro y robusto del proceso donde será aplicado.

A partir de la implementación de estrategias APC en el Sistema Básico de Control de Proceso (BPCS) de una planta, es posible desarrollar estaciones o módulos de entrenamiento donde se pueden simular condiciones o estados inseguros en el proceso de producción a los cuales el personal operativo podría verse enfrentado, registrando las acciones que el operador ejecuta a través de la interfase hombre-máquina (HMI) para su corrección antes que el riesgo pudiese materializarse.

2 Modelo y simulación del proceso

El modelo del proceso es el elemento clave para el desarrollo efectivo de estrategias APC. Inicialmente es necesario construir sobre un simulador de proceso el modelo en estado estacionario a partir de los P&ID, PFD, composiciones y condiciones de las corrientes de entrada y salida, históricos de variables, entre otros.

A parir del modelo en estado estacionario, previamente validado frente a los datos con los cuales fue construido, se procede a integrar las curvas características de elementos finales de control, bombas, compresores, entre otros, planteando el balance en estado no estacionario. Este balance debe ser complementado con ecuaciones constitutivas tales como relaciones PVT, expresiones cinéticas, etc., y condiciones iniciales y de frontera para desarrollar el modelo dinámico.

Como es bien sabido, la complejidad matemática incrementa considerablemente al pasar del estado estacionario al estado dinámico, en esencia porque el modelo cambia de ecuaciones algebraicas a ecuaciones diferenciales (ordinarias y parciales).

A partir de lo anterior, se deduce que la construcción de un modelo de simulación para el desarrollo de estrategias APC debe hacerse sobre un simulador comercial que soporte el estado estable y dinámico.

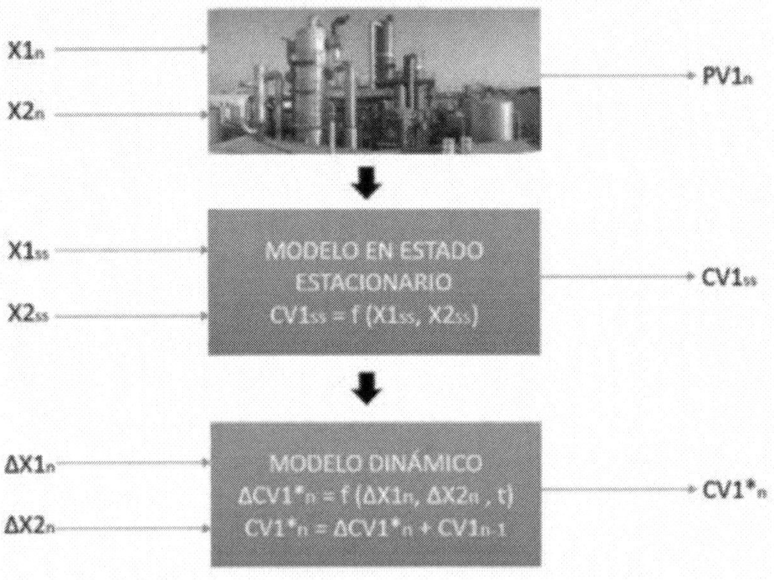

Figura 1. Desarrollo del modelo dinámico del proceso

3 Estrategias de control avanzado

Normalmente los proveedores de sistemas de control que ofrecen soluciones APC soportan sus aplicaciones por encima del Sistema Básico de Control de Proceso (BPCS), es decir, agregando componentes de hardware y aplicativos de software dentro de la pirámide de automatización de una planta, haciendo que el costo de implementación y el desarrollo de estaciones o módulos de entrenamiento incremente sustancialmente.

A través de desarrollos de aplicativos APC en el BPCS por medio de lenguajes de programación de acuerdo a IEC 61131-3 [3], es posible implementar no solamente la estrategia de control APC, sino también derivar de forma directa y flexible, estaciones o módulos de entrenamiento para asegurar del desempeño de los operadores.

La selección de la estrategia APC que será implementada en el BPCS de una planta o proceso, depende del objetivo que se quiera alcanzar, el cual estará orientado normalmente a la robustez y al desempeño. Control Multivariable (MVC), Modelo de Control Predictivo (MPC), Lógica Difusa (FL), Gain Scheduling o controladores H-infinito, son algunas de las estrategias que pueden ser implementadas de forma directa en el BPCS de una planta para posteriormente desarrollar las estaciones o módulos de entrenamiento.

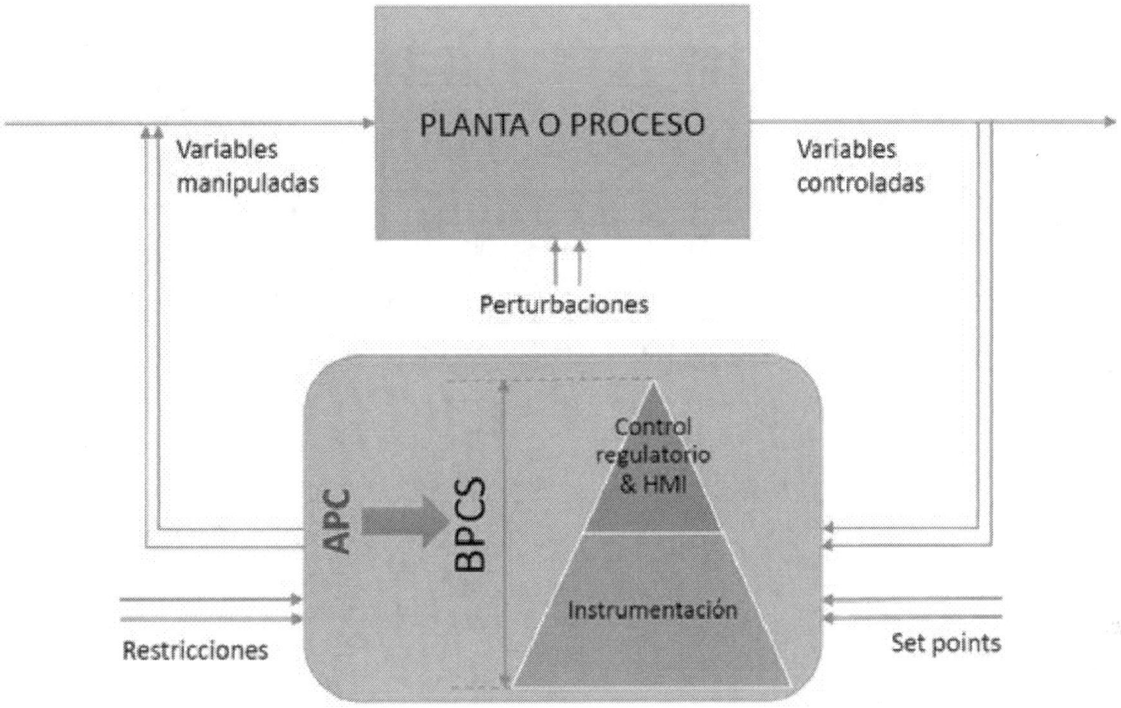

Figura 2. Diagrama de bloques simplificado de control avanzado

4 Simulador de entrenamiento

Cada estación de entrenamiento estará conectada a un emulador BPCS de la planta, el cual a través de una interface *OLE for Process Control©* (OPC) y un *script* para transferencia de datos hace posible la conexión con el modelo dinámico del proceso. El ambiente de la estación de entrenamiento será exactamente igual al del HMI de la planta.

Directamente sobre el modelo dinámico se podrán simular condiciones o estados inseguros a los cuales el personal operativo podría verse enfrentado, registrando las acciones que el operador ejecuta a través del HMI para su corrección antes que el riesgo pudiese materializarse.

Otros parámetros que deberán ser configurados durante el desarrollo de la estación de entrenamiento, será el tiempo de seguridad de proceso (*Process Safety Time* - PST), los modos de operación, entre otros.

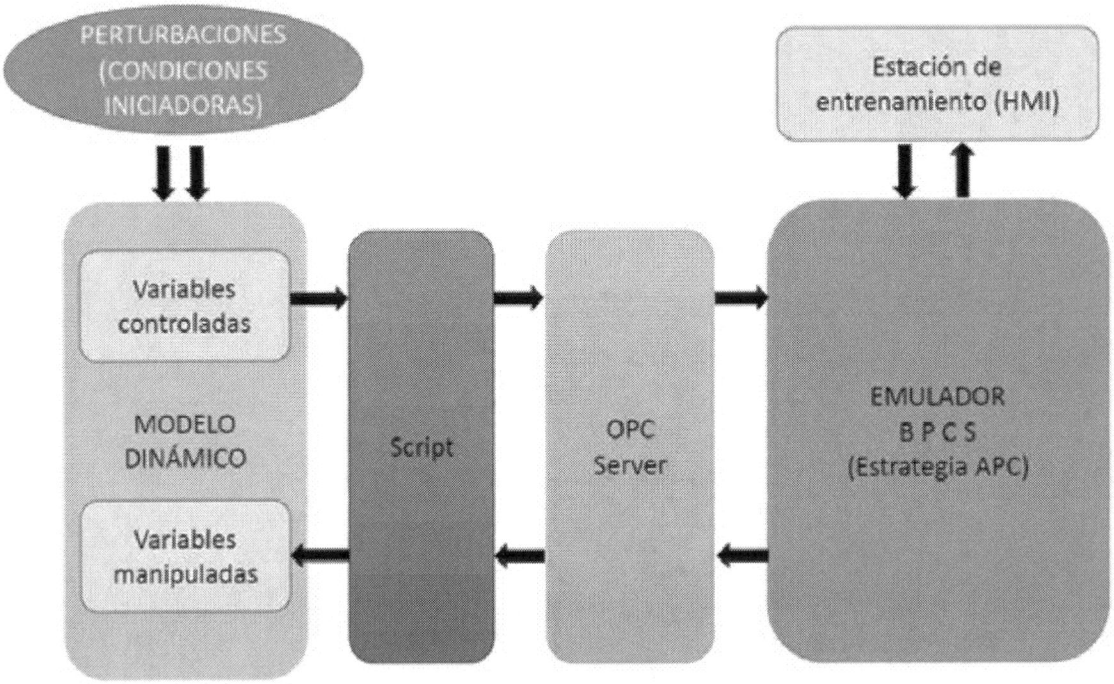

Figura 3. Arquitectura de una estación de entrenamiento

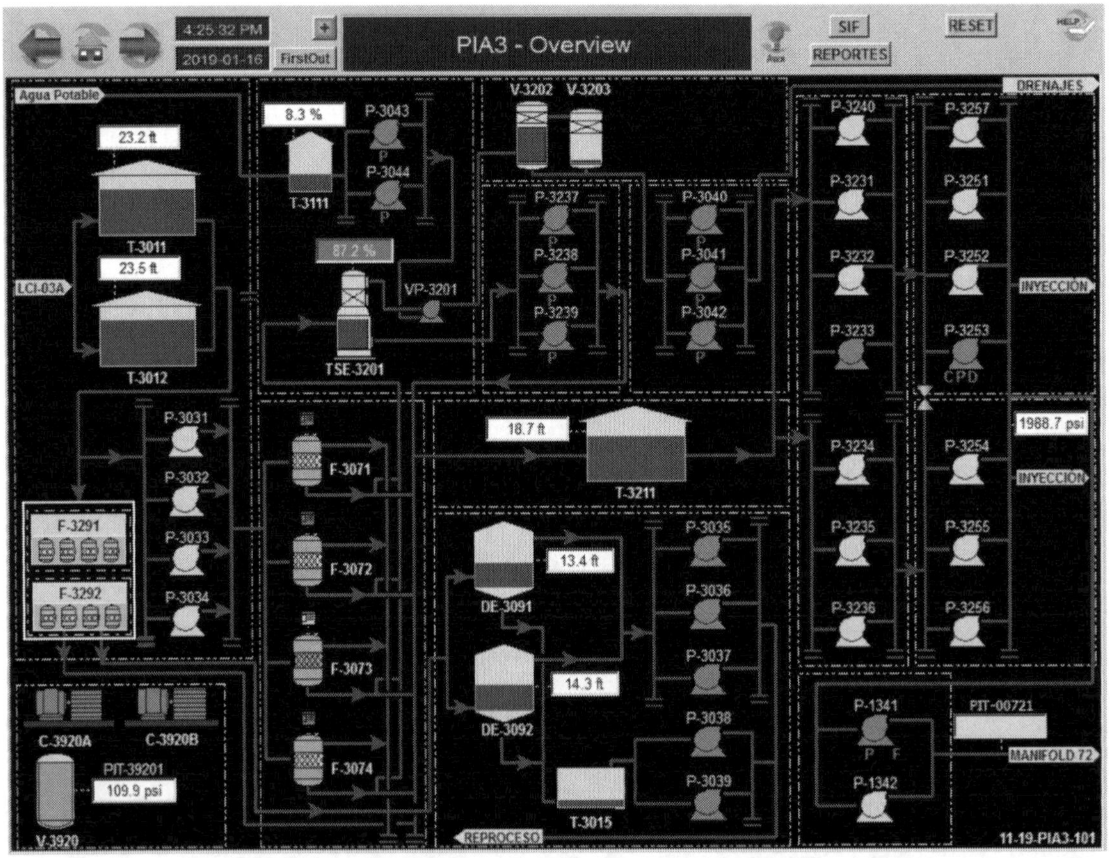

Figura 4. HMI en estación de entrenamiento

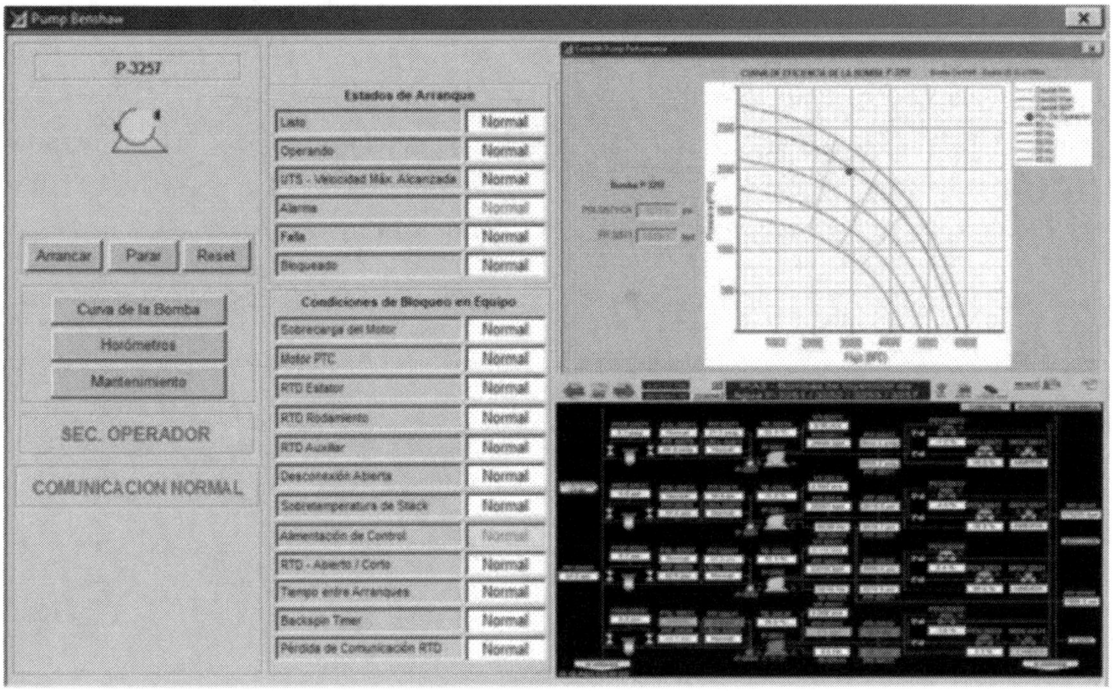

Figura 5. Interface para generación de perturbaciones sobre el modelo dinámico del proceso

5 Tiempo de seguridad de proceso (PST)

Para cada escenario de riesgo a simular es necesario definir el PST. Esencialmente, el PST corresponde al periodo de tiempo entre la ocurrencia de una falla en el proceso y la ocurrencia o materialización del evento peligroso. En otras palabras, el PST determina el tiempo máximo en el cual el operador podrá tomar una acción correctiva para evitar que el evento peligroso pueda ocurrir.

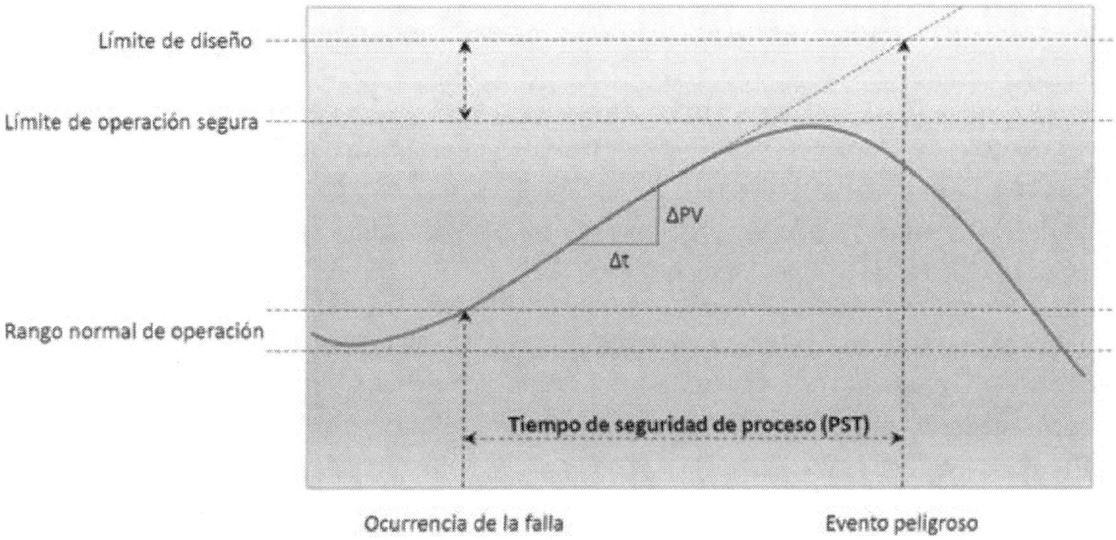

Figura 6. Línea de tiempo de un escenario peligroso [4]

6 Modos de operación

Teniendo en cuenta que cerca del 70% de los accidentes ocurren durante el desarrollo de operaciones no rutinarias [5], las estaciones o módulos de entrenamiento deberán incluir los diferentes modos de operación de un proceso, tales como el *start-up*, operación normal, *shutdown*, *process upset*, mantenimiento, *shutdown* de emergencia.

Cuando se trata de un proceso *batch*, es necesario incluir sus estados inactivos, transitorios y finales.

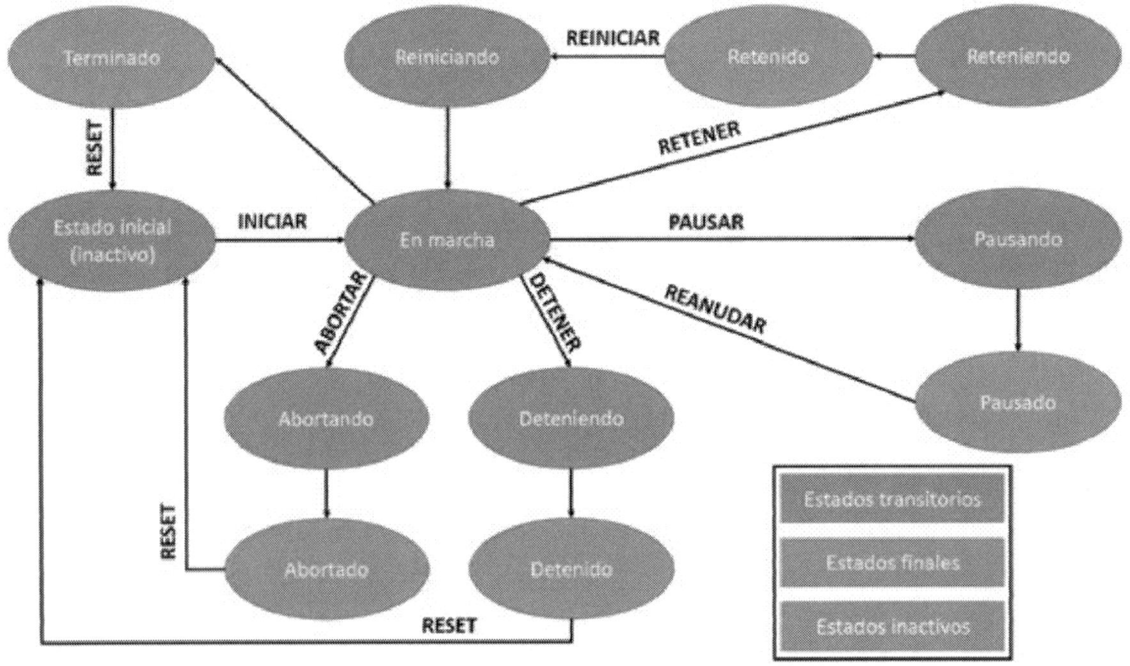

Figura 7. Diagrama de transición de estados (tomado de ISA-88.01-1995 (R2006))

7 Seguimiento

Los instrumentos de seguimiento de entrenamiento y aseguramiento del desempeño de operadores deben estar orientados a modelos que permitan establecer la curva de aprendizaje de cada uno de los participantes en las actividades de capacitación sobre las estaciones o módulos de entrenamiento.

A partir de los manuales de fabricantes y licenciatarios de equipos, y junto con los conceptos y argumentos de operadores senior de la facilidad simulada en la estación de entrenamiento, es posible desde la misma estación, calcular *Key Performance Indicators* (KPI) para establecer métricas y estadísticas que alimenten las curvas de aprendizaje individuales.

8 Conclusión

Para el entrenamiento y aseguramiento del desempeño, a partir de la implementación de estrategias APC en el Sistema Básico de Control de Proceso (BPCS) de una planta por medio de lenguajes de programación de acuerdo a IEC 61131-3 [3], es posible desarrollar de forma directa y flexible estaciones o módulos de entrenamiento donde se pueden simular condiciones o estados inseguros en el proceso de producción a los cuales el personal operativo podría verse enfrentado, registrando las acciones que el operador ejecuta a través de la interfase hombre-máquina (HMI) para su corrección antes que el riesgo pudiese materializarse.

El aseguramiento del desempeño a través de estaciones de entrenamiento puede desarrollarse desde operaciones unitarias simples hasta procesos complejos; sin importar el caso, siempre el modelo del proceso será el elemento clave para el desarrollo efectivo de estrategias APC.

9 Referencias

[1] A. Sepeda. "Understanding Process Safety Management", CEP Magazine, August (2010) 26-33.

[2] CCPS. "Guidelines for the Management of Change for Process Safety", Center for Chemical Process Safety. Wiley-Interscience, 2008.

[3] IEC 61131-3:2013. "Programmable Controllers – Part 3: Programming languages", International Electrotechnical Commission.

[4] G. Barnard and W. Creel. "Impacts of Process Safety Time on Layer of Protection Analysis", 11th Global Congress on Process Safety. AIChE, 2015.

[5] PHA of Non-Normal Operating Modes, available at http://www.process-improvement-institute.com/consulting-services/pha-of-non-normal-operating-modes/ accessed on January 22, 2019.

La Gestión del Cambio del Sistema Instrumentado de Seguridad y su Impacto en el Riesgo

Elio Gómez, CFSP
exida Consulting
Amores 1029 – 201, Mexico City 03100
egomez@exida.com

Alejandro Esparza, CFSE
exida Consulting
Amores 1029 – 201, Mexico City 03100
aesparza@exida.com

Prepared for Presentation at
American Institute of Chemical Engineers
2019 Spring Meeting and 15th Global Congress on Process Safety
New Orleans, LA
March 31 – April 3, 2019

La Gestión del Cambio del Sistema Instrumentado de Seguridad y su Impacto en el Riesgo

Elio Gómez, CFSP
exida Consulting
egomez@exida.com

Alejandro Esparza, CFSE
exida Consulting
aesparza@exida.com

Keywords: Identificación de Peligros y Análisis de Riesgo, Sistemas Instrumentados de Seguridad, SIS, SIL, Seguridad Funcional

Abstracto

Los Sistemas Instrumentados de Seguridad (SIS) contribuyen de manera muy importante a mantener la seguridad de un proceso peligroso. A lo largo de la operación del proceso, lo cual puede ser durante varios años, es común que el SIS sufra modificaciones por muy diversas razones, lo cual en si no es un problema, siempre y cuando se lleve a cabo una adecuada gestión del cambio (MOC, por sus siglas en inglés). De no llevarse a cabo adecuadamente la mencionada gestión del cambio, se puede incrementar la probabilidad de que la protección provista por el SIS se vaya degradando, aumentando consecuentemente el riesgo del proceso peligroso.

1 Introducción

Los Sistemas Instrumentados de Seguridad (SIS), cuando son diseñados, construidos, instalados y operados de manera correcta, son un contribuyente muy importante en el manejo del riesgo del proceso, es decir, colaboran en gran medida a la operación segura una planta. Ese diseño, construcción, instalación y operación "de manera correcta" suele conseguirse siguiendo procesos de ingeniería como el llamado Ciclo de Vida de Seguridad Funcional, de acuerdo con estándares como IEC 61511 e ISA 84.

Al momento de la validación de un sistema de seguridad, es decir, cuando se demuestra que han sido satisfechos exitosamente todos los requisitos de seguridad que hubiesen sido definidos durante la etapa de análisis de riesgo del proceso, nos aseguramos de haber obtenido seguridad funcional en dicho sistema, reduciendo significativamente la posibilidad de que el SIS falle, de modo que no actúe cuando sea necesario para evitar un accidente. Cabe destacar que la validación del sistema de seguridad debe realizarse previo a poner en operación la planta por primera vez. Pero ¿qué tan rápido se puede degradar la seguridad provista por el SIS?

No deberíamos asumir que el sistema de seguridad permanece sin disminución de su desempeño a lo largo del tiempo, existen varios factores que degradan la seguridad funcional, es decir, que aumentan la probabilidad de falla del sistema de seguridad, por ejemplo:

- Desgaste de los componentes del sistema de seguridad (sensores, válvulas, etc.), inclusive de aquellos componentes electrónicos que no tienen partes móviles.
- Fallas sistemáticas (por ejemplo, mantenimiento deficiente, errores en la configuración, etc.)
- Modificaciones al SIS … de no realizarse de manera adecuada

Respecto al último punto, en el estándar de seguridad funcional IEC-61511, desde la cláusula 9 hasta la 11, se muestran actividades relacionadas con los requisitos específicos para la gestión de los SIS, que incluyen específicamente a la gestión del cambio relacionada directamente con el SIS y sus componentes de hardware y software.

2 La Gestión del Cambio del SIS

La llamada gestión del cambio no consiste solamente de un procedimiento sistemático para registrar las modificaciones al hardware y software del SIS, sino que también implica la necesidad de evaluar la solicitud de cambio, determinando si realmente es necesario, conveniente y posible.

Las solicitudes de cambio al SIS pueden deberse a varias razones. Estas solicitudes de cambio en el SIS podrían tener un impacto negativo en su desempeño de seguridad, es decir, en el Nivel de Integridad de Seguridad (SIL por sus siglas en inglés) que puede ser alcanzado, por lo que habría que revisar detalladamente que dicho SIL, definido durante la fase de análisis del ciclo de vida de seguridad, siga siendo alcanzado y mantenido.

3 Razones del Cambio del SIS

Existen múltiples razones para realizar cambios, y por ello hay que analizar cada caso detenidamente, para determinar si realmente hay alguna debilidad en el diseño, o si es que existe alguna variable en el proceso que no se está considerando como debiera, o si realmente el cambio representará una mejora. También se debe incluir un análisis del impacto del cambio.

Normalmente, las razones específicas para realizar un cambio en el SIS son:

- Cambios en los procedimientos de operación. Por ejemplo, la gerencia general de la planta ha determinado que se debe incrementar la producción
- Cambios al proceso. Por ejemplo, el departamento de proceso requiere incrementar el nivel normal de operación en un tanque de almacenamiento

- Cambios en las Especificaciones de Requisitos de Seguridad (SRS). Por ejemplo, durante las revisiones periódicas (típicamente cada 5 años) al Análisis de Peligros y Riesgos de la planta, se decide aumentar el requisito de nivel SIL de alguna función instrumentada de seguridad (SIF, por sus siglas en inglés)
- El departamento de operaciones requiere un cambio en una SIF, por ejemplo por que el SIS esté presentando una tasa de fallas en falso mayor a lo esperado
- Incremento en la demanda de actuación del SIS. Por ejemplo, ha ocurrido un incidente y se piensa que podría utilizarse al SIS para evitar que la falla se repita nuevamente
- El departamento de mantenimiento necesita modificar algún equipo que esté obsoleto, ya sea porque haya fallado y requiere reparación y/o porque haya llegado al fin de su vida útil
- Correcciones de errores en software o firmware
- Corrección de fallas sistemáticas
- Cambios en las aplicaciones del software
- Cambios en la legislación

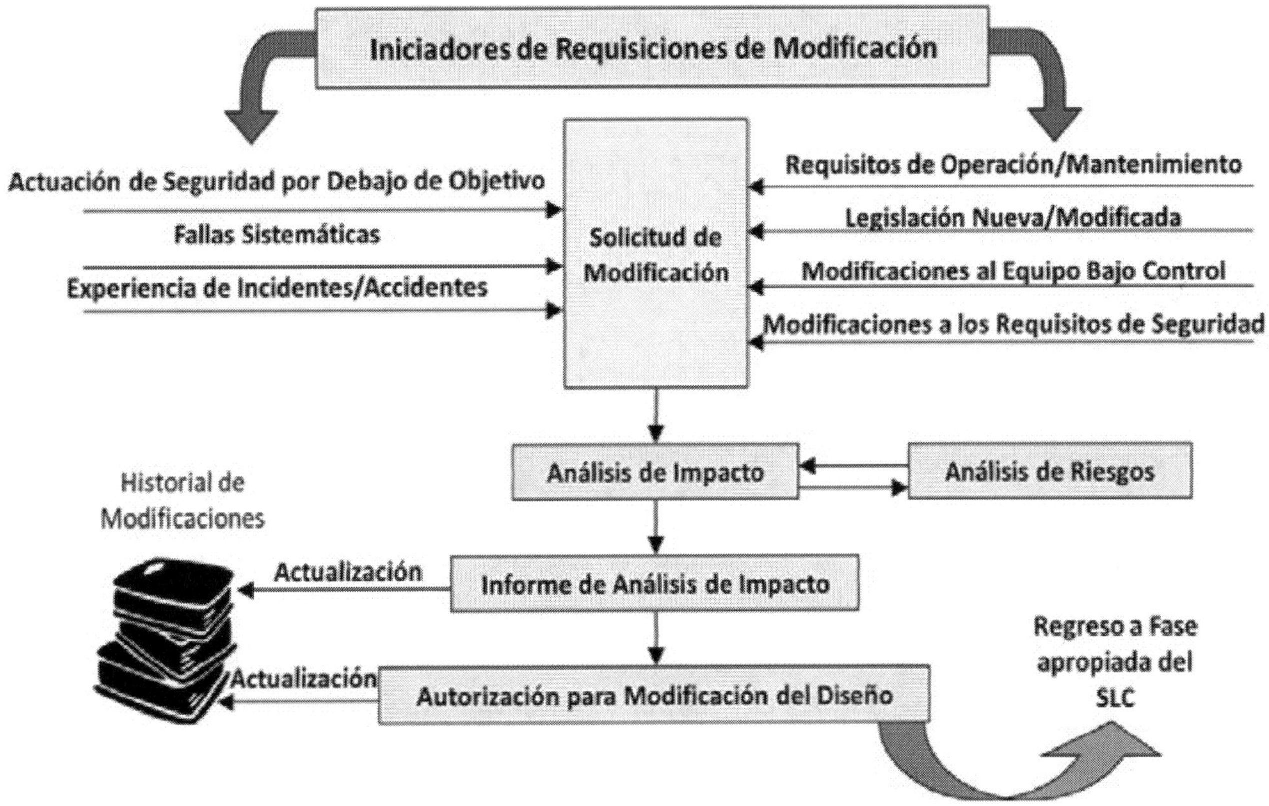

Figura 1. Causas Típicas de Solicitudes de Modificación del SIS

4 Análisis de Impacto del Cambio del SIS

Existen varios aspectos que deben ser incluidos en las obligaciones al realizar un análisis de impacto del cambio. Se recomienda que:

- El periodo de tiempo necesario para implementar el cambio sea indicado de forma clara y precisa.
- La disponibilidad de espacio de memoria (en el CPU del SIS) sea verificada.
- Sean determinados los efectos sobre el tiempo de respuesta de la SIF
- Se hayan definido los cambios en el CPU on-line versus off-line y los riesgos involucrados
- Las pruebas para comprobar si el cambio se llevó a cabo correctamente y que el SIS sigue su desempeño según el diseño requerido, tal y como se ha especificado en las SRS
- La configuración adecuada del histórico de eventos
- Las pruebas funcionales para comprobar que el cambio realizado no ha repercutido negativamente en otras partes del SIS que no fueron modificadas

"Cuando el análisis muestra que la modificación propuesta podría impactar a la seguridad, entonces debería haber un retorno a la primera fase del ciclo de vida del SIS afectado por la modificación" (IEC 61511, Cláusula 17.2.3). Lo anterior implicaría la revisión del análisis de riesgos del proceso (por ejemplo, del HAZOP). En la Figura 2 se muestra, como ejemplo, que la salida de la actividad de Modificación del SIS se vuelve la entrada a la actividad de Análisis de Peligros y Riesgos del Proceso.

Es importante hacer notar que la planeación de seguridad para la modificación y actividades de re-verificación, son específicamente requeridas en el estándar IEC 61511 (Cláusula 17.2.4). La Evaluación de la Seguridad Funcional (FSA por sus siglas en inglés) es ampliamente recomendable después de la modificación (Cláusula 17.2.6).

También se menciona en el estándar que "Toda la documentación afectada por la modificación debería ser actualizada" (IEC 61511, Cláusula 17.2.5), además de que es imprescindible llevar una bitácora de modificaciones (IEC 61511, Cláusula 17.2.7).

Los cambios en el SIS pueden realizarse sobre su hardware (por ejemplo, componentes), software (por ejemplo, configuración) y/o procedimientos relacionados (por ejemplo, pruebas periódicas).

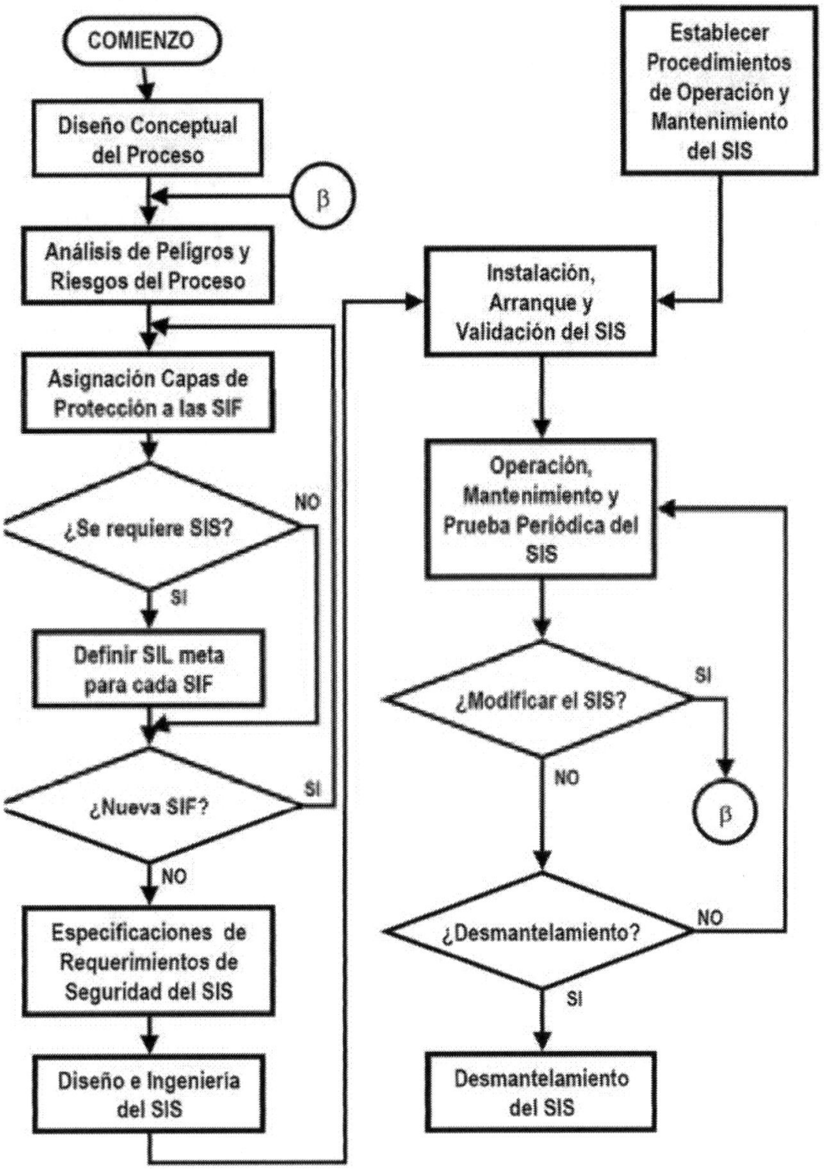

Figura 2. Ciclo de Vida de Seguridad Funcional

5 Procedimientos de Gestión del Cambio del SIS

Es común que las compañías cuenten con procedimientos propios de gestión del cambio en sus plantas, los cuales en ocasiones están enfocados solamente al proceso sin necesariamente abarcar al SIS. A continuación se muestran algunos puntos a considerar en la generación de procedimientos específicos de gestión del cambio del SIS:

- Los procedimientos deben incluir un método claro de identificar y solicitar el trabajo que hay que realizar, así como los peligros que podrían verse afectados
- Deberá especificarse la realización de un análisis para determinar el impacto en la seguridad funcional como resultado de las modificaciones propuestas
- Deberá indicarse que la actividad de modificación solo comienza si existe la apropiada autorización
- Deberá mantenerse la información de todos los cambios al SIS, incluyendo una descripción del cambio, razón del cambio, así como las aprobaciones requeridas para cada cambio
- El documento de gestión del cambio del SIS debería continuamente estar siendo mejorado, de manera que lo aprendido durante la aplicación de dicho procedimiento pueda ser utilizado para mejorar su efectividad
- Criterios para definir cuándo el remplazo de un componente por otro de similares características no es considerado un cambio propiamente.

6 Conclusión

Los cambios pueden ser generados desde múltiples áreas y con diversas intenciones, pero lo importante es que, una vez más, se pueden apoyar en las fases tempranas Ciclo de Vida de Seguridad, donde se realizan los análisis de peligros y riesgos, y de ser necesario, se determina un nuevo requisito de Nivel de Integridad de Seguridad (SIL) de la Función Instrumentada de Seguridad, mejorando los componentes de esta, así como su prueba y puesta en servicio. Para mayor detalle verifique la cláusula 17 del estándar IEC 61511.

7 Referencias

[1] IEC 61511:2016. "Functional Safety – Safety instrumented systems for the process industry sector". International Electrotechnical Commission. 2016

[2] Jonathan De Los Santos & Alejandro Esparza. "Seguridad Funcional, Incluyendo Ciberseguridad y Administración de Alarmas" exida, 2013.

[3] Iwan Van Beurden & William M. Goble. "Safety Instrumented System Design: Techniques and Design Verification" ISA, 2018.

[4] Manual de curso "FSE 102: Diseño y Operación del SIS", exida, 2019

¿Cómo implementar un Sistema de Gestión de Competencias de Seguridad Funcional en la industria de procesos?

Romel Rodríguez
CSF Consultoría en Seguridad Funcional
CC El Ingenio, Primer Piso, Oficina 24, Nueva Barcelona – Barcelona, CP 6010 - Edo. Anzoátegui, Venezuela
rodriguezrx@csf.com.ve

Eliana Berroteran
CSF Consultoría en Seguridad Funcional
CC El Ingenio, Primer Piso, Oficina 24, Nueva Barcelona – Barcelona, CP 6010 - Edo. Anzoátegui, Venezuela
berroterane@csf.com.ve

Prepared for Presentation at
American Institute of Chemical Engineers
2019 Spring Meeting and 15th Global Congress on Process Safety
New Orleans, LA
March 31 – April 3, 2019

¿Cómo implementar un Sistema de Gestión de Competencias de Seguridad Funcional en la industria de procesos?

Romel Rodríguez
CSF Consultoría en Seguridad Funcional
CC El Ingenio, Primer Piso, Oficina 24, Nueva Barcelona – Barcelona, CP 6001 - Edo. Anzoátegui, Venezuela
rodriguezrx@csf.com.ve

Eliana Berroteran
CSF Consultoría en Seguridad Funcional
CC El Ingenio, Primer Piso, Oficina 24, Nueva Barcelona – Barcelona, CP 6001 - Edo. Anzoátegui, Venezuela
berroterane@csf.com.ve

Keywords: competencias, sistema de gestión de competencias, seguridad funcional, sistemas instrumentados de seguridad.

Abstract

Ser competente es un requisito esencial para todos los involucrados en el diseño, instalación, operación y mantenimiento de sistemas relacionados con la seguridad; pero, a menudo, la palabra competencia es utilizada sin entender su significado. Competencia implica conocimiento, entrenamiento, experiencia, comportamiento, actitudes y habilidades para cumplir una tarea según un estándar definido. Por lo tanto, al gestionar las competencias se debe proveer un marco en el que se defina el perfil requerido en cada actividad de forma que se pueda asignar efectivamente cada rol y medir el desempeño del personal.

En este trabajo se muestra cómo implementar un sistema de gestión de competencias de seguridad funcional, desde el enfoque de un consultor de ingeniería, considerando el diseño del sistema, los criterios de evaluación y la identificación de necesidades de capacitación en materia de seguridad funcional.

1 Introducción

En forma general, la competencia, trabajando individualmente o en equipos de trabajo, está asociada a la idoneidad para hacer algo o intervenir en un asunto determinado. Es una combinación de habilidades que va más allá de la formación técnica para adquirir conocimientos. Implica el entendimiento de la operación los sistemas con los que se trabaja (procesos, sistemas de control, sistemas instrumentados de seguridad, etc.); la comprensión de procedimientos y prácticas de trabajo; así como comportamiento, actitudes y habilidades para asumir responsabilidades y cumplir una tarea de acuerdo con procedimientos, normas o estándares definidos.

Si trasladamos este concepto a la industria de los procesos, vemos que la competencia es un requisito esencial para aquellos que realizan actividades relacionadas a sistemas de seguridad, cualquier persona con cierto grado de incompetencia actuará como un posible eslabón débil que eventualmente podría conducir a un accidente.

2 ¿Por qué es importante gestionar las competencias en materia de seguridad funcional?

Debemos reconocer cuán importante es el tema de la seguridad funcional y la tendencia de procesos productivos utilizando tecnología cada vez más compleja; sistemas de seguridad con un mayores rendimientos y migraciones de funciones de seguridad mecánicas a unidades con controles electrónicos. Pero, trabajar en seguridad funcional es algo más que un trabajo rutinario, es una integración entre la comprensión de las normas relacionadas, la gestión y planificación de un ciclo de vida de seguridad y el manejo de parámetros relacionados con la seguridad en varios niveles.

Como vemos en los siguientes ejemplos, cualquier actividad relacionada con la seguridad funcional que sea realizada por personal con conocimiento limitado, sin la experiencia adecuada o una actitud displicente, puede resultar en sistemas con una integridad de seguridad inadecuada, debido a que, dada la relación entre cada actividad, las tareas que no se realicen de forma apropiada pueden afectar la seguridad a nivel global:

- Análisis de peligros y riesgos sin la identificación adecuada de escenarios peligrosos y estimación inadecuada del riesgo (con riesgos mayores a los que realmente pueden ser tolerados) o generación de recomendaciones que no son necesarias.
- Identificación inadecuada de capas de protección independientes (IPL) durante las sesiones de LOPA resultando en una estimación errónea del nivel de integridad de seguridad (SIL) requerido.
- Omisión de parámetros claves en el desempeño del sistema (por ejemplo, tiempo de actuación de la función instrumentada de seguridad (SIF)), durante la especificación de los requerimientos de seguridad.
- Falta de conocimiento sobre el tratamiento que debe ser aplicado a las IPL consideradas en el diseño para garantizar la reducción de riesgo que debe ofrecer cada SIF.
- Instalación y/o calibración inadecuada de equipos o instrumentos asociados al sistema instrumentado de seguridad (SIS) por desconocimiento del personal encargado.
- Degradación del desempeño del SIS por falta de comprobación de las premisas de diseño durante la operación y mantenimiento o incumplimiento de los plazos estimados de pruebas.
- Modificaciones en las condiciones de proceso, hardware, software, procedimientos, etc. sin evaluar el impacto sobre la integridad de la seguridad del SIS.

3 Sistemas de gestión de competencias

Un sistema de gestión de competencias involucra políticas, procedimientos y prácticas diseñadas para desarrollar, medir, mantener y mejorar continuamente las competencias del personal dentro de una organización.

Al desarrollar un sistema de gestión de competencias y definir el perfil de competencia mínimo que debe ser cubierto, se puede asegurar que los trabajadores tienen los atributos necesarios para realizar sus tareas, incluida la posibilidad de demostrar a reguladores, auditores y, en ocasiones a clientes, que las competencias del personal se están gestionando de forma efectiva. Así, es posible evaluar de forma individual las capacidades de cada integrante del equipo, asignar quienes podrán desarrollar efectivamente una función y, en los casos que sea necesario, implementar un plan de capacitación para desarrollar el potencial de cada persona.

Normalmente, la gestión de competencias es considerada como parte del sistema de gestión de una organización y se recomienda su integración con los procesos, procedimientos y estructuras organizativas, así como un monitoreo constante que permita su mantenimiento a largo plazo. En líneas generales, en un sistema de gestión de competencias se debe considerar:

- Definición de las áreas que debe cubrir el sistema de gestión de competencias para prevenir incidentes mayores y garantizar la seguridad.
- Relación entre competencias, roles y niveles de competencia.
- Sistemas que permitan el desarrollo de las competencias y la mejora continua del sistema (procedimientos de trabajo, entrenamiento, procesos de evaluación de competencias, indicadores de desempeño).
- Documentación de las competencias requeridas en cada área de trabajo.

4 Desarrollo de un sistema de gestión de competencias en seguridad funcional

La norma de seguridad funcional para la industria de los procesos IEC 61511 establece que las organizaciones que tengan responsabilidad de una o más fases del ciclo de vida de seguridad del SIS deben desarrollar y disponer de un mecanismo para gestionar las competencias de su personal, sin embrago, no se muestra orientación sobre cómo cumplir estos requisitos.

La normativa también establece que los involucrados en alguna de las fases del ciclo de vida de seguridad de los SIS deben ser competentes para realizar las actividades de las que son responsables. Por tanto, un modelo de gestión de competencias en seguridad funcional debe especificarse cuidadosamente y mostrar las habilidades y conocimientos técnicos particulares (medidos por niveles de competencia) de las tareas que corresponden a las diferentes actividades que se realicen.

El sistema de gestión de competencias de CSF, como empresa de consultoría en seguridad funcional, está enfocado en las primeras fases del ciclo de vida de seguridad de los SIS:

- Análisis de Peligros y Riesgos.
- Asignación de las Funciones de Seguridad a las Capas de Protección.
- Especificación de los Requerimientos Funcionales y de Seguridad para el SIS.
- Diseño Básico de las Funciones Instrumentadas de Seguridad.

(La instalación, operación y mantenimiento, no son parte de su proceso medular por lo que estas actividades no fueron consideradas).

En este trabajo solo se describen las tareas asociadas a la actividad de Análisis de Peligros y Riesgos que forman parte de los objetivos de la normativa de seguridad funcional, a saber:

- Realizar y documentar el análisis de peligros y riesgos.
- Identificar eventos peligrosos.
- Identificar causas y consecuencias de cada evento peligroso.
- Estimar el riesgo de cada evento peligroso.
- Identificar salvaguardas y funciones de seguridad requeridas.
- Verificar, Evaluar y Auditar el análisis de peligros y riesgos.

Una vez claras las actividades a realizar, se identificaron los roles requeridos. En seguridad funcional existen, al menos, cinco roles asociados a actividades particulares: Elaborador o Desarrollador, Verificador, Evaluador, Auditor y Gerente de la Seguridad Funcional.

Los mismos son definidos en términos genéricos, pero indicando las responsabilidades que corresponden a cada uno, lo que permite asignar las competencias más fácilmente. Aunque se debe aclarar que un individuo podría realizar varios roles (en diferentes actividades), ya sea individualmente o como parte de uno o más equipos de trabajo.

- Gerente de la Seguridad Funcional: es el responsable de la seguridad funcional a nivel global y encargado de planificar y coordinar actividades, recursos y de la asignación de personal. Coordina las evaluaciones y auditorias de la seguridad funcional incluyendo la comunicación y la integración de la documentación.

- Elaborador o Desarrollador: este rol aplica a personas, grupos de personas, departamentos, organizaciones o unidades responsables de ejecutar las actividades técnicas. Este rol podrá estar representado por las figuras de Líder, Secretario Técnico o Grupo de Análisis dependiendo de la actividad.

- Verificador: es responsable de ejecutar actividades técnicas para demostrar el cumplimiento de los objetivos y requisitos establecidos en cada una de las fases del ciclo de vida de seguridad (sin estar involucrado en el desarrollo de la actividad).

- Evaluador: persona encargada de juzgar y determinar, por medio de la revisión de toda la documentación, si la seguridad funcional ha sido alcanzada. Es encargado de revisar lo que el gerente de la seguridad funcional gestiona.

- Auditor: es responsable de determinar, a partir de registros, si se han aplicado los procedimientos adecuados, por personas con la competencia necesaria y con la frecuencia especificada.

Para los diferentes roles, se establecieron niveles de competencia para conocer el requisito mínimo que debe ser cubierto:

- Aprendiz: personal en proceso de aprendizaje y formación. Tiene conocimiento de la teoría y muestra la comprensión conceptual sobre el tema y los términos asociados. Puede realizar tareas rutinarias con supervisión significativa y sabe dónde obtener más información en caso de no tener una respuesta apropiada a determinado tema.

- Practicante Supervisado: personal con suficiente conocimiento y comprensión de las mejores prácticas, normativas y procedimientos de la organización para llevar a cabo tareas generales relacionadas con su trabajo. Realiza tareas fundamentales y rutinarias sin imponer una carga al Practicante o Experto que pudiese comprometer el rendimiento de la organización. Requiere supervisión ocasional, trabaja con otros y se encuentra en proceso de aumentar su funcionalidad, experiencia y capacidad.

- Practicante: personal con suficiente conocimiento y comprensión de las mejores prácticas, normativas y procedimientos de la organización para realizar tareas con una dirección mínima y sin necesidad de supervisión detallada. Tiene suficiente experiencia para completar las tareas asignadas de forma independiente, integra el trabajo con otras disciplinas y evalúa o compara diversas opciones o alternativas. El Practicante puede supervisar el trabajo realizado por un Practicante Supervisado.

- Experto: personal con habilidades y formación especializada que aplica soluciones creativas a problemas complejos y se mantiene al tanto de las tecnologías, soluciones, normas, y estándares relacionados. Posee amplia experiencia y conocimiento para trabajar en situaciones novedosas, hacer frente a múltiples problemas bajo presión y asumir responsabilidades.

- Líder: personal con amplia experiencia y conocimiento para coordinar recursos y asignar tareas y realizar de forma efectiva una actividad en particular.

Finalmente, fueron definidos los criterios de competencia (correspondientes a cada tarea) que buscan un estándar de desempeño. Para esto fue necesario analizar la naturaleza de la actividad, definiendo criterios medibles según los cuales evaluar o juzgar las competencias del personal.

A las tareas específicas de cada actividad se le atribuye un criterio de competencia por cada rol, ofreciendo una visión de la forma correcta de realizar una tarea, las habilidades técnicas, conocimientos y atributos personales requeridos, debido a que el nivel de competencia requerido en cada tarea puede ser diferente, incluso en un mismo rol.

Como vemos en la tabla 1, la asignación del personal está basada en el conocimiento de los requisitos de competencia de cada rol para las tareas específicas asociadas a una actividad (a efectos de este trabajo la actividad de Análisis de Peligros y Riesgos).

Tabla 1. Criterios de Competencia

Criterios de Competencia Análisis de Peligros y Riesgos (PHA)		Nivel de Competencia					
		Elaborador					
		Líder	Secretario	Verificador	Evaluador	Auditor	Gerente de la Seguridad Funcional
Realizar y documentar el análisis de peligros y riesgos	Conocimiento y aplicación de técnicas de análisis de peligros y riesgos como HAZOP, HAZID, WHAT IF, APP y Check List	3	2	3	2	NA	2
	Capacidad para explicar al grupo de trabajo la técnica de análisis de peligros y riesgos a utilizar	3	1	1	2	1	2
	Conocimiento práctico de la herramienta de software a utilizar (hojas de cálculo tipo MS Excel, procesadores de texto tipo MS Word, software Hazard Leader Review, PHA-Pro, PHA-Works, Open PHA, etc.)	2	3	2	2	NA	1
	Conocimiento sobre normas, regulaciones y procedimientos de trabajo que apliquen	3	1	2	4	4	3
	Capacidad para analizar y organizar información técnica, planos y documentos de ingeniería	3	2	2	3	4	3
	Capacidad para dividir el proceso a evaluar en secciones que faciliten el análisis	3	2	2	2	NA	NA
	Capacidad para sintetizar la información técnica manejada durante las sesiones de trabajo	3	3	2	2	NA	NA
	Capacidad para documentar los resultados obtenidos durante las sesiones de trabajo o cualquier otra actividad relevante	3	3	2	2	3	2
Identificar eventos peligrosos	Conocimiento, formación y experiencia para la aplicación de proceso y tecnología utilizada	3	2	2	3	NA	3
	Conocimiento sobre sistemas de control de procesos y sistemas de seguridad	3	1	2	2	NA	2
	Capacidad para identificar los eventos peligrosos que pudiesen iniciarse en el proceso, ambiente, falla de equipos, errores humanos o violaciones de seguridad a nivel de hardware y software	3	1	2	3	NA	2

Criterios de Competencia Análisis de Peligros y Riesgos (PHA)		Nivel de Competencia					
		Elaborador					
		Lider	Secretario	Verificador	Evaluador	Auditor	Gerente de la Seguridad Funcional
Identificar causas y consecuencias de cada evento peligroso	Capacidad para identificar las causas y consecuencias potenciales de un acontecimiento peligroso relacionado con los procesos a ser analizados	3	2	2	3	NA	3
	Capacidad para identificar los eventos peligrosos que pudiesen iniciarse en el proceso, ambiente, falla de equipos, errores humanos o violaciones de seguridad a nivel de hardware y software	3	1	3	3	NA	3
Estimar el riesgo de cada evento peligroso	Capacidad para determinar el nivel de daño causado a personas, medio ambiente y activos asociado a cada evento peligroso	3	2	2	3	NA	3
	Capacidad para determinar la reducción de riesgo requerido según los criterios de tolerabilidad de riesgo definidos	3	2	3	3	NA	3
Identificar salvaguardas y funciones de seguridad requeridas	Capacidad para identificar las funciones de seguridad, capas de protección y salvaguardas necesarias para controlar los riesgos	3	2	3	3	NA	3
	Capacidad para determinar la aplicabilidad de capas de protección independientes (IPL) y Funciones Instrumentadas de Seguridad (SIF)	3	2	3	3	NA	3
Verificar, Evaluar y Auditar el análisis de peligros y riesgos	Capacidad para determinar el cumplimiento de los requisitos del análisis de peligros y riesgos según la norma IEC-61511	3	2	3	3	3	3
	Capacidad para definir los procedimientos que deben ser utilizados al desarrollar cada actividad	3	2	2	4	3	4
	Capacidad para determinar si los procedimientos utilizados para realizar el análisis de peligros y riesgos fueron utilizados adecuadamente	3	2	3	4	4	4
	Capacidad para determinar si el personal asignado al proyecto tiene las competencias adecuadas según las tareas realizadas	NA	NA	NA	4	4	4
	Conocimiento sobre sistemas de gestión, guías y procedimientos de trabajo de la organización incluyendo las tareas de verificación y evaluación (assessment)	2	2	3	4	4	4

Criterios de Competencia Análisis de Peligros y Riesgos (PHA)		Nivel de Competencia					
		Elaborador		Verificador	Evaluador	Auditor	Gerente de la Seguridad Funcional
		Líder	Secretario				
Atributos Personales	Habilidad para comunicarse claramente dentro del grupo de trabajo	4	2	2	3	3	4
	Facilidad para relaciones interpersonales	4	2	2	3	3	4
	Capacidad para formar parte de un grupo de trabajo, colaborar y cooperar con los demás	4	2	2	3	3	4
	Capacidad de liderazgo, orientando la acción del grupo de trabajo en una dirección determinada para conseguir consenso	4	2	2	3	3	4
	Capacidad de negociar imparcialmente y dirigir o controlar una discusión utilizando técnicas ganar-ganar	4	2	2	3	3	4
	Capacidad de crear conocimiento nuevo mediante la improvisación, experimentación, creatividad o nuevas combinaciones del conocimiento existente	3	2	2	4	2	4
	Capacidad para identificar y planificar las actividades necesarias para cumplir con el alcance del proyecto	3	2	2	2	2	4
	Capacidad para cumplir con los tiempos de entrega establecidos como parte del proyecto	3	2	2	2	2	4
	Capacidad para coordinar las actividades necesarias para cumplir con el alcance del proyecto	3	2	2	2	2	4
	Capacidad para establecer eficazmente las metas y prioridades de sus tareas (considerando las acciones, plazos y recursos requeridos)	3	2	2	2	2	4
	Capacidad para identificar, seleccionar y dirigir recursos (personal, herramientas, equipamiento, etc.) para alcanzar objetivos en función de las prioridades de la organización y el desempeño del equipo	3	2	2	2	2	4
	Capacidad para trabajar bajo presión y superar problemas o dificultades laborales, comprometiéndose con el logro de los objetivos de la organización	3	2	2	2	2	4

Leyenda: Aprendiz: 1, Practicante Supervisado: 2, Practicante: 3, Experto: 4, Líder: 5.

Una vez especificados los criterios de competencia, es posible evaluar a una persona, y si cumple con el perfil especificado, la persona se considera competente para el rol en particular. En caso que la persona no cubra las competencias para ejercer un rol, no podrá ser asignada a esa tarea. De igual forma si una persona demuestra un nivel de competencia más alto que el requerido para el trabajo actual, debe ser considerado en futuras evaluaciones.

Si una vez evaluada una persona, se considera necesaria la formación de competencias, las deficiencias deben ser comunicadas a fin de proporcionar una visión clara de qué competencias necesitan, evitar la sobrestimación de habilidades y promover la aceptación de procedimientos y medidas de capacitación.

Finalmente, considerando que pueden haber cambios de tecnologías, normas, condiciones operacionales, etc. que pueden hacer que las competencias pueden cambiar con el tiempo, todas las personas deben ser evaluadas al menos una vez al año y se deben manejar indicadores de rendimiento para la implementación del sistema (por ejemplo, % de roles entrenados, % de evaluación de competencia completada, número de personas con déficit de competencia, etc.) para reducir la probabilidad de que se pase por alto el desempeño deficiente del personal, permitir que se tomen medidas que eviten el deterioro de las potencialidades de cada integrante del equipo y garantizar la mejora continua del sistema de gestión de competencias.

5 Conclusiones

- Para un consultor de ingeniería, un sistema de gestión de competencias puede proporcionar personal competente y motivado, asegurar que el personal está en capacidad de asumir cada rol y demostrar que los productos son entregados de conformidad con la norma IEC 61511.

- Es importante demostrar y evaluar la competencia de las personas, en lugar de suponer que son competentes al completar cursos de capacitación y tener cierto grado de experiencia, sin cuestionar si está en capacidad de ejercer el rol asignado, si fue evaluado el resultado del adiestramiento o si su desempeño en el trabajo fue satisfactorio. La evaluación de competencias puede revelar si el personal cumple con los requisitos de la norma IEC 61511 y reduce la probabilidad de que se pase por alto un rendimiento inferior al requerido.

- Mantener y desarrollar una competencia depende de la experiencia que se genere de la ejecución de una tarea en forma regular, por lo que las competencias deberán ser evaluadas, a través de revisiones periódicas y planificadas que permitan verificar el desempeño del personal.

6 Referencias

[1] International Electrotechnical Commission. Functional safety – Safety instrumented systems for the process industry sector. Geneva, Switzerland, 2016.

[2] CCPS. Guidelines for Defining Process Safety Competency Requirements, American Institute of Chemical Engineers, New York, NY, 2015.

[3] HSE. Managing competence for safety-related systems, Part 1 & 2, 2007.

[4] European Process Safety Centre. Process Safety Competence: How to Set up a Process Safety Competence Management System, 2013.

[5] IChemE Safety Centre. Guidance Process Safety Competency – a Model, 2015.

[6] P. Baybutt. “Competency requirements for process hazard analysis (PHA) teams”. Journal of Loss Prevention in the Process Industries, 2014.

[7] C. Sandom. “Competence Considerations for Systems Safety”.

Risk Assessment Activities Deployed by JXTG

Presenter/Tomoya Ohtani
JXTG Nippon Oil & Energy Corporation
Safety, Health & Environment Department
Staff General Manager
1-1-2, Ote-machi, Chiyoda-ku, Tokyo, Japan
Ohtani.tomoya@jxtg.com

Other Author/Hisashi Shibuya
JXTG Nippon Oil & Energy Corporation
Refining & Manufacturing Department
Process Technology & Engineering Group
Manager
1-1-2, Ote-machi, Chiyoda-ku, Tokyo, Japan
Shibuya.hisashi@jxtg.com

Prepared for Presentation at
American Institute of Chemical Engineers
2019 Spring Meeting and 15th Global Congress on Process Safety
New Orleans, LA
March 31 – April 3, 2019

Risk Assessment Activities Deployed by JXTG

Presenter/Tomoya Ohtani
JXTG Nippon Oil & Energy Corporation
Safety, Health & Environment Department
Staff General Manager
1-1-2, Ote-machi, Chiyoda-ku, Tokyo, Japan
Ohtani.tomoya@jxtg.com

Other Author/Hisashi Shibuya
JXTG Nippon Oil & Energy Corporation
Refining & Manufacturing Department
Process Technology & Engineering Group
Manager

Keywords: Risk Assessment, Hazard Identification, Line by Line HAZOP, Procedure-based HAZOP

Abstract

JXTG Nippon Oil & Energy Corporation (JXTG Energy) was formed by the merger of JX Energy and TonenGeneral Sekiyu K.K. in April 2017. A risk-based safe operations management system (SOMS) for process safety, personnel safety, health and environment has been structured based on the heritage-TonenGeneral safe operations management system to achieve a higher level of safe and stable operations.

Based on the following, I will introduce risk assessment activities now being implemented as an element of SOMS.

- Safe Operations Management System and Risk Assessment
- Activities for Process Safety
- Risk Assessment Execution Structure and Training
- Conclusion

1 Safe Operations Management System and Risk Assessment

A serious incident in refining or petrochemical production operations can threaten the lives and health not only of employees, but of the local community as well, thereby having a significant impact on society at large. As our company mission is to provide a safe and stable supply of oil and petrochemical products, it is very important for us to maintain safe operations and eliminate serious incidents, which threaten the safety and peace of society. To this end, we think risk assessment and management which includes hazard identification and risk mitigation is extremely important for maintaining safe and stable operations and is the foundation for the existence of our company's business.

Our safe operations management system (SOMS) is a risk-based process safety management system that has been implemented at all 16 of our refineries and chemical plants. The core elements structure of the system is shown in Fig. 1.

- Commitment to a high level of safety
- Understanding of hazards and risks
- Risk management
- Learning from experience

The core elements structure of the system comprises 12 pillars, called "elements'', on 4 main principles, which form the foundation.

Risk management is an important principle (foundation) in hazard identification and risk evaluation.

Fig.1 Safe Operations Management System – Core Elements Structure

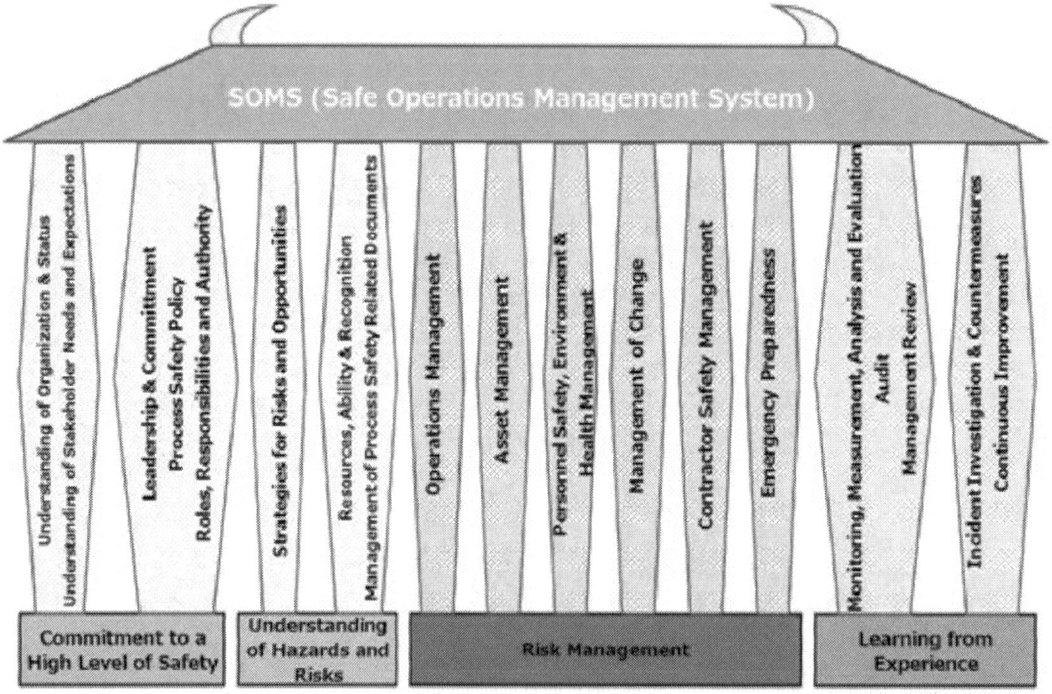

2 Activities for Process Safety

2.1 Risk Matrix and Risk Assessment Resources

We execute various kinds of process safety activities to mitigate the consequences in the event of a serious incident and loss of containment.

There is a tendency in Japan to advocate “zero risk” and set ideal targets of “absolute safety”, but it is impossible to reduce all risks to zero.

However, the mindset in Japan is gradually changing from “zero risk'' to “tolerable risk”. In the international safety standard ISO/IEC guide 51, safety is not defined as “zero risk'' or “absolute safety'', it is prescribed as follows: “The goal is to achieve tolerable risk for people, property and the environment and minimize adverse effects on the environment.''

Resources (staff, materials, money) are limited in a company. First, we think it is necessary to develop a risk assessment training structure and trained risk assessment leaders and screeners who can promote risk assessment activities effectively and in a timely manner, focusing on reducing serious risks. The key factor for success is maintaining the quality of risk assessment and risk management at the site and in the field.

We have built operating systems to enable us to conduct comprehensive, high-quality risk assessments for various work activities. Evaluation standards, procedures and guidelines have been established for the “Understanding of Hazards and Risks” principle using various risk assessment techniques.

In addition, we assign qualified risk assessment leaders and risk screeners at each site in all areas of our operations to promote process safety and ensure the quality of risk assessments through structured risk assessment training and qualification for site members (refer to Fig. 4).

Adopting a risk-based process safety mindset, we evaluate risk levels using a company-wide risk matrix, and determine action to be taken based on risk levels.

We have defined five consequence levels for four areas: safety and health, environment, society, and financial. We have also defined five levels of probability. Through a combination of consequence levels and probability levels, we define the four risk categories as follows in a 5-by-5 risk matrix. We also define approval levels and action to be taken at each risk level as shown below.

Category 1: Unacceptable risk. Category 1 risks must be mitigated immediately and the risk level must be reduced by at least one level.

Category 2: Unacceptable risk in principle. Category 2 risks must be mitigated in order of priority and the risk level must be reduced by at least one level.

Category 3: Acceptable risk. However, the acceptability of Category 3 risks is determined by whether or not they can be mitigated through a relatively low investment of costs and resources.

Category4: Acceptable risk. Further risk mitigation is usually unnecessary.

Fig. 2 Risk Matrix with Approval Levels

A	3	2	1	1	1
B	3	3	2	1	1
C	4	3	2	1	1
D	4	4	3	2	1
E	4	4	4	3	2
	V	IV	III	II	I

Risk Category	Continue Operations Until Resolution		Residual Risk
	Approval	Notify	
1	NA	Director	NA
2	3rd level	Site Manager	Site Manager
3	4th level	3rd level	3rd level
4	No need	No need	GM

2.2 Hazard Identification

In our company, risk assessments to identify hazards are categorized as comprehensive risk assessments and ad-hoc risk assessments. Comprehensive risk assessments include line-by-line HAZOP, procedure-based HAZOP, potential high-consequence risk assessment, risk assessment for safety valves and firefighting facilities, etc. Comprehensive risk assessments are managed by the site and the timing is scheduled in a 10-year plan. Ad-hoc risk assessments include risk assessments for routine work, such as operation and maintenance activities, management of change, patrol findings, etc., equipment repair planning, incident investigation inside and outside the company, and SOMS audit follow-up.

2.3 Line-by-line HAZOP and Procedure-based HAZOP

Line-by-line HAZOP: First, we split the P&ID into multiple nodes. Afterwards, we evaluate the effects of operational deviation for each node using guide words. One feature of our HAZOP technique is that our guide word list includes not only operational deviation-related words, but also safety engineering standards, so we also evaluate the effects of deviation from safety engineering standards. In addition, we assign HAZOP leaders from among personnel who have specialized in-house training and qualifications, and also belong to an organization outside the organization in charge of the operation for which the HAZOP is being conducted. This is to maintain the quality of HAZOP. We also conduct field surveys for fireproofing and water spray facilities, safety critical devices bypass control, etc. that cannot be identified by documents (P&ID and related design documents).

Procedure-based HAZOP: Procedure-based HAZOP is conducted for procedures involved in start-up, shut-down, emergency shut-down and potentially high-consequence routine operations. The objective of procedure-based HAZOP is to ensure that non-routine and risky operations like start-up, shut-down, emergency shut-down and potentially high-consequence routine operations are executed more safely and appropriately through procedure improvement. First, we evaluate the potential risk level by scoring the level of risk in terms of both fire/explosion and toxicity, and determine whether to conduct a detailed procedure review depending on the total score of the risk level. The procedure-based HAZOP teams, led by the leader, review the procedures step by step to improve the procedures as well as the warning and caution indications, and to include potential risks when errors occur and suggest double checks by operators for the safety-critical steps in the procedures. Similar to line-by-line HAZOP leaders, procedure-based HAZOP leaders have the necessary qualifications and are selected from outside the organization where the HAZOP is being conducted.

2.4 Ad-hoc Risk Assessment

Fig. 3 Ad-hoc Risk Assessment Work Flow

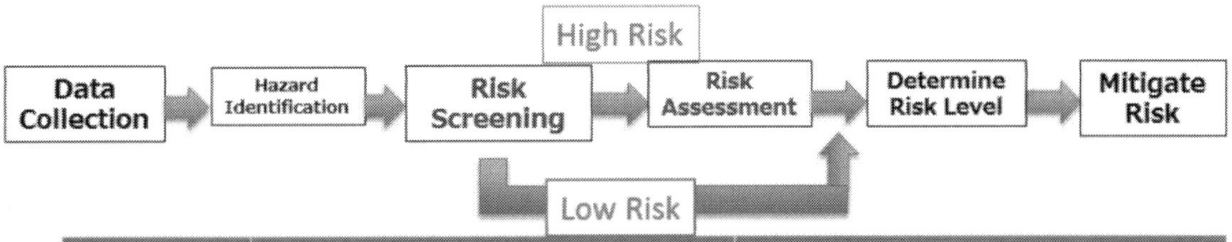

	Risk Screening	Risk Assessment
Objective	Evaluate formal RA necessity (evaluate risk category as potential 1 or 2)	Thorough evaluation pf risk category by RA team
Scope	All hazards	Hazards with potential risk category 1 or 2
Procedure	Develop scenarios → Evaluate potential consequence and probability → Evaluate potential risk category	For potential risk category 1 or 2, same procedure done by RA team
Members	Risk screener	RA team led by RA leader

Risk Assessment

- ✓ Scopes of consequences : Safety, health, environment, society, finance etc.
- ✓ Probability : Databases based on credible data
- ✓ Risk category : Defined by colored risk matrix for company-wide use

Ad-hoc risk assessment consists of two steps, risk screening and risk assessment, as shown in Fig. 3. First, the risk screener evaluates the risk level for all potential hazards. The formal risk assessment by the team is executed for potential category 1 or 2 items. The company-wide risk matrix is used both for risk screening and risk assessment. Once the risk level has been determined, we examine reasonable risk mitigation and get the appropriate approval for continuance of operation until resolution or residual risk. One key factor for success is appropriately communicating hazard information to risk screeners and risk assessment leaders, and another is the quality of risk screeners and risk assessment leaders.

We also have a training and qualification system for risk screeners and risk assessment leaders to maintain and improve their skills and knowledge. In addition, in the selection of risk assessment personnel, we have requirements for knowledge and experience.

3 Risk Assessment Execution Structure and Training

3.1 Risk Assessment Execution Structure

The risk assessment execution structure is shown in Fig. 4. Each site has at least one process safety specialist (PSS), who promotes site process safety activities including ensuring the quality of site risk assessments. For ad-hoc risk assessments, risk assessment leaders and risk screeners are key personnel, and for comprehensive risk assessments such as line-by-line HAZOP, procedure-based HAZOP and potential high-consequence risk assessment, site process safety specialists and comprehensive risk assessment leaders have key roles.

Fig. 4 Risk Assessment Execution Structure

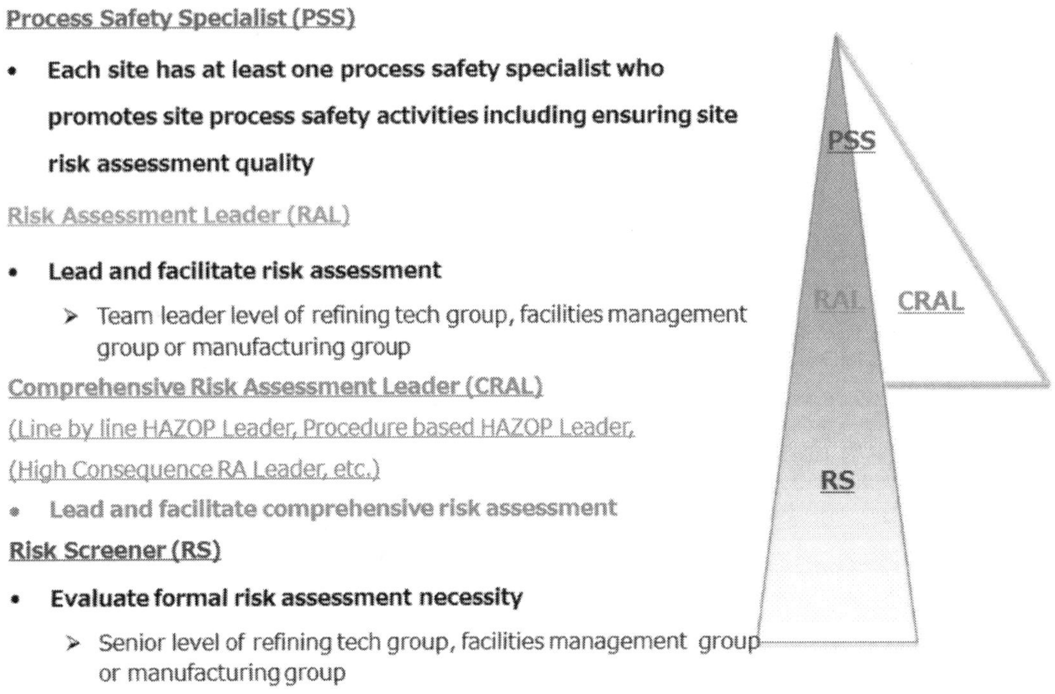

The main roles of a process safety specialist are as follows.

- Develop plan for, lead and facilitate comprehensive risk assessment
- Lead, facilitate and support ad-hoc risk assessment in the event of MOC such as design change, operation change, etc.
- Support development of risk mitigation plans based on risk assessments
- Investigate process safety incidents and support development of countermeasure plans
- Support development of firefighting tactics for process safety incidents such as BLEVE, VCE, toxic gas release, tank fires, etc.
- Implement enlightenment and training activities for process safety

The risk assessment leader (RAL) plays a lead role for ad-hoc risk assessments with the support of RA personnel, and is generally selected from team leaders of the refining tech group, asset management group and operation group.

The comprehensive risk assessment leader (CRAL) plays a lead role in comprehensive risk assessments such as line-by-line HAZOP, procedure-based HAZOP, potential high-consequence risk assessment, etc.

Risk screeners (RS) are generally selected from staffs of the refining tech group, facilities management group and manufacturing group.

Requirements for training, operations experience and qualifications have been established for all risk assessment key positions.

3.2 Required Training for Risk Assessment Core Positions

Risk Assessment training is required for all risk assessment key positions such as risk screeners and risk assessment leaders. In addition, risk assessment leaders must have at least 10 years' experience in the refining or petrochemical industries, as well as RA participation experience as a member of the RA team.

HAZOP leader candidates are required to have safety engineering basic training and HAZOP method training, and must also lead a line-by-line HAZOP under the observation a qualified HAZOP leader, after which they are qualified with the approval of the qualified HAZOP leader.

A HAZOP leader's qualifications are not permanent. In order to maintain his or her qualifications, a HAZOP leader must lead a HAZOP at least once every two years and must also learn about revisions in safety engineering standards in a timely manner.

Fig. 5 Qualification Criteria for Risk Assessment Key Positions

Main qualification criteria for risk screener

	Risk Screener	RA Leader
Basic RA training	○	○
10+ years experience in petro/petrochem industries	—	○
Formal RA participation	—	○

Main qualification criteria for HAZOP leader

	HAZOP Leader
Basic safety design training	○
SS/Non-SS HAZOP training	○
Experience leading HAZOP supported by certified HAZOP leader	○

4 Conclusion

Here we have touched on the risk assessment activities currently being implemented at JXTG Nippon Oil & Energy Corporation.

Since the April 2017 company merger, we have developed and introduced risk assessment using the company-wide risk matrix. As of now, all risk assessment key positions have been trained and assigned. All 16 sites have started two types of risk assessments: comprehensive risk assessment and ad-hoc risk assessment.

All risk assessment activities are core elements of the risk-based safe operations management system (SOMS). Under the SOMS PDCA cycle, we aim to achieve continuous improvement in our risk assessments and prevent serious process safety incidents.

Thank you for your attention.

Analysis of Spontaneous Ignition in Various Polymer Resin

Takashi Odagiri
Mitsubishi Chemical Corporation
1 Toho-cho, Yokkaichi, Mie, Japans
odagiri.takashi.mb@m-chemical.co.jp

Kouya Murai
Mitsubishi Chemical Corporation
1 Toho-cho, Yokkaichi, Mie, Japans
murai.kouya.mc@m-chemical.co.jp

Shuichi Tokutake
Mitsubishi Chemical Corporation
1 Toho-cho, Yokkaichi, Mie, Japans
tokutake.shuichi.mn@m-chemical.co.jp

Keywords: Self-heat, spontaneous ignition, polymer

Abstract

In this study, two types of polymers were evaluated for their spontaneous ignition behavior through thermal analysis and bulk volume analysis. Thermal analysis of polymers using differential scanning calorimetry (DSC) in presence of oxygen showed exothermic onset after melting point. Wire basket (WB) experiments showed that bulk volume of polymers can ignite at much lower temperature than DSC exothermic onset. The critical conditions of spontaneous ignition were determined through sets of wire basket experiments of varying sizes in accordance with European standards DIN EN 15188: determination of the spontaneous ignition behavior of dust accumulations. By understanding the critical condition, we were able to validate the safe condition to avoid hazards of spontaneous ignition.

1 Introduction

A bulk volume of combustible material such as polymers can self-heat through slow reaction with oxygen in its void. Such polymers when handled in large industrial quantity may accumulate enough heat to undergo spontaneous ignition leading to undesirable fire hazards. Polymers in industrial manufacturing are packaged, stored and processed in various volume and temperature conditions that such hazards need to be considered. The condition governing self-heating is the heat balance in the bulk material to generate heat through oxidation reaction and release heat to the surrounding. Assuming that sufficient oxygen is available in the bulk volume [1], oxidation reaction on the surface of polymer follows Arrhenius type reaction and the heat generation is proportional to the volume of the bulk. Heat dissipation from the bulk to the surrounding atmosphere takes place across the bulk surface through heat transfer and is proportional to the surface area. From the balance of heat generation and heat dissipation, it is clear that the volume (V) or the volume to surface area ratio (V/A) is the key factor affecting hazards in the bulk material. In Japan there are various standard tests available to evaluate material undergoing spontaneous ignition, however these test are designed for evaluation of pyrophoric substances that ignite within minutes from contacting oxygen and does not require heat accumulation from slowly self-heating of the bulk material. UN recommendation for the transportation of dangerous goods, Division 4.2, Self-heating substance [2] does consider volume of the material however its purpose is to classify the potential hazard at the transport condition and it is not meant to determine the critical condition of spontaneous ignition. The European standard DIN EN 15188 [3], takes into account the slow self-heating of the bulk material and determine the critical conditions necessary for spontaneous ignition. In this study, spontaneous ignition characteristics of self -heating polymers are determined by following DIN EN 15188, and the results are used to set procedures in safe handling of the polymers.

2 Experiment

2.1 Model substance

Two model case polymers are selected in this study. Polymer A, is a white powder resin that have no anti-oxidation stabilizer and is known to have fuming incidents due to self-heating during storage. Based on experience, package volume was reduced to minimize hazards of self-heating, however the effect of volume change had not been evaluated. Polymer B is more stable than Polymer A that it hardly self-heat at normal storage condition. However when it is treated in hot air dryer for post processing, cases has been reported for Polymer B to self-heat and degrade. The post processing procedure required to set safe treatment condition.

2.2 Analysis method

Thermal analysis is performed firstly using differential scanning calorimetry (DSC) to evaluate the polymers. Wire basket (WB) experiment is then performed to evaluate for the bulk material characteristics.

2.2.1 DSC Analysis

Thermal analyses on Polymer A were performed using High Pressure DSC (HP-DSC) under nitrogen atmosphere and air atmosphere. Figure 1 shows the DSC curves obtained from the two measurements. The two DSC curves show a common melting point at 163℃. Above the melting point, the curve measured under nitrogen atmosphere (dotted line) does not show any peaks until it reaches 400℃, where it start thermal degradation. The curve measured under air atmosphere (solid line) shows a sharp exothermic onset at 164℃ and the exothermic signal continues up to 600℃. The difference in the two curves is due to presence of oxygen in the system, and that the Polymer A can undergo oxidation reaction when heated above 164℃.

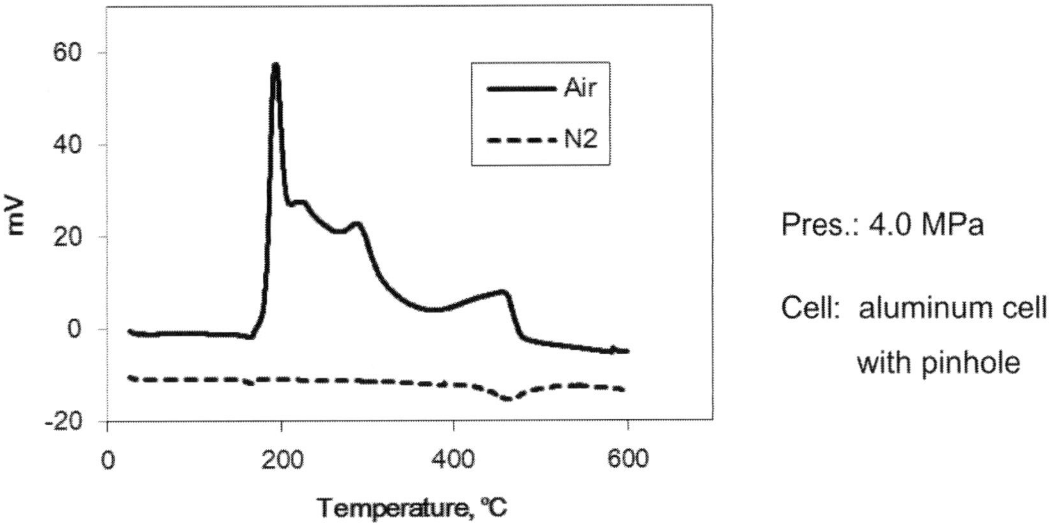

Figure 1. HP-DSC curve for Polymer A, under air and nitrogen atmosphere

2.2.3 Wire basket test

Wire basket (WB) experiment based on DIN EN 15188 consists of temperature controlled oven and wire basket of various volumes as sown in Figure 2.

Figure 2. Wire basket of volume 137, 237, 703 and 1140mL and wire basket suspended in hot storage oven.

Basket is made of wire mesh that can contain the sample and air can pass through. Oven temperature is measured at half the distance between the basket and the oven wall. Sample temperature is measured at the center of the basket.

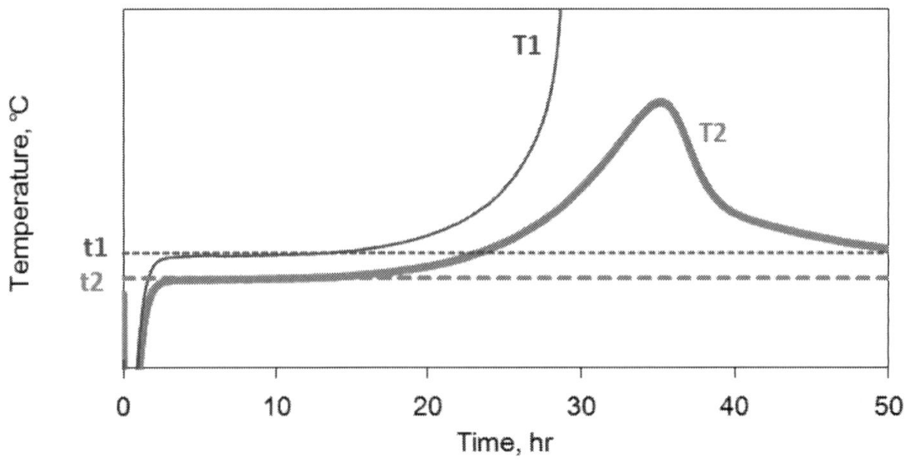

Figure 3. Temperature profile at center of sample (solid line) and oven temperature (dotted line).

In an experiment, oven is kept at constant temperature and sample temperature is monitored to see if the samples ignite or not. Figure 3 shows ideal temperature profile from two sets of experiments performed with same basket size at different oven temperatures. Dotted line shows the oven temperature of the two experiments. Solid line shows the sample temperatures of two respective oven temperatures. In DIN EN 15188 standard, ignition is defined with temperature rise greater than 60℃ and/or the temperature profile having an inflection point. If the oven temperature is set higher than self-ignition temperature (T_{SI}), sample temperature will rise rapidly, indicating ignition as in T1. If the temperature is at T_{SI} or lower, sample temperature will rise above the oven temperature until the heat dissipation overcome heat generation, then temperature will come down to slightly higher temperature and reach steady state as in T2. In order to ensure that highest non-igniting oven temperature is T_{SI}, the experimental oven temperature between t1 and t2 should be less than 2℃.

WB experiments for Polymer A was performed using four different size baskets as shown in Figure 4. For each sizes, at least three runs were performed to determine T_{SI}. Oven temperatures and respective sample temperature profiles are indicated using same color. Blue is lowest oven temperature that the Polymer A ignited. Red is highest oven temperature that the sample did not ignite, defined as self-ignition temperature (T_{SI}). Orange is lower non-ignition temperature.

Ignited temperature profiles in blue have inflection point in which the temperature raised rapidly; however, the temperature rise stops shortly after. It can be assumed that the bulk core has reached melting point and the resin fused together. With closing of the void, the oxidation reaction stops and the temperature drop back to the oven temperature. The sample may not have ignited as defined in DIN EN 15188, however we can see from the picture of the tested sample in Figure 5 that it is no-longer acceptable as an industrial product, and such conditions should be avoided.

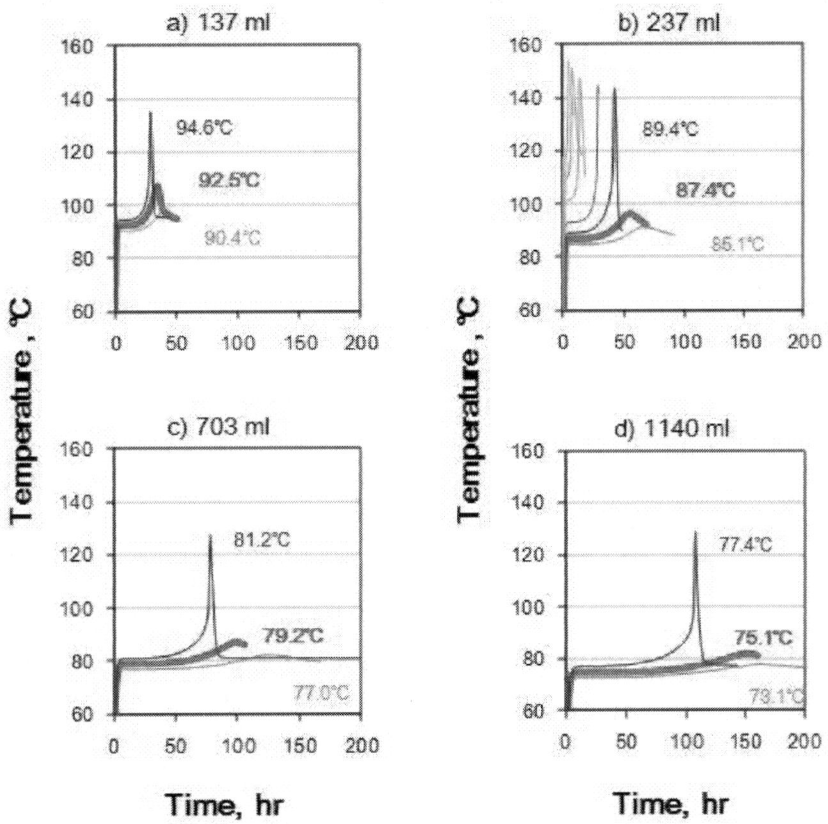

Figure 4. Temperature profile of Polymer A in hot storage and corresponding storage temperatures; ignition temperature, highest non-ignition temperature (T_{SI}) and non-ignition temperature. Wire basket size a) 137mL, b) 237mL, c) 703mL and d) 1140mL.

Figure 5. WB tested sample after ignition.

Experimental results are summarized in Table 1. It can be seen from the table that as the sample volume increased from experiment 1 through 4, T_{SI} reduces and takes longer time to ignite.

Tabe 1. Summary of wire basket experimental results: Polymer A sample volume, self-ignition temperature and induction time

Experiment Run	Volume V (mL)	Self-Ignition Temperature T_{SI} (℃)	Induction Time θ (hr)
1	137	92.5	24.8
2	237	87.4	35.3
3	703	79.2	68.8
4	1140	75.1	96.5

3. Result and discussion

In WB experiment, we cannot determine the heat generation (Q) and heat dissipation (q), however for any given temperature conditions, Q and q are proportional to the volume (V) and surface are (A) respectively (Eqn. 1 and 2). Heat balance can then be expressed as volume to surface area ratio (V/A) that can be determined from the basket size and geometry (Eqn. 3).

$$Q \propto V \text{ and } q \propto A \qquad \text{(Eqn. 1 and 2)}$$

$$Q/q \propto V/A \qquad \text{(Eqn. 3)}$$

T_{SI} results are plotted on pseudo-Arrhenius plot (Figure 6). This is a practical way to understand the critical condition of the experimental results. From Figure 6, it can be seen that T_{SI} for $1m^3$ package is 30℃, and therefore it is very likely that it can self-heat and ignite if it is stored in ordinary storage condition. Based on this result, it can be seen that T_{SI} for reduced size package of 20kg bag (about 40L) is 50℃ and much less likely to self-heat in ordinary storage.

According to thermal explosion theory developed by Frank-Kamenetzkii [4], correlation between characteristic length of the sample (r) and its $1/T_{SI}$ is given the following equation (Eqn. 4);

$$\ln\left(\delta_{cr} \cdot \frac{T_{SI}^2}{r^2}\right) = \ln\left(\frac{E \cdot H_0 \cdot \rho \cdot k_0}{R \cdot \lambda}\right) - \frac{E}{R} \cdot \frac{1}{T_{SI}} \qquad \text{(Eqn. 4)}$$

Figure 7 shows an Arrhenius plot with critical Frank-Kamenetzkii parameter (δ_{cr}), corrected to Thomas for cube. From the plot, the activation energy (E) and the pre-exponential factor (k_0)

can be determined as 76,555 J/mol and 1.14E+4 s^{-1} respectively. These kinetic parameters can be used in more detailed simulation that can consider actual shapes and conditions of polymers handling in industry.

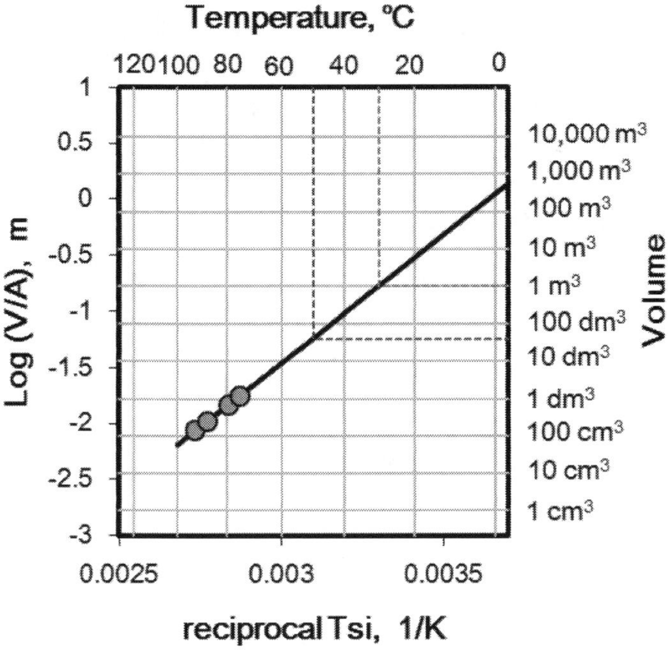

Figure 6. Pseudo-Arrhenius plot according to DIN EN 15188.

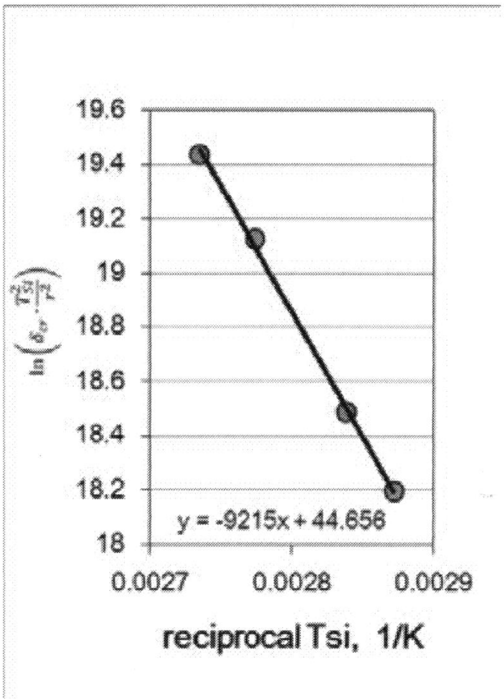

Figure 7. Arrhenius plot according to DIN EN 15188.

Induction time is another factor that need to be considered if the condition allows enough self-heating and ignite. If the storage temperature is above T_{SI}, meaning that heat generation is greater than heat dissipation, temperature of the core is increased with oxidation reaction without influence of heat dissipation. Induction time of ignition for the cases that ignited is independent of the sample size. The induction time of ignition is plotted on Figure 8, in which all data lie on single line regardless of the size.

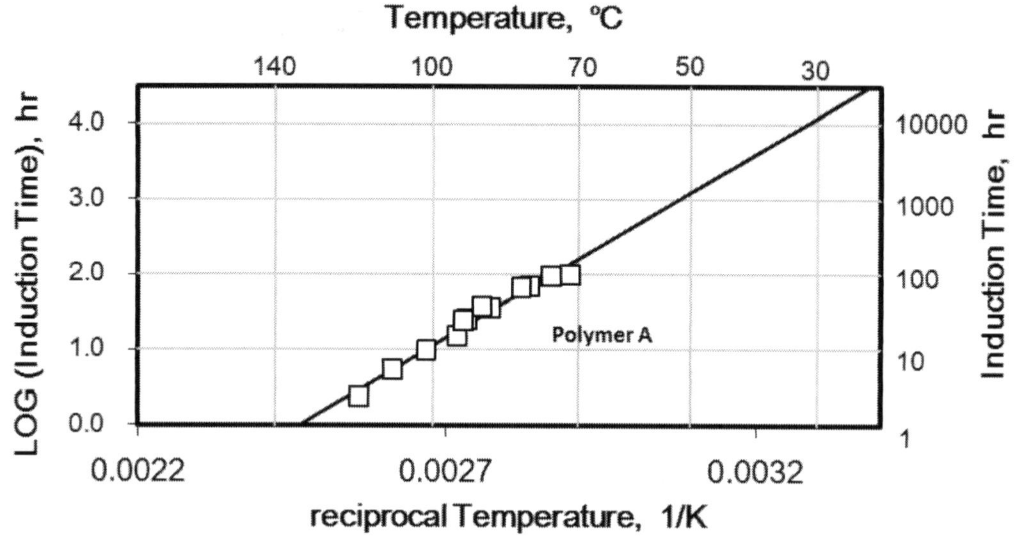

Figure 8. Induction time of ignition. Result from all experiment at oven temperature T_{SI} and higher.

Wire basket experiment for model Polymer B was also performed in comparison. Polymer B is known to be thermally stable and it hardly self-heat at ambient condition. However, when Polymer B is heated in hot air dryer for prolonged duration, it can self-heat and possibly ignite. Temperature profile from WB experiment for Polymer B, using three different size baskets are shown in Figure 9. It can be seen from the profile of 134mL sample, Polymer B does not have "sharp" distinction of the temperature profile between ignition run and no ignition run. Experiments with larger sample size did not show the distinction either. If conditions for Polymer B to reach steady state without ignition do exist, it would require much bigger sample, lower temperature and/or longer time.

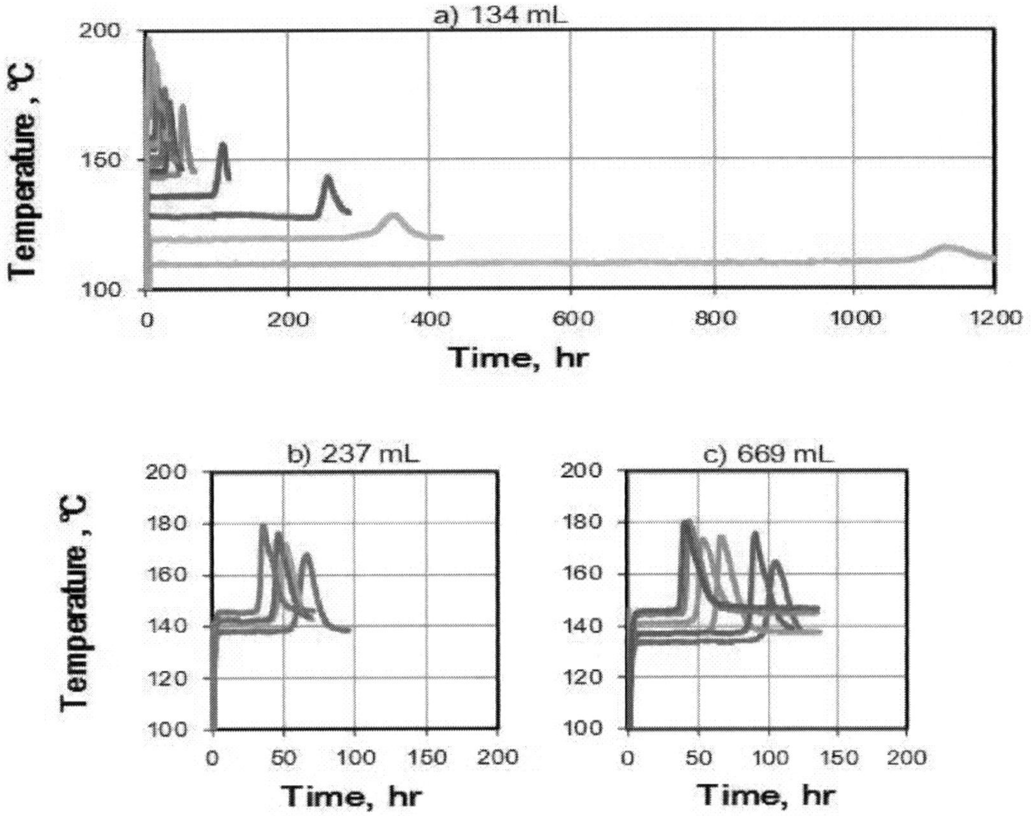

Figure 9. Temperature profile of Polymer B in hot storage.

Induction time of Polymer B is plotted together with Polymer A in Figure 10. Polymer B is more stable than Polymer A, that induction time is two orders magnitude longer. Storage of Polymer B at normal condition is not important because even if it could self-heat and ignite, the induction time is estimated to be in the order of few years. However, because Polymer B needs to be dried in hot air dryer, operation time and temperature must be kept below the temperature - induction time line. Based on the induction time, safe drying condition for Polymer B was set at 140℃, and 4 hours (Figure 10, filled circle)

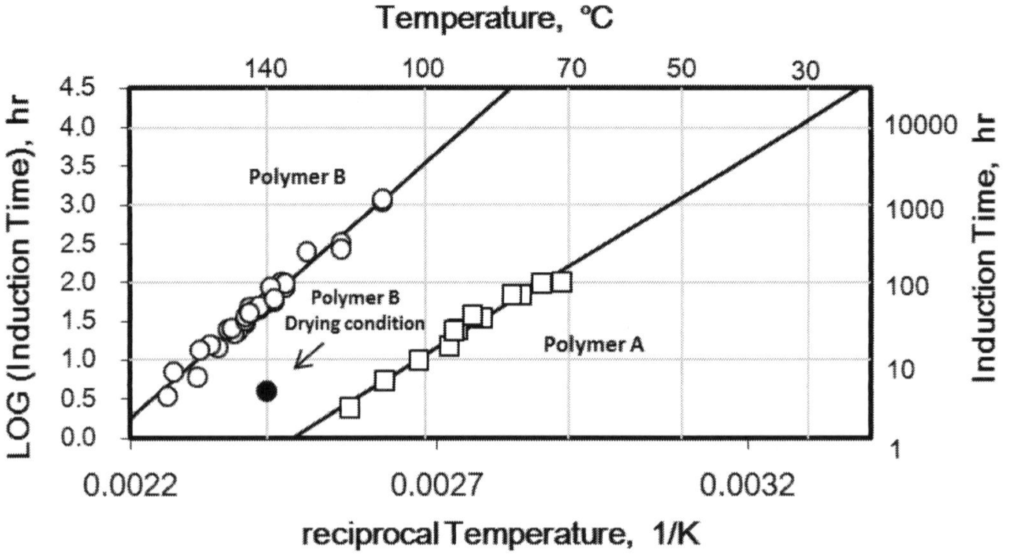

Figure 10. Induction time of ignition. Result from all experiment at oven temperature T_{SI} and higher. Safe drying condition for Polymer B (filled circle)

4. Conclusion

In this study, self-ignition characteristics of the two polymers A and B were studied by applying thermal explosion theory by following DIN EN 15188 standards. For Polymer A, we were able to determine the size dependent T_{SI} and confirm the danger in $1m^3$ package and reducing the package size to 40L can reduce the danger significantly. For Polymer B, we were not able to determine the critical condition that separate ignition and non-ignition conditions. However, by determining the correlation between temperature and induction time, we were able to recommend safe hot air drying condition by limiting time and temperature.

Standard tests for evaluating self-ignition characteristics had not been adopted in Japan, however can be very useful in safe handling of polymer and other combustible material.

5. Acknowledgment

The authors wish to thank Dr. Schmidt for his effort in generating data and having many discussions to deepen our understanding.

6. References

[1] K. Takeda and K. Akita. J. of Japan Soc. for Safety Engr., Vol 14, No. 3, 1975, pp131-135.

[2] UN ESC: Recommendations on the Transport of Dangerous Goods 14th ed.

[3] European Standard DIN EN 15188: 2007-11, Determination of spontaneous ignition behavior of dust accumulation

[4] W. Hensel. "The dependence of self ignition temperatures of dust upon grain size", Archivum Combustions, Vol. 7 1987 No. 1-2

19 AIChE Spring Meeting
15TH GLOBAL CONGRESS ON PROCESS SAFETY
NEW ORLEANS • 2019

Improvement of Piping Integrity Management with Advanced Technology

Suguru Koyanagi
Idemitsu Kosan Co., Ltd.(Manufacturing&Technology Dept.)
Safety&Environmental Section in Tokuyama Complex
1-1, Shingu-cho,shunan-shi, Yamaguchi, 745-8613, Japan
suguru.koyanagi@idemitsu.com

Yuma Okamoto/ Suguru Shojiguchi
Idemitsu Kosan Co., Ltd.(Manufacturing&Technology Dept.)
Engineering & Manegement Section in Tokuyama Complex
1-1, Shingu-cho,shunan-shi, Yamaguchi, 745-8613, Japan
yuma.okamoto@idemitsu.com / suguru.shojiguchi@idemitsu.com

Prepared for Presentation at
American Institute of Chemical Engineers
2019 Spring Meeting and 15th Global Congress on Process Safety
New Orleans, LA
March 31 – April 3, 2019

Improvement of Piping Integrity Management with Advanced Technology

Suguru Koyanagi
Idemitsu Kosan Co., Ltd.(Manufacturing&Technology Dept.)
Safety&Environmental Section in Tokuyama Complex
1-1, Shingu-cho,shunan-shi, Yamaguchi, 745-8613, Japan
suguru.koyanagi@idemitsu.com

Yuma Okamoto/ Suguru Shojiguchi
Idemitsu Kosan Co., Ltd.(Manufacturing&Technology Dept.)
Engineering & Manegement Section in Tokuyama Complex

Keywords: Asset Integrity, Advanced technology

Abstract

Asset Integrity takes precedence over anything as the plant ages and deteriorates. IDEMITSU KOSAN Co., Ltd. is adopting advanced technology aiming at high level Asset Integrity. In this presentation, we introduce two examples of Asset Integrity management that utilize advanced technologies.

- Entire-length inspection of the underground pipeline between complexes using "Intelligent PIG"
- Ultrasonic wall thickness continuous monitoring system

As a result, we were able to obtain higher asset integrity by applying them to underground pipeline integrity management. Furthermore, we also confirmed that these technologies can reduce inspection cost and workload.

1 Introduction

In IDEMITSU Tokuyama complex, there are many aged plants and pipelines over 40 years since construction. We have experienced some unpredictable defects on them due to the change of circumstances surrounding us.
Therefore, IDEMITSU has adopted advanced technology of inspections aiming at higher Asset Integrity. In this paper, we introduce two examples of asset integrity management concerning underground pipe corrosion.

[Two examples for an underground pipe corrosion inspection]
(1) Entire-length inspection of the underground pipeline between complexes using "Intelligent PIG"
(2) Ultrasonic wall thickness continuous monitoring system

2 Pipe corrosion management with advanced technologies

2.1 Entire-length inspection of a pipeline between complexes using "Intelligent PIG"

There are long-distance underground pipelines between factories in the Shunan petrochemical complex and the asset integrity management of the underground pipeline is important. If the underground pipeline suffers damages from such as corrosion, necessary oil or gas cannot be pipelined in the complexes and some factories could not be able to continue the operation.

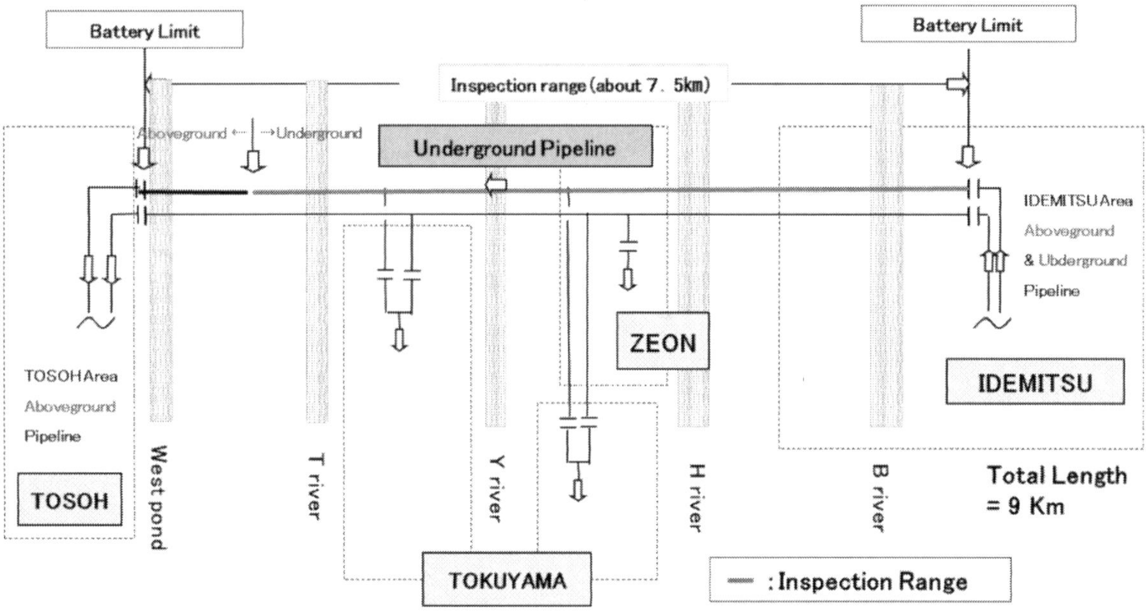

Figure 1. The long distance pipeline of Syunan petrochemical complex

As shown in Figure 1, most part of the pipeline lays underground. The external surface corrosion for aboveground part of the pipeline can be inspected visually. On the other hand, as shown in Figure 2, indirect investigation has been being employed for the underground part to estimate the corrosion tendency of entire-length of the underground pipeline by monitoring the state of the cathodic protection and direct visual inspection results by partially digging of pipeline as a representative data.

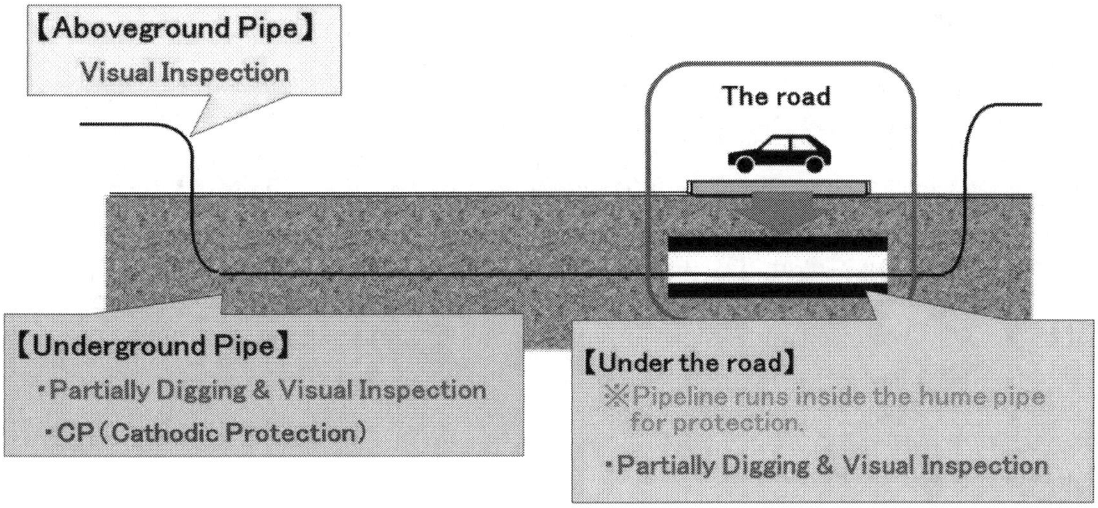

Figure 2. Conventional management method of outer surface corrosion

If it could be possible to inspect an entire-length pipeline directly, asset integrity would improve remarkably and risk of disaster would be lower. In order to realize that the "Direct inspection of whole pipeline" is possible, we have adopted "Intelligent-PIG technology" instead of "Digging & Direct inspection". By adopting this technology, we aimed at lower cost and lower workload.

2.1.1 Inspection Method

Inspection instrument (Pig for inspection) and necessary temporary equipment are shown in Figure-3.

The inspection instrument consists of a position identifier, Odometer, Data processor & a storage device, an ultrasonic wave sensor, battery. The pipeline is inspected by Ultrasonic flow detection (UT) method. This inspection instrument is moved in the pipeline with water pressure and inspects the corrosion of internal and external surface of the pipe.

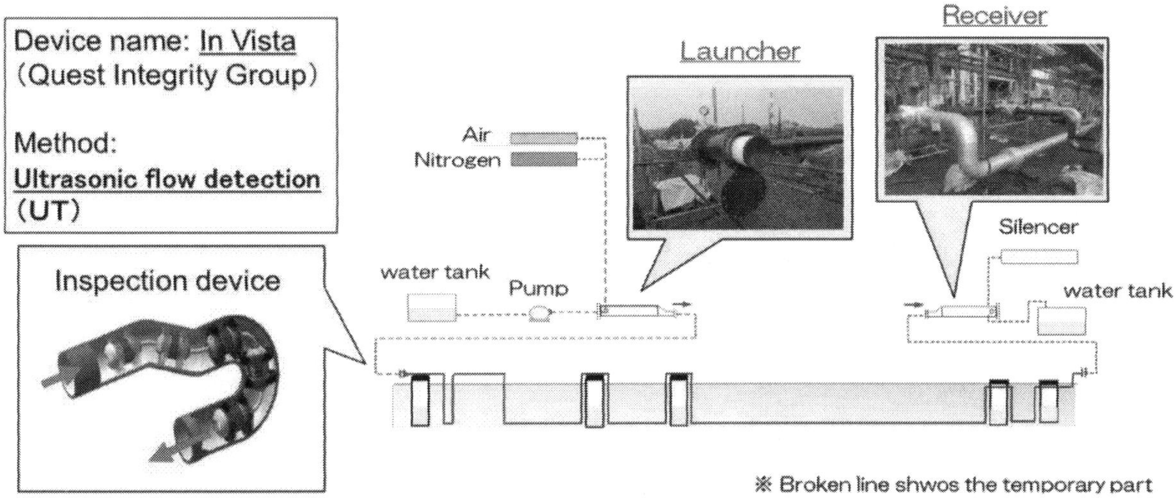

Figure3. Inspection device and temporary facility

2.1.2 Verification of the inspection result

The advanced inspection technology (Intelligent-PIG) was verified as effective by comparing with the conventional digging visual inspection as shown in Figure4.

Figure 4 shows test results' comparison in a development of a certain length of the pipe (3 m). Upper development shows the test results by new inspection technology and lower one is the conventional inspection technology (digging visual inspection) at the same point. We could confirm that "Intelligent PIG inspection results" is almost consistent with "Digging visual inspection result" (Red circle part) regarding maximum thinning portion, distribution of thinning point/ depth of thinning.

As a result, we have been convinced that the new inspection technology has same accuracy as conventional inspection technology.

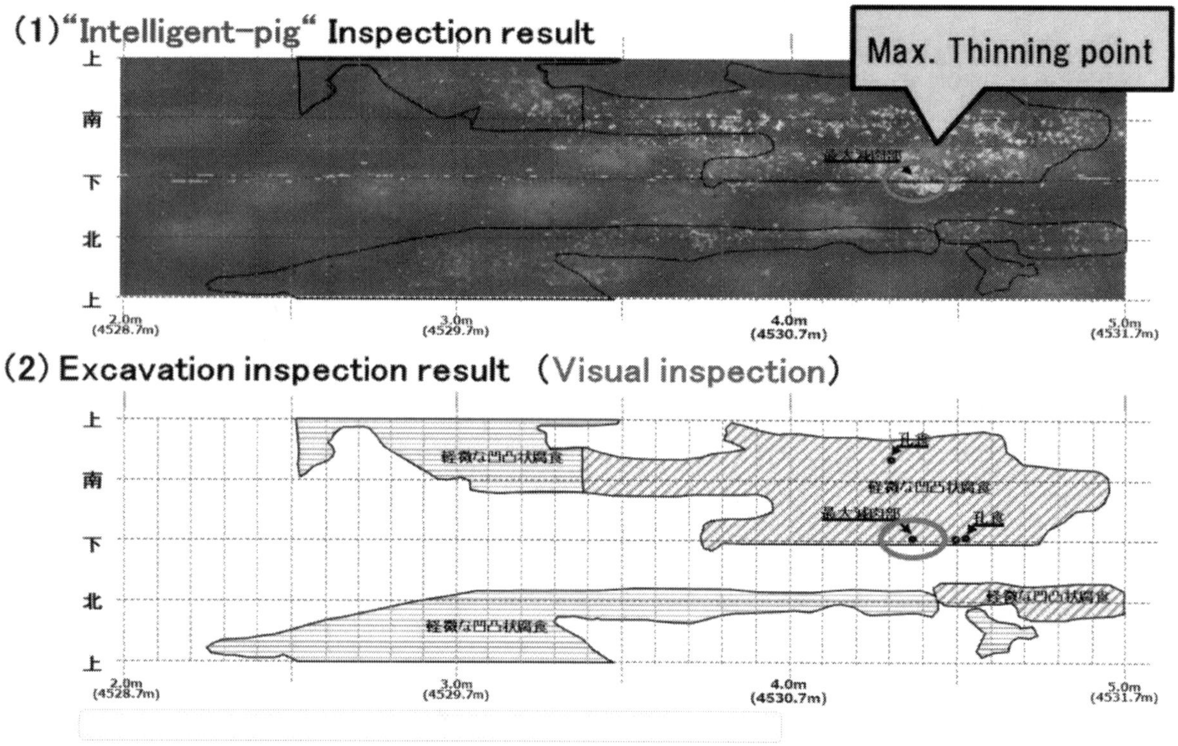

Figure 4. Verification result of "Inspection-PIG" inspection

2.1.3 Inspection result (Entire-length inspection)

Table1 shows the inspection results of entire length of the pipeline with new inspection technology. Thanks to this inspection, we were able to identify the corrosion situation.

Table1. Identified thinning point with the inspection result

No.	Distance from the start (m)	Place	Corrosion situation
#1	1.3km	Ground	Spot(Piping support part)
#2	4.4km	Under Ground	Uneven thinned scattered
#3	4.5km	Under Ground	Uneven thinned scattered
#4	5.5km	Boundary	Spot thinning near the boundary of ground surface
#5	6.9km	Ground	Spot(Piping support part)
#6	7.2km	Ground	Spot(Piping support part)
#7	7.2km	Ground	Spot(Piping support part)
#8	7.2km	Ground	Spot(Piping support part)
#9	7.6km	Under Ground	Uneven thinned scattered

Following items are confirmed thanks to adoption of new inspection technology, then we were able to achieve more accurate inspection and higher asset integrity;

- We were able to confirm that the pipeline was healthy as a general and did not have any critical problem.
- We were able to inspect and grasp the local corrosion situation of "the underground pipe" where it was not able to inspect by conventional inspection technology without digging
- Furthermore, It became possible to grasp the corrosion situation at "hard-to-inspection" part such as soil-to air interface portion, pipe support contact portion.

2.2 Ultrasonic wall thickness continuous monitoring system

Recently, our petrochemical plants have been diversifying the raw materials (imported naphtha), and the type of naphtha to be processed is switched from day to day. As a result, the operating condition of the plants changes accordingly.
Consequently, even if we learned the progress of thinning by periodic inspection (wall thickness measurement), it was difficult to identify when defect will be occurred by what mechanism, influencing factors.

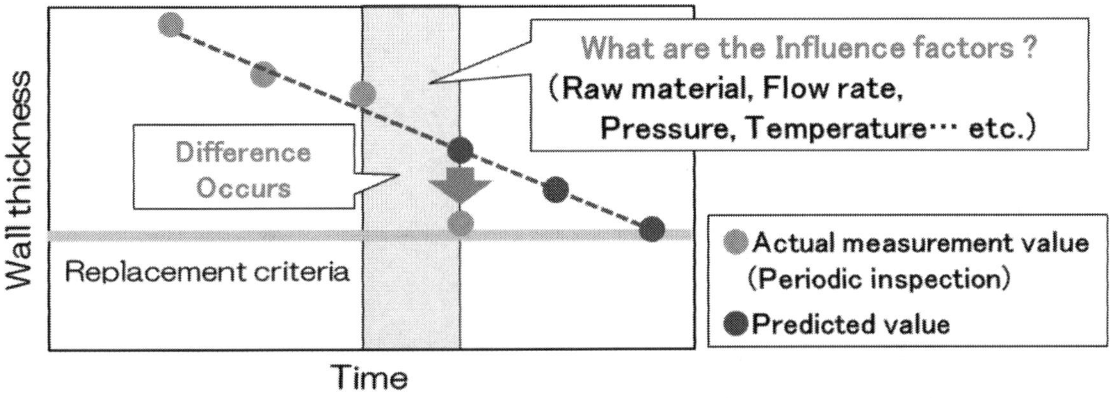

Figure 5. Problem of periodic inspection with frequent operation change

If "Continuous inspection" is possible and we can detect the thinning phenomena continuously, it is possible to find out the thinning at its early appropriate stage, carry out early repair and learn the correlation between the thinning phenomena and operating conditions even though operating conditions change frequently. We examined the possibility of adoption of "Ultrasonic wall thickness continuous monitoring system" for the purpose of "the reduction of the leak risk" and "improvement of the prediction accuracy" that are expected by the utilization.

2.2.1 Inspection Method

The inspection instrument is shown in Figure6.

The inspection instrument, consisting of DSI (Display Serial Interface) of the controller and probe installing to the pipe, gathers data with a mobile device. The data gathering interval is adjustable and 6 hours' interval was employed for the trial use.

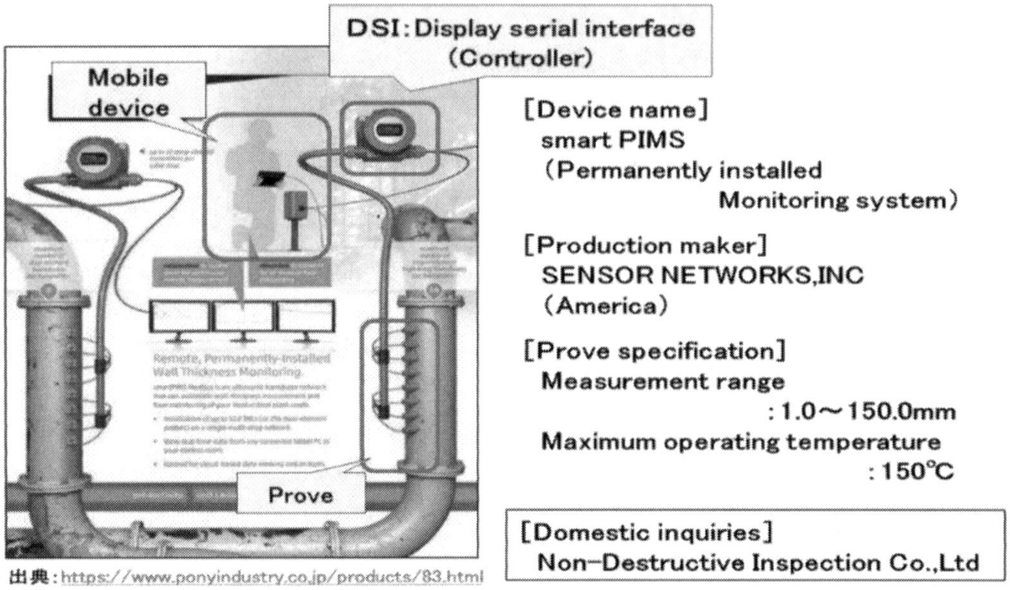

Figure.6 Inspection device (continuous monitoring system) [1]

2.2.2 Installation location of inspection devices

This inspection instruments were installed at the overhead pipe of the naphtha splitter in our complex , as shown in Figure7.
Among these 4 points, Point③ (Off-gas Flare Line) have been showing severe thinning tendency from the past. Other 3 points were added for comparison purpose.

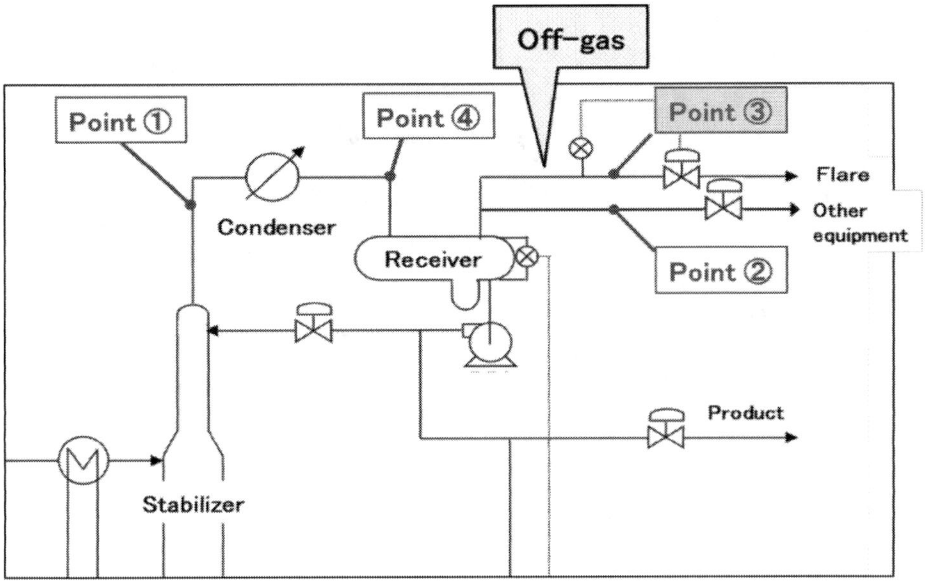

Figure.7 Installation location of inspection devices (4 places)

2.2.3 Installation result

Inspection results for approximately one month time are shown in figure.8.
There was no thickness change at Point ①, Point ② and Point ④, while Point ③ showed anticipated thinning phenomena. Especially, the thinning rate at Point③ was confirmed irregular, discontinuous as Figure 8 shows.

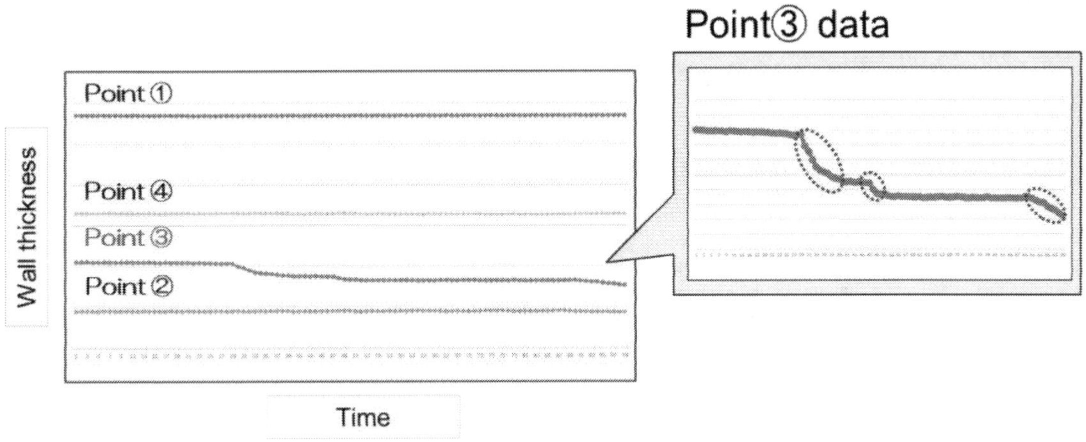

Figure.8 Inspection result with Ultrasonic wall thickness continuous monitoring system

2.2.4 Consideration

Inspection results in Figure8 shows irregular, sudden thinning in a specific period. We tried to identify the cause of thinning by comparing the thickness monitoring data with the various operation data in chronological order.

As a result, we found that off-gas flow rate was increasing always during the specific period when sudden thinning occurred. We have learned that sudden thinning may be suppressed by reducing the off-gas flow rate from these analysis data.

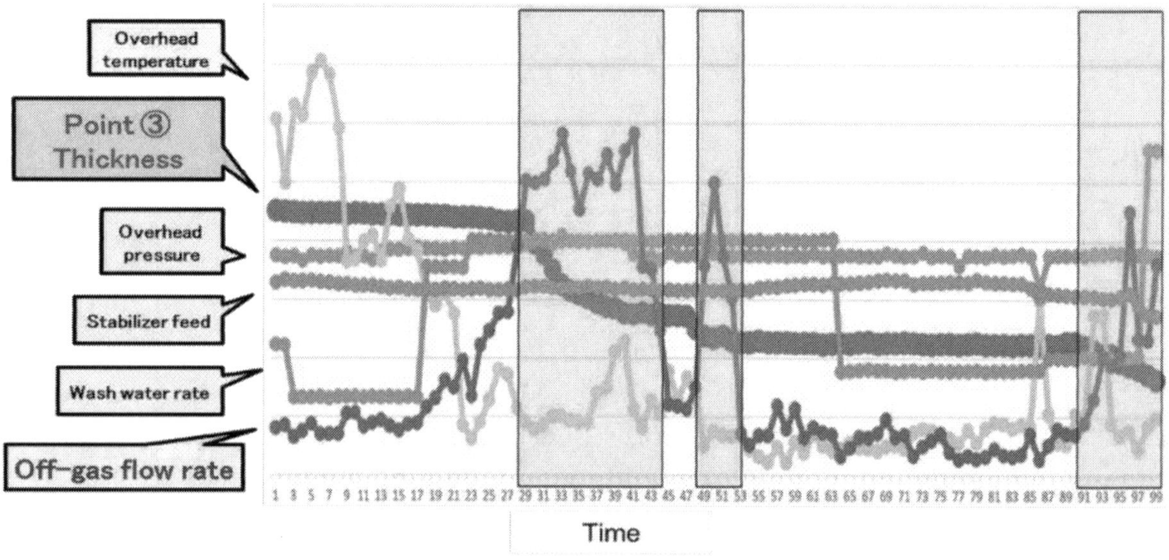

Figure.9 Comparison of "Wall thickness" with "Various operation data"

However, there was some periods when sudden thinning did not occur even though off-gas flow rate was high. Consequently, we considered that there would be another intrinsic cause other than the flow rate for the sudden thinning phenomena.

Based on these facts, we have investigated operation conditions, facility conditions and process conditions for Point② piping, which has no thinning tendency even though the fluid (off-gas) is same as point③.

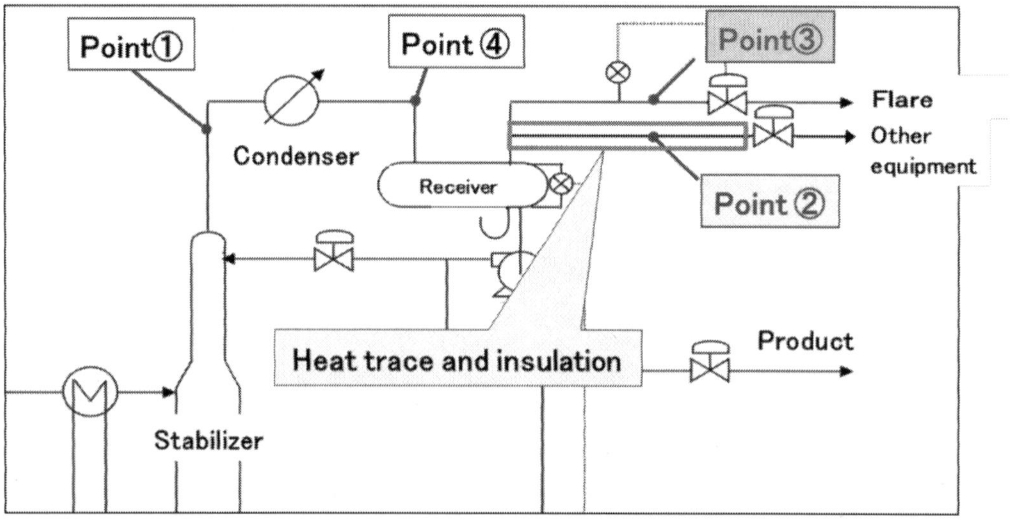

Figure.10 Difference of "Point②" and "Point③"

Off-gas piping of Point ② has heat trace and insulation on it while no heat trace and insulation on Point③ piping as indicated in Figure10. Since the temperature at Point ③ piping is atmospheric temperature and moisture in the off-gas would be condensed. It was speculated that the condensed water might cause corrosion in the Point③ piping.

Based on these investigation, we are planning to install "insulation + heat trace" on the Point③ off-gas piping.
If thinning is suppressed successfully by this installation, we don't need to reduce off-gas flow rate anymore for Point③ flare off-gas piping. This means that more flexible operation becomes possible.

X Conclusion

High level asset integrity (healthiness of our facilities) has been achieved in our complex by adopting two advanced technologies to our long-distance underground pipelines. Optimization of inspection period and the facility replacement timing will be realized by the inspection accuracy improvement. Also, it was confirmed that these technologies were able to lead to economical improvements such as "Reduction of inspection cost and workload" and "Expansion of flexibility in operation conditions".

Furthermore, we believe that following secondary effects are expected.

(1)Elimination of degradation, damage accompanied by inspection preparatory work such as digging

(2)Improvement of safety for preparatory work and inspection work at height and/or in the confined space

(3) Improvement of persuasiveness of accountability to stakeholders

We would like to promote high level Asset Integrity as the advancement of Process Safety Management by utilizing an advanced technology positively.

3 References (examples given to show format).

[1] Ultrasonic wall thickness continuous monitoring system, Available at https://www.ponyindustry.co.jp/products/83.html on December 17, 2018.

Capacity building of risk communication with the local community in RMP (Risk Management Plan)

Hwajin HEO
National Institute of Chemical Safety
90, Gajeonbuk-ro, Yuseong-gu, Daejeon, 34111, Republic of Korea
un1115@korea.kr

Junheon YOON
National Institute of Chemical Safety
90, Gajeonbuk-ro, Yuseong-gu, Daejeon, 34111, Republic of Korea
soiljh@korea.kr

Yeon-Kie RYOO
National Institute of Chemical Safety
90, Gajeonbuk-ro, Yuseong-gu, Daejeon, 34111, Republic of Korea
yk1234@korea.kr

Prepared for Presentation at
American Institute of Chemical Engineers
2019 Spring Meeting and 15th Global Congress on Process Safety
New Orleans, LA
March 31 – April 3, 2019

Capacity building of risk communication with the local community in RMP (Risk Management Plan)

Hwajin HEO
National Institute of Chemical Safety
90, Gajeonbuk-ro, Yuseong-gu, Daejeon, 34111, Republic of Korea
un1115@korea.kr

Junheon YOON
National Institute of Chemical Safety

Yeon-Kie RYOO
National Institute of Chemical Safety

Keywords: Risk management plan (RMP), Chemicals control Act (CCA), Chemical accidents, Precautionary Chemicals of Accident (PCAs), NICS public notification system, Risk communication

Abstract

Cooperative relationship with local community is a very important factor to respond to chemical accident effectively. To that end, the Risk Management Plan (RMP) system requires chemical companies to prepare emergency response plan to inform local communities. When a company get approved its RMP, it should disseminate the information of RMP including the company's name, location (e.g. address), phone number, information on hazardous substance it handles (e.g. type, hazard, and risk of chemical accident), and emergency response information (e.g. consequence range, accident-control equipment, how to deliver alarms, what people should do, a list of emergency agencies and their phone numbers). The National Institute of Chemical Safety (NICS) public notification system (http://icis.me.go.kr) is a useful tool to disseminate such information to the public. The company should post their information on the system within one month after approval. If the company does not want to use the system, it has to prepare a report for local residents within three months after RMP approval by using one of the following ways; letter (via either post or email), public hearing or presentation, information delivered to representatives of community.

However, it is just one-way communication, so it is not easy for local communities to find and use accurate information. The appropriateness of the emergency response plan in the community is always controversial with low credibility.

To develop an emergency response plan that can be implemented effectively, reliable and specific information on the region must be necessary. That's why the role of local communities and local governments is important. Since the introduction of the RMP, some local governments have been working with the central government to increase their capacity building of risk communication with the local communities for emergency response. However, it will take a long time to ensure that all regions with chemicals handling facility have effective emergency response system. However, the central government plans to improve an information-sharing system regarding emergency response plan, and establish a strategy to communicate well with local governments and local communities. This study shows several capacity building activities for risk communication in Korea.

1 Introduction

After a leak of hydrogen fluoride in Gumi in 2012, the Chemicals Control Act (CCA) was implemented in 2015 to effectively prevent and swiftly respond to chemical-related accidents. However, there is still an indication of insufficient measures to cope with accidents. There is growing concerns over negative effect of chemical accident as they occur frequently, and the public raises their voice to ask safety-related information disclosure for the right to know. (Y.K. Jung et al., 2018).

Recent hydrofluoric acid leak in Geumsan (June 6, 2016), fire in Incheon (Apr. 13, 2018), Tungsten hexafluoride (WF_6) leak in Yeongju (Apr. 13, 2018) and silicon tetrachloride ($SiCl_4$) leak in Gunsan (Nov. 13, 2018) showed that emergency management system has some problems such as damaged mutual relationship between local residents and concerned company, inadequate response at the beginning of the accident, inaccuracy of provided information, and insufficient evacuation information for residents. These are typical cases in which local residents, companies, local governments and the central government have not been able to communicate and share information properly.

Chemical accident is a problem not only at the companies but also in the local community. In order to prevent accidents and to cope with them in a timely manner, members of local community should be able to know about chemicals in their area, and be able to continue to participate in establishing an emergency response plan. In this study, based on the Risk Management Plan (RMP) of the CCA, we will seek to strengthen emergency response system in local community and revitalize risk communication.

2 Risk Communication with Risk Management Plan (RMP)

2.1 Public notification of RMP

Korea's RMP is a system to minimize damage to the local community and the environment by promptly responding to chemical accidents. Any company where uses, manufactures or stores one of the Chemical Substances Requiring Preparedness of Accident under the CCA more than certain limits should get approved its RMP and notify the local community of emergency response information at least once a year, including company name and location, hazard information on chemical substance, range of consequence, accident-control equipment, a way to deliver alarms, code of conduct, a list of agencies for emergency contact.

The National Institute of Chemical Safety (NICS) has established a public notification system in 2014 and operates it in order to help companies fulfill their obligation of public notification and to ensure the right to know of local residents. The NICS public notification system is built for easy access to chemical safety information by local residents. However, the system is not used well due to its one-way communication method, insufficient user convenience, passive participation of company, and insufficient legal management measures.

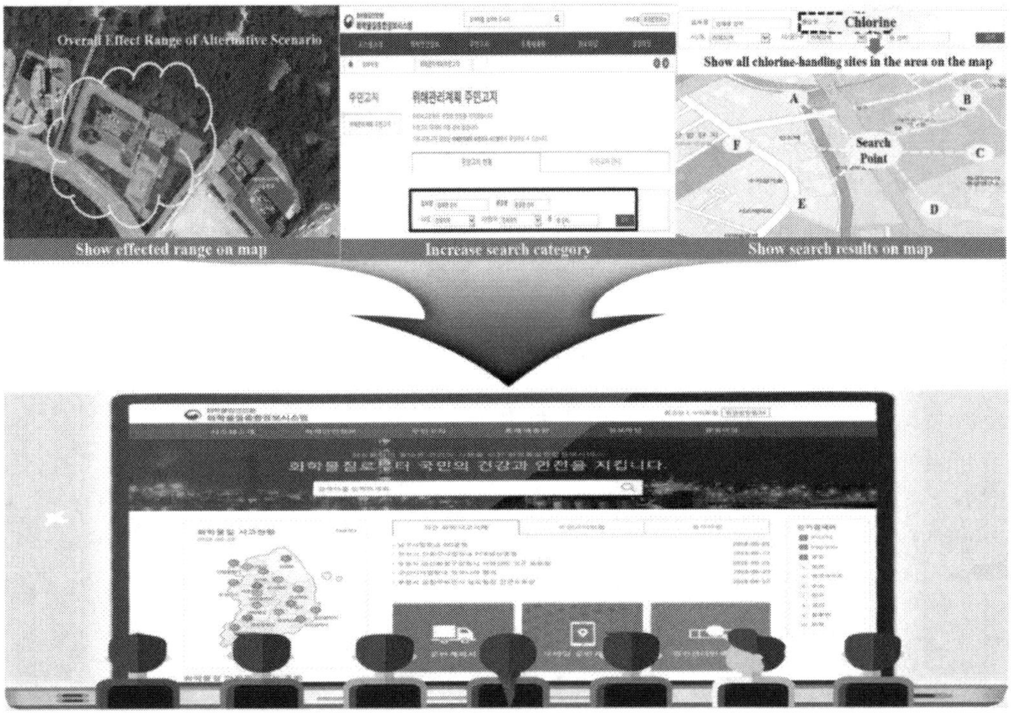

Figure 1 Improvement of NICS public notification system

Therefore, the Ministry of Environment (MOE) and the NICS try to find and implement various measures to operate effective RMP notification system for local communities. First, the CCA was amended in 2019, stating that the companies handling substances requiring preparedness of accident should notify local residents of relevant information through the NICS public notification system, and also inform them with a written notice of the fact that it registered in the system. Second, we try to upgrade the function of public notification system so that users can easily access emergency response information. Fig 1 shows major upgraded function of the system.

2.2 Local Government's Participation in the RMP Approval

Since the CCA has enacted, the MOE and the NICS have sought and supported the ways for local governments to participate in emergency response planning of RMP and RMP information provision to local residents. As a result, some local governments have established and are operating their own ordinance on chemical management including RMP review and information provision to local community. The CCA amendment requires local governments and the NICS to review RMP emergency response program in a joint manner from May 30, 2017. Local government reviews the adequacy of emergency response program in emergency response summary of RMP prepared and submitted by company, and then informs the NICD of the review result. In the review, local government focuses on protection targets in the rage of consequence, accident reporting system, evacuation alarm, alarm delivery method through local government, resident evacuation site, resident notification method, emergency rescue plan.

Despite these efforts, information sharing is not enough to establish an effective emergency response plan. First, the information provided to local government is limited to effectively perform its duty such as situation dissemination, resident protection and evacuation, and control activities in the event of a chemical accident. Second, RMP information submitted by companies before the regulation on involvement of local government came into effect on May 30, 2017 is not shared with local governments, so they do not have the information available to respond to chemical accidents that occurs in such companies. Lastly, there is a lack of interest of local government. Since chemical management works were transferred to the central government in 2015, most of the local governments have employees with less expertise in chemical management and emergency response system or sometimes have no staff to arrange chemical management works. For above-mentioned reason, some local governments have no response when the NICS requests them to review an emergency response program of RMP.

To overcome these limitations, the central government is seeking ways to improve the information sharing system in order to strengthen capacity building of local governments for emergency respond. First, the CCA will be amended to give a right to access a full RMP or to provide additional information upon request by the local government. Second, we will develop online RMP system where local governments can access necessary information readily and swiftly for its chemical management and its response to chemical accident. Lastly, the central government will set up a guideline to enhance experts who review RMP emergency response program in local governments, and will hold regular meetings with the experts to guide the review process, and listen their opinions to make the review process more effective.

2.3 Study on Emergency Response Information-Sharing System in RMP

The CCA requires RMP-approved companies to implement the approved RMP's emergency response plan in the event of a chemical accident. Approximately 1,200 RMPs were approved from 2015 to 2018, but it is questionable whether the emergency response plan set by each company is good enough to be used in real chemical accidents. In particular, small-scale companies lack competency to prepare RMP, and it is difficult to cooperate with local government or adjacent other companies. With these reasons, emergency response plan prepared by small-scale company has low on-site applicability and credibility. Korea witnessed several chemical accidents where concerned company got approved RMP but failed to implement the RMP emergency response plan such as damage control and restoration measures. At the time, the companies were blamed by the press, environmental groups and so on.

It is difficult for small-scale companies to share the information needed for chemical accident prevention, preparedness and response with the local community because of lack of human and material resources, and it is also difficult to establish feasible emergency response system. In general, small industrial complex is more vulnerable to accidents, but its capacity is poor relatively to deal with chemical accident properly. That could lead to severe damage caused by chemical accident. Therefore, it is necessary to equip with emergency response system enabling joint response and share the information with each other. As a matter of fact, fire accident in Incheon (Apr. 13, 2018) was a massive disaster with property damage of approximate 2.3 billion won, which started from fire in one company but spread to neighboring companies and vehicles. At the time, emergency responders were unable to confirm the information needed to deal with the accident, such as materials handled by the company, point site of the accident, and accident control information. They also failed to notify the nearby residents of accident alarm and evacuation promptly.

In order to solve these problems, the NICS plans to carry out an annual project to establish a joint emergency response system and an information sharing system based on the characteristics of each region or industrial complex, taking into account emergency response system of RMP. For the project, the NICS will select companies which are small-sized and hard to set up an emergency response system by themselves, and are located in industrial complex of the same type of business such as plating and dyeing. With this project, companies are able to prepare for potential chemical accidents and improve their capacity to cope with accidents by establishing a joint emergency response system. Furthermore, they can relieve anxiety of local residents by sharing information on the emergency response system, which leads to restore mutual trust.

2.4 Review on System for Prevention, Preparedness and Response of Chemical Accidents

As the CCA came into effect in 2015, chemical management works were transferred from the local government to the central government. In the event of chemical accident, the local government sometimes failed to play its role such as accident notification, protection and evacuation of residents, and accident response supporting. In order to build the local government's capacity, the MOE has conducted a project to build a local-level chemical accident preparedness system for three years starting from 2016.

In 2016, the central government developed a Korean model for building local-level chemical accident preparedness system. This model proposed phase-in development of local preparedness system. Phase 1 is to enact an ordinance by the local government. The objective of this model is to enact and implement ordinance through discussions by stakeholders. Phase 2 is to have a system of ordinance implementation in accordance with the enacted ordinance. A key element for ordinance implementation system is to establish a committee where needs participation of local government, NGOs, and companies. Phase 3 is to set up an emergency preparedness plan with the committee. In order to establish an effective emergency preparedness plan, it is necessary to provide the information including the current status of chemical substances handled by companies in the region and estimated off-site consequence of chemical accidents. In addition, companies also need to communicate with residents on a regular basis.

In 2017, based on the results of previous research, the central government has standardized the process of building local preparedness system. It has selected four areas nationwide first, and enacted an ordinance and formed a committee to set up local preparedness system in each area. They have developed a web site for dissemination and information-sharing of local preparedness system.

Korea has a total of 226 basic local governments. It is difficult for the central government to promote local preparedness system of all basic local governments. Therefore, the central government has selected some of the basic local governments in 2018 that need to build a local preparedness system first, based on the population, chemicals, and the sales volume of industrial complexes. For the selected governments, roadmap has been developed to build local preparedness system within the next five years. Important element of the roadmap is funding and nationwide networking. For expansion of local preparedness system nationwide, Korea needs to carry out project to build local preparedness system, capacity building with national network and quality assurance with funding assistance.

When comparing the MOE's "Project on Local Preparedness System for Chemical Accident" with the NICS's "Study on RMP Emergency Response Sharing System", both have a similar goal to build a system of prevention, preparedness and response for chemical accident based on characteristics of each local communities. Local governments and NGOs are the main targets of the MOE project, while the main targets of the NICS project are chemical companies. What is common to succeed both projects is that the risk communication among local community members is critical.

With the above-mentioned project of the NICS, it wants voluntary cooperation of chemical companies to build a joint emergency response system and get the impetus to persuade local governments and communities together. This is because they are experts who can best prevent, prepare, and respond to chemical accidents in the local community. In the end, the two projects will have to be interconnected. Since the emergency response system of the local community is only possible when all local governments, residents and companies work together.

3 Conclusion

Over the last four years, chemical companies have extended their protection to nearby residents and surrounding environment through the RMP, rather than focusing on prevention of chemical accident in their place of business only. They also have strengthened preparedness and response as well as prevention of chemical accidents. In the early days of the system implementation, many companies thought that the RMP was merely annoying task only for business license, overlapping with the chemical accident management system of other ministries. However, not only companies but also residents and local government think that the RMP is essential information for chemical accident prevention, preparedness and response.

If the RMP was operated as intended, it would inform local residents and responding agencies of hazards and risks of concerned substance before the accident, and make them well understand how they act in case of chemical accident. If chemical accident occurs, they would deliver alarms quickly and cooperate with the local government to safeguard residents according to the resident evacuation plan. It would be also possible to minimize, remove and restore the damage of accidental substance according to the emergency action plan.

The NICS is currently developing and implementing measures to overcome the limitations of the RMP since the companies who submitted the RMP in the first cycle should re-submit the RMP in 2020. In the end, the core of RMP is that the businesses, residents, local governments and the central governments cooperate to establish emergency response system that are tailored to regional characteristics and use them for prevention, preparedness and response to chemical accidents. Risk communication is significant to establish effective emergency response system.

Risk communication means the process of information exchange and communication between stakeholders. The RMP conducts risk communication with local residents by sharing the information and RMP emergency response program, and then establishes a framework with mutual trust and cooperation.

To do this, first we will strengthen the company's notification obligations and improve the NICS public notification system so that users can easily use the system. Second, we will improve the information sharing system so that the RMP information can be efficiently used for local government's prevention, preparedness and response to chemical accidents. Lastly, we will carry out annual project to build joint emergency response system and information sharing system specifically for small industrial complex with the same type of business so that small companies can set up an emergency response system and respond quickly to accidents. These improvements will contribute greatly to achieve the original objective of RMP to protect people and the environment from chemical accidents if local emergency response system is settled and information sharing with the local community is invigorated.

4 References

[1] National Institute of Chemical Safety, Available at http://icis.me.go.kr/rmp_notice/Main.do Accessed on February 3, 2017

[2] National Institute of Chemical Safety, Available at http://icis.me.go.kr/ Accessed on April 1, 2019

[3] J.H. Yoon, B.C. MA, Y. Yoon, "Risk Management Program (RMP) for chemical industry in Korea", 13th Global congress on process safety, 2017

[4] Y.K. Jung, M.J. Lee, H.J. Heo, J.D. Park, J.H. Yoon, C.H. Shin, "A Study on the Way to Revitalize Resident Notification System of Hazardous Chemicals", Crisisonomy Vol.14 No.6, pp. 41-50, 2018

[5] Ministry of Environment, Establishment of Local System for Chemical Accident Prevention, Preparedness and Response, 2016-2018

A decision making model for chemical accident responses based on machine learning

Chuntak Phark
Department of Environmental Engineering, Ajou University
Suwon, Gyeonggi-do, 16449, Republic of Korea
Phark.c.t@gmail.com

Seungho Jung
Department of Environmental Engineering, Ajou University
Suwon, Gyeonggi-do, 16449, Republic of Korea
processsafety@ajou.ac.kr

Prepared for Presentation at
American Institute of Chemical Engineers
2019 Spring Meeting and 15th Global Congress on Process Safety
New Orleans, LA
March 31 – April 3, 2019

A decision making model for chemical accident responses based on machine learning

Chuntak Phark
Department of Environmental Engineering, Ajou University
Suwon, Gyeonggi-do, 16449, Republic of Korea
Phark.c.t@gmail.com

Seungho Jung
Department of Environmental Engineering, Ajou University

Keywords: Emergency evacuation orders, Decision making, Chemical accident responses, Deep neural network, Machine learning

Abstract

Chemical accidents are one of the biggest obstacles to the development of the chemical industry. Prevention and mitigation should be considered to reduce the damage of chemical accident ahead. When they are failed emergency response should play an important role. An emergency evacuation order is one of the emergency response method to mitigate the damage after the accident occurred. In order for a chemical emergency response method such as emergency evacuation order to be effective, its accuracy and quickness must be secured.

In the previous study, a model of predicting issuing emergency evacuation order using machine learning was created. The previous prediction model studied 61,563 chemical accident data in NTSIP(National Toxic Substance Incidents Program) from 1996 to 2009 in the United States, and the learning objective was the issuance of emergency evacuation orders under certain accident situation. However, incorrect evacuation order could have been issued in its history and the methodology needed to be improved. In this study, the necessity of emergency evacuation orders was determined by using logical judgment and machine learning in order to solve the problems using the same database. For logical judgment, information on issuing emergency evacuation orders and civilian victim information in the chemical accident database were used. The accuracy of the model was increased and it was confirmed that it could be used significantly in the chemical accident responses.

1 Introduction

As the amount and type of chemicals being used increase, the patterns of chemical accidents are also diversifying. Initial responses are important to reduce damage due to chemical accidents. Emergency evacuation is one of initial responses to chemical accidents. It is a way to reduce casualties by moving people in the areas near the accident. Therefore, quick and accurate decision making is necessary. ALOHA, which is used to analyze the extent of damage, is limited in the consideration of the detailed terrain, and numerical analysis such as computational fluid dynamics is more accurate but takes much time for analysis. In order to solve the problems of promptness and accuracy at the same time, this study constructed a decision-making model using machine learning.

Phark.et.al. (2018) is a previous study on the foregoing. The previous study that used ATSDR's NTSIP database and HSEES database, which recorded chemical accidents occurred in the United States from 1996 to 2014, showed that the prediction of the issuing of emergency evacuation orders using machine learning was significant. However, in this study, machine learning was carried out using whether emergency evacuation order was issued or not without judging the effectiveness of the orders. Therefore, it is possible that cases of wrong decision making for emergency evacuation orders have been learned. In addition, this study has a limitation that the sensitivity indicating the accuracy of the accidents at which the emergency evacuation order was issued is somewhat low. Therefore, in the present study, the emergency evacuation order decision-making model was improved by solving the problems of the previous study using the Deep Neural Network (DNN), one of machine learning algorithms.

2 Methodology

The database used in the present study is the NTSIP database and the HSEES database of ATSDR in the United States, which recorded chemical accidents that occurred in the US from 1996 to 2014. The present study is largely consists of a stage to judge the necessity of emergency evacuation orders, a stage to modify the databases, and a stage to construct a decision-making model using machine learning.

2.1 Judgement of the necessity of emergency evacuation orders

The progression of a chemical accident generally follows the following process.

When a chemical accident has occurred, the consequences of the accident are determined by the internal and external factors of the chemical facility. The decision maker for responses to chemical accidents estimates the consequences of the chemical accident to make decisions on the emergency evacuation order. The interactions among the consequences of the chemical accident, the result of decision-making on the emergency evacuation order, and the implementation methods may or may not result in victims in the vicinity. Since these events occur in a chronological sequence, it can be judged whether the emergency evacuation order was effective or not only after the victims of the chemical accident occurred.

When an emergency evacuation order has been issued, victims may or may not occur depending on the method of implementing the order. If an emergency evacuation order was issued and there is no victim due to the chemical accident, it is difficult to judge the effectiveness of this emergency evacuation order based on the victim alone. On the other hand, if no emergency evacuation order has been issued, the necessity of issuing an emergency evacuation order can be judged directly according to whether any victim has occurred or not. Table 1 shows this relationship as a matrix.

Table 1. Action-Victim matrix

		Emergency Evacuation Order Information	
		Issued	**NOT Issued**
Victim	**0**	**Required or Not**	**Not required**
	0<	**Required**	**Required**

In the present study, the necessity of an emergency evacuation order was judged using Table 1 and accidents at which emergency evacuation orders were issued but there was no victim, were approached conservatively so that the emergency evacuation orders were judged to have been necessary. According to the results of judgment of the necessity of emergency evacuation orders, 10,084 cases, which are 16.4% of the entire 61,564 chemical accidents, required the issuing of an emergency evacuation order, while 51,480 cases of chemical accidents, which are 83.6% of the entire chemical accidents, did not require any emergency evacuation order. In addition, among the 10,084 cases of chemical accidents judged to have required an emergency evacuation order, an emergency evacuation order was issued in 6,438 cases (63.8%).

2.2 Modification of the databases

Phark.et.al. (2018) modified the NTSIP database and the HSEES database before they conducted their study. The most significant characteristic of the database used in Phark.et.al. (2018) is that substances that cause chemical accidents were replaced with a NFPA rating. Their study is also characterized by the fact that machine learning was carried out excluding data that can be identified only when the chemical accident has ended such as the cause of the chemical accident and victims due to the chemical accident.

The present study was conducted after an additional modification of the databases used in Phark.et.al. (2018). A total of 61,564 chemical accidents were recorded in the databases. Whereas whether emergency evacuation orders were issued or not was learned in the previous study, the results of judgment of the necessity of an emergency evacuation order was learned using Table 1, thereby minimizing the possibility to have learned wrong decision-making on emergency evacuation orders in the present study. In addition, industrial codes were not included in the learning to reduce the noises generated from the sparse data distribution. The databases used in the present study express chemical accidents using the 14 attributes listed in Table 2.

Table 2. Descriptions of the Chemical accident database

Attribute Class	Description
EVNTTYPE	Type of event
THRTACTU	Was the release actual or threatened
SEASON	Season when event occurred
WEEKDAY	Portion of week when event occurred
TIME	Time range that event occurred
AREATYP	Description one of type of area where event occurred
AREA_RES	Residential area within ¼ mile of event
INST_TYPE	Accident equipment type
LIVEQTR	Number of people living within ¼ mile of event
SUB_CAT	Substance category
H	NFPA rating; Health hazard(0–4)
F	NFPA rating; Fire hazard(0–4)
R	NFPA rating; Instability hazard(0–4)
RELS1CHEM1	First type of release for Chemical
Fianl_Judge	Emergency evacuation required or not

2.3 Deep-Neural-Network overview

Machine learning is one of big data processing technologies and is largely classified into supervised learning, unsupervised learning, and reinforcement learning. Supervised learning is used to estimate given target values and labels while unsupervised learning is used to classify data for which no target value has been given. The present study, which should learn the situations of chemical accidents to estimate the label termed the necessity of emergency evacuation orders, can be regarded to be an example of supervised learning.

The machine learning technique used in the present study is the Deep Neural Network(DNN). Its basic structure is a perceptron that simulates a single nerve cell of a living organism. Equations (1) and (2) are calculated for each perceptron. The results of the calculation are passed through the activation function to judge whether or not to transmit the information to the subsequent stage. Figure 1 is a mimetic diagram that represents the foregoing.

$$u_k = \sum_{j=1}^{p} w_{kj} x_j \qquad \text{Equation (1)}$$

$$y_k = \varphi(u_k - \theta_k) \qquad \text{Equation (2)}$$

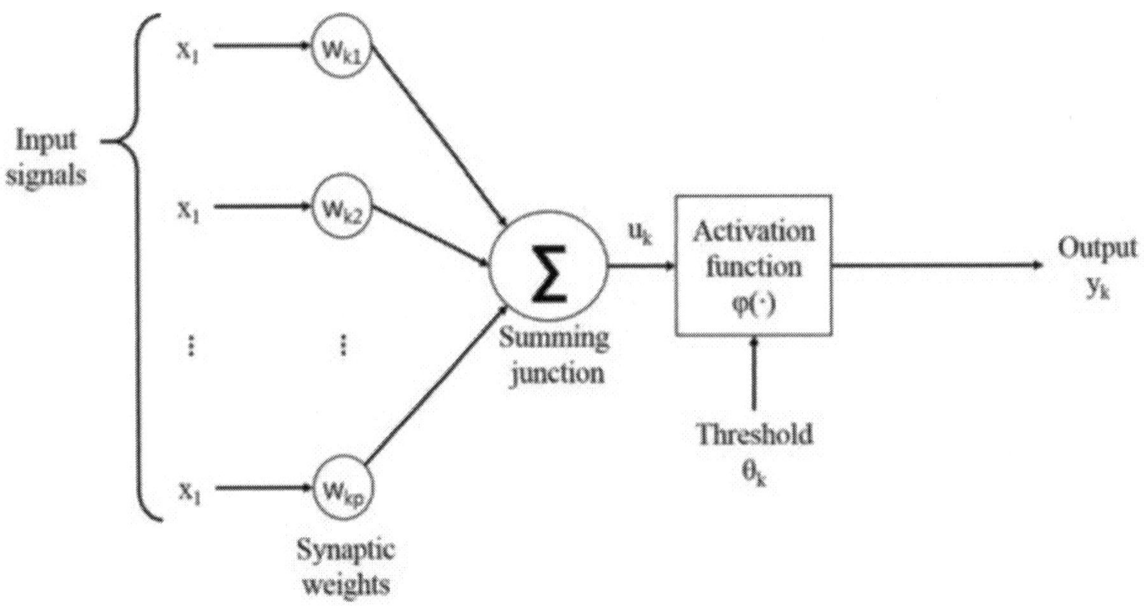

Figure 1. Perceptron model

The DNN is a way to derive the optimal decision boundary by connecting multiple perceptrons and repeating linear fitting and nonlinear transformation. Figure 2 depicts the general structure of DNN.

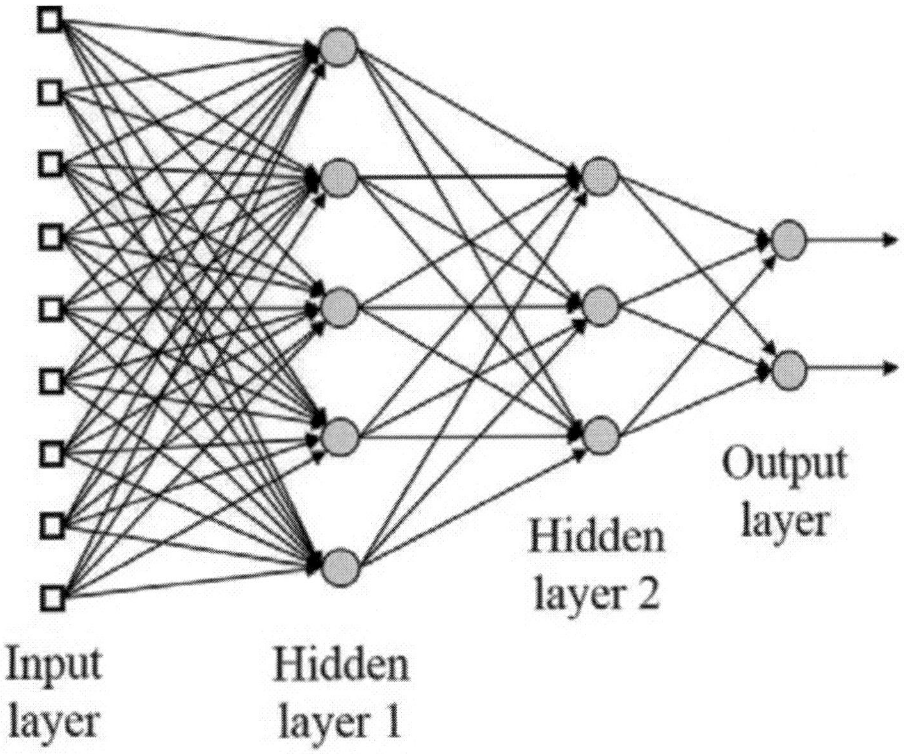

Figure 2. DNN structure

Each circle in Figure 2 is a perceptron. The data are entered into the input layer of the neural network, calculations are carried out while the data go through the hidden layers, and the calculation results come out from the out layer. The learning of the DNN can be regarded to be a process of calculating the loss using the calculation results and modifying the weights of the perceptrons in the reverse direction of the data flow. Those that greatly affect the accuracy of the DNN include the structure of the hidden layers, the activation function, loss calculations, and the optimization function for modifying weights.

Although there is no formula for determining the size or number of hidden layers, it is known that the highest accuracy is shown when the number of perceptrons in the hidden layers is about 1.5 times the size of the input layer. In the present study, the size of the input layer is the number of attributes used to represent chemical accidents in the database used.

The labels of the databases used in the present study are biased toward one side. The asymmetric data were learned effectively by normalizing the meaning of the result of each chemical accident using the ratios of the labels and using the results of normalization as weights in loss calculation.

2.4 DNN design

Tensor flows were used for the implementation of the DNN. The input layer was designed with 14 attributes, which were used to describe chemical accidents in the databases used in the study. The hidden layers were designed in two stages in total and each hidden layer was composed of 11 perceptrons. Since the emergency evacuation order is either necessary or unnecessary, the output layer has two perceptrons. Figure 3 is a mimetic diagram of the foregoing.

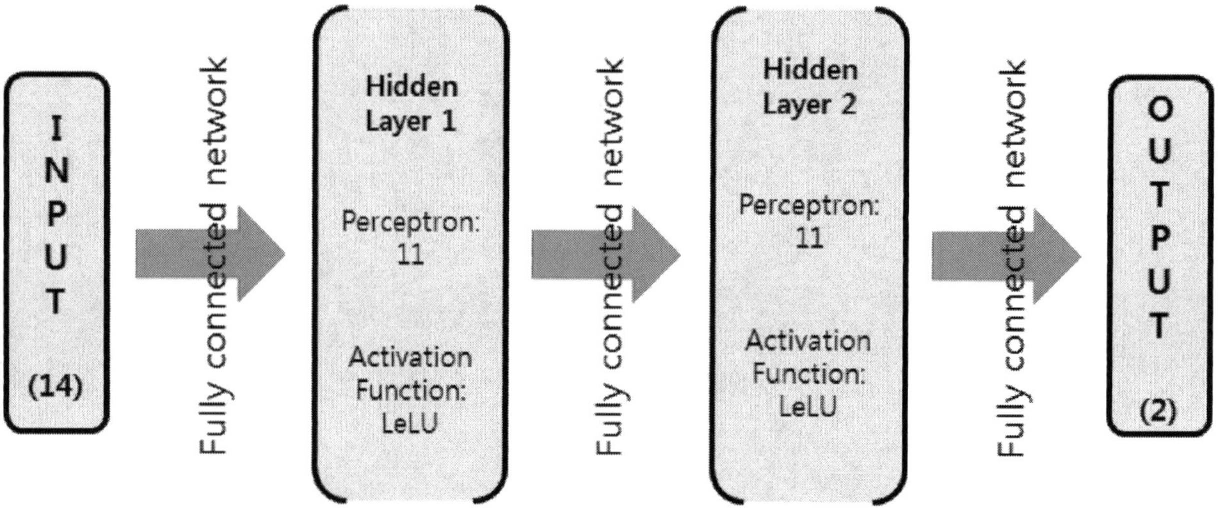

Figure 3. Structure of Decision making model for Chemical accident response

The weights according to the necessity of emergency evacuation orders were reflected in the label data, and LeLU was used as the activation function of each perceptron. The cross entropy was used for the calculation of losses. The Adam optimizer was used for the weight learning. In order to prevent over fitting, the drop out method to remove perceptrons with low weights was used in the learning process. Over fitting is a phenomenon in which data for learning are perfectly learned so that the accuracy is low in data that were not learned.

2.5 Assessment of model accuracy

Out of the 61,564 cases of chemical accident data, 60%(36,918 cases) were used for learning by the DNN and 40%(24,646 cases) were used to assess the accuracy.

The 'Overall Percentage Agreement (O.P.A)', 'Sensitivity' and 'Specificity' were used to evaluate the accuracy of the emergency evacuation order decision-making model. O.P.A means the accuracy of the entire data labels. The sensitivity means the accuracy of chemical accidents requiring an emergency evacuation order, and the accuracy of chemical accidents that do not require any emergency evacuation order is expressed as Specificity. Table 3 shows the matrix used for the calculations of the Sensitivity, Specificity, and O.P.A , and Equations (3), (4), and (5) are the calculation formulas.

Table 3. Confusion matrix

		Prediction	
		Positive	**Negative**
Actual	**Positive**	**TP (True Positive)**	**FN (False Negative)**
	Negative	**FP (False Positive)**	**TN (True Negative)**

$$Sensitivity = \frac{TP}{TP + FN} \quad (3)$$

$$Specificity = \frac{TN}{TN + FP} \quad (4)$$

$$Overall\ Percentage\ Agreement = \frac{TP + TN}{TP + FN + FP + FN} \quad (5)$$

3 Results

An emergency evacuation order decision-making model using a DNN that learned learning data 3,000 times was assessed and according to the results, the O.P.A for the 24,646 test data was 93.5% and the loss was 0.63. The time spent for the learning was about 2 minutes. Figure 4 is a graph of O.P.A and Loss according the increases in the number of times of learning.

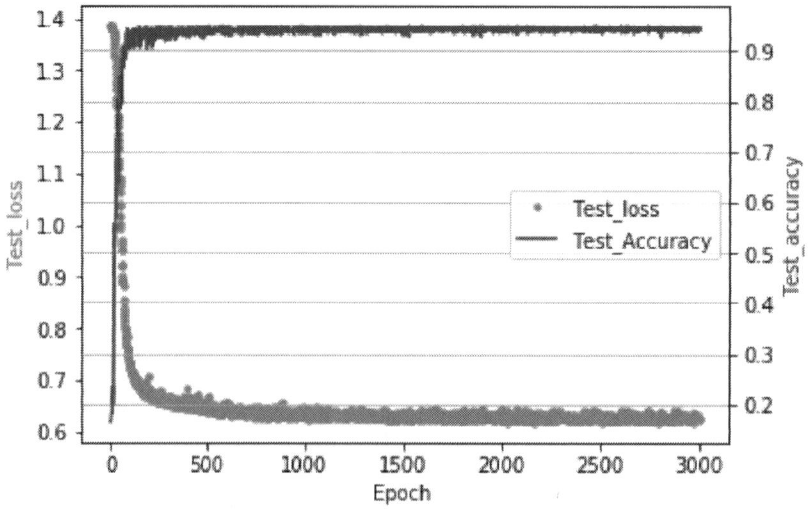

Figure 4. Test loss and test accuracy

<Table 4> shows the predicted values of the decision-making model and the actual values as a confusion matrix. The sensitivity and the specificity were evaluated as 74.6% and 98.4%, respectively. Figure 5 shows a comparison with the study findings of Phark.et.al. (2018). Overall, the accuracy of the model has increased.

Table 4. Prediction results

		Prediction	
		Required	**Not required**
Actual	**Required**	**3,006 cases**	**1025 cases**
	Not required	**331 cases**	**20,284 cases**

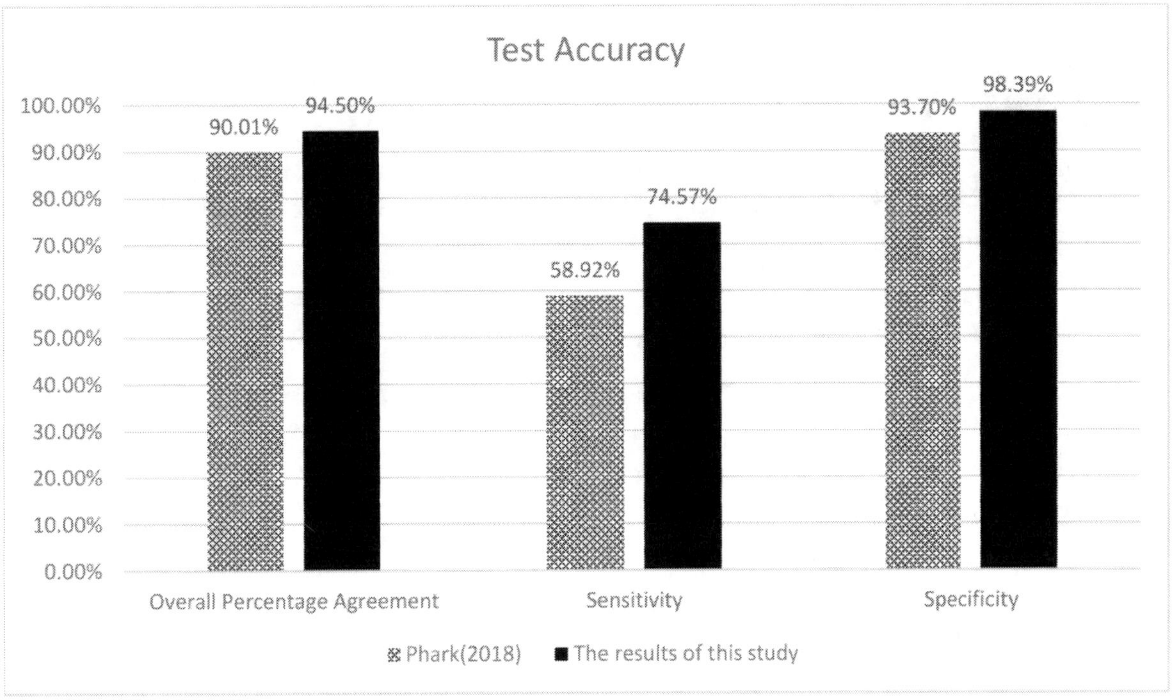

Figure 5. Test accuracy

4 Conclusion

It was identified that judging the necessity of issuing an emergency evacuation order using the information in the NTSIP / HSEES databases is effective. In addition, a model that can be used to judge the necessity using the DNN was constructed and the necessity of an emergency evacuation order could be judged at quite high accuracy using the model. This seems to be attributable to the normalization of asymmetric data using weights.

Although the overall accuracy of the model is important, in the case of responses to accidents, accurately identifying accidents at which emergency evacuation orders are necessary is particularly important. Of the 10,084 chemical accidents judged to have required an emergency evacuation order in the Modified NTSIP / HSEES databases, an emergency evacuation order was 'actually' issued in 6,438 cases, which are about 63.8%. On the other hand, the ability of the decision making model derived through the present study to identify accidents that require an emergency evacuation order, that is, the sensitivity was evaluated to be 74.6%. These results suggest that the decision making model derived through the present study can significantly help decision-making in responding to actual chemical accidents.

5 References (examples given to show format).

[1] C. Phark, W. Kim, Y. S. Yoon, G. Shin and S. Jung. Prediction of issuance of emergency evacuation orders for chemical accidents using machine learning algorithm, Journal of Loss Prevention in the Process Industries 56(2018), 162-169

[2] U.S. Department of Health and Human Services. Available at http://www.atsdr.cdc.gov/HS/HSEES/index.html access on December 1, 2018.

[3] U.S. Department of Health and Human Services. Available at http://www.atsdr.cdc.gov/NTSIP access on December 1, 2018.

[4] S. O. Haykin. Neural Networks: a Comprehensive Foundation, first ed. Prentice Hall, New Jersey, 1994.

[5] S. Karsoliya. Approximating Number of Hidden Layer neurons in Multiple Hidden Layer BPNN Architecture, International Journal of Engineering Trends and Technology 3(6) (2012), 714-717

[6] S. Marshland. Machine Learning: an Algorithmic Perspective, second ed. Chapman & Hall / CRC, 2014

[7] J. A. Swets. Measuring the accuracy of diagnostic systems. Science 240(4857), 1285-1293

550365 The Changes of PSM(Process Safety Management) System in Republic of Korea

The Changes of PSM(Process Safety Management) system in Republic of Korea.

The chemical industries has been a lower frequency of accidents than other kind of industries, but it can cause enormous damage with one accident and it takes a considerable period of time to recover the damage.

As major industrial accidents had taken place all over the world, and international organizations strengthened the system for prevention, Korean government jointly organized an international seminar on prevention of industrial accidents in cooperation with the ILO to introduce a process safety management system in October 1992.

In accordance with the ratification of the ILO Convention for the Prevention of Industrial Accidents (C-174) (1993), the basis of process safety management system was established on January 5, 1995.
Since January 1, 1996, the process safety report review and confirmation policy implemented and has contributed to the prevention of chemical accidents.

At the beginning of the introduction of the process safety management system under Article 49-2 of the Industrial Safety and Health Act was revised on January 5, 1995, seven hazardous industries and 21 kind of hazardous materials were managed under of this Act.

The Industrial Safety and Health Act revised on January 7, 2000 abolished the 5-year process safety report re-submission and review rules for existing facilities as the regulations were relieved in accordance with the Special Measures for Deregulation of Business Activities.

After 2001, the implementation of the process safety report was verified on site. However, due to the increased social interest in the major industrial accident, the new law was introduced on March 31, 2005, Article 49-2, Paragraph 8 of the same Act, to reinforce the requirement to resubmit the process safety report if the complementary status was poor through the evaluation.

The Ministry of Labor visits workplaces where reports have been reviewed and confirmed, evaluates the level of implementation of the 12 elements of the process safety report, classifies them into 4 grades(P, S, M + and M- grades)

In the initial management system, the P-grade business sites were exempted the performance status check for 4 years and the M-grade business sites were repeatedly inspected every year.

However, P-grade business sites were inspected the performance status check once for 4 years from 2014 and if the level of implementation is insufficient, it can be reevaluated for downgrade.

In the event of a result of large-scale chemical accidents such as the hydrogen fluoride leakage in 2012, the kind of chemical substances to submit process safety report were increased from 21 to 51.

When Emergency Response Goes Fatally Wrong: Lessons Learned from Investigating Two Major Incidents

Scott Davis
Gexcon US
4833 Rugby Avenue Suite 100
Bethesda, MD
sgdavis@gexcon.com

Prepared for Presentation at
American Institute of Chemical Engineers
2019 Spring Meeting and 15th Global Congress on Process Safety
New Orleans, LA
March 31 – April 3, 2019

When Emergency Response Goes Fatally Wrong: Lessons Learned from Investigating Two Major Incidents

Scott Davis
Gexcon US
4833 Rugby Avenue Suite 100
Bethesda, MD
sgdavis@gexcon.com

Keywords: emergency response planning, lessons learned, case studies, major explosion incidents

Abstract

Emergency response plans are an essential, yet oftentimes overlooked, layer of protection for facilities where all other safeguards failed to prevent an incident. While catastrophic accidents such as large releases of chemicals, fires or explosions are devastating for the process industry, experience investigating numerous incidents has shown that a lack of an effective emergency response plan can lead to an unnecessary and tragic escalation of the initial incident. More specifically, investigation of two recent incidents: (1) 2013 West Texas Explosion and (2) 2015 explosion on the FPSO Cidade São Mateus; will demonstrate how fatal consequences resulted from the lack of emergency planning in the first case study and from deficiencies in the emergency response plan or implementation of the plan in the second case study. More specifically, the FPSO explosion resulted in 9 fatalities, eight of which were responding to a leak of a light-end hydrocarbon, and the West Texas explosion that resulted 15 fatalities, most of whom were responding to the fire that preceding the devastating explosion. Lessons will show that the fatal consequences could have been avoided with the proper emergency planning. In addition, while emergency response plans consider "maximum credible" scenarios, past events have shown that low probability high consequences should also be at least considered.

1 Introduction

Emergency response plans are an essential, yet oftentimes overlooked, layer of protection for facilities where all other layers of protection failed to prevent an incident. An emergency plan specifies procedures for handling sudden or unexpected situations, where the ultimate objectives are to be prepared to: (1) prevent fatalities and injuries; (2) reduce damage to buildings, stock, and equipment; (3) protect the environment and the community; and (4) accelerate the resumption of normal operations [1]. Developing the emergency response plan consists of five major steps (see Figure 1). The first step is to assess the vulnerability of the site and determine what hazards pose a threat to the installation. Such an exercise will help determine not only what "credible" events can occur at a given facility, but also what catastrophic or major events can occur. It should be noted that major emergencies are generally rare events.

Next, it is necessary to develop your emergency response plan and determine what means are currently available to execute the plan and what is still needed. At the planning stage, it is important that several groups be asked to participate, such as: subject matter experts and process engineers; the health and safety group; and even appropriate municipal officials should also be consulted since control may be exercised by the local government in major emergencies and additional resources may be required. Once the emergency response plan is developed, the necessary facilities and equipment need to not only be provided but tested and maintained. Communication, training and periodic drills will ensure adequate performance if the plan must be carried out. Lastly, there should be a periodic review of the emergency response plan.

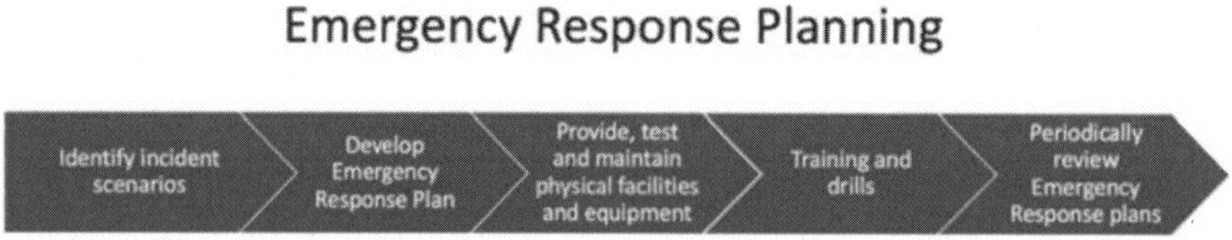

Figure 1: Steps of Emergency Response Planning

While the major benefit is to provide guidance during an emergency, developing the plan has other advantages. For example, previously unrecognized hazardous conditions, which would aggravate an emergency situation, could be discovered during the emergency planning and means put in place to mitigate or eliminate these hazards. The planning process may also identify deficiencies, such as the lack of resources (equipment, trained personnel, supplies), which can be easily corrected before an emergency occurs.

While catastrophic accidents such as large releases of chemicals, fires or explosions are devastating for the process industry, experience investigating numerous incidents has shown that a lack of an effective emergency response plan can lead to an unnecessary and tragic escalation of the incident. Chaos can quickly occur during an emergency if there are lack of resources and trained personnel coupled with an urgent need for rapid decisions and a very short period of time. The stress of the situation can lead to poor judgment resulting in severe losses, and normal channels of authority and communication cannot be relied upon to function routinely. Therefore, it is imperative to eliminate the "it can't happen here" mentality, as emergencies can occur and preplanning is necessary.

The importance of emergency planning will be discussed within the context of two recent incidents: (1) 2013 West Texas Explosion and (2) 2015 explosion on the FPSO Cidade São Mateus. These two case studies will demonstrate how the lack of emergency planning resulted in devastating consequences in the first case study and from deficiencies in the emergency response plan or implementation of the plan in the second case study.

2 Background of Incidents

2013 West Texas Explosion

On April 17th, 2013, an explosion occurred at the West Fertilizer Company storage and distribution facility in West, Texas. The explosion at West Fertilizer resulted from an intense fire in the seed and fertilizer area of the facility that led to the detonation of approximately 30 tons of ammonium nitrate stored inside a wooden receiving bin. The explosion occurred while emergency services personnel were responding to a fire at the facility. Fifteen people were killed, more than 250 were injured, and numerous buildings were damaged or destroyed.

The seed and fertilizer building (Figure 2) was built in 1958 and stored fertilizer and seed, which were distributed to farmers. Figure 3 shows the layout of the seed and fertilizer building, which included an ammonium nitrate (AN) storage bin neighboring the seed room, as well as an empty bin used for ammonium nitrate storage, and bins containing ammonium nitrate, K-Mag (fertilizer mix of potassium, magnesium and sulfur), Potash, ammonium sulfate, DAP (diammonium phosphate), and a bin for potassium chloride along the west wall. This building was capable of receiving rail shipments directly from the rail car by means of a rail spur junction and an underground conveyor belt and bucket elevator; the bucket elevator dumped directly into the northern AN storage bin, or onto a conveyor belt feeding the bins along the west wall.

Figure 2: West Fertilizer Facility – viewed from the west (upper) and east (lower)

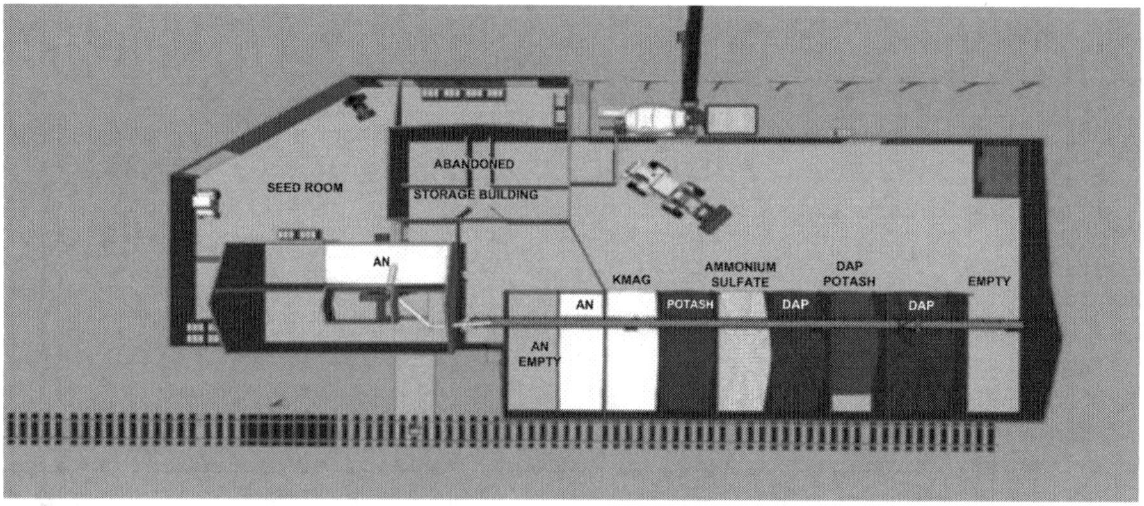

Figure 3: Layout of the Seed and Fertilizer Building.

On April 17th, 2013, at approximately 7:29 pm, a fire was reported, and fire fighters responded to the fertilizer plant [1]. All operations were already stopped for the day and employees had left the plant at about 5:30-6:00 pm. GexCon obtained several videos that recorded the progression of the initial fire [3]. Figure 4 shows the initial fire as viewed from the east and west; note that the highest point on the building represents the location of the bucket elevator by the northern AN storage bin. The image shows that the fire is isolated to the northern section of the building, which includes the seed room and portions of the northern AN storage bin containing about 30 tons of ammonium nitrate.

At approximately 7:50 pm, a massive explosion occurred at the fertilizer plant. Figure 5 shows images from a video taken from the high school parking lot to the southeast and Figure 6 the ensuing blast damage and crater from the detonation. Based on the videos from the day of the incident, there were actually two explosions: one smaller blast followed by a larger blast roughly 300 msec later.

Figure 4: Close up of the Seed and Fertilizer Building during fire as viewed from east (left image) and viewed from the west (right image)

Figure 5 – View of the explosion (viewed from southeast)

Fifteen people were killed as a result of the explosion, which included voluntary firefighters, and over 250 people were injured [3]. Damage was extensive throughout the neighboring community, which included homes, schools, an apartment building and a nursing home.

Figure 6: Blast damage and resulting crater

2015 explosion on the FPSO Cidade São Mateus

Due to the confidentiality of matters in this case, all references are made to two publicly available reports regarding the incident [5], [6]. The Cidade São Mateus was a Floating, Production, Storage and Offloading of Oil and Gas Export (FPSO), with a capacity to produce 4,000m³/day of oil and 10,000,000m³/day of gas as shown in Figure 7.

Figure 7: 2015 FPSO Cidade São Mateus.

On February 11, 2015 there was a leak of condensate (light-end hydrocarbon liquid) discovered in the pump room at a spade blind flange connection. Figure 8 shows that the pump room is situated just forward of the engine room and aft of the storage tanks. The layout and levels of the pump room is given in the left image of Figure 9, while the leaking spade blind flange is in the right image.

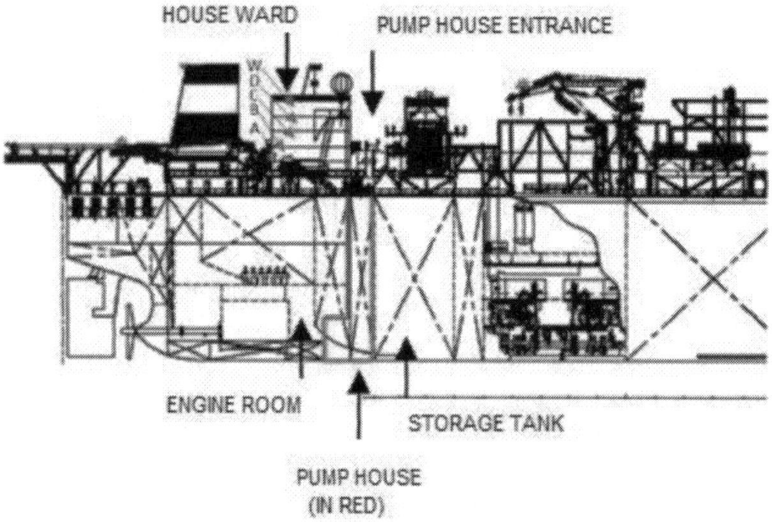

Figure 8: Cutaway side view of the pump room on the FPSO Cidade São Mateus.

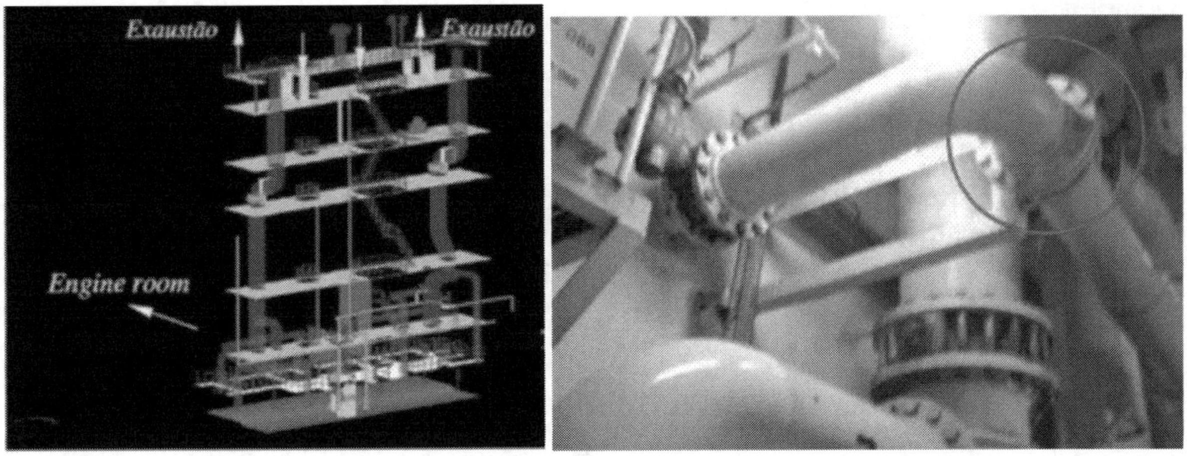

Figure 9: Model of the pump room on the FPSO Cidade São Mateus (left) and spade blind flange (right).

Just prior to the leak, the operations crew was attempting to drain condensate center cargo tank 6 (6C) into an alternate tank and they used an alternate pump alignment (stripping pump). However, the valve on discharge side of stripping pump was closed and they were pumping against a spade blind that was inadequate for such pressures, which ultimately failed the blind and seals. The gas detectors in the pump room began to alarm, and ultimately all three were alarming.

The ventilation was interlocked to shutdown per "confirmed gas" in the pump room. Emergency response teams entered the pump room to attempt clean up and repair the flange. There were three separate entries, and on during the third entry the explosion occurred. Figure 10 through Figure 12 show the level of blast damage to the pump room. Nine people were killed, of which eight were first responders.

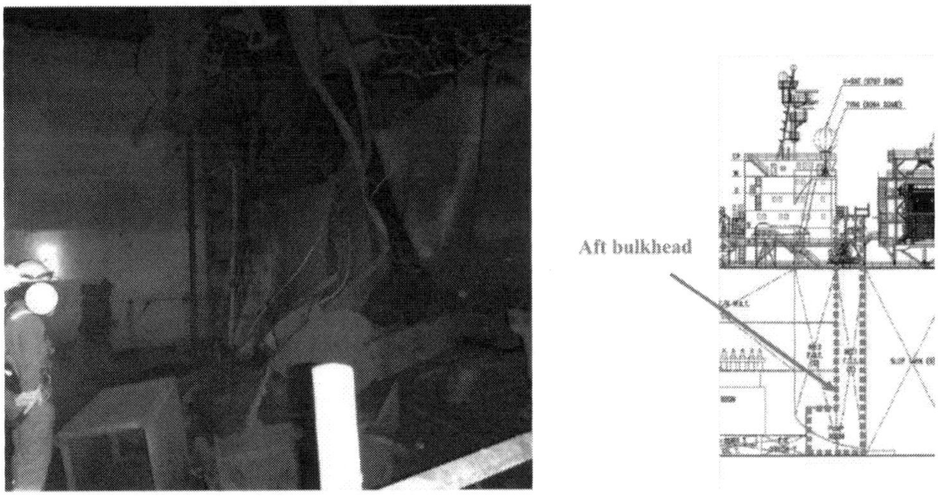

Figure 10: Damage to level above the pumps – complete bulkhead failure.

Figure 11: Damage to the engine room.

Figure 12: Damage to upper level of pump room.

3 Emergency Response

Response during the 2013 West Texas Explosion

As the fire occurred 20 minutes prior to the explosion, fire personnel were deployed to scene. Of the 15 people killed, 12 were first responders (8 volunteer firefighters and emergency services), and the other individuals were good Samaritans who supported the emergency response at the fertilizer plant. One of the volunteer fire fighters, who died in the blast, was employee of West Fertilizer. In fact, he was the plant foreman and knew what was stored in the facility. The emergency response is outlined in Figure 13.

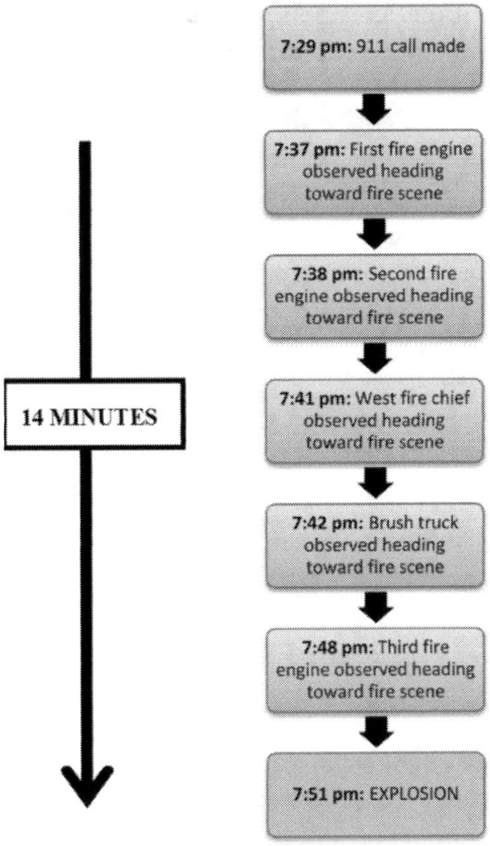

Figure 13: Emergency Response Plan at West Texas

Plant personnel and management did not understand the unique explosive nature of ammonium nitrate. More specifically, while the conditions necessary for detonation are not precisely known, the data suggest that detonations are promoted by severe shock or by heating with confinement, such as the massive fire at the plant. In fact, pools of molten ammonium nitrate may form in the event of fire, and if the molten mass becomes confined (e.g., in the elevator pit) it could explode. These hazards were not known nor understood by plant personnel and emergency responders. Therefore, responders were unaware of the potential of ammonium nitrate to detonate, had no knowledge of how long the fire had been burning, no robust incident pre-planning process in place, and no previous ammonium nitrate-related fire emergency training or drills. Hence, the strategy and tactics utilized by the West Volunteer Fire Department were not appropriate for the rapidly developing situation and they were exposed to unnecessary and extreme risk that proved fatal.

Response during 2015 explosion on the FPSO Cidade São Mateus

The first detector alarmed at 11:31, and by 11:38 all three detectors reached the over limit levels. The 1st team entered the pump room between 11:49-11:54 to assess the gas alarms, and they reported a leak at the flange and a 2-meter diameter pool under the flange. A 2nd team entered the pump room around 12:07 to assess the damage and whether repairs could be made. Despite the confirmed gas on all three detectors, the 2nd team still entered, and a personal gas detector alarmed at 100% LEL. The 2nd team exited around 12:15 and the ship was partially demobilized around 12:20 for lunch. The 3rd team enters around 12:27 to clean and repair, with ladders, wrenches, fire hose and absorbent blankets. The explosion occurred around 12:38 and killed four members in the pump room and four members at the top of the pump room on watch. Due to the failure of the wall into the engine room, the utilities operator was fatally injured in the engine control room.

There was a detailed Emergency Response Plan on the FPSO, along with the muster points for each person involved in the emergency response as shown Figure 14.

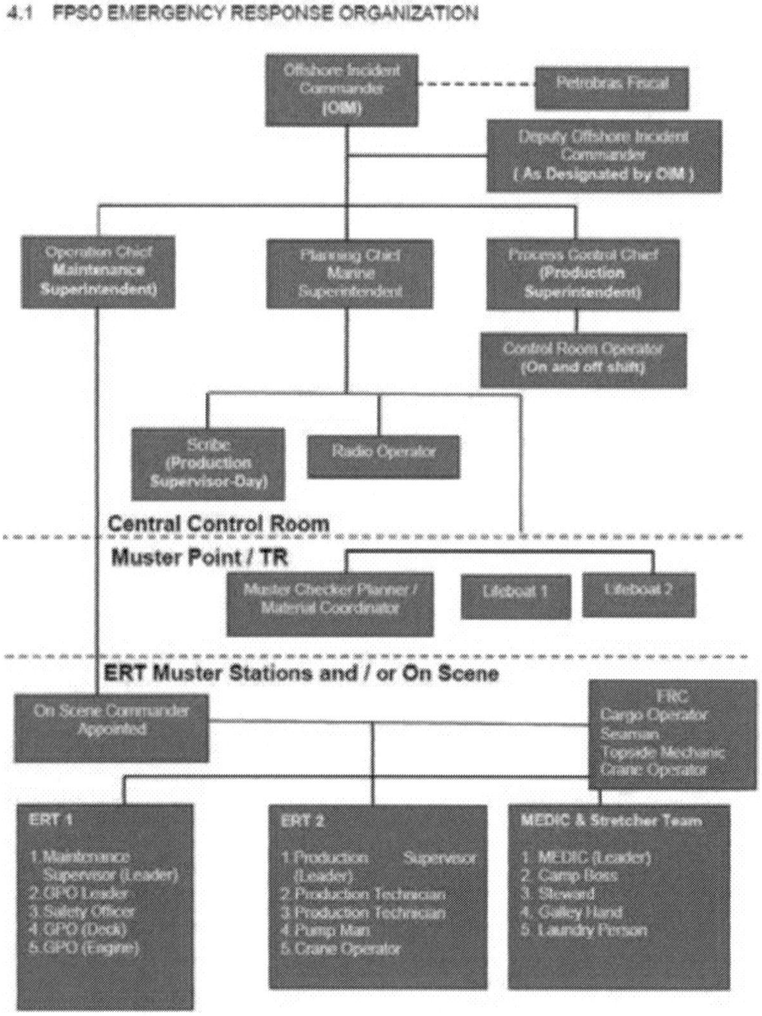

Figure 14: Organization Chart and muster points of the emergency response team

The Offshore Incident Commander is the Offshore Installation Manager (OIM) and would have the roles of implementing a command structure that would satisfy the organizational needs to protect life, environment and the installation, as well as deploying quick and efficiency actions to control incidents, as well as managing response actions to ensure that the escalating potential, risks of injury to personnel and damages to the installation are minimized. In the event of a leak in potentially explosive atmospheres, the emergency response must consider the possibility of the event escalating to a fire or explosion scenario. Thus, systems and procedures should be provided for the crew to control the situation and mitigate the risks of escalation from explosive atmospheres.

Investigation showed that the vessel was not designed for condensate, but for crude and gas. Yet despite the fact that there was confirmed gas detection within the pump room and on personal gas detectors, the crews entered three times. On the third entry, the explosion occurred. It was also reported that an explosive atmosphere in the pump or engine room was not included as an accident scenario in the emergency response plan, and there were no clear instructions of this particular scenario. Despite the knowledge that there was confirmed gas detection on the fixed sensors installed at the bottom of the pump room and also on personal detectors, Emergency Response Teams were sent into the pump room to evaluate the scenario in an explosive atmosphere.

The actions of Command essentially played down the risk of the condensate leak and allowed personnel to enter the pump room on multiple occasions to clean the spill without first performing a risk analysis, and without additional monitoring the room's atmospheric conditions. This noncompliance with the applicable emergency procedure was not only observed during entry operations, but also through the partial or total demobilization of the people irrespective of their muster points with an ongoing confirmed gas detection in the pump room. This demobilization resulted in two members being injured as they were found in a damaged elevator shaft after the blast.

4 Analysis

While deficiencies in the emergency response plans resulted in unnecessary and tragic consequences, the reasons each were lacking are very different and distinct.

For the West Texas Explosion, the Texas State Fire Marshal incident investigation [2] concluded that neither West Fertilizer nor West Fire Department had pre-fire and pre-incident plan at the facility. -Therefore, the recommendations provided were to ensure that local fire departments implement an Emergency Response Plan, with a particular focus on high-hazard occupancies. It was identified that fire departments visit and pre-incident plan responses to commercial structures in their jurisdiction. Further, they concluded that it is imperative that high-risk occupancies containing hazardous materials are identified and appropriate emergency response strategies and tactics are developed through pre-plan documents and SOP directives. Despite the extremely low likelihood of ammonia nitrate detonations, the recommendations state that high-hazard and high-risk occupancies containing heavy fire loads and hazardous materials should be identified by Fire Departments and all members made aware of their potential dangers. This will better prepare them for not only credible hazards, but also those high consequence hazards which may be found in the event of a fire and also provide them with insight as to when it is prudent to attack a fire or evacuate the area. More specific recommendations regarding the emergency response planning included:

- Develop a detailed Emergency Response Plan and Evacuation Plan for the area within 1 mile of existing ammonia nitrate buildings
- Develop standard operating procedures/guidelines to train and educate members on firefighting and emergency scene strategies and tactics for high-hazard occupancies
- Develop pre-incident plans by visiting the commercial and high-risk properties within the response jurisdiction.
- Utilize risk management systems on all emergency responses to determine when to employ alternative tactics.

- Incorporate incident management systems in every emergency incident, training exercise and basic operational process of the fire department to provide appropriate command and control for conducting operations.
- Implement a strategic incident action plan (IAP) for all emergency operations through a coordinated application of appropriate emergency scene tactics for the presented dangers
- Ensure that an incident safety officer (ISO), a person who is independent from the incident commander, is appointed and effectively utilized at every incident that would require an ISO.
- Establish a hazardous materials program and actively train members on the tactics to use when an incident involving hazardous materials occurs

In contrast, for the 2015 FPSO event, there was a detailed Emergency Response Plan for the vessel. Despite this plan, investigations concluded that there were significant deficiencies in the plan that needed to be remedied, and largely included reference to personnel exposure, failure to identify the accident scenarios for the emergency plan, training and mobilization. The execution of the plan failed to properly identify the potential high consequence and catastrophic result of the actions of the emergency responders. More specifically:

- Exposure to personnel
  - In the event of a leak in potentially explosive atmospheres indoors, the plan must consider the possibility of event escalation to a fire and/or explosion. Thus, systems and procedures should be provided once there is the formation of an explosive atmosphere. To ensure the success of mitigation actions, emergency procedures and resources for emergency response should be pre-planned, drawing on the use of all available resources, including the time of scenario assessment.
  - To determine that all functions involved in emergency response, including land-based support, have periodic theoretical training in emergency response procedures.
  - Set in the emergency response plan the available material and human resources to the response of each accidental
- Failure to identify accident scenarios or not included from the risk scenarios
  - Establish criteria and procedures for the migration of risk scenarios identified in the risk analysis and safety studies for emergency response plans. In addition to the selected scenarios, the procedures should contain scenarios that represent challenges to response actions, such as fatalities, and multiple injuries, among others.
  - Establish periodically simulated emergency covering all identified accident scenarios, including scenarios with multiple injuries and fatalities. Such simulations should include the activation of external resources to inform platform that response structure is effectively tested and that corrective measures arising from simulated ensure continuous improvement of the response actions of all entities involved.
- Demobilization of meeting points
  - Periodically coordinate all staff regarding the hazards involved in the assumptions made in the response procedure to the emergency unit.

  - Prohibit the release of people to normal activities before the full control of the emergency.
- Ignition of the explosive environment
  - As the ignition was introduced by the action of the emergency responders within explosive atmospheres, review the procedures related to emergency response and safe work practices considering the possible ignition sources identified in existing rules

5 Conclusion

The lessons from two specific incidents show that critical deficiencies in the emergency response plan or implementation of the plan can result in fatal consequences that could have been avoided. The 2013 West Texas explosion shows how important it is to not only recognize high risk events, but also have the appropriate plans in place to avoid unnecessary escalation of an already tragic incident, where 12 of the 15 people fatally injured were first responders. By preparing first responders for not only credible hazards, but also those high consequence hazards, can provide them with crucial insight as to when it is prudent to attack a fire or evacuate the area. In addition, not properly following or playing down the risk of a high consequence event was observed in the 2015 FPSO event, where eight of the nine crew members killed by the blast were first responders. The failure to recognize the potentially high consequence and catastrophic event, led to an emergency response that unnecessarily exposed responders and ultimately led to an exacerbated event. This study confirms that emergency response plans are an essential layer of protection for facilities where all other safeguards failed to prevent an incident, which could have prevented unnecessary and tragic escalation of the incident.

6 References

[1] Preparedness for Emergency Response - FEMA Training - FEMA.gov

[2] Texas State Fire Marshal, "Firefighter Fatality Investigation FFF FY 13-06," West, TX, 2013

[3] Davis, SG, DeBold, TF, & Marsegan, C (2017). Investigation Findings and Lessons Learned in the West Fertilizer Explosion. Journal of Fire Science. Volume 35, Issue 5, September 5, 2017: pp. 379-395.

[4] "A Public Health Report on Injuries Related to the West (Texas) Fertilizer Plant Explosion," June 24, 2014.

[5] Brazilian Navy Directorate of Ports and Coasts, Maritime Accident Safety Investigation Report, FPSO Cidade São Mateus - Explosion followed by flooding and casualties, February 11th, 2015.

[6] Investigation report of the explosion incident occurred on 11/02/2015 in the FPSO Cidade São Mateus, Brazilian National Agency of Petroleum, Natural Gas and Biofuels, Superintendence of Operational Safety and the Environment, August 2015.

19 AIChE Spring Meeting
15TH GLOBAL CONGRESS ON PROCESS SAFETY
NEW ORLEANS • 2019

Major Explosion at a Chemical Company in China: the Causes and Lessons Learned

Wei Wang
China University of Petroleum (East China)
No. 66, West Changjiang Road, Huangdao District, Qingdao, China
15621097585@163.com

Yi Liu
China University of Petroleum (East China)
No. 66, West Changjiang Road, Huangdao District, Qingdao, China
liuyi@upc.edu.cn

Dongfeng Zhao
China University of Petroleum (East China)
No. 66, West Changjiang Road, Huangdao District, Qingdao, China
zhaodf@vip.sina.com

Yanchun Wang
China University of Petroleum (East China)
No. 66, West Changjiang Road, Huangdao District, Qingdao, China
Wychun25@163.com

Jian Chen
China University of Petroleum (East China)
No. 66, West Changjiang Road, Huangdao District, Qingdao, China
836771528@qq.com

Hanhai Dong
China University of Petroleum (East China)
No. 66, West Changjiang Road, Huangdao District, Qingdao, China
donghanhaiupc@126.com

Prepared for Presentation at
American Institute of Chemical Engineers
2019 Spring Meeting and 15th Global Congress on Process Safety
New Orleans, LA
March 31 – April 3, 2019

Major Explosion at a Chemical Company in China: the Causes and Lessons Learned

Wei Wang
China University of Petroleum (East China)
No. 66, West Changjiang Road, Huangdao District, Qingdao, China
15621097585@163.com

Yi Liu
China University of Petroleum (East China)

Dongfeng Zhao
China University of Petroleum (East China)

Jian Chen
China University of Petroleum (East China)

Yanchun Wang
China University of Petroleum (East China)

Hanhai Dong
China University of Petroleum (East China)

Keywords: Fishbone Diagram; Analytic Hierarchy Process; Process Safety Management; Incident Investigation

Abstract

On July 12, 2018, a devastating explosion incident occurred in Yibin City Hengda Technology Co., Ltd. in Sichuan Province, resulting in 19 deaths and 12 injuries, and impacting an area of about 1,800 square meters. This was one of the most disastrous explosions that occurred in a chemical plant in China in the past five years. The incident reflects the general deficiencies of process safety management in small and medium-scale companies. Therefore, it is significant to analyze the causes of the incident and put forward countermeasures to avoid such kind of incidents.

According to the characteristics of complexity, diversity, and uncertainty about the influencing factors of the explosion, this paper presented a systematic analysis from qualitative and quantitative perspectives. The fishbone diagram model was used to identify influence factors from six aspects of people, design, management, environment, safety prevention, and materials. Furthermore, an assessment index including 26 factors was determined. The analytic hierarchy process method was used to analyze the importance degree of each factor quantitatively, and the key factors leading to the incident were obtained. The results showed that the sum of weights of three main aspects, including people, design, and safety prevention is nearly80%, so they were the dominant causes leading to this incident. Lack of chemical industry experience, non-standard process design, and incomplete automatic control system were the focus of prevention among 26 factors.

Process safety management is the application of management systems to the identification, understanding, and control of process hazards to prevent incidents from occurring. This analysis considered process hazard analysis, pre-startup safety review, operation procedures, training, and emergency plan and response, which were the main elements in process safety management framework. Finally, we proposed safety precautions for the identification, storage, production, and use of hazardous chemicals, as well as recommendations for process equipment management.

1 Introduction to the event

At 18:42:33 on July 12, 2018, an explosion occurred in Yibin Hengda Technology Co., Ltd. in the industrial park of Jiang'an County, Sichuan Province, causing 19 fatalities and injuring 12 people. The explosion caused the second and third workshops to catch fire, and the combustion area is $1800m^2$ [1]. The first workshop, warehouse, drying room, analysis room, office building, and other buildings had been damaged to some extent. The fire department dispatched 42 fire engines and more than 200 firefights. It took more than two hours to extinguish the fire on the spot as shown in Figure 1. It was estimated that the total energy of the explosion in this incident was equivalent to 235 kg of TNT. The direct economic loss was 41.42 million yuan.

At the time of the incident, the company was producing imazamox and triazole. Triazole is a pesticide, and imazamox is an herbicide. The process of the incident was the dehydration of butyramide in the process of synthesis of imazamox. At about 11:30 on July 12, the company collected 2 tons of chemicals labeled as raw materials (actually the explosive chemical sodium chlorate, not the raw material required for the production of butyramide). At around 14:00, the deputy director of the company's second workshop received raw materials from the warehouse. At around 15:30, the workers in the second workshop raised the raw materials for production through the lift to the third floor of the second workshop. At around 16:00, the 2R301 kettle for dehydration of butyramide was completed, and the feed was in an empty state. Before 17:20, the 2R301 kettle was completed. At about 6 pm, the dehydration completed. At 18:42:33, it was due to the on-site handover time. A chemical explosion occurred in the 2R301 kettle on the third floor of the second workshop. The direct cause of the incident was that the operator replaced the unlabeled sodium chlorate with butyramide and put it into the 2R301 kettle for dehydration, which caused the explosion and fire. Sodium chlorate is a kind of explosive chemical, which is easy to form an explosive mixture when mixed with butyramide. The explosion led to the disintegration of 2R301 kettle. The high-temperature toluene vapor rushed out rapidly formed an explosive mixture with the external air and produced a secondary explosion, which also caused the explosion and combustion of the materials such as sodium chlorate, toluene, and methanol stored at the workshop site.

Figure 1. Extinguishing at the scene.

According to the investigation, in this incident, the second workshop was the core area of the explosion, and the overall structure was severely damaged as shown in Figure 2. The platform on the third floor collapsed. The reactor with a height of nearly two meters and a diameter of about one meter fell from the upper level.

Figure 2. Severely damage to the second workshop.

2 Incident evaluation index system based on the fishbone diagram

Fishbone Diagram is also named as cause and branches figure analysis. Kaoru Ishikawa, a Japanese management master, invented it for finding out the root cause of incident [2]. It indicates the relationship between a problem and its underlying causes. It is a qualitative analysis tool to help identify incident causes [3]. With step by step in-depth study to analyze its influencing factors,

Fishbone Diagrams can systematize complicated incident causes. When starting the specific report, we first analyze the factors which affect the safety most and then followed by departure from the overall reason to find the basic reason and to detect and determine the main reason [4,5,6].

After the incident, the government set up a team of experts to identify risk factors from six aspects: human, design, management, environment, safety prevention, and material and establish an incident evaluation index system with 26 elements. The fishbone diagram that induced the explosion is shown in Figure 3.

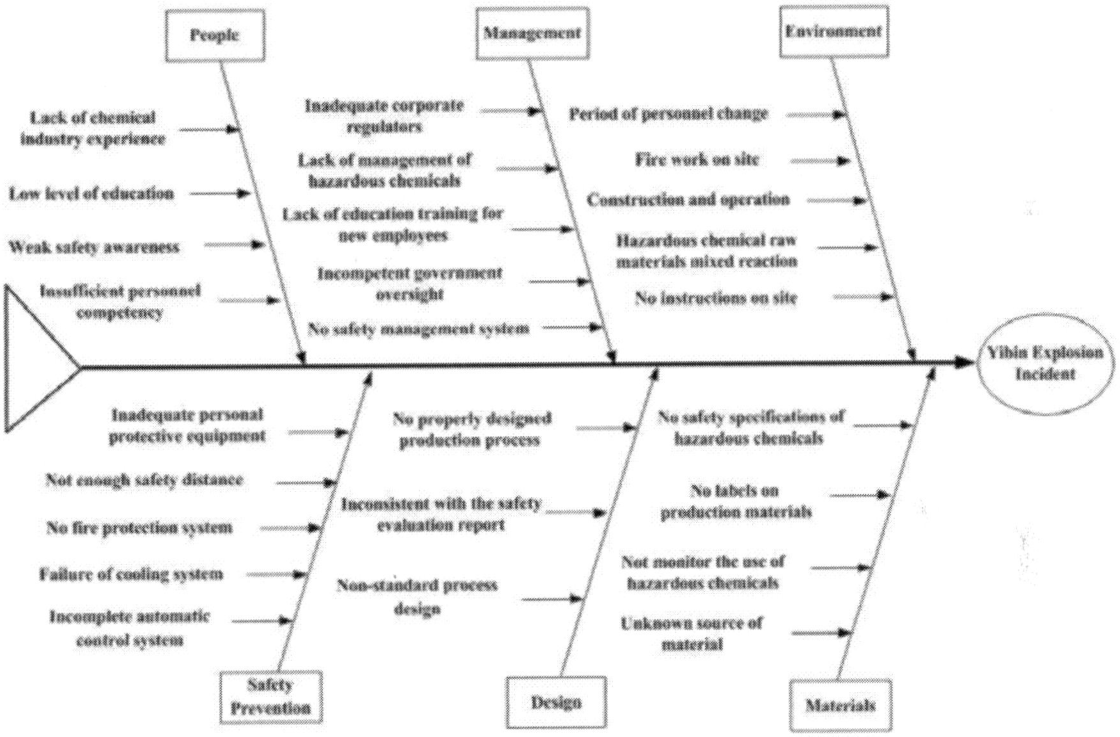

Figure 3. Fishbone diagram of Yibin explosion incident [7].

2.1 People

Neither the safety managers nor the operators demonstrated proficiency in the requirements of process safety competency. There were two main aspects below:

- Corporate management - The general manager of the enterprise had responsibility for the mechanical processing and production contract. However, he did not have appropriate experience in the chemical industry. The technical director did not have experience with management of process safety and production systems. The corporate legal person responsible for finance and logistics also had no experience in chemical engineering.

- Operators - Among the 19 people who died in the incident, 16 were operators, and most of these operators were local farmers. They lacked the basic knowledge of chemical process safety production, and they did not understand the process safety risks in the production process.

2.2 Design

The second workshop where the incident occurred was not constructed and installed strictly by the design drawings, and it was originally planned to have nine sets of 5000L tanks and one set of 3000L tanks and produce 5-nitroisophthalic acid. As built, the workshop had three sets of 5000L tanks, seven sets of 3000L tanks, and eight sets of 2000L tanks were set up. The product was also different from the original, but imazamox and triazole. A similar situation existed in the third workshop. The production technology of imazamox and triazole used by the company had no formal technical source, and the enterprise also did not entrust professional institutions for process calculation and construction drawing design. The technical director of the enterprise only planned the required production equipment and operation process according to his experience, but he knew nothing about the process safety risks of related technologies. It was during the production of imazamox that the distillation kettle used for the dehydration of butyramide by toluene first exploded. According to the *Notice on Further Strengthening the Safety Design Management of Hazardous Chemicals Construction Project*, Hengda company needed to start construction in April 2018, but the company had already started construction in July 2017. Also, the fire separation design of each building did not meet the requirements of the *Petrochemical Enterprise Design Fire Code (GB50160)*.

2.3 Management

The lack of process safety culture is the main cause of management failure. Safety culture is developed through the interactions of factors such as safety policies and safety goals, the discipline of the organization, the behavior and attitude of the company staff, and the competence of workers. Improper operating discipline and procedures can lead to a negative process safety culture and an increased likelihood of process safety incidents. The enterprise had not formulated the post safety responsibility system, safety management rules, and regulations. Furthermore, the enterprise manager and the person in charge of each department only had the oral authorization, and there was no formal appointment document to clarify the post responsibility. The safety management organization was not complete. The company's safety and environmental protection department had only one safety officer who did not have safety management capabilities. There were also no full-time or part-time safety officers in each workshop. There was a lack of management of hazardous chemicals in the process safety management. The technology provider concealed some chemical name marks for confidentiality, and there was no safety specification. Required knowledge and competence connected with safety for workers were the pivotal content of safety training. New employees did not have security education directly work or internship. The government had not strictly regulated the safety departments of local key hazardous chemicals companies.

2.4 Environment

The factory built the workshop while producing, and used the polypropylene pipe which was easy to produce static electricity to transport flammable and explosive liquid materials. The building materials, hardware, and chemicals in the warehouse were mixed and piled with hidden dangers. The incident occurred during shift time, resulting in a large number of casualties. In the on-site operation workshop, there was neither operation instructions nor warning signs for hazardous chemicals. At the time of the incident, the second workshop produced two products at the same time. This practice of producing multiple produced at the same time in the same workshop has excellent security risks.

2.5 Safety prevention

Compared with the 20 elements of CCPS, China drew on the process safety management experience of the Occupational Safety and Health Administration (OSHA) and further optimized the formation of 12 elements for the production of Chinese chemical companies. After that, a series of process safety technologies were developed and promoted for the key elements of risk management and equipment integrity. In 2013, the SAWS issued *Guidance for enhancing chemical process safety management*. The Ministry of Emergency Management, which was integrated by the State Administration of Work Safety(SAWS) in 2018, attaches great importance to the supervision of hazardous chemicals, hazardous chemical processes, and major hazardous sources, referred to as "Two Priorities and One Major." If an enterprise violates The relevant provisions, it shall be punished according to the upper limit of economic punishment according to law.

In the production process of imazamox and triazole, the nitrification, oxidation, diazotization, hydrazine hydrate, hydrogen peroxide, methanol, and other dangerous chemicals and processes involved were all "Two Priorities and One Major" projects, which needed strict supervision. However, the company's automatic control system, flammable and toxic gas alarm system and fire water system and other safety facilities were not installed. Once there was a danger, the enterprise itself could not deal with an emergency. The lack of an automated control system led to more than ten operators performing manual operations around the reactor, which was an important cause of major casualties.

2.6 Materials

According to the investigation, the safety evaluation report made by the enterprise before the incident did not fully consider the process safety information, such as the lack of description of dangerous chemicals and the risk identification of the process, so the enterprise had great potential risks. Workers didn't know the source of raw materials, and many chemical production raw materials were not marked. Workers knew nothing about the risks of these dangerous chemicals, and company officials failed to communicate information about the hazards of the process to employees. The workers compiled codes to replace the names of products and raw materials, and these numbers were not abbreviated as specifications. The raw materials used by Hengda Company, such as hydrazine hydrate, nitric acid, toluene, hydrochloric acid were dangerous chemicals or precursor chemicals, but the company still purchased them illegally without relevant licenses. In the production process, regulators did not supervise the management of hazardous chemicals.

3 Determination of the weights of the incident incentive index based on the analytic hierarchy process

The Analytical Hierarchy Process (AHP) is a decision-making method proposed and used by American operations researcher T.L.Saaty in the mid-1970s [8]. AHP is a process that breaks down a problem in a hierarchical matrix with increasing level of detail to determine the most significant causes of an incident. According to the obtained fishbone diagram of the Yibin explosion incident, an incremental layer structure for analytic hierarchy analysis was constructed, as shown in Fig. 4, where the A, B, and C layers illustrated the primary effect and hierarchy of causes.

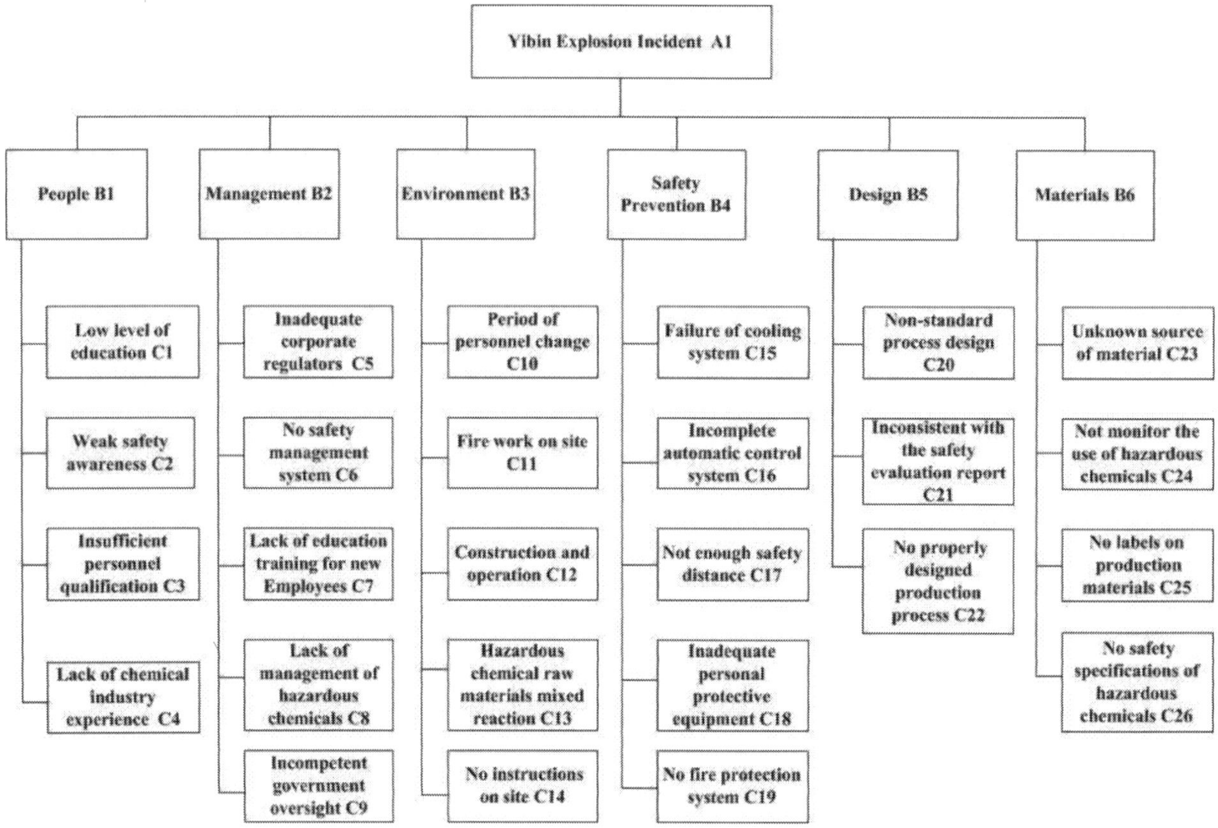

Figure 4. Hierarchical structures of safety evaluation system for Yibin explosion incident.

3.1 The single-layer index weights of the influencing factors

According to the actual data collected and expert opinions, we used 1-9 scale to construct the judgment matrix (A) of the first-level evaluation index, as shown in table 1, and obtained the influence of various factors in layer B on layer A. λ_{max} is the maximum eigenvalue, CI is the consistency index, RI is the average random consistency index, and CR is the random consistency ratio; When CR<0.1, it passed the consistency test [9]. According to table 1, the weights of design is the largest, followed by people and safety prevention, with the sum of the weights of the three as high as 0.8. It can be seen that the leading cause of this incident is due to the low level of process design, operators' insufficient qualification of chemical safety production and non-compliant protection measures. Therefore, enterprises should pay attention to intrinsic safety design, improve the safety awareness of staff by popularizing safety knowledge and operational guidelines, and finally achieve the goal of safe production.

Table 1. Judgment matrix of primary evaluation indicators and weights results

A	B1	B2	B3	B4	B5	B6	Weights	Consistency	Ranking
B1	1	2	3	1/2	1/3	7	0.1645	λ_{max}=6.4698 CI=0.094 RI=1.26 CR=0.0746<0.10 Satisfy consistency	3
B2	1/2	1	2	1/3	1/4	6	0.1060		4
B3	1/3	1/2	1	1/4	1/5	5	0.0700		5
B4	2	3	4	1	1/2	8	0.2545		2
B5	3	4	5	2	1	9	0.3810		1
B6	1/7	1/6	1/5	1/8	1/9	1	0.0239		6

3.2 The comprehensive weights of the influencing factors

We built each judgment matrix of layer C to layer B and calculated the single layer weights of layer C, and finally, we obtained the comprehensive weights of incidents caused by bottom influencing factors. We considered the product of these two layers as the comprehensive weights of the sub-factor. For example, the comprehensive weights of C1 is equal to the single-layer weights of C1 to B1 multiplied by the single-layer weights of B1 to A. The comprehensive weights result and ranking of each sub-factor were shown in table 2.

Table 2. Single and synthetic weights of C layer indexes

		Single weights	Consistency	Synthetic	Ranking
B1	C1	0.0843	λ_{max}=4.1375 CI=0.0458 RI=0.9 CR=0.0515<0.1 Satisfy consistency	0.0139	16
	C2	0.2120		0.0349	8
	C3	0.1544		0.0254	12
	C4	0.5492		0.0904	3
B2	C5	0.1556	λ_{max}=5.2550 CI=0.0638 RI=1.12 CR=0.0569<0.1	0.0165	15
	C6	0.0566		0.0060	21

	C7	0.4735	Satisfy consistency	0.0502	5
	C8	0.2578		0.0273	11
	C9	0.0566		0.0060	22
B3	C10	0.6253	λmax=5.3598 CI=0.08995 RI=1.12 CR=0.0803<0.10 Satisfy consistency	0.0438	7
	C11	0.0901		0.0063	19
	C12	0.0435		0.0030	23
	C13	0.1976		0.0138	17
	C14	0.0435		0.0030	24
B4	C15	0.0781	λ_{max}=5.1717 CI=0.042925 RI=1.12 CR=0.0383<0.1 Satisfy consistency	0.0199	13
	C16	0.5895		0.1501	2
	C17	0.1271		0.0324	9
	C18	0.0781		0.0199	14
	C19	0.1271		0.0324	10
B5	C20	0.6833	λ_{max}=3.0246 CI=0.0123 RI=0.58 CR=0.0236<0.1 Satisfy consistency	0.2604	1
	C21	0.1168		0.0445	6
	C22	0.1998		0.0761	4
B6	C23	0.1242	λ_{max}=4.1891 CI=0.0630 RI=0.9 CR=0.0708<0.1 Satisfy consistency	0.0030	25
	C24	0.2626		0.0063	20
	C25	0.5465		0.0131	18
	C26	0.0667		0.0016	26

As can be seen from table 2, the full weights of C16 (Incomplete automatic control system) and C20 (Non-standard process design) was greater than 0.1. The full weights of C4 (Lack of chemical industry experience) and C22 (No properly designed production process) were very close to 0.1. The combined weights of these four factors reached 0.577, which was more than half of the total. Therefore, these four influencing factors were the primary causes of the explosion incident in Yibin, Sichuan, as strict control and prevention were given priority.

Among them, both C20 and C22 belong to the design aspect, and enterprises should entrust qualified formal companies to design the process and production plan. In the method of safety design, the operation and design should be analyzed in real time to check the status of dangerous materials, effectively reduce the operating temperature and pressure, and control the influence of hazards factors in the process on chemical equipment. The equipment used in production shall meet the strength and anti-corrosion performance required by the process, and shall be designed and manufactured by qualified factories.C4 is a human problem. Therefore, it is necessary to conduct chemical safety education and operation training for operators before entering the plant. Many of Hengda's workers were the relatives and friends of management personnel. They were generally poorly educated and skilled in operation, and had not received any training in chemical safety. The workers did not go through systematic learning, but the workshop director directly taught them operating skills. C16 is a problem of safety prevention. The automatic control system can not only improve the operation efficiency of production equipment but more importantly, it can protect the safety of chemical personnel and reduce the loss caused by incidents. Enterprises should follow the safety production regulations, install automatic control systems, toxic and harmful gas alarm systems, and fire protection systems to effectively prevent incidents.

The weights of C7 (Lack of education training for new employees), C10 (Period of personnel change) and C21 (Inconsistent with the safety evaluation report) were all around 0.05. The weights of C2 (Weak safety awareness), C19 (No fire protection system) and C17 (Not enough safety distance) also exceeded 0.03. These five factors were important causes of incidents and should be effectively controlled.

C7 is a new employee training management problem. Enterprises should pay attention to the educational training of employees before entering the plant and implement the three-level safety training education of the plant, workshop, and team. C10 involves change management. At the time of the incident, it was during the shift time of employees, the change of personnel and the change of materials and equipment, which had great potential risks. Enterprises should establish a change management system, conduct comprehensive training for the staff who play a vital role in the change management process and clarify the position and responsibility of each person in the change system.

4 Lessons learned from the causes of the incident

The Yibin explosion was a pharmaceutical chemical incident. Through the investigation and analysis of historical disasters in the chemical industry in China, it was found that the proportion of pharmaceutical chemical incidents in the chemical industry was huge [10]. It was characterized by a random change of chemical process, small scale, low level of process control, involving dangerous materials and complex chemical reaction. Especially in small and medium-sized chemical enterprises, the lack of process safety management is an essential cause of such incidents. From the perspective of process safety management, this paper summarized and put forward suggestions for preventing such incidents.

- Process Hazard Analysis - In the design stage, the risk identification and assessment tool system should be applied to identify the risks of materials in the process of technology. In the review stage, HAZOP, SIL are used to review the process, layout and material flow. Companies should use scientific models to calculate the explosion-proof grade and safety distance of buildings. By adding the explosion-proof wall and the explosion-proof film, the building close to the installation area can reach the required safety factor. After the device is put into operation, it needs to be compared with the hazard analysis in the design stage, and the hazard that is inconsistent with the design should be analyzed again. Hengda's actual operating equipment, raw materials, and products were inconsistent with the original plan, and the company had not discovered and controlled the real hazards [11].

- Pre-startup Safety Review - The company shall organize a safety inspection team, whose task is to confirm whether the improvement measures and safety guarantee measures in the process hazard analysis report have been implemented as required; Whether adequate training of personnel in operation, maintenance, and management of process equipment has been completed; Whether necessary safety information of process equipment is completed and updated; Whether necessary equipment testing and checking have been completed and confirmed; Whether the process equipment is constructed and installed according to the design requirements [12].

- Operation Procedures - The incident occurred during the change management phase, and there was no review or other risk assessment during the shift. Staff should review the operating procedures as often as possible to ensure that the operating procedures reflect current operating conditions, including changes in chemicals, process equipment, and facilities. The enterprise shall confirm the adaptability and effectiveness of the operating procedures and ensure that the operators understand and master the operating systems [13].

- Training - A complete process safety management is inseparable from the participation of operators, and the human factor is the most unstable factor in process safety management. Therefore, Enterprises should establish strict production operation disciplines and require operators to master process safety information, equipment integrity management points, and emergency response policies and finally improve workers' ability to control equipment safety operations [14].

- Emergency Plan and Response - Enterprises need to establish a set of overall emergency plans, which usually stipulate in written form how the factory should deal with abnormal or urgent situations. Also, enterprises should establish emergency response teams, traditionally composed of enterprise personnel, but also include external staff; The responsibilities of each team member should be clear, ensuring that members have no questions about accountability and delegation. In case of an emergency, the person in charge may determine the safe area according to the emergency response manual and direct the person to evacuate to a safe place [15].

5 Conclusion

The circumstances of the incident described in this article are typical in China, and the incident reflects the common safety problems in small and medium-sized chemical enterprises. Managers should understand these lessons we learned from this incident. By using fish-bone diagram method, the causes of the explosion incident were obtained through qualitative analysis, including six aspects, such as people, design, management, environment, safety prevention and materials, and 26 factors influencing the incident were further analyzed. Then the analytic hierarchy process was used for quantitative analysis, and the results showed that lack of chemical industry experience, non-standard process design, and incomplete automatic control system were the main factors affecting the incident, so the enterprise should strictly control and prevent these factors. Finally, we summarized the causes of such incidents and put forward the measures from five aspects of process safety management, including process hazard analysis, pre-production safety inspection, operation procedures, training, and emergency plan and response plan. An incident brings not an only economic loss but also bad social influence to the factory. It is our hope that this paper will help prevent future accidents.

6 References

[1] Sichuan Provincial Emergency Management Office, Yibin Hengda Technology Co., LTD. "7.12" Major Explosion Fire Accident Investigation Report, Available at http://www.chinasafety.gov.cn/gk/jsts/201902/t20190214_224397.shtml, Accessed on February 14, 2019.

[2] Tongyuan Luo, Chao Wu and Lixiang Duan, Fishbone diagram and risk matrix analysis method and its application in safety assessment of natural gas spherical tank, Ind. Eng. J. 25(02), 2007, 138-141.

[3] Xing, W., QC Team Based Materials, China Society Press, Beijing, China, 2004.

[4] T Deng, X., Chen, Q. and Jiang, D., Application of fishbone chart analysis method in prevention for steam turbine overspeed incident, Guangdong Electr. Power, 20(02), 2013, 73-77.

[5] Changlong Zhu, Jiayong Zhu, Li Wang and M. Sam Mannan. Lessons learned from analyzing a VCE incident at a chemical plant, Journal of Loss Prevention in the Process Industries,50,2017, 397-402.

[6] Dennis P. Nolan, Chapter 5: Specialized Reviews—CHAZOP, EHAZOP, Bow-Tie Analysis, Layers of Protection Analysis, Safety Integrity Level, FishboneDiagram, and Cyber Security Vulnerability Analysis, Safety and Security Review for the Process Industries (Fourth Edition), Gulf Publishing Company,2015,Pages 17-27.

[7] Xinhua News Agency, Emergency management department: 7·12 incident enterprises have no law and discipline, and government supervision is lax, Available at http://www.gov.cn/xinwen/2018-07/17/content_5307196.htm?_zbs_baidu_bk, Accessed on July 17, 2018.

[8] Seong KonLee, GentoMogi and Jong WookKim. Decision support for prioritizing energy technologies against high oil prices: A fuzzy analytic hierarchy process approach, Journal of Loss Prevention in the Process Industries,22(6), 2009, 915-920.

[9] Saaty T.L., A scaling method for priorities in hierarchical structures, Journal of Mathematical Psychology, 15 (3),1977, 234-281.

[10] Heda Zhang and Xiaoping Zheng, Characteristics of hazardous chemical incidents in China: A statistical investigation, Journal of Loss Prevention in the Process Industries, 25(4), 2012, 686-693.

[11] Hao WU, Igor PEÑARRUBIA, Lin CUI and Jinsong ZHAO, Process safety management considerations for biofuel production, Higher Education Press, 4(3), 2017, 357-367.

[12] Yuan Gao, Yunxiao Fan, Jing Wang, Xi Li and Jingjing Pei, The mediating role of safety management practices in process safety culture in the Chinese oil industry, Journal of Loss Prevention in the Process Industries, 57, 2019, 223-230.

[13] Paul R.Amyotte and Cathleen S.Lupien, Chapter Three - Elements of Process Safety Management, Methods in Chemical Process Safety, 1, 2017,87-148.

[14] T. Kontogiannis, M.C. Leva and N. Balfe, Total Safety Management: Principles, processes and methods, Safety Science, 100(B), 2017, 128-142.

[15] Hao WU, Igor PEÑARRUBIA, Lin CUI and Jinsong ZHAO, Process safety management considerations for biofuel production, Front. Eng. Manag, 4(3), 2017, 357- 367.

Taming the Wild River Rapids – How Process Safety Can Apply Outside the Process Industries

Trish Kerin
IChemE Safety Centre
Level 7, 455 Bourke St, Melbourne, Victoria, Australia
tkerin@icheme.org

Prepared for Presentation at
American Institute of Chemical Engineers
2019 Spring Meeting and 15th Global Congress on Process Safety
New Orleans, LA
March 31 – April 3, 2019

Taming the Wild River Rapids – How Process Safety Can Apply Outside the Process Industries

Trish Kerin
IChemE Safety Centre
Level 7, 455 Bourke St, Melbourne, Victoria, Australia
tkerin@icheme.org

Keywords: lessons learned, process safety, design, maintenance

Abstract

On the 25th October 2016 the Thunder River Rapids ride at Dreamworld on Australia's Gold Coast suffered a pump failure, resulting in the tragic deaths of four riders. This incident was investigated by the police and later a coronial inquest to determine how a tragedy like this could occur. As a process safety professional, it was interesting to follow the coronial inquest findings and then think about how process safety management may have helped to prevent this incident. Assuming that a process safety event can be defined as a high consequence or potential consequence low likelihood loss of control event [1], this incident had all the hallmarks. Therefore, it is possible to apply the IChemE Safety Centre (ISC) framework [2] for process safety to identify where the system failed and understand how it could be averted in future. This paper will explore various elements of the incident and how process safety management techniques may have helped, including that the ride had been modified to be different from all other similar rides, that the staff did not know how to stop the ride in an emergency and that all means to stop the ride required manual detection and intervention, with no alarms or shut down systems installed.

1 The Incident

The Thunder River Rapids ride at Dreamworld in Queensland, Australia, was a water ride, simulating white water rafting. Up to six riders sit in a circular raft, which flows along a turbulent river, including travelling up a conveyor to ride down a waterfall. The water flow is driven by pumps and made turbulent by the installation of underwater obstacles. It was in full operation on 25th October 2016, when the pump managing the water flow tripped out. This resulted in the water level on the ride dropping. This drop in water level caused a raft to become stuck on the rails at the conveyor leading to the disembarkation platform. This raft remained stationary for some 57 seconds while the ride continued to run before it was impacted by the next raft in the sequence. This following raft contained 6 passengers, four adults and two children. When this raft struck the stationary one, it flipped over, killing the four adult passengers. The two children managed to escape the raft [3]. This incident is currently under a Coronial Inquest. As such this paper is written from several media outlets that have been reporting the proceedings and based on the evidence reported, not the findings of the Coroner. The Coronial Report is due to be published in mid 2019.

2 History of the Ride

The Thunder River Rapids ride was completed in 1986 and was likely an in-house design and construction [4]. Sometime after construction, the ride was modified to remove some of the timber slats that form the base of the conveyor [4]. Two in every three slats were removed, though is unclear why this was done. This did create significant gaps in the base of the conveyor belt. The ride was also refurbished some months prior to the incident. [5].

Over the history of the ride there had been a number of incidents that had occurred with the rafts and the pumps. This in itself is not unusual, in that pumps can trip out from time to time, however there had been an incident in 2001 where the pump had tripped, and a raft became stuck [6]. This was then struck by a second raft, which flipped over. In this incident there were no passengers in the raft, and therefore no injuries. The investigation found that the ride operator had failed to adequately monitor the ride and he was dismissed. Details of this incident were not shared broadly and the operators and maintainers of the ride in 2017 had no knowledge of it. At the time an email was written by a park engineer discussing the incident and stated "I shudder when I think if there had been guests on the ride" [3]. This demonstrates that someone at some time recognized the significance of this incident, but it was then lost in history.

3 Technical and Organizational Causes of the Incident

There is one key immediate cause of this incident, however there are many different and wide-ranging causal factors. The immediate cause has been determined to be the tripping of the water pump, which in turn allowed the water level to drop, causing the first raft to become stuck. This provided the obstacle that the second raft, with six passengers, struck, causing it to then flip over. But let us examine the broader aspects of this incident, from the pump reliability to the safety systems and organizational capability. The ISC framework [2] provides a useful template to examine the causes. This framework in represented by Figure 1. Fundamentally leadership underpins all aspects of process safety. There are six functional elements that must be robust in an organization to manage major hazard safety. These are; knowledge & competence; engineering & design; systems & procedures; assurance; human factors and culture. In this incident there were clear failures in each of these elements.

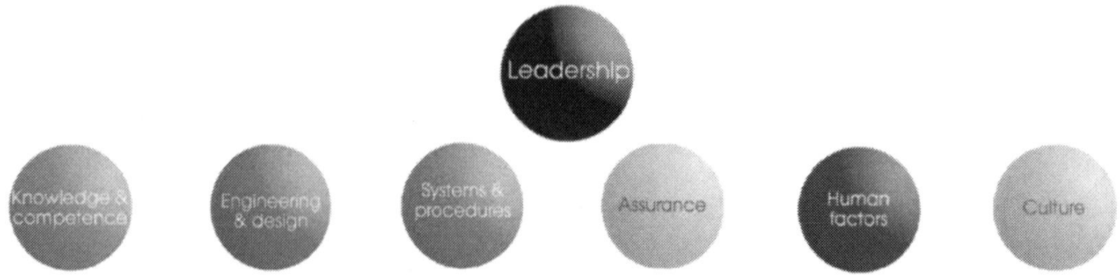

Figure 1. ISC Framework for Process Safety

3.1 Knowledge & Competence

The training of the ride operators took only 90 minutes and was unstructured with no evidence of competence consistently recorded [7]. There was a procedure manual however it was thought to be confusing by the ride operators [8]. The operator running the ride on 25th October 2016 had been trained for 90 minutes that morning, as it was her first day running the Thunder River Rapids. The junior operator stated that the trainer "showed me the front page of the manual and just flipped through it. I had to go through it myself. She said just sign off on it when you are happy." [9] The trainer stated that she sometimes had to "invent" training techniques because of the complexity of the control panel [8]. There was no clear guidance on how and when to use the emergency stop for the ride. The new operator testified that she was told "don't worry about it, you don't need to use it" with regard to the emergency stop button [10]. This is disputed by the person who delivered the training. The new operator believed it was the role of the senior operator to shut the ride down. It should be noted that the emergency stop would shut down the moving equipment with in two seconds, where the normal shutdown took around eight seconds. The ride operator that day stated that she "would have waited for the senior operator to tell her to hit the e stop" [10]. The senior operator that day was "surprised" to learn there was an emergency stop that would have shut the ride down within two seconds [7].

3.2 Engineering & Design

Intamin Amusement Rides designed and built the first river rapids rides in the world, and following this tragedy released a statement stating that the ride was "neither built by Intamin nor supplied by Intamin". The ride at Dreamworld was a custom build, and this was said to be common for rides built in the 1980's in the Gold Coast[4].

3.2.1 The Timber Slats

—

A few years after the ride was opened, a modification was made to the timber slats in the conveyor system [4]. The slats appear to serve at least two safety purposes. The first was to provide a solid platform for a passenger if they fell from the raft, essentially isolating them for the moving machinery and providing a platform to stand on [11]. The second was so that the rafts did not get caught in gaps and become stuck or indeed flip over. The Thunder River Rapids ride had two of every three slats removed. This provided large gaps between the remaining slats. There does not appear to be a theme park anywhere else that has made this change. There also appears to be no evidence as to why the change was made, nor was there a considered Management of Change (MoC) review undertaken. It is not clear if the organization had a structured and systematic MoC process. The safety regulator did not approve the modification and believed it added in additional hazards [4, 12]. In this incident, the raft became lodged in the gap, and as the conveyor continued to move forward, the raft was driven up vertically. When the passengers then fell from the raft, there was no solid base and they landed in the mechanism of the conveyor. There had also been conveyor framework modifications made to the ride in early 2016 in an attempt to stop rafts flipping over [13]. It is unclear what this modification was or indeed why it was made.

Figure 2. Aerial view of Thunder River Rapids showing the large gaps in the timber slat conveyor after the incident [10].

Figure 3. Aerial view of Infinity Falls, Orlando, showing close spacing of the boards on the conveyor [14].

3.2.2 The Pump

The water pumps had failed three times in the days leading up to the incident, and then twice on 25th October 2016, before the final failure that lead to the tragedy [15]. The pump's purpose was to keep the water flowing through the ride to keep the rafts moving. If the pumps stopped the water not only stopped flowing, it also drained back, and the water level dropped such that rafts could become stuck on the underwater obstacles that produced the rapids [16]. The reason for the pumps tripping out was not determined, the protocol seemed to be to just reset them to get the ride operating again [3]. There also is some confusion as to whether an alarm was sounded when the pumps tripped out, with an old procedure having the comment about an alarm crossed out [17]. At the time of the incident the pumps were approximately 10 years old, and the manufacturer stated that they had a life expectancy of about a decade. There did not appear to be any plans to replace the pumps, however there was a maintenance request into the pump supplier in the week prior, but this visit did not eventuate [18].

3.2.3 Low Water

Low water in the ride was a critical hazard, as the rafts could become stuck or indeed flip. There had been previous incidents where low water level hazard led to rafts flipping, notably in 2001. There was no low water level switch that would shut down the ride if the water level became too low [3]. The only indication the operators had was to notice the water stain level on the wall of the ride canal to determine if the water was too low [19]. When the water level dropped on 25th October, one empty raft became stuck on the rails near the base of the conveyor. As the second raft containing four adults and two children engaged with the conveyor it was then driven into the stuck raft. A safety review conducted after the event stated that the primary cause was a lack of a safety rated water level detection system. It was believed that such a device could have been fitted for as little as AUD 3,000. The ride had been refurbished some seven months prior to the incident, but a device was not fitted. The nearby log ride at the park had such a safety device fitted [18].

3.2.4 Emergency Stop

Around two years prior to the incident an external safety report was written suggesting the emergency stop system should be simplified, however this was rejected by the company as the emergency system in place was seen to be an "acceptable risk". The stopping procedure for the ride contained several individual steps. There was a single button which would stop all components of the ride, however this was located at the unloading area and not the control panel where the operators worked. [20]

There was a great amount of confusion regarding the emergency stop button and what it actually did. The senior operator stated that they were "surprised" to learn there was an emergency stop that would stop all components and the junior operator on her first day was unsure what the button did [7]. She stated she was told to not worry about the button because she did not need to use it and testified, she "would have waited for my senior operation to tell me to press it". She also did not know the button would stop the conveyor from moving. [10] The maintenance workers who performed the daily checks on the ride also stated they did not know what the emergency stop button did as they never tested it [21].

3.2.5 Maintenance

Seven months prior to the incident a directive to cut all repairs and maintenance spending was ordered by the owners. At the time of the incident the facility was approximately AUD125,000 over maintenance budget for the year to date. Capital expenditure could continue to be spent during this time [16]. One maintainer stated that he performed comprehensive daily, monthly and annual checks on the ride, which include consumable replacements such as nuts and bolts [22]. When the regulator inspected the ride after the incident, he noted it was in "generally poor condition" with excessive corrosion and rust on the electrical cables [23].

The maintenance planner noted that if spending could be deferred it would be. He also noted that the ride was shut annually for review, but on the nine occasions he witnessed a shutdown, he never saw an engineer inspect the ride nor were any risk assessment carried out [24].

3.2.6 Seatbelts

The Thunder River Rapids rafts had seat belts fitted for the safety of the riders. It should be noted that in May 2017, some seven months after the Dreamworld incident, that an 11 year old girl died on a similar ride, Splash Canyon, at Drayton Manor in the United Kingdom. The Splash Canyon ride rafts had no seatbelts installed at all, and as the child stood to swap seats with a fellow rider she fell overboard and struck her head. She died days later in hospital. This was not the first time a rider had fallen from the Splash Canyon ride [25]. This incident highlights the importance of proper restraints on these types of rides.

The seatbelts provided on the Thunder River Rapids rafts consisted of Velcro straps. It was vital that these were engaged correctly to provide the necessary restraint. Unfortunately, the Velcro was worn, and this made correct engagement more difficult. It is thought that at least one of the four riders may have survived the incident if the rafts had a better quality seatbelt, such as a mechanical buckle type that are on aircraft. This is because as the raft was shaken by the mechanism the belt failed to hold and the rider dropped to the conveyor system below [23, 26].

3.3 Systems & Procedures

Dreamworld has a series of operating procedures for the Thunder River Rapids ride however they were thought to be confusing [8]. During the investigation into the 2001 incident where empty rafts flipped, it was determined that the operator failed to follow the startup and emergency procedures. The investigation also reviewed the operating procedures and determined that the chance of a similar incident occurring with passengers on the rafts "is nil" [6]. A senior engineer believed that the cause of the 2016 incident was the operator not following procedures and that if procedures were followed the "system is a safe system" [27, 28].

There was also a lack of adequate emergency response procedures. Operators had discussed wanting better procedures and undertaking emergency drills, however, even two years after the incident, these have not been conducted at all [29].

3.4 Assurance

—

Over the history of the ride there had been a number of third-party reviews, inspections or audits carried out. Three years prior the incident a safety audit was conducted, and it recommended replacing the complicated emergency shut down process with a single emergency stop button, but this recommendation was never actioned [30].

There was also a legal requirement for the rides to have a "certificate to operate" and the Thunder River Rapids ride certificate was due to expire in January 2016. The company sought at extension as they were unable to find a "competent person" as defined under the legislation to inspect the ride [16]. Eight day before the incident the ride was inspected by a third-party engineer. He declared the ride as "structurally sound and properly functioning" even though he did not complete the full safety audit and merely did a visual inspection. He also did not inspect the log book as was required by law in a push to have the ride certified before it expired, though he admitted that he could have refused to certify the ride because of a "lack of documentation" [31, 32].

3.5 Human Factors

The control panel for the ride was confusing [3] and the operators needed to perform 36 different tasks in a one minute period, consistently [10]. This made for a system that encouraged human error, as it was not possible for the operators to be effective under these conditions. In addition to this, the engineering team believed the operators needed to simply follow procedures for the ride to be safe, not taking into account how difficult the procedures were or the actual workload. A senior engineer stated "In my view if procedures are followed the system is a safe system, you may as well not open the park. Procedures are there to be followed, if we don't follow procedures in everyday life, things happen." This blind focus on procedures does not take into account how the human interfaces with the machine and how the system advantages or disadvantages human performance.

3.6 Culture

A maintenance worker stated that he believed that the "safety culture at Dreamworld is very sound" [22] however this is in contrast to other workers who testified that familiarity had bred contempt with respect to the Thunder River Rapids ride [16], where new attractions were given more focus that existing ones [17]. Junior operators believed there was a culture of fear and had engaged with their union over their working conditions and the demands put on them regarding lack of rest breaks [33].

The culture also appeared to not learn from incidents, with the 2001 incident of flipping rafts said to be a one off and having "nil" chance of occurring with riders on board [6]. even though the consequence appeared to be understood at the time with an engineer writing in an email "I shudder when I think if there had been guests on the ride." [3].

The park relied on its apparent good safety history rather than challenging how and when an incident could occur [11]. They did not have an effective incident tracking system to

—

help with the learning process. They introduced a database but there was "general reluctance to use what was a fairly complex database" by the operators [34].

Safety did not appear to be a value in the company culture. The former safety manager stated that "everyone thought it was someone else's responsibility". He believed that deficiencies in the corporate structure led to a lot of problems. In 2018 his corporate safety role was made redundant.

Around five months prior to the incident, a memo was issued to staff to 'remain vigilant when monitoring conveyor movements" for items that might interfere with the rafts. This suggests there was some understanding of hazards and consequences, but it does to appear to have driven any improvement activities or modifications to prevent an incident [24].

4 Ramifications

The full ramifications of this incident are not yet clear. The Queensland government has initiated some legislative improvements to manage theme parks, but there is likely to be more recommendations coming from the Coronial Inquest when it releases its findings. Aside from the tragic loss of four lives, from the Dreamworld perspective, their guest numbers have remained low since the incident occurred, and it is struggling to attract patrons. This has led to their parent company suffering financial losses for the past two years and 50% decreases in the share price over the two year period [35].

The incident also provided an interesting lesson in crisis management. The company was scheduled to hold their annual general meeting in the days following the incident. The meeting went ahead and in it the CEO was awarded a performance bonus of $840,000 including both shares and a cash component despite the four deaths days before [36]. From a public relations perspective this was not a successful event. The CEO also stated they had been in contact with the victim's families, but this was publicly disputed by the families [37]. This further damaged the company reputation at a difficult time.

This incident will not only have ramifications for the families of the deceased, but for the survivors and the staff who responded due to the nature of injuries sustained by the victims. Four former staff members are suing Dreamworld for psychological damage [38].

These ramifications show how a process safety type incident can have significant impact on a business and its ability to continue operations in a positive way.

5 Could the Principles of Process Safety Have Helped?

As had been shown, there were clear and systematic failures across the whole organization that led to this tragedy. Starting with knowledge and competence, which relates to organizational capability, had the operators been trained better in the safe operation of the ride as well as in emergency response, their performance that day may have been improved. From an engineering and design perspective, the design and installation of the ride was inadequate, it underwent undocumented changes which introduced hazards, and it was not adequately maintained. The systems and procedures for the operation and maintenance the ride was confusing and did not appear to contain necessary information. The assurance processes conducted did not seem to be thorough, rather they focused on keeping the ride operating and recommendations were not actioned. The ride was not designed with human factors in mind, from the high workload to the confusing operating panel. The organization culture did not seem to place a value on safety, it only focused on cost reduction. It also did not appear to be a learning culture, where lessons from the past are understood and embedded. Lastly at the time of crisis, the overall leadership displayed by the parent company was publicly found to be lacking.

There are transferrable learnings for the process industries from this incident, with some clear questions to ask. How are you managing your competency program to ensure you have the right capability? Are your designs adequate, maintained and any changes managed? Are your procedures thorough, identifying hazards, consequences and controls and easy to comprehend? How are you monitoring the operation of your facility? Does the design and operation of your facility enhance human performance or make it more difficult? Does your culture support safe operations?

6 Conclusion

A focus on the elements of process safety can help identify what is important in a business. In this instance there had been recommendations for a low-cost instrumented safety system to be installed that would have prevented the incident from occurring. The importance of this was not understood by the organization, and the opportunity to upgrade the ride during its recent refurbishment was ignored. It is clear that the hazard and potential consequences and controls were not understood. As is often the case, when the hazard is not understood and therefore not managed, people suffer tragic consequences.

7 References

[1] Kerin, T. *Managing Process Safety. In the Core Body of Knowledge for Generalist OHS Professionals*. Safety Institute of Australia, Tullamarine, VIC. 2017

[2] ISC (IChemE Safety Centre). Safety Centre Framework, Available at https://www.icheme.org/knowledge/safety-centre/framework/, Accessed on January 2, 2019

[3] ABC News, Available at https://www.abc.net.au/news/2018-06-18/dreamworld-fatal-accident-ride-had-broken-down-inquest-told/9871898, Accessed on January 2, 2019

[4] Australian Financial Review, Available at https://www.afr.com/news/dreamworlds-rapids-ride-different-from-others-around-the-world-20161101-gsfd07, Accessed on January 2, 2019

[5] 9 News, Video 11, Available at https://www.9news.com.au/2018/12/06/03/33/dreamworld-inquest-enters-final-days, Accessed on January 2, 2019

[6] news.com.au, Available at https://www.news.com.au/national/queensland/courts-law/coroner-james-mcdougall-requests-proof-all-dreamworld-rides-were-certified/news-story/e77275b97b0bbee9c1c541bda6df0035, Accessed on January 2, 2019

[7] ABC News, Available at https://www.abc.net.au/news/2018-06-19/dreamworld-operator-on-first-day-when-four-people-died-on-ride/9885214, Accessed on January 2, 2019

[8] ABC News, Available at https://www.abc.net.au/news/2018-06-27/dreamworld-trainer-promoted-since-four-killed-on-ride/9914188, Accessed on January 2, 2019

[9] 9 News, Video 5, Available at https://www.9news.com.au/2018/12/06/03/33/dreamworld-inquest-enters-final-days, Accessed on January 2, 2019

[10] ABC News, Available at https://www.abc.net.au/news/2018-06-20/dreamworld-operator-impossible-load-responsibilities-manage-ride/9889776, Accessed on January 2, 2019

[11] 9 News, Video 1, Available at https://www.9news.com.au/2018/12/06/03/33/dreamworld-inquest-enters-final-days, Accessed on January 2, 2019

[12] 9News, Available at https://www.9news.com.au/2018/11/20/03/32/dreamworld-ride-passed-wphs-safety-audit, Accessed on January 2, 2019

[13] ABC News, Available at https://www.abc.net.au/news/2018-10-08/dreamworld-inquest-hears-total-failure-to-ensure-safety/10350034, Accessed on January 2, 2019

[14] Intamin Worldwide, Available at https://www.intaminworldwide.com/infinity-falls-orlando/ , Accessed on December 12, 2018

—

[15] 9 News, Video 8, Available at https://www.9news.com.au/2018/12/06/03/33/dreamworld-inquest-enters-final-days, Accessed on January 2, 2019

[16] ABC News, Available at https://www.abc.net.au/news/2018-06-25/dreamworld-inquest-told-of-ride-maintenance-cutbacks/9906214, Accessed on January 2, 2019

[17] Australian Financial Review, Available at https://www.afr.com/real-estate/commercial/hotels-and-leisure/dreamworld-prioritised-new-attractions-over-maintenance-of-old-rides-inquest-20180626-h11urb, Accessed on January 2, 2019

[18] 9 News, Video 10, Available at https://www.9news.com.au/2018/12/06/03/33/dreamworld-inquest-enters-final-days, Accessed on January 2, 2019

[19] ABC News, Available at https://www.abc.net.au/news/2018-06-21/dreamworld-ride-operator-highly-distressed-give-evidence-inquest/9893308, Accessed on January 2, 2019

[20] news.com.au, Available at https://www.news.com.au/national/queensland/courts-law/dreamworld-engineer-tells-families-i-apologise-for-not-being-to-make-eye-contact-with-you/news-story/40545dc747c95eb96b0293c888d77d96, Accessed on January 2, 2019

[21] Sunshine Coast Daily, Available at https://www.sunshinecoastdaily.com.au/news/dreamworld-thunder-river-rapids-button-wasnt-regul/3543997/, Accessed on January 2, 2019

[22] ABC News, Available at https://www.abc.net.au/news/2018-06-26/dreamworld-ride-operator-says-conveyor-was-never-stopped/9910192, Accessed on January 2, 2019

[23] news.com.au, Available at https://www.news.com.au/national/queensland/courts-law/safety-report-shows-simple-change-on-dreamworld-ride-could-have-saved-lives/news-story/9c1fa2c195efd12f86a3e8ad516ea1b8, Accessed on January 2, 2019

[24] 9 News, Video 20, Available at https://www.9news.com.au/2018/12/06/03/33/dreamworld-inquest-enters-final-days, Accessed on January 2, 2019

[25] news.com.au, Available at https://www.news.com.au/travel/travel-updates/incidents/girl-11-dies-after-falling-into-the-water-on-splash-canyon-ride-during-school-trip-to-uk-theme-park/news-story/f83330b0ce12ea11df2d089ece676519, Accessed on January 2, 2019

—

[26] 9News, Available at https://www.9news.com.au/2018/12/06/03/33/dreamworld-inquest-enters-final-days, Accessed on January 2, 2019

[27] 9News, Available at https://www.9news.com.au/2018/10/16/14/43/dreamworld-inquest-thunder-river-rapids-ride-engineers-give-evidence, Accessed on January 2, 2019

[28] 9News, Available at https://www.9news.com.au/2018/10/17/10/49/dreamworld-inquest-engineer-to-appear-again, Accessed on January 2, 2019

[29] Australian Financial Review, Available at https://www.afr.com/news/dreamworld-ride-trainer-promoted-after-tragedy-staff-wanted-emergency-training-20180628-h11yke, Accessed on January 2, 2019

[30] 9 News, Video 12, Available at https://www.9news.com.au/2018/12/06/03/33/dreamworld-inquest-enters-final-days, Accessed on January 2, 2019

[31] ABC news, Available at https://www.abc.net.au/news/2018-10-09/dreamworld-ride-declared-safe-days-before-tragedy/10352164, Accessed on January 2, 2019

[32] ABC News, Available at https://www.abc.net.au/news/2018-10-17/dreamworld-supervisor-sympathy-has-families-in-tears-inquest/10387186, Accessed on January 2, 2019

[33] 9 News, Video 9, Available at https://www.9news.com.au/2018/12/06/03/33/dreamworld-inquest-enters-final-days, Accessed on January 2, 2019

[34] 9 News, Video 16, Available at https://www.9news.com.au/2018/12/06/03/33/dreamworld-inquest-enters-final-days, Accessed on January 2, 2019

[35] Australian Financial Review, Available at https://www.afr.com/research-tools/AAD/share-prices?days=1095, Accessed on November 26, 2018

[36] The Guardian, Available at https://www.theguardian.com/australia-news/2016/oct/27/deborah-thomas-set-to-receive-840000-in-bonuses-despite-dreamworld-deaths, Accessed on December 19, 2018

[37] ABC News, Available at https://www.abc.net.au/news/2016-10-27/dreamworlds-shareholders-ask-parent-company-to-support-victims/7970104, Accessed on January 2, 2019

—

[38] news.com.au, Available at https://www.news.com.au/lifestyle/real-life/news-life/dreamworld-first-responders-to-sue-theme-park-for-psychological-damage/news-story/e8fd7399eec9d165c657dbdb3ac35968, Accessed on January 2, 2019

Fatal Toxic Chemical Release at DuPont

Tamara Qureshi
U.S. Chemical Safety and Hazard Investigation Board
1750 Pennsylvania Ave, NW Suite 910 Washington, DC 20006
tamara.qureshi@csb.gov

Dan Tillema, Steve Cutchen, Lauren Grim, and Chris Lyon
U.S. Chemical Safety and Hazard Investigation Board
1750 Pennsylvania Ave, NW Suite 910 Washington, DC 20006

Prepared for Presentation at
American Institute of Chemical Engineers
2019 Spring Meeting and 15th Global Congress on Process Safety
New Orleans, LA
March 31 – April 3, 2019

Fatal Toxic Chemical Release at DuPont

Tamara Qureshi
U.S. Chemical Safety and Hazard Investigation Board
1750 Pennsylvania Ave, NW Suite 910 Washington, DC 20006
tamara.qureshi@csb.gov

Dan Tillema, Steve Cutchen, Lauren Grim, and Chris Lyon
U.S. Chemical Safety and Hazard Investigation Board

Keywords: DuPont, toxic release, gas detection, alarm

Abstract

On November 15, 2014, four workers were killed after approximately 24,000 pounds of highly toxic methyl mercaptan released from an insecticide production unit (Lannate® Unit) at the E.I. du Pont de Nemours chemical manufacturing facility in La Porte, Texas. This paper will discuss the events leading to the methyl mercaptan release, the deficiencies in DuPont La Porte's toxic gas detection system, and offer lessons on how to prevent some of the deficiencies in the toxic gas detection system from occurring. An effective toxic gas detection system can protect workers by alerting them of a potentially dangerous situation and preventing them from unknowingly entering a potential toxic chemical release.

1 Introduction

On November 15, 2014, approximately 24,000 pounds of highly toxic methyl mercaptan released from an insecticide production[1] unit (Lannate® Unit) at the E.I. du Pont de Nemours chemical manufacturing facility in La Porte, Texas. Four workers at the DuPont La Porte facility entered the Lannate® Unit's manufacturing building (manufacturing building) and the highly toxic methyl mercaptan release. Three of these workers were responding to a coworker's distress call from inside the manufacturing building. The fourth worker entered after all the workers inside the manufacturing building stopped communicating with workers inside the control room. The Lannate® Unit's methyl mercaptan toxic gas detection system did not effectively warn workers of the large chemical release occurring within the manufacturing building. Additionally, this toxic gas detection system did not provide emergency responders with sufficient information to effectively evaluate the potential harm to the public.

[1] The Insecticide Business Unit (IBU) was divided into two primary insecticide product lines: Lannate®, which is DuPont's trade name for methomyl; and Agricultural Products Intermediates (API), which produced insecticides and nematicides known as Vydate®, the trade name for oxamyl [3, p. 11].

The DuPont La Porte methyl mercaptan toxic gas detection system had several deficiencies. First, the methyl mercaptan detector system did not meet its design intent. The alarm set point for methyl mercaptan concentration and the method of conveying alarms did not forewarn workers of the immediately dangerous to life and health (IDLH) conditions they would encounter by entering the manufacturing building. The methyl mercaptan gas detectors only alarmed on the Lannate® control room operator's computer screen. There was no audible or visible alarm provided in the field for outside personnel. Second, the placement of the methyl mercaptan toxic gas detectors did not fully portray the extent of the November 15, 2014 methyl mercaptan release. Third, the DuPont La Porte site normalized the hazards associated with methyl mercaptan. Finally, the alarms did not prompt protective action or investigations into the sources of alarms. During the hours prior to the November 15, 2014 incident, multiple highly toxic chemical gas detectors alarmed (sounded). However, the DuPont emergency response team was not notified nor was the area cleared of personnel. In addition, methyl mercaptan releases on November 13 and 14, 2014, were identified by methyl mercaptan detectors, but were never reported as releases nor investigated as serious process safety incidents. Personnel believed they knew the source of the releases and normalized their responses to these alarms. Consequently, there was a delayed awareness of the release by operations, and workers who entered the manufacturing building after the release began were not alerted of nor prepared for the lethal concentrations of methyl mercaptan being released, resulting in the death of two of these workers.

The DuPont La Porte incident is the CSB's third investigation of a fatal DuPont incident in five years.[2] On September 30, 2015, the CSB issued a presentation, safety video, and an Interim Recommendations report [1],[3] which detailed needed safety improvements at the facility and issued recommendations to the DuPont La Porte facility to be implemented prior to resuming operations of the Lannate® Unit.[4] Due to the amount of methyl mercaptan stored at the DuPont La Porte site, the Lannate® process was covered by Occupational Safety Health Administration's (OSHA) Process Safety Management (PSM) program and the Environmental Protection Agency's (EPA) Risk Management Plan (RMP) Rule [2].

An effective toxic gas detection system can make the difference between personnel being prepared for a potentially dangerous situation or entering a potential toxic chemical release. To be effective, a toxic gas detection system needs to consider the chemical's hazards, the gas detector's

[2] The CSB also investigated an incident at a DuPont plant in Belle, West Virginia (three incidents that occurred in a 33-hour period over January 22-23, 2010, resulting in one fatality), and another in Buffalo, New York (a single incident that occurred on November 9, 2010, resulting in one fatality and one serious injury) [16], [17].

[3] In the Interim Recommendations, the CSB identified significant process safety management deficiencies at the DuPont La Porte facility, including delayed maintenance of safety-critical equipment, lack of written procedures, poor hazard analysis practices, lack of implementation of important inherently safer design concepts, and a lack of hazard recognition.

[4] In response to the CSB findings and recommendations, DuPont committed to making changes at the DuPont La Porte facility. DuPont La Porte, however, was not able to address all of these findings and recommendations, because in the spring of 2016 DuPont announced that it would not reopen the Lannate® Unit. The company had determined that "significant changes in market conditions during the period of the shutdown [would] persist over the long term [14]" and the cost required for the restart "[was] not a long-term viable and cost-efficient option for the DuPont Crop Protection business [15]." In 2017, DuPont dismantled and removed the buildings and equipment from the DuPont La Porte facility that were associated with its insecticide and herbicide business units, which included the Lannate® Unit.

capabilities, the location of the gas detectors, and the alarms associated with the gas detectors. This paper will discuss some of the events leading to the methyl mercaptan release, the deficiencies in DuPont La Porte's toxic gas detection system and offer lessons on how to prevent some of the deficiencies in the toxic gas detection system from occurring.

2 Events Leading to the Incident

On Monday, November 10, 2014, an inadvertent chemical dilution caused operating difficulties that forced a shutdown of DuPont La Porte's Lannate® Unit [3, pp. 20-21]. On Wednesday, November 12, 2014, operators tried to restart the Lannate® process. During the shutdown, the methyl mercaptan piping to the reaction section of the process plugged due to the formation of a clathrate hydrate (hydrate),[5] halting restart. At that time, operators did not know the source of the plugging and began troubleshooting the process to try to clear the piping, unaware that the solid hydrate had formed.

Two days later, on Friday, November 14, 2014, the troubleshooting efforts to clear the plugging were still ongoing. That morning, a technical team composed of engineers and other employees experienced in the Lannate® process, met with operations personnel to discuss troubleshooting options to clear the plugging. This meeting identified the likely scenario that water had entered the methyl mercaptan system and a solid hydrate formed as a result. Based on their understanding of the hydrate,[6] the technical team asked operators to put hot water on the outside of the methyl mercaptan piping, under the insulation, to warm the piping and its contents, to break-up (dissociate[7]) the plugging. The team also realized that when heated, methyl mercaptan would expand, requiring a safe place to vent to avoid over-pressuring the piping. To address this concern, operations personnel opened valves between the methyl mercaptan piping and a waste gas vent header, not recognizing that this alignment created a pathway for liquid methyl mercaptan to eventually release from the piping inside the manufacturing building when drain valves were opened. In addition, when operations personnel opened other valves while troubleshooting, such as drain valves, methyl mercaptan released to the atmosphere causing a strong methyl mercaptan odor[8] that could be smelled by site personnel.

[5] At low temperatures (≤ 52 °F) water and methyl mercaptan form a solid, ice-like material called methyl mercaptan clathrate hydrate [10]. Temperatures in the Houston area for the 24 hours preceding the incident averaged approximately 40 °F and had been consistently below 55 °F since Tuesday, November 11, 2014. Although the potential for methyl mercaptan hydrate formation had been identified in DuPont's methyl mercaptan technical standard and in a process hazard analysis years earlier, DuPont never implemented safeguards, such as heat tracing, or developed a procedure to dissociate the hydrate safely [3, p. 22].

[6] DuPont's methyl mercaptan technical standard identifies the potential to form a hydrate at low temperatures. The technical standard states, "[methyl mercaptan] will form a hydrate with water, which is a solid below 40 deg F [degrees Fahrenheit] per information provided by a [methyl mercaptan] supplier."

[7] Methyl mercaptan clathrate hydrates dissociate back to the liquid phases of methyl mercaptan and water at approximately 52 °F [10].

[8] Methyl mercaptan has a low odor threshold. It gives off a smell like rotten cabbage. It can be smelled at only two parts per billion (ppb) [9, p. 1]. OSHA's permissible exposure limit (PEL), however, is 10 parts per million (ppm), which is 5,000 times the minimum concentration it can be smelled. The low odor threshold of mercaptans makes them useful as fuel gas odorants [9, p. 2].

During this troubleshooting process, methyl mercaptan released both outside and inside the manufacturing building, triggering 32 methyl mercaptan gas alarms on the control panel throughout the 17 hours preceding the incident. While methyl mercaptan is a toxic and flammable chemical, and at high enough concentrations lethal, operations personnel did not respond in accordance with the nature of the emergency circumstances, because they associated the alarms with ongoing troubleshooting efforts. They did not perceive the methyl mercaptan alarms as signifying a serious hazard because they normalized[9] both the methyl mercaptan odor within the Lannate® Unit as well as the detector alarms associated with them.

Early in the morning on Saturday, November 15, 2014, the hot water from the hoses warmed the hydrate plugging, causing it to dissociate, clearing the plugging. Due to the valve alignment, liquid methyl mercaptan flowed into the waste gas vent header, located inside the manufacturing building [3, pp. 25-26]. This vent header, however, was not intended for liquid methyl mercaptan.

At 2:51 am alarms began to sound on the control system indicating high pressure in equipment located inside the manufacturing building. Operations personnel did not realize that liquid methyl mercaptan was causing the high pressure. Instead, they attributed the high pressure to a common, long-standing problem with process condensate (that DuPont La Porte personnel believed to be mostly water) accumulating in the vent header piping [3, p. 28]. The Lannate® control room operator (Board Operator) contacted the night shift supervisor (Shift Supervisor) and another operator (Operator 1) separately, asking them to help troubleshoot the high-pressure situation. The Shift Supervisor and Operator 1 each went separately to the manufacturing building, likely with the intent to drain liquid from the vent header piping manually.[10] Sometime between 3:01 am and 3:13 am, two sets of drain valves on the vent header piping, located on the third floor of the manufacturing building were opened (Figure 1). But instead of being comprised mostly of water as workers believed, liquid methyl mercaptan that had filled the piping escaped out of the valves, vaporized, and killed the Shift Supervisor.

Between 3:24 am and 3:26 am, three methyl mercaptan detectors inside the manufacturing building sensed at least 25 parts per million (ppm) methyl mercaptan[11] and triggered alarms at the control panel. The Board Operator—still focused on reducing the high pressure—did not realize a major chemical release inside the manufacturing building was occurring.

[9] Normalization of deviance is a phenomenon in which individuals or work teams gradually accept a lower standard of performance until the lower standard becomes the norm. It is typically the result of conditions slowly changing and eroding over time [11, p. 4].

[10] In 2011, DuPont installed a NOx (nitrogen oxides) reducing scrubbed (NRS) incinerator system. The waste gas vent header piping to the NRS Incinerator was designed without sufficient consideration of liquid accumulation. It contained low points where liquid could accumulate, and there was no engineered equipment provided, such as a knock out drum, to remove liquid safely from the waste gas vent header piping system. The liquid accumulation in the piping would create pressure in process equipment. Pressure buildup within the equipment was resolved by manually draining liquid from the waste gas vent header. See DuPont La Porte, Texas Chemical Facility Toxic Chemical Release: Interim Recommendations [3, p. 18].

[11] The methyl mercaptan detectors alarm at 25 ppm–well above the National Institute of Occupational Safety and Health (NIOSH) recommended exposure limit (REL) of 0.5 ppm and the OSHA enforceable (PEL) ceiling limit of 10 ppm. OSHA recommends that employers consider using the alternative occupational exposure limits because the agency believes that exposures above some of these alternative occupational exposure limits may be hazardous to workers, even when the exposure levels are in compliance with the relevant PELs [5]. The Cal-OSHA permissible exposure limit (PEL) is an 8-hour time-weighted average (TWA) of 0.5 ppm and NIOSH recommended exposure limit (REL) is a 15- minute ceiling limit of 0.5 ppm [7], [6].

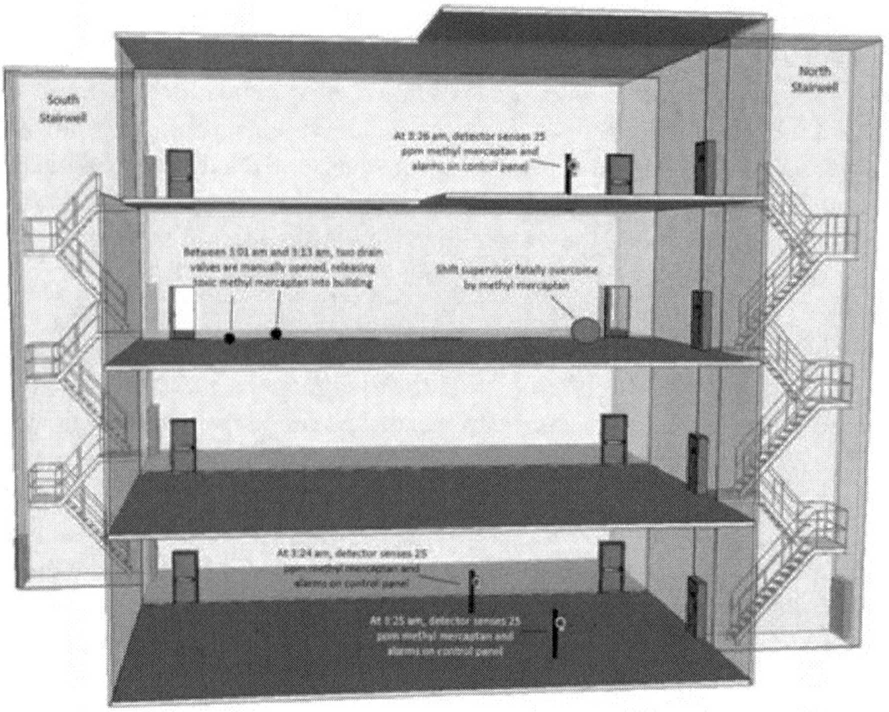

Figure 1. Depiction Showing Location Where Drain Valves Were Opened. These drain valves released toxic methyl mercaptan into the manufacturing building. Methyl mercaptan detectors on the first and fourth floors detected high concentrations of methyl mercaptan shortly after the release began. This depiction is a simplified graphic of the manufacturing building and does not show the location of Operator 1. Source: CSB

At about 3:30 am, Operator 1 made an urgent call for help over the radio. Various personnel interpreted her communication in different ways, generally as either "We need help!" or "I need help on the fourth floor!" The Board Operator tried to get more information through radio communication, but neither the Shift Supervisor nor Operator 1 ever responded [3, p. 29].
Two operators (Operator 2 and Operator 3) who were in the control room and heard the distress call ran to the manufacturing building to help. Another operator (Operator 4) saw Operator 2 and Operator 3 running to the manufacturing building, and he followed them. None of these operators knew there was a major release of toxic methyl mercaptan inside of the manufacturing building, and therefore they did not wear any respiratory protection when they ran into the manufacturing building. The manufacturing building lacked automatic visual or audible alarms to alert field workers or prevent them from entering a potentially toxic atmosphere.

Operator 2, Operator 3, and Operator 4 each entered the south stairway and took different routes inside the manufacturing building to try to find Operator 1. Operator 2 went to the third floor, where he was fatally overcome by methyl mercaptan [3, p. 29]. Operator 4 went to the second floor. He walked about ten feet and hit what he later described as a "wall" of methyl mercaptan, but he managed to retreat to the stairwell [3, p. 29]. Operator 3 went to the fourth floor but did not find anyone. He then announced on the Lannate® Unit's public-address system that he did not see anyone on the fourth floor. The Board Operator responded that the Shift Supervisor and Operator 1 may be on the third floor. Operator 3 then began to feel light headed. He made his way to the stairwell and lost consciousness while descending from the fourth floor [3, p. 29].

When Board Operator could not reach people inside the manufacturing building, an operator in the control room, Operator 6[12] (the brother of Operator 2), then grabbed three 5-minute escape respirators. Other operators in the control room warned him not to enter the manufacturing building because they did not know what was going on or where the non-responsive operators were. Operator 6, however, took the escape respirators and rushed to the manufacturing building [3, p. 29].

At about 3:40 am, Operator 6 encountered Operator 4 in the south stairway of the manufacturing building and put an escape respirator on him. The breathing air helped Operator 4 recover, and he exited the manufacturing building safely. A worker—likely Operator 6—then manually activated the manufacturing building fume release alarm, an alarm intended to alert area workers of a toxic chemical release in the building. Operator 6 then went to the third floor of the manufacturing building. At some point inside the manufacturing building after helping Operator 4, Operator 6 retrieved a self-contained breathing apparatus (SCBA). He tried to rescue his brother (Operator 2) by putting one of the escape respirators on his face. Operator 6 was then fatally overcome by methyl mercaptan before he could connect his own respirator mask to his SCBA tank.

3 Lessons from the DuPont La Porte Incident

3.1 Ensure Toxic Gas Detector Systems Operate as Intended

DuPont La Porte's methyl mercaptan technical standard stated, "[a]utomatic or continuous specific chemical detection systems *must* be installed where a [methyl mercaptan] release could have any off-site impact […] [Emphasis in the original]." DuPont La Porte's methyl mercaptan toxic gas detection system consists of methyl mercaptan detectors, alarms, distributed control system (DCS), control panel, and a control room operator. The manufacturing building is equipped with gas detectors for methyl mercaptan in various locations. The third floor, where the release occurred, however, did not have any methyl mercaptan detectors. The detectors inside the manufacturing building were also set to alarm on the control panel when they detected a 25 ppm concentration of methyl mercaptan and were intended to protect workers from exposure to methyl mercaptan.[13]

DuPont used hydrogen sulfide (H_2S) detectors to identify methyl mercaptan. DuPont records indicate that the manufacturer has stated the detector should be capable of detecting methyl mercaptan within 1 ppm.[14] The alarms activated when the detectors sensed the hydrogen sulfide equivalent to 25 ppm methyl mercaptan.[15] The implementation of the methyl mercaptan toxic gas

[12] Operator 6 was a board operator for another process in the Lannate® Unit.

[13] The 25 ppm alarm set point is 50 times the NIOSH's REL of 0.5 ppm and more than double OSHA's ceiling limit of 10 ppm [7]. Therefore, alarming at this detection level could not protect workers from exposure to methyl mercaptan.

[14] DuPont La Porte's detectors for methyl mercaptan are calibrated for 0-50 ppm using a hydrogen sulfide (H_2S) calibration gas.

[15] The methyl mercaptan detectors were evaluated post-incident. Although the detectors inside the building have the ability to span 0-50 ppm, these detectors were set to span 0-25 ppm. Post-incident, the span was increased to 50 ppm. This configuration of the alarms limited the amount of information communicated to DuPont La Porte personnel.

detection system inside the manufacturing building, however, did not meet the design intent, federal safety regulations, or DuPont corporate standards.

The methyl mercaptan detectors outside and inside the manufacturing building were installed for two different purposes—the outside detectors were geared towards an offsite release and the inside detectors were meant to protect workers from a hazardous environment. DuPont La Porte documentation stated that the basis for the alarms outside the manufacturing building was to provide an early warning mechanism for a potential offsite methyl mercaptan release. The alarm setpoint was based on the ERPG (Emergency Response Planning Guidelines)-2[16] concentration[17] for methyl mercaptan which is 25 ppm. DuPont La Porte installed the methyl mercaptan detectors inside the manufacturing building after the methyl mercaptan detectors outside the building. Unlike the outside detectors, there was no documentation for the basis for setting the alarm setpoint to 25 ppm for the methyl mercaptan detectors inside the manufacturing building nor was there evidence that these setpoints were set at 25 ppm to prevent methyl mercaptan alarms from becoming a nuisance. It appears that the setpoints were copied from the outside methyl mercaptan detectors even though the inside detectors served a different purpose—worker protection. The methyl mercaptan detectors alarmed at 25 ppm—well above the NIOSH recommended exposure limit (REL) of 0.5 ppm and the OSHA enforceable PEL ceiling limit of 10 ppm.

Depending on the purpose of a toxic gas detection system alarm, there are different sets of values for facilities to consider. For instance, ERPG values are designed for the protection of the public during an emergency and not intended for the everyday routine protection of workers [4].[18] For worker protection, OSHA recommends that employers consider using the lower alternative occupational exposure limits because the agency believes that exposures above some of these alternative occupational exposure limits may be hazardous to workers, even when the exposure levels are in compliance with the relevant PELs [5]. For example, the Cal-OSHA permissible exposure limit (PEL) is an 8-hour time-weighted average (TWA) of 0.5 ppm and NIOSH's REL is a 15- minute ceiling limit of 0.5 ppm [6], [7].

Additionally, information from the methyl mercaptan detectors inside the manufacturing building only displayed on the DCS monitor and was not recorded by the site's DCS and data historian [3, p. 37]. CSB investigators, however, were able to obtain information from alarm logs about the detectors.

DuPont La Porte's management of change documentation for installation of methyl mercaptan detectors inside the manufacturing building stated the condition prompting the change was the

[16] ERPG-2 is the maximum airborne concentration below which nearly all individuals could be exposed for up to 1 hour without experiencing or developing irreversible or other serious health effects or symptoms which could impair an individual's ability to take protective action. At concentrations above ERPG-2, some members of the community may experience significant health effects which could impair them from taking protective action. Emergency responders often alert the community to take protective action in areas susceptible to the ERPG-2 concentration and above [19].

[17] The ERPG values for methyl mercaptan in parts per million (ppm) are: ERPG-1, 0.005; ERPG-2, 25; ERPG-3, 100 [18, p. 6].

[18] Emergency Response Planning Guidelines (ERPGs) are chemical concentration levels that are used during emergency response and when preparing for emergencies. They are designed to anticipate public health effects not occupational health effects from exposure to airborne chemicals [4].

possibility that personnel could enter a potentially hazardous location without knowing methyl mercaptan had been released:

> *Present Conditions and Description of Proposed Change:*
>
> *Personnel can enter a potentially hazardous location without knowing MeSH [methyl mercaptan] has been released to the environment.*

The methyl mercaptan gas detectors, however, only alarmed on the Lannate® control room operator's computer screen.

DuPont La Porte site did not have audible or visual alarms on or in the manufacturing building to alert field personnel of a methyl mercaptan release automatically. DuPont relied on verbal communication of alarms that automatically displayed on a continuously manned control board. Without being made aware of the major chemical release inside the building, Operator 2, Operator 3, and Operator 4 entered without respiratory protection. As a result, Operator 2 was killed from methyl mercaptan exposure, and Operator 3 and Operator 4 were exposed to dangerous concentrations of methyl mercaptan.

Facilities should make strategic decisions when choosing a toxic gas detector to ensure that a toxic gas detection system operates as intended. Considerations should include regulatory requirements, NIOSH recommendations, chemical interferences, and utility (worker protection, release modeling, etc.). The methyl mercaptan detectors at the DuPont La Porte facility had a span of 0-50 ppm. This range may be appropriate for early detection of a toxic leak when coupled with alarms and clear instructions to evacuate the premises at lower concentrations. If the goal is to understand other aspects of a release, like potential lethal concentration (150 ppm is methyl mercaptan's IDLH concentration) or modeling an offsite release, then a wider concentration range should be considered.

Additionally, facilities should also consider the use of automatic alarms designed to alert personnel of hazardous conditions—such as a release of a toxic chemical or explosive atmosphere—which can relay critical safety information immediately to personnel, without potential delays inherent in a system that solely relies on a control room operator to verbally communicate safety alarms to operators in the field.

3.2 Ensure Toxic Gas Detector Placement is Strategic

Although the DuPont La Porte site had 13[19] fixed methyl mercaptan detectors inside and outside the manufacturing building, there were no fixed methyl mercaptan detectors located on the third floor inside the manufacturing building where a majority of the November 15, 2014 methyl mercaptan release occurred. Additionally, no workers were provided with either portable or personal methyl mercaptan monitors. Part of the Lannate® process occurred inside the closed manufacturing building, a four-story building containing a large amount of piping and process equipment, dividing walls, multiple stairwells, and an elevator. The Lannate® Unit had methyl

[19] The site had a total of 15 methyl mercaptan detectors—there were two additional methyl mercaptan detectors installed in another process building adjacent to the manufacturing building.

mercaptan process piping both inside and outside the manufacturing building. A facility siting study and a process hazards analysis had identified the necessity of having methyl mercaptan detectors in the Lannate® manufacturing building. These studies identified the need for protecting workers from toxicity and explosivity hazards related to potential methyl mercaptan releases.

DuPont La Porte installed 10 methyl mercaptan detectors near process piping outside the manufacturing building. The design basis for the outside methyl mercaptan detectors was to provide an early warning of a methyl mercaptan release. DuPont La Porte placed these detectors around its methyl mercaptan storage tank and the area where methyl mercaptan could be unloaded. There were, however, no fixed methyl mercaptan detectors along the fence line that could relay whether there was an offsite release.

DuPont La Porte installed three methyl mercaptan detectors inside the manufacturing building. The stated design purpose for the methyl mercaptan detectors inside the building was worker protection. The inside detectors were located within the building on the first and fourth floors. These detector locations were determined based on potential release scenarios from a 2001 process hazard analysis (PHA).

The methyl mercaptan release occurred on the third floor where there were no methyl mercaptan detectors. Although the 2001 PHA did not identify methyl mercaptan detector locations on the third floor, DuPont La Porte routed the waste gas vent header piping along the third floor and this piping was directly connected to the top of the methyl mercaptan storage tank. During abnormal or upset conditions, the waste gas vent header could have been a potential source for a methyl mercaptan release. For example, ineffective hazard management while troubleshooting the plugged methyl mercaptan feed piping during the November 15th incident allowed liquid methyl mercaptan to flow into the waste gas vent header piping towards the NRS incinerator—a location where it was never intended to go.

Additionally, the routine draining of liquid from the waste gas vent header should have triggered a management of change review and considered whether additional safeguards like methyl mercaptan detectors were necessary. As discussed earlier, DuPont La Porte never fully resolved liquid accumulation in the waste gas vent header piping to the NRS incinerator with an engineered solution. Instead, operators were directed to drain the system by area management through written instructions in a daily instruction log book to operations staff [8, p. 18].[20] This routine draining presented potential dangers because DuPont La Porte personnel never sampled or analyzed the liquid workers routinely drained from this system [8, p. 18]. Daily instructions before the incident requested the operators to sample the uncharacterized liquid. DuPont La Porte management wanted samples of the liquid for a subsequent laboratory analysis to identify its chemical composition. Workers reported that the liquid had a strong, noxious odor of sulfur compounds [8, p. 18]. Even though DuPont La Porte was aware that the liquid likely contained more than just water and was unaware of the actual chemical composition [8, p. 18], instructions to workers did

[20] Beginning January 14, 2014, and continuing until the incident, the daily instructions directed operators to drain the methyl mercaptan waste gas vent header's "low point" on the 3rd floor once per shift. After the installation of a flexible hose from the waste gas vent header drain valve to the floor, the instructions changed to include directing the liquid through the hose under a running safety shower until the liquid stops flowing from the hose [8, p. 18].

not specify that additional precautions—like wearing additional PPE—were needed when draining the liquid from the waste gas vent header to the floor inside the manufacturing building.

Consequently, DuPont La Porte workers dealt with the common problem of liquid accumulation in the waste gas vent header on a routine basis by draining the liquid (line breaking) without engineered solutions or ensuring the use of safety procedures or personal protective equipment. On November 15, 2014, when the liquid drain valves were opened this time, flammable and highly toxic methyl mercaptan flowed onto the floor and filled the manufacturing building with toxic vapor and there no methyl mercaptan detectors on the third floor to alert personnel of the hazardous conditions that existed there.

As with toxic detector selection, the placement of toxic gas detectors must be strategic in order to capture the potential areas where a release could occur and the extent of the release. The methyl mercaptan detector placement both externally and internally at the DuPont La Porte facility did not operate as intended in part because the detectors did not fully cover the areas where the majority of the methyl mercaptan was released. For example, there was no real time data for emergency responders to use to evaluate the potential for offsite concentrations during the methyl mercaptan release. Although emergency responders conducted offsite monitoring hours after the initial release, the information obtained from air monitoring could not give an accurate portrayal of the methyl mercaptan release. Furthermore, the methyl mercaptan monitors inside the building did not provide accurate information about the release because they were not located near the open drain valves.

Facilities can adopt a variety of methods to select the location of toxic gas monitors with the understanding that they should be prepared for scenarios where the monitoring location may not be effective. One technique that a facility could use is to develop a series of release scenarios and based on those scenarios place monitors in the path of those releases (including by the potential sources of a leak). This technique, however, may not cover all potential release areas. Additionally, facilities should evaluate additional locations for toxic gas detectors if there is a change in process that creates new potential release scenarios. Another technique, if focusing on worker protection, is to give workers personal monitors when they enter toxic gas hazardous areas, especially if these areas are not normally occupied or congested with equipment. At DuPont La Porte, commercially available personal hydrogen sulfide monitors could have been issued to workers. Additionally, if using fixed toxic gas monitors for personnel protection, they should be installed near the breathing zone.

3.3 Ensure Personnel Understand Toxic Gas Hazards

DuPont La Porte personnel did not fully appreciate the effects of methyl mercaptan's physical characteristics. Although methyl mercaptan[21] is a highly toxic, flammable chemical, its low odor

[21] Personnel at DuPont La Porte would refer to methyl mercaptan as "MeSH." MeSH is shorthand based upon methyl mercaptan's chemical formula: CH_3SH. "Me" represents the methyl group (CH_3) and "SH" represents the mercaptan group (SH)—and when combined, these initialisms forms MeSH.

threshold[22] and ability to cause olfactory fatigue in personnel can generate a false sense of security when working with the chemical.

Although DuPont's methyl mercaptan technical standard states, "[a]ll DuPont sites *will* protect the site and neighboring businesses and communities from the escape of a highly toxic material [,]" methyl mercaptan detectors provided essentially no value to DuPont personnel [Emphasis in the original]. DuPont operations staff became accustomed to the odor of methyl mercaptan being present in their work environment. The low odor threshold of methyl mercaptan resulted in a false perception of safety. Consequently, the workers did not associate detecting the odor of methyl mercaptan as representing an immediate hazard.

> *"And I smell MeSH [methyl mercaptan] all the time ... but it never ... I smelled MeSH but I never felt I couldn't breathe. Like whether it's, there's -- whether it's a small leak or anything, it just never -- I didn't think MeSH was that dangerous. I thought it, you know, but it is."*

As seen by this comment from a DuPont La Porte worker, smelling methyl mercaptan at low concentrations without having any observable, adverse effects can potentially normalize the danger of a highly toxic chemical.

Furthermore, extended exposure to methyl mercaptan could potentially eliminate the perception of a methyl mercaptan release because olfactory fatigue will prevent workers from being able to smell the chemical. The odor threshold for methyl mercaptan is approximately two parts per billion (2 ppb). However, olfactory fatigue is known to occur with methyl mercaptan, resulting in the exposed person to no longer being able to detect the chemical by smell. As a result, using odor as an indicator of a release may not provide adequate warning of hazardous concentrations [9].

In the events leading to the major methyl mercaptan release, there were many smaller releases, filling the area with methyl mercaptan odor. In part because personnel believed they understood the source of releases and normalized the risk associated with these smaller releases, personnel did not apply additional safeguards like clearing the area or wearing the required personal protective equipment required by the site when dealing with a methyl mercaptan release.

Facilities should ensure that employees understand the hazards of chemicals that they are working with and may be potentially exposed to if there is a release. In cases where there are multiple hazardous chemicals, such as methyl isocyanate[23] (MIC) and methyl mercaptan (as was the case at DuPont La Porte), it is important that companies effectively train workers on the hazards of each chemical at their facility. Furthermore, companies should train workers that handle multiple highly toxic chemicals, such as methyl mercaptan and MIC, to treat these chemicals with equal

[22] Odor threshold, in general, is the lowest concentration of a gas or a material's vapor that can be detected by odor. Odor threshold values are not fixed physiological parameters or physical constants but are statistical points representing the best estimate value from a group of individual responses [12, p. 2].

[23] Methyl isocyanate (MIC) is a raw material used in the production of insecticides at La Porte including the Lannate® process. In December 3, 1984, a methyl isocyanate release at the Union Carbide insecticide plant in Bhopal, India resulted in an estimated 3,800 people that died within days, and tens of thousands that were injured. Eventually, the release killed tens of thousands of people. See CSB Safety Message on 30th anniversary of fatal chemical release that killed thousands in Bhopal, India [13].

importance. This training is important for workers to help ensure they have a working knowledge of those chemical hazards.

3.4 Ensure Response to a Toxic Gas Detector Alarm Is Appropriate

The toxic gas detection system at DuPont La Porte had a history of false readings. Because of these false readings, personnel supplanted their ability to observe and detect a leak over the toxic gas detection system. Personnel adapted their responses to a potential toxic gas release to include determining whether the toxic gas detectors were functioning properly by using their perception of smell before implementing the site's leak response procedures.

For example, the Lannate® Unit used chlorine in addition to methyl mercaptan. Workers could observe from a distance a chlorine leak by the green color of the gas or the immediate adverse effects (i.e., difficulty breathing) of the gas. Workers, similarly, would also use their sense of smell to detect methyl mercaptan. Methyl mercaptan gas, however, is colorless and the odor is not a reliable indicator of a release (for the reasons noted above). In both situations where workers used their physical observations to determine where there was a leak, workers were at risk for being overcome by a toxic chemical release because they were not using additional safeguards like personal protective equipment (PPE).

DuPont's highly toxic material standards require that toxic gas detection systems be treated the same as safety interlocks. DuPont defines a safety interlock as a system or function that detects an out-of-limits (abnormal) condition or improper sequence and brings it to a safe condition. In this case, an alarm should trigger a predetermined response from operators to bring the system to a safe condition. In practice, the methyl mercaptan detectors were not treated as a safety interlock.

Multiple DuPont employees informed the CSB that odor was the primary method used to locate a methyl mercaptan leak. DuPont personnel routinely responded to methyl mercaptan detector alarms without any additional respiratory protection. When they responded, if a significant leak was identified, these personnel consistently stated they would leave to get a SCBA and then return.

On the night of the incident, multiple methyl mercaptan detector alarms activated, but because the workers were aware that methyl mercaptan was being released from their activities associated with clearing the hydrate, these alarms were normalized and not treated as a safety interlock. The alarms were not treated as a near miss and there was no actual investigation into the source or how to prevent these releases. No additional personal protective equipment, such as air-purifying respirators or SCBAs was used. In addition, DuPont did not order a unit evacuation and did not notify the emergency response team until hours later, after Operator 1 made the distress call.

When a toxic detection system has a history of false readings it creates an opportunity for workers to normalize toxic gas detector readings and their response, making the toxic gas detection system ineffective. Facilities should take steps to address the potential for false readings and reduce opportunities for workers to normalize false readings and their alarms. One method is to treat toxic gas detector alarms as a near miss and investigate them even if there are no injuries or the readings are determined to be false. This method should trigger an effort to understand alarm management and safety management system issues and address them. Additionally, facilities should ensure that workers are using leak response procedures to prevent personnel from entering into potential hazardous situations without proper safeguards, like PPE.

4 Conclusions

The Lannate® Unit's toxic gas detection system did not effectively warn workers of the November 15, 2014 methyl mercaptan release occurring inside the manufacturing building. The manufacturing building, including the building's overall construction, detectors, and alarms, was not designed such that personnel could be readily notified of a major chemical release in the manufacturing building. Additionally, the methyl mercaptan toxic gas system did not provide real-time, accurate information to warn of potential offsite consequences.

There are several lessons that facilities can learn from the incident to help improve their toxic gas detection systems. Gas detectors must operate as intended to prevent workers from normalizing the hazard the detectors are protecting workers from. For toxic gas detection system to be effective, the detector selection and placement have to be strategic to identify and quantify releases. Finally, personnel have to understand and appreciate the hazards associated with the toxic gases they work with to help ensure they respond appropriately to potential toxic gas releases.

5 References

[1] U.S. Chemical Safety and Hazard Investigation Board (USCSB), "DuPont La Porte Facility Toxic Chemical Release," [Online]. Available: https://www.csb.gov/dupont-la-porte-facility-toxic-chemical-release-/. [Accessed 1 March 2019].

[2] U.S. Environmental Protection Agency (USEPA), "List of Regulated Substances under the Risk Management Plan (RMP) Program," [Online]. Available: https://www.epa.gov/rmp/list-regulated-substances-under-risk-management-plan-rmp-program. [Accessed 2017 August 28].

[3] U.S. Chemical Safety and Hazard Investigation Board (USCSB), "DuPont La Porte, Texas Chemical Facility Toxic Chemical Release: Interim Recommendations," 30 September 2015. [Online]. Available: https://www.csb.gov/assets/1/20/dupont_la_porte_interim_recommendations_2015-09-30_final1.pdf?15526. [Accessed 28 February 2019].

[4] National Oceanic and Atmospheric Administration (NOAA) Office of Response and Restoration, "Emergency Response Planning Guidelines (ERPGs)," [Online]. Available: https://response.restoration.noaa.gov/oil-and-chemical-spills/chemical-spills/resources/emergency-response-planning-guidelines-erpgs.html. [Accessed 10 March 2019].

[5] United States Department of Labor (USDOL) Occupational Safety and Health Administration (OSHA), "Permissible Exposure Limits - Annotated Tables," [Online]. Available: https://www.osha.gov/dsg/annotated-pels/. [Accessed 28 February 2019].

[6] United States Department of Labor (USDOL) Occupational Safety and Health Administration (OSHA), "Annotated Table Z-1," [Online]. Available: https://www.osha.gov/dsg/annotated-pels/tablez-1.html. [Accessed 28 February 2019].

[7] Centers for Diseases Control and Prevention National Institute of Occupational Safety and Health (NIOSH), "NIOSH Pocket Guide to Chemical Hazards - Methyl Mercaptan," 11 April 2016. [Online]. Available: https://www.cdc.gov/niosh/npg/npgd0425.html. [Accessed 28 February 2019].

[8] U.S. Chemical Safety and Hazard Investigation Board (USCSB), "DuPont La Porte, Texas Chemical Facility Toxic Chemical Release: Interim Recommendations," 30 September 2015. [Online]. Available: https://www.csb.gov/assets/1/20/dupont_la_porte_interim_recommendations_2015-09-30_final1.pdf?15526.

[9] Agency for Toxic Substances and Disease Registry (ATSDR), "Methyl Mercaptan," [Online]. Available: https://www.atsdr.cdc.gov/MHMI/mmg139.pdf. [Accessed 28 February 2019].

[10] A. H. Mohammadi and D. Richon, "Equilibrium Data of Sulfur Dioxide and Methyl Mercaptan Clathrate Hydrates," *Journal of Chemical & Engineering Data,* vol. 56, pp. 1666-1668, 2011.

[11] Center for Chemical Process Safety (CCPS), Recognizing Catastrophic Incident Warning Signs in the Process Industries, Hoboken: Wiley, 2012.

[12] A. H. Lehocky, S. S. Murnane and D. Patrick, Odor Thresholds for Chemicals with Established Occupational Health Standards, Falls Church: AIHA, 2013.

[13] USCSB, "On 30th Anniversary of Fatal Chemical Release that Killed Thousands in Bhopal, India, CSB Safety Message Warns it Could Happen Again," [Online]. Available: http://www.csb.gov/on-30th-anniversary-of-fatal-chemical-release-that-killed-thousands-in-bhopal-india-csb-safety-message-warns-it-could-happen-again-/?pg=4.

[14] K. Baker, "DuPont Closing Texas Plant Where 4 Died," The News Journal, 1 April 2016. [Online]. Available: https://www.usatoday.com/story/money/2016/04/01/dupont-closing-texas-plant-where-4-died/82512270/. [Accessed 29 June 2017].

[15] DuPont, "Form 8-K," 31 March 2016. [Online]. Available: https://www.sec.gov/Archives/edgar/data/30554/000003055416000115/dd33116form8-k.htm.

[16] U.S. Chemical Safety and Hazard Investigation Board (USCSB), "E.I. DuPont de Nemours & Co., Inc. Belle, West Virgina Investigative Report," September 2011. [Online]. Available: https://www.csb.gov/assets/1/20/csb%20final%20report.pdf?13966. [Accessed 1 March 2019].

[17] U.S. Chemical Safety and Hazard Investigation Board (USCSB), "E. I. DuPont De Nemours Co. Fatal Hotwork Explosion," [Online]. Available: https://www.csb.gov/e-i-dupont-de-nemours-co-fatal-hotwork-explosion/. [Accessed 1 March 2019].

[18] American Industrial Hygiene Association (AIHA), "Current ERPG Values," 2014. [Online]. Available: https://www.aiha.org/get-involved/AIHAGuidelineFoundation/EmergencyResponsePlanningGuidelines/Documents/2014%20ERPG%20Values.pdf.

[19] F. Cavender, S. Phillips and M. Holland, "Development of Emergency Response Planning Guidelines (ERPGs)," *Journal of Medical Toxicology,* vol. 4, no. 2, pp. 127-131, 2008.

553063 Hurricane Harvey 20 Months Later - Insights on the Arkema Crosby Flooding

On August 24, 2017, Hurricane Harvey, a Category 4 hurricane, made landfall in southeast Texas producing an unprecedented amount of rainfall. Hurricane Harvey killed 68 people and flooded over 300,000 structures, forcing roughly 40,000 people to evacuate their homes. Hurricane Harvey required a large emergency response effort supported by local, state, and Federal officials.

At Arkema's Crosby site, Harvey delivered an unprecedented amount of rainfall. The National Oceanic and Atmospheric Administration estimates show that the rainfall during Harvey had less than a one-in-one-thousand-year likelihood of occurring. Harris County has estimated that at Crosby, Harvey was a 5,000 year to 20,000 year event. The plant's hurricane preparedness procedure was followed and resulted in a safe shutdown and hurricane preparedness of the manufacturing units. However, finished goods inventory remaining in the warehouses was overwhelmed by high water and overcame multiple layers of protection used to safely store organic peroxide finished goods.

This presentation will explain the sites preparedness ahead of time, the events leading up to the flooding, the incident itself, and the emergency response.

The presentation will focus on the impacts, the main lessons learned and the key recommendations from the CSB report.

AUTHOR INDEX

AUTHOR INDEX

AIChE
120 Wall Street, FL 23
New York, NY 10005-4020

ISBN 978-1-5108-8690-2
9 781510 886902

CURRAN ASSOCIATES INC.
proceedings.com